PLANT LIFE IN
ANAEROBIC ENVIRONMENTS

PLANT LIFE IN
ANAEROBIC ENVIRONMENTS

edited by

DONAL D. HOOK
Professor and Director
The Belle W. Baruch Forest Science Institute of
Clemson University
Georgetown, South Carolina

R. M. M. CRAWFORD
Professor, Department of Botany
St. Andrews University
Scotland

ANN ARBOR SCIENCE
PUBLISHERS INC / THE BUTTERWORTH GROUP
P.O. BOX 1425 ■ ANN ARBOR, MICHIGAN 48106

Second Printing 1980

Copyright © 1978 by Ann Arbor Science Publishers, Inc.
230 Collingwood, P. O. Box 1425, Ann Arbor, Michigan 48106

Library of Congress Catalog Card No. 77-85085
ISBN 0-250-40197-5

PREFACE

This book is designed to provide the reader with an understanding of the influence of hypoxia and anoxia on the physiology and adaptations of higher plants and the chemical and microbiological factors which affect plant growth in wetland soils.

Most higher plants are sensitive to conditions which limit aeration of the roots, stem or the whole plant. Yet, rice, swamp and marsh species thrive with their lower stems and roots submerged in poorly aerated-inundated soils. Such seemingly atypical growth led researchers to ask "What mechanisms and/or adaptations do these plants possess which enable them to live and thrive in the harsh wetland environments?" Although scientists have sought answers to this and similar questions for over a century, the total effort expended on this subject has been relatively small, sporadic and disjointed in nature. As a consequence an overall understanding of the biology and problems of plant growth in anaerobic environments has not materialized.

The first part of the book has authoritative chapters on the influence of hypoxia and anoxia on the ultrastructure of plant organelles, energetics, metabolism, gaseous diffusion and gas exchange, growth and development of higher plants. It also contains chapters on the chemical and microbiological properties of short- and long-term flooded soils.

The second part of the book contains recent contributions on cytoenzymology, nitrate metabolism, alcohol and lactate dehydrogenase identification and lipid production of higher plants under hypoxia.

The interested reader may choose to read through the book in chronological order to obtain a thorough understanding of anaerobiosis or pick and choose chapters according to his or her particular area of interest.

Donal D. Hook
R. M. M. Crawford

ACKNOWLEDGMENT

The editors wish to acknowledge the efforts and contributions of Professor B. B. Vartapetian and his colleagues in the USSR for organizing and bringing together the scientists who attended the first symposium on "Anaerobiosis and Related Plant Adaptations" at the 12th Botanical Congress in Leningrad, USSR, in July 1975. This book grew out of the contacts and intercommunication that developed among those attending the symposium.

Also we wish to acknowledge the financial support that The Belle W. Baruch Foundation provided Professor Hook during the editing and publishing process. Without this support and the time allotted the senior editor by Clemson University the book could not have been completed in a timely manner.

CONTENTS

SECTION I
PROCESSES IN ANAEROBIOSIS

SECTION II

RECENT CONTRIBUTION ON ANAEROBIOSIS

SECTION I

PROCESSES IN ANAEROBIOSIS

INTRODUCTION–LIFE WITHOUT OXYGEN

B. B. Vartapetian

K. A. Timiriazev Institute of Plant Physiology
Academy of Sciences
Moscow 127106 USSR

Life under the conditions of oxygen deficiency, or even in an environment completely devoid of molecular oxygen, is widespread on this planet. The phenomenon of anaerobiosis, permanent or temporary, complete or partial, can be observed at virtually every level of evolution of the biological system, from the primitive unicellular organisms up to man.

This property is particularly conspicuous in microorganisms, among which there are species such as the facultative anaerobic yeast *Saccharomyces cerevisea*, that can easily change from an oxygen-rich environment to one totally devoid of O_2 and retain their normal life function.

Strict anaerobes rank even higher in this sense. For example, the butyric fermentation bacteria *Clostridium Pasteurianum* not only lack any enzymatic mechanism for interaction with oxygen, but moreover, molecular oxygen is toxic to them. This species stands at the opposite extreme: it lives without oxygen but dies in its presence.

We find the phenomenon of anaerobiosis even in so highly organized a creature as man. The organism of a newborn baby at the moment of birth, being in a state of very active metabolism, nevertheless does without oxygen for as long as 30 minutes; then normal oxygen respiration and normal aerobic metabolism begin.

Still more striking cases of adaptation to anaerobiosis are observed in the animal kingdom, among worms especially parasites such as flat and round worm species, insects and even higher animals, reptiles and mammals, particularly the sea-living species. The champion is the turtle who can stay underwater for hours.

Some other sea animals, such as seals, dolphins and whales are also remarkably adapted to an oxygen-free environment. These animals, like man, breathe through their lungs, so they should not be compared to fish who live underwater and whose cells do not suffer from oxygen deficiency because their gills easily extract oxygen from the medium. By contrast, when a sea mammal takes a long dive, its external oxygen supply is close to that of a human diver.

Remarkable examples of life under an oxygen shortage or in a complete oxygen deprivation can also be cited from botany, especially of higher plants. Since many chapters in this book deal with adaptation of higher plants to oxygen deficit, only one example of higher plants' ability to live without oxygen will be cited here: rice seedlings. It is well known that rice seeds germinate readily under a layer of water. Recent experiments have shown them to germinate even under a strict anoxia, in a vacuum, when all molecular oxygen was eliminated from the environment, actively growing rice coleoptiles preserved a fine cellular organization in the intact, functionally active state.

Today no one would deny, or even be surprised, that life can exist in an environment devoid of molecular oxygen. But a century ago, it was believed that without breathing oxygen, there could be no life.

It was only after the brilliant studies by Louis Pasteur, who demonstrated that the fermentation process is carried out by microscopic organisms in the absence of oxygen, and after a heated debate that the possibility of life without oxygen was considered.

Pasteur's work on anaerobiosis is a perfect example of what might now be called science in action. He has probably been credited with more discoveries of practical value achieved by profound theoretic, sometimes very academic, laboratory experiments than any other scientist in biology, but apart from its importance for microbiological theory, Pasteur's work on anaerobiosis revolutionized industrial microbiology—a very practical field.

Since the discovery of anaerobiosis and the emergency of the theory of life without oxygen, it has been intimately linked with practical human needs and has also opened a new stage in the history of biological science. With the further progress of science, in particular of the theory of the origins of life on earth—also founded by the classical works of Louis Pasteur and developed by Alexander Oparin—there came the amazing conclusion that those recently discovered anaerobic organisms existed on this planet long before the first oxygen-respiring ones. Anaerobes originated in the early history of the earth when there was still no molecular oxygen on the planet and reductive conditions prevailed. It was only much later, when photosynthesis had developed, that molecular oxygen accumulated, and the reductive medium gave way to an oxidative one, so that aerobes could come into existence. In evolutionary terms, anaerobes preceded aerobic organisms.

All the foregoing examples of life without oxygen are remarkable in this respect—facultatively and obligatively anaerobic yeast or bacteria, turtle, neonate humans and rice seedlings. Some of them can live permanently without molecular oxygen, the others must again return to an aerobic environment to compensate "oxygen debt" following prolonged anaerobiosis. These are outstanding illustrations that higher organisms can be in the oxygen-free environment in the state of active life, for long periods—a fact of great scientific and practical importance.

These organisms are "prodigies" of anaerobiosis. What about other organisms, which do not belong in that category but are the majority on this planet? What about the strictly aerobic microorganisms, the higher plants, the animals and the man? Have the cells of these organisms lost all capacity for survival without oxygen? Do they inevitably and immediately degrade when placed in an anoxic medium? Experience and experiment show that this is not the case. Some faculty for anaerobiosis— complete or partial, short-term or prolonged—is possessed by most living organisms, including such strict aerobes as the green leaves of higher plants and the cells of any human organs.

Consider human muscle under physical stress. When at rest, the muscle is engaged in normal aerobic respiration fully meeting its needs with the energy released in mitochondria during oxidative phosphorylation. When muscle stress is increased abruptly, for instance under great strain, the muscle is no longer satisfied with the amount of oxygen conveyed through the circulatory system, and the high-energy phosphate compounds generated in the mitochondria no longer meet the energy needs of the heightened muscular activity. The reserve of high-energy compounds, stored in the muscle in the form of creatine phosphate, runs out very soon. There is a sharp deficit of oxygen. The possibilities of aerobic respiration are exhausted. At this point, the mechanism of anaerobic breakdown of glucose is switched on in the muscle and starts working. Glycolytic reactions provide energy for the organism to cope with the increased stress. With muscles, the physiological expediency for the organism as a whole is evident in these potential reserves of energy generation to be tapped in time of need.

But the cells of other organisms and organs, which, unlike muscles, do not frequently experience an oxygen shortage, can also exist for a limited time without oxygen. Leaves of the higher plants are a case in point. Owing to its great number of stomata and intercellular spaces, the leaf is perfectly aerated. Nevertheless, when placed in an anoxic environment, such as an atmosphere of gaseous nitrogen, it shows no evident pathological alteration of the fine cellular organization for about one hour, and sometimes for several hours. Biochemical analysis indicates that intensive metabolic and even synthetic processes occur in the leaf during that time on the basis of anaerobic metabolism.

One may ask what kind of material base in the cell enables aerobic organisms to survive the lack of oxygen in the environment. Does an aerobic organism, when experiencing anaerobiosis or adapting itself to it, drastically rebuild its metabolism, for example, by creating a new energy structure of the cell? Or does it just utilize something preexisting in the cell before the onset of anaerobiosis? The fact is that the mechanism of anaerobic generation of energy (glycolysis) is universal to living beings. It is possessed by every cell, including a strict aerobe.

As has been mentioned, aerobic organisms came into being after molecular oxygen appeared on the earth as a result of photosynthesis, and after the reductive conditions on the planet were replaced with oxidative ones. The cells of these organisms developed new and more efficient engines for extracting energy from organic matter. It was the mechanism of oxygen respiration, *i.e.*, the cycle of ditricarboxylic acid, the mitochondrial respiratory chain, and the enzymatic system accumulating the energy released during the transport of electrons from an organic compound to oxygen.

The engine of aerobic respiration succeeded the anaerobic one in the process of evolution. It did not supplant or even parallel the old system of energy extraction but was built as a superstructure on the anaerobic base. The aerobic cell retained the memory of the older system of energy generation and incorporated it into the new one. The result was an integrated energy system combining the ancient and the new mechanism.

In the absence of oxygen, the superstructure is readily switched off, *i.e.,* the mechanisms of aerobic energy release (the ditricarboxylic acid cycle and the mitochondrial transport of electrons) cease to work while the base, the enzymatic systems of anaerobic energy generation, continues to function independently of the superstructure.

Consequently, the anoxia medium does not spell immediate death to the aerobic cell, nor does it compel this cell to create some stopgap system to extract energy from organic compounds in the new situation. Indeed, it has inherited such a system from its ancestors. Imperfect as it is, in case of emergency it is automatically switched on and begins to function. Faced with a critical oxygen deficit, any aerobic cell can thus continue to function for a certain, longer or shorter, lapse of time. On subsequent appearance of molecular oxygen in the environment the superstructure, that is, the enzymatic systems of cell aerobic respiration, easily comes out of the latent state and again resumes functioning.

The foregoing is no more than a general and simplified scheme. In the real cell and the real situation the relationship between the aerobic and anaerobic systems of energy generation is more complex and is regulated by more intricate mechanisms.

A direct regulatory link between the glycolytic system and the oxygen respiration is created because glycolysis precedes the aerobic stage and prepares the material for the oxygen phase. The connection and interaction of the two systems is obvious, with the anaerobic system determining and coordinating the aerobic one.

However, the cell also possesses other much more perfect mechanisms which are involved in the control of aerobic and anaerobic breakdown of glucose and regulate the interaction between these two systems of energy generation. This is elucidated by the following example.

When a cell deprived of molecular oxygen switches off its aerobic respiration, the breakdown of the respiratory substrates in glycolysis is greatly stimulated. Conversely, when the same cell is transferred from anaerobic conditions into a medium containing oxygen, the rate of organic matter expenditure is substantially reduced. The phenomenon is known as Pasteur's effect. The overall physiological essence of this effect is evident: in the presence of molecular oxygen, energy is extracted from the oxidized substances more efficiently so that organic matter is spent more economically. But the fine mechanisms of this regulation have been the subject of wide speculation and hypotheses for a very long time. Only recently the real picture of the interaction between respiration and glycolysis began to be elucidated. Glycolytic enzyme phosphofructokinase is of primary importance in this process to catalyze the nonequilibrium reaction of conversion of fructose-6-phosphate into fructose-diphosphate. The activity of this enzyme in determining the rate of glycolysis is controlled by almost ten metabolites among which are citrate, ATP and inorganic phosphate. It has been found that citric acid, which is synthesized under aerobic conditions in the cycle of ditricarboxylic acids, is an allosteric inhibitor of the key enzyme of glycolysis, phosphofructokinase. In the presence of oxygen, the Krebs cycle functions and the cell continues to synthesize the allosteric inhibitor—the citric acid. As a result, glycolysis is greatly suppressed. When there is no ambient oxygen, the Krebs cycle is naturally switched off, citric acid is no longer synthesized, and the blocking of phosphofructokinase is removed. This opens up an intensive phosphorylation of fructose-6-phosphate.

Apart from citric acid, ATP is also a powerful allosteric inhibitor of phosphofructokinase. So when the main line of the cell's energy generation (mitochondrial respiration) is switched off in the anoxic medium, the concentration of ATP, this second allosteric inhibitor of phosphofructokinase, drops sharply, and this, too, must step up the glycolytic reactions.

Inorganic phosphate is an allosteric activator of phosphofructokinase. An increase of its amount in the cell under the conditions of oxygen deficiency should also stimulate, through activation of phosphofructokinase, breakdown of substrate in the glycolytic reactions.

When the cell is again supplied with oxygen, the Krebs cycle is restarted as is the mechanism of mitochondrial transport of electrons and coupled phosphorylation; as a result, the concentration of allosteric inhibitors (ATP and citric acid) rises and that of the activator, inorganic phosphate, decreases substantially and lowers the rate of glycolysis. Another important enzyme responsible for regulation of glycolysis rate is pyruvatekinase which, as well as phosphofructokinase, catalyzes the nonequilibrium reaction and has strategic location in metabolism.

We have only discussed the extreme states of the two systems: a cell deprived of oxygen and one provided with sufficient quantity of oxygen, but the entire spectrum of intermediate states can occur. The controlling influence exercised on the cell by molecular oxygen is more diversified than in the foregoing example with allosteric inhibition and de-inhibition. In fact, oxygen affects many other cardinal cell processes and systems.

For example, deprivation of the cell of oxygen affects the cytoplasmic and mitochondrial protein-synthesizing system as well. This has been graphically demonstrated on the cells of lower organisms, especially facultative anaerobic yeast.

In the absence of oxygen, for example, the production of an important respiratory enzyme of mitochondria, cytochromeoxidase, (in the synthesis of whose polypeptides both mitochondrial and the cytoplasmic protein-synthesizing systems take part) is discontinued. Along with the repression of the synthesis of this enzyme in the absence of oxygen, as has been clearly shown in plants, the synthesis on the cytoplasmatic ribosomes of the glycolytic enzyme, alcohol dehydrogenase (ADH), is induced and its concentration grows sharply as anaerobiosis continues.

If the cell is then returned into an oxygen-rich environment, the reverse process is clearly observed by the experimentor-induction of cytochrome oxidase synthesis and repression of ADH synthesis.

Hence, molecular oxygen not only controls the processes of aerobic and anaerobic energy generation, but also specifically regulates cellular processes related to the functioning of nuclear and mitochondrial genetic systems. Therefore, the transition of aerobic organisms into oxygen-free conditions and the adaptation of the organisms to hypoxia and anoxia should essentially affect the fundamental processes of cellular metabolism. In addition, it should be emphasized that molecular oxygen acts as a substrate for the synthesis of the important chemical compounds of cellular membranes such as unsaturated fatty acids, phospholipids and sterines.

After this general view of anaerobiosis as a universal biological phenomenon and consideration of the interrelation between the aerobic and anaerobic systems of energy generation and the regulatory role of oxygen, let us now proceed to the problem of oxygen deficit experienced by the higher plants which is the main subject of the book.

Anatomical and morphological features of a higher plant facilitate gas exchange of its tissues with the surrounding atmosphere. At first, it might seem that higher plants should easily maintain aerobic metabolism and never suffer from an oxygen shortage. However, a closer look shows that higher plants often experience situations strongly limiting the access of atmospheric oxygen to the plant as a whole or to its parts, causing a temporary or prolonged oxygen starvation.

For example, the cells of the succulent organs (fruit, tuber) sometimes (particularly during their long-term storage) are short of oxygen because of the organ's specific structure. Anaerobiosis is frequently experienced by roots and seeds both in nature and in cultivation. Sometimes it leads to a mass destruction of the plants and entails grave economic losses. Seeds of plants can suffer from lack of oxygen due either to poor permeability of their coats or to the "soaking injury" during germination in waterlogged soils or in years with cold, wet springs.

The supply of external oxygen is limited for the cells of the meristematic zone of the roots owing to the peculiarities of the morphological structure of these compact tissues with reduced intercellular spaces. Even under conditions of normal aeration, the cells of the meristematic zone can experience partial anaerobiosis.

Anaerobiosis is a threat to roots in the vast regions of the world where fields are periodically flooded for irrigation. In areas with a long and severe winter, hypoxia frequently affects the roots of perennial and winter crops when an ice crust seals the soil surface and makes it impervious to the diffusion of oxygen and carbon dioxide. Problems with atmospheric oxygen supply are experienced by roots of wild and cultivated plants, including trees, in water-logged swampy soils. Since excessively wet soils cover vast territories in many countries, the issue is very important in practical terms. Finally, anaerobiosis may be experienced by the roots of plants cultivated without soil in nutrient solutions, which is now commercial practice as well as a laboratory research technique.

However, one can notice some contradiction in all that precedes. Thus the cell has a finely regulated mechanism which enables it to survive and function without oxygen; however, we now speak of the noxious effect of anaerobiosis on many organisms. Apparently the real situation with anaerobiosis is more involved than the preceding description. Although the aerobic cell can switch to anaerobic generation of energy, this is known to be considerably less effective than aerobic energy generation. If the cell does not possess the appropriate mechanisms for a sharp increase in the rate of glycolytic reactions in the oxygen-free environment, the organism will suffer from energy starvation.

Furthermore, in anaerobiosis there is an accumulation in the tissues of the end-products of glycolysis (ethyl alcohol, lactic acid) which are toxic for the organism. The enhancement of glycolysis during anaerobiosis accelerates the accumulation of these toxic compounds in the organism.

On the other hand, anaerobiosis, by stimulating the breakdown of organic compounds in the cell, naturally leads to their more rapid depletion in the tissues. Therefore, oxygen starvation during anaerobiosis in a number of cases is followed by even more harmful carbon starvation if the tissues do not possess sufficient reserves of organic compounds. The occurrence under oxygen starvation of noticeable inhibition of the transport of organic compounds from some organs into others (for instance, from the leaves into the roots or from the grain into the seedling) intensify the situation.

Finally, the regulatory mechanisms responsible for the cell's transition from aerobic to anaerobic energy generation and matabolism probably do not work ideally in all cases when the level of oxygen supply is changed. In all likelihood, there are many other still unknown factors and mechanisms that determine a cell's capacity for switching over the metabolism mode in response to a changed oxygen supply. Some organisms may not have such mechanisms.

Indeed, some organisms adapt themselves to oxygen deficit quite easily, while the cells and tissues of other organisms and organs, which obviously lack the appropriate regulatory mechanisms, soon perish when partially or completely deprived of oxygen.

This is illustrated by human muscular or cerebral cells. Minutes after the heart stops and the oxygen supply is cut off, the brain cells undergo an irreversible pathological change leading to the death of the organism. By contrast, the muscular cells, as we have seen, endure an oxygen deficit relatively well and even continue to function for a long time.

Similar examples are found among higher plants. As mentioned, rice seedlings germinate readily in an anoxic medium and grow a longer coleoptile than under aerobic conditions. Meanwhile, the seeds of most other plants cannot germinate without oxygen; they never start to grow and soon perish.

For the investigators studying the regulatory mechanisms of the cell in relation to anaerobiosis, of particular interest are naturally the organisms in which these mechanisms have been well established by historical evolution so that a lack of oxygen in the environment does not lead in them to a disruption of metabolism and degradation of the cell's structure and function. Such biological objects are convenient models for studies of molecular and cellular mechanisms of self-regulation under an oxygen shortage.

As for adaptation of higher plants to oxygen deficiency, it is realized not by a single mechanism but by various mechanisms depending on the species and the conditions of the environment. The adaptation can occur through significant enhancement of anaerobic generation of energy owing to the ability to activate glycolytic breakdown of substrate in the oxygen-free environment. Indeed, the Pasteur effect that was detected in most plants studied in this respect (except buckwheat) is most pronounced in the plants capable of active metabolism and growth under the conditions of anoxia (rice seedlings). By virtue of a sharp increase of glycolysis under anoxia in the cells of these plants, a high level of energy charge, indicative of the energy supply level of the cell, is maintained.

Adaptation to anaerobic environments can occur by means of using other external acceptors of electrons in addition to molecular oxygen, for example, nitrates. Induction of nitratereductase in the roots of plants under anaerobic conditions has been shown experimentally.

Plants are differently adapted to the accumulation and detoxication of the end-products of glycolysis (fermentation). Some of them excrete these toxic products (for example, ethyl alcohol) directly into the environment. The others transport them from the poorly aerated organs (roots) to well aerated organs (leaves) which detoxicate the end-products by oxidative conversions.

A peculiar way to avoid the formation under anaerobic conditions of compounds harmful to the organism is diversification of the end-products of glycolysis, that is, the formation of neutral or less toxic compounds— for example, alanine or malate instead of toxic ethanol.

So far we have spoken mainly of the biochemical mechanisms of adaptation to oxygen deficit; but many plant species in the course of evolution have developed a basically different adaptation to life on a soil with impeded gas exchange: they transport oxygen from the subaerial organs to the roots. The morphological features of such plants (hydrophytes) facilitate inflow of oxygen to the rhizosphere. The root cells of these species inhabit a medium devoid of oxygen, but they do not suffer from any oxygen deficit. That explains why these root cells have not evolved any specific biochemical mechanism of adaptation to a shortage of oxygen and are no more resistant to anoxia than those of the plant species growing in well aerated soil.

One supposes that the transport of molecular oxygen from the green organs may play a protective part in the mesophytes, whose roots also frequently undergo conditions of oxygen deprivation. The problem has been studied for the past two decades in many of the world's laboratories and vividly discussed in the literature. The interest in this issue is heightened by its implications for farming and forestry on soils with poor gas exchange.

Different species of plants possess (or are devoid of) to a variable degree the above-mentioned devices and mechanisms which permit them to experience long-term or short-term anaerobiosis in the environment.

The author suggests that plants should be classified according to the character of their resistance to anoxia in the following categories: (1) truly resistant to anoxia, (2) resistance to anoxia is apparent rather than true and (3) nonresistant to anoxia.

As for the plants possessing the traits of true physiological resistance to anoxia, their ability to experience the absence of molecular oxygen in the environment is biochemical adaptation in nature. This is due to reformation in metabolism of the cell experiencing oxygen deficit and primarily in its protein metabolism and energetics. Under conditions of complete exclusion of oxygen in the environment, such organisms can preserve for a comparatively long period the fine organization of the cells in the intact, functionally active state and are even capable of growing. Representative of this category are young rice seedlings which can serve as a splendid biological system for studying the molecular, subcellular and cellular mechanisms of adaptation of higher plants to oxygen deficiency.

The plants whose resistance to anoxia is considered to be apparent rather than true also grow freely on waterlogged marshy soils with extremely poor gas exchange. Moreover, such plants (adult hydrophytes) can grow on anaerobic soils for much longer periods and even within the whole ontogeny than the plants falling into the first category. Nevertheless, the nature of the resistance of these organisms to anaerobic environments differs radically from that of the plants of the first category, that is, truly resistant plants. Indeed, as the recent studies have shown, when molecular oxygen is excluded from the environment a fine organization of the cells of the roots of these organisms does not exhibit signs of resistance to anoxia greater than those of the organisms not capable of growing on anaerobic soils. In the course of evolution the cells of hydrophytes' roots did not develop biochemical mechanisms of adaptation to anaerobiosis. The capacity of these organisms to grow long or permanently on soils practically devoid of the possibility of gas exchange with the external atmosphere is accounted for by the ability of these plants to transport oxygen from aerial parts to underground organs (roots) rather than by the peculiarities of root metabolism. This transport is strongly facilitated by the morphological organization of hydrophytes, whose bodies are penetrated with large intercellular spaces. Although the cells of their roots are constantly in the anaerobic environment, they do not experience shortage of molecular oxygen. In the oxygen-free environment the roots of hydrophytes are capable of active growing on the basis of aerobic metabolism. Hence, the organisms mentioned living permanently in the anaerobic environment nevertheless avoid anaerobiosis.

Finally, the category of plants nonresistant to anoxia should include those plants—mesophytes—which do not possess the traits of either apparent or true resistance or possess them to much lower degree. This category unfortunately comprises the great majority of agricultural plants; their characteristic of nonresistance to anoxia is a source of great economic loss.

These three categories should not be considered in all cases as isolated, clear-cut groups of plants. In fact, when dealing with a great variety of plant species and conditions of their environment one can find plants which in their relation to anaerobiosis do not exhibit "pure" traits of one single category but possess to a variable degree traits inherent to different categories. Nevertheless, even in these cases the advanced classification may be of certain usefulness.

Thus, in the biology of plants as well as in medicine and microbiology, theoretical studies of anoxia and hypoxia and of the mechanisms of adaptation of higher plants to oxygen deficit are essential for, and stimulated by, practical human needs.

The study of the structure, function and metabolism of plants under oxygen deficiency and also elucidation of the mechanisms of the adaptation will give better orientation to plant-breeders, crop specialists and other practical workers in agriculture and forestry in selecting the appropriate species of plants and in creating favorable conditions both for cultivation of plants and for storage of agricultural products.

On the other hand, it should be emphasized that the problem is one of basic theoretical importance, since the mechanisms of anaerobic energy generation and the regulatory mechanisms ensuring the cell's functioning at various levels of oxygen supply affect the fundamental metabolic processes of the cell.

The expediency of carrying out such investigations on the molecular and cellular levels as well as on the level of the whole organism of plants is evident.

2

PLANT CELLS UNDER OXYGEN STRESS

B. B. Vartapetian, I. N. Andreeva and N. Nuritdinov

K. A. Timiriazev Institute of Plant Physiology
Academy of Sciences
Moscow 127106 USSR

INTRODUCTION

This chapter summarizes and evaluates 12 years of research by the authors on anaerobiosis and adaptation of higher plants to oxygen deficiency with regard to the current literature on the subject.

The studies were carried out on two main lines: (1) the investigation of ultrastructure, functions and development of subcellular structures, mainly mitochondria, in plants grown under conditions of strict anoxia. The studies were carried out on young rice seedlings (*Oryza sativa* L.), whose coleoptiles have a remarkable ability to grow under conditions of total absence of molecular oxygen in the environment. Because of this trait, the rice seedling proved a very convenient object to study cellular and molecular mechanisms of the adaptations of higher organisms to oxygen deficiency; and (2) the investigation of the physiological role of molecular oxygen transported from shoots to roots under conditions of root anaerobiosis. In these experiments, the representatives of two ecologically opposite plant groups served as test objects: mesophytes, mainly pumpkin (*Cucurbita peop* L.) and cotton plants (*Gossypium hirsutum* L.), cultivated usually on easily aerated soils; and hydrophytes, chiefly adult rice plants, cultivated on soils with an extremely restricted gas exchange.

The main results of the studies on two lines mentioned are described below.

STRUCTURE AND FUNCTIONS OF MITOCHONDRIA OF ANAEROBICALLY GROWN RICE SEEDLINGS

Among higher plants and maybe even among higher organisms, it is difficult to find a more interesting object than rice seedlings for studying the phenomenon of adaptation to anaerobiosis. This species is remarkable not only from the scientific point of view, but also regarding its practical value, since more than half the population of the world uses rice as its main source of nutrition.

Rice seeds germinate at extremely low oxygen content in the environment.[1-4] Seeds of some other plants (cucumber, wheat, celosia) can germinate at decreased oxygen content in an environment (0.5-5%) but not in environments totally devoid of molecular oxygen.[5-9]

Our own attempts to germinate seeds of some cultivated plants (bean, haricot, pea, pumpkin, sunflower, wheat, lettuce and other) under strict anaerobic conditions have been unsuccessful. Rice seeds under the same anaerobic conditions germinated vigorously. Figure 1 presents a photograph of 6-day-old rice seedlings grown in a nitrogen atmosphere and normal aerobic conditions.[10,11] Only coleoptiles can grow under anaerobic conditions, while roots and leaves grow only if oxygen is present in the environment. Coleoptiles grown under anaerobic conditions are longer than those grown in the presence of oxygen.

The ability of coleoptiles to grow in nitrogen atmosphere cannot be explained by the presence of traces of molecular oxygen in gaseous nitrogen. Rice seeds germinate easily not only in gaseous nitrogen but under vacuum (Figure 2).[12]

From the beginning of our studies with rice seedlings, we were interested in the fine structure of cells of coleoptiles grown under conditions of strict anoxia. In this regard, mitochondria are of particular interest, as they are organelles of the cell oxygen metabolism. Do the cells of anaerobically grown rice coleoptiles contain mitochondria? If so, what is its ultrastructure? Do such mitochondria have respiratory enzymes that are able to function in the presence of oxygen in the environment? Finally, can the mitochondria develop, under conditions of a total exclusion of oxygen in the environment?

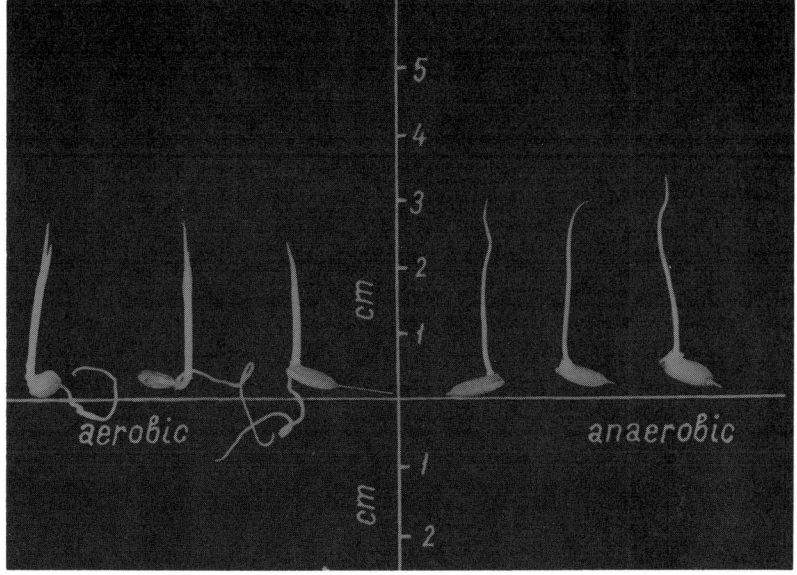

Figure 1. Six-day-old rice seedlings grown under aerobic and anaerobic conditions.

Figure 2. Six-day-old rice seedlings grown under vacuum.

These questions arose when some specialists in the biochemistry and physiology of rice concluded that anaerobically grown rice coleoptiles have no enzymatic systems of aerobic respiration and, therefore, are not capable of metabolizing atmospheric oxygen.[13] The investigators mentioned classified rice seedlings as obligatory anaerobes.

Electron microscope studies[14] as well as the results of Ueda and Tsuji[15] and Öpik[16] have led to a contrary conclusion. It has been demonstrated that mitochondria are present in the cells of anaerobically grown rice coleoptiles and that anaerobically grown coleoptiles have the ability to carry on respiration.[10,11,16-18] Finally, the presence of cytochromes in mitochondria isolated from the cells of such coleoptiles[19,20] and the ability of anaerobic mitochondria to transport electrons to molecular oxygen and coupled phosphorylation[21] adequately substantiate the capabilities of anaerobic mitochondria to carry on aerobic metabolism when returned to an aerated environment. Here, and in the following account, coleoptiles grown under anaerobic and aerobic conditions will be identified as anaerobic coleoptiles and aerobic coleoptiles, and mitochondria isolated from anaerobic and aerobic coleoptiles will be called anaerobic mitochondria and aerobic mitochondria, respectively.

On the whole, the results of these studies led us to conclude that rice coleoptiles are facultative anaerobes, an extremely rare phenomenon among higher organisms in general, and among higher plants in particular.

Mitochondrial Ultrastructure in Anaerobically Grown Coleoptiles

The first observations under the electron microscope on anaerobic coleoptiles grown under nonsterile conditions and strict anoxia show that the cells of such coleoptiles contain all the organelles typical for aerobic coleoptiles including mitochondria.[14] Such mitochondria were larger in size (1.5-2 μm in diameter) than mitochondria of the aerobic coleoptile cells (0.5-0.6 μm) and were irregularly shaped. The matrix of such mitochondria was clarified: mitochondria had low numbers of cristae and the intracristal space was dilated.

The ability of anaerobic mitochondria to restore their ultrastructure within 1-3 hr after transfer of the anaerobic seedlings into an ordinary atmosphere suggests that the mitochondria are not degraded. Upon return to an aerated environment, anaerobic mitochondria decrease in size to 0.6-0.8 μm in diameter. Also, they form an oval shape, the matrix becomes finely granular and electron dense, and cristae take on the appearance of elongated tubes. Such mitochondria become similar in ultrastructure to mitochondria of aerobically grown coleoptiles.

Simultaneously, anaerobic mitochondria were identified in rice coleoptiles by Japanese investigators Ueda and Tsuji,[15] who described it in detail in the process of anaerobic growth of coleoptiles. Somewhat later, Öpik[16] published an article that confirmed both the results of Ueda and Tsuji and our findings.

Subsequent investigations were carried out on rice coleoptiles grown under anaerobic conditions in an aseptic medium that excluded the influence of the products of microorganisms' metabolism on the ultrastructure of cellular organelles.[10]

Results of the studies (Figure 3) show that the changes of mitochondrial ultrastructure under nonsterile anaerobic conditions observed in the first experiments are chiefly retained by coleoptile germination in an aseptic medium.

After six days of germination under conditions of anoxia, coleoptile cells (Figure 3b) contain complete sets of organelles, although their number is small and the cytoplasm is of low electron density. The endoplasmic reticulum has an appearance of elongated profiles, and vesicles and the number of ribosomes on the membranes are small. Free ribosomes are scarce in the cytoplasm and dictyosomes are infrequent. Number of mitochondria in the cells is also small and their ultrastructure has characteristic distinctions from the mitochondria of the aerobic variant (Figures 3a and 3b).

To compare quantitatively some parameters of mitochondria (area, length of the outer membrane and number of cristae) of aerobically and anaerobically grown coleoptiles, as well as the changes of mitochondrial ultrastructure during aerobic adaptation after preliminary anaerobic germination of coleoptiles, morphometric investigations (according to Wiebel[23]) were carried out (Table I). Most mitochondria of anaerobic coleoptiles are larger, area and length of the outer membrane are increased, and the mitochondrial matrix is less electron dense than in mitochondria of aerobic coleoptiles (Figure 3b). Cristae are also changed: their number in anaerobic mitochondria decreases, and the intracristal space is so dilated that the majority of cristae are of rounded shape. In some mitochondria, great cavities formed due to an enlargement of the intracristal space in one or several cristae. Along with "altered" mitochondria, there are fewer changed mitochondria similar to the controls, regarding sizes of the organelles and the shape of cristae (Figure 3c).

As soon as 30 min after transfer of anaerobic seedlings into an ordinary atmosphere, a decrease of sizes of mitochondria and an increase of electron density of the matrix occur. Three to six hours after transfer into air, all the mitochondria become similar to the mitochondria of aerobic coleoptiles in ultrastructure, shape and sizes (Figure 3d). Number of cristae

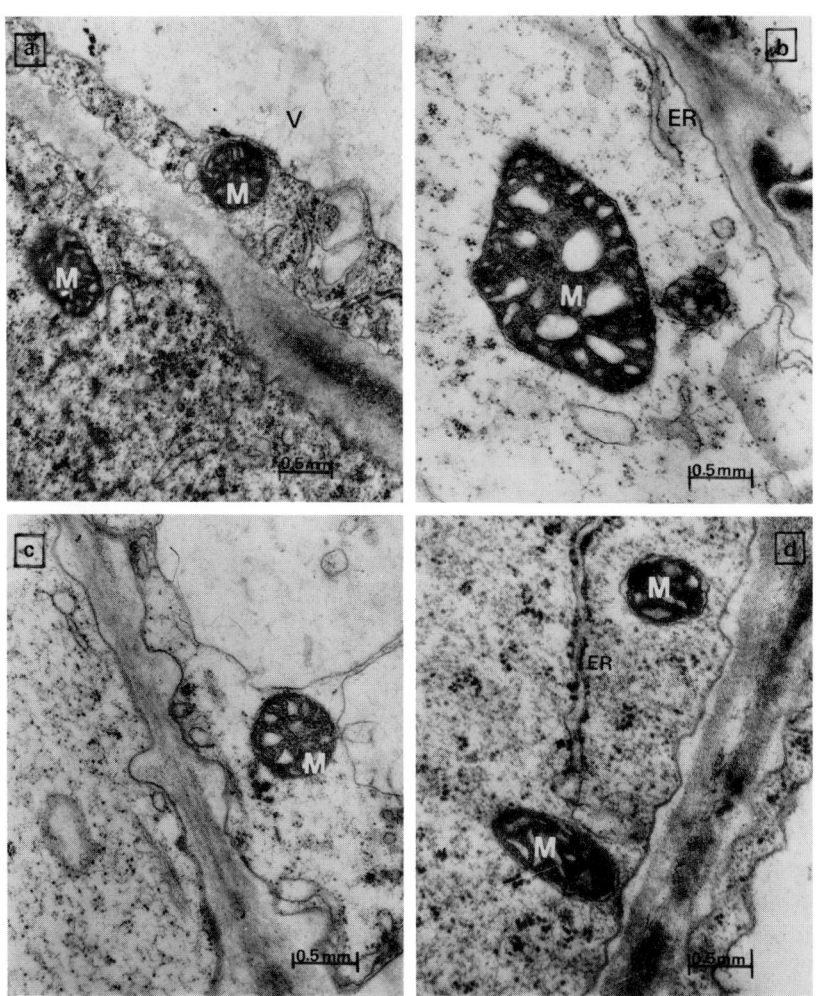

Figure 3. Ultrastructure of mitochondria in rice coleoptile cells, grown (a) aerobically, 6 days; (b, c) anaerobically, 6 days; and (d) anaerobically, 6 days, then 6 hr in air. Here and hereafter, the material for electron microscope investigations was fixed according to Karnovsky,[22] embedded in EPON 812, and the specimens were examined in an electron microscope JEM-100B.

Table I. Some Parameters of Mitochondria From Coleoptiles Grown Under Aerobic and Anaerobic Conditions and Transferred to Air

Coleoptile Treatment	Length of Outer Membrane (μm)	Area of One Mitochondrion (μm^2)	Number of Cristae Per Mitochondrion
Aerobic (6-day-old)	2.122 ± 0.085	0.304 ± 0.015	18
Anaerobic (6-day-old)	2.895 ± 0.156	0.475 ± 0.031	11
Anaerobic (6-day-old, after 30 min in air)	2.366 ± 0.107	0.344 ± 0.019	10
Anaerobic (6-day-old, after 1 hr in air)	1.914 ± 0.127	0.264 ± 0.025	10
Anaerobic (6-day-old, after 6 hr in air)	2.024 ± 0.086	0.286 ± 0.017	13

increases but does not reach the number of mitochondrial cristae in aerobic coleoptiles.

Thus, in the course of adaptation to oxygen, mitochondria of anaerobic coleoptiles are capable of prompt morphological rearrangement.

Respiratory Activity and Cytochromes of Anaerobic Coleoptiles and Mitochondria

Since mitochondria have been found in anaerobically grown coleoptiles, it is critical to ascertain whether their cells are capable of aerobic metabolism, *i.e.*, whether they have enzymatic systems capable of utilizing molecular oxygen. An integral idea can be developed in the elucidation of the ability of the coleoptile cells to absorb oxygen from the environment. Experiments conducted using a Warburg apparatus have shown (Figure 4) that coleoptiles grown under strictly anaerobic conditions are able to activate and to absorb molecular oxygen.[10] Although the intensity of oxygen uptake by anaerobic coleoptiles (calculated in fresh weight) was considerably lower than in aerobic, the former consumed oxygen in notable quantities. Similar results were obtained by others.[11,16] Tsuji[11] has studied in detail the respiratory activity of anaerobically grown coleoptile

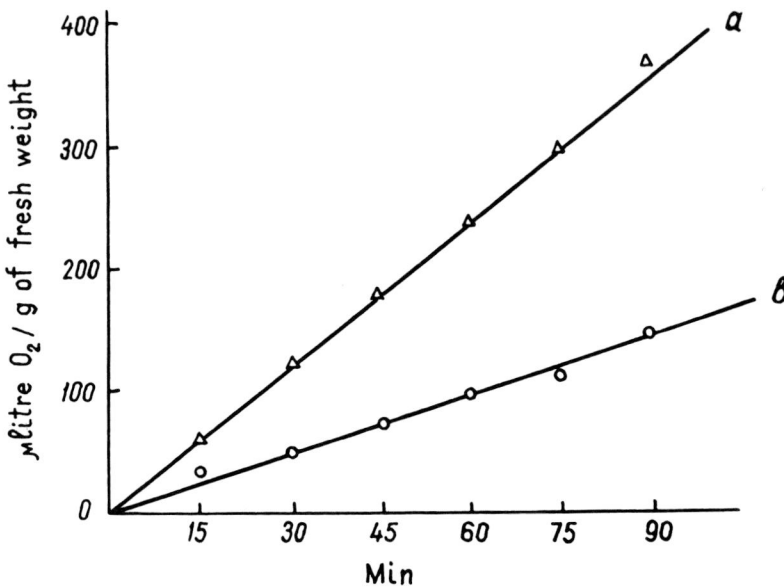

Figure 4. Rate of O_2 absorption by 6-day-old rice coleoptiles grown (a) aerobically, and (b) anaerobically.

after transfer into an ordinary atmosphere and has shown that the intensity of respiration of the coleoptiles increased, reaching and even exceeding the level of aerobically grown coleoptiles.

However, both in the experiments of Tsuji and Öpik[11, 16] and in ours, the possibility of synthesis of respiratory enzymes in a time inverval after the contact of anaerobic coleoptiles with molecular oxygen and before the measurement of its respiration (a phenomenon of oxygen induction) could not be excluded. In a Warburg apparatus, one cannot begin to measure oxygen absorption by coleoptiles sooner than 10-20 min after the contact of anaerobic coleoptile cells with atmospheric oxygen. In the subsequent experiments, we therefore tried as much as possible to shorten the time between the coleoptile contact with an ordinary atmosphere and the measuring of its oxygen uptake by polarographic techniques.[18,21] Figure 5a shows the results of one such experiment in which the time between the contact of a seedling with oxygen and the beginning of measuring was 35 sec. The data support the theory that the respiratory apparatus pre-exists in the coleoptile cells under anaerobic conditions, and after transfer of the coleoptiles into air, this apparatus begins to function at once.

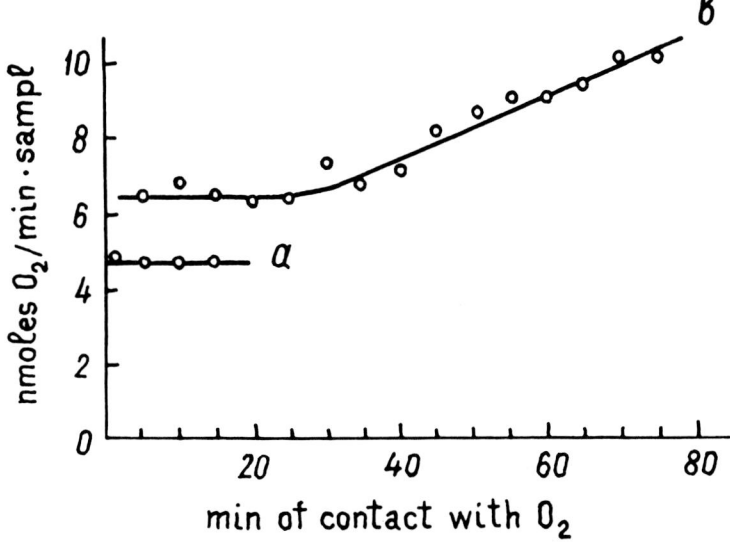

Figure 5. Oxygen absorption by 6-day-old anaerobic coleoptiles after transfer into a polarographic cell: (a) 15-min continuous recording; (b) 75-min recording (during the intermeasurement intervals the suspension was aerated).

The intensity of respiration remains at the same level during the first 30 min after the coleoptile contact with oxygen. After 30 min, a gradual increase of the intensity of respiration is noted (Figure 5b). These observations imply that in the initial period, a respiratory apparatus of the cells pre-existed before the contact of the seedlings with oxygen and is capable of functioning immediately upon contact with oxygen. Later, the oxydative enzymes appear as a result of oxygen induction.

Mitochondria isolated from anaerobic coleoptiles were examined to ascertain whether they had cytochromes, the most important enzymes of the respiratory chain, and whether such mitochondria are able to transport electrons to molecular oxygen and to accomplish coupling phosphorylation.

A direct spectrophotometric determination of cytochromes in mitochondria isolated from 6-day-old coleoptiles (Figure 6) has shown the presence of cytochromes of all three types: a, b, c.[19,21] The number of cytochromes calculated according to the spectra of cytochrome absorption in the course of reduction by dithionite and oxidation by ferricyanide in anaerobic mitochondria proved to be notably lower than in aerobic mitochondria. The content of cytochrome oxidase (aa_3) in anaerobic mitochondria is 3 times less, and the amount of b and c

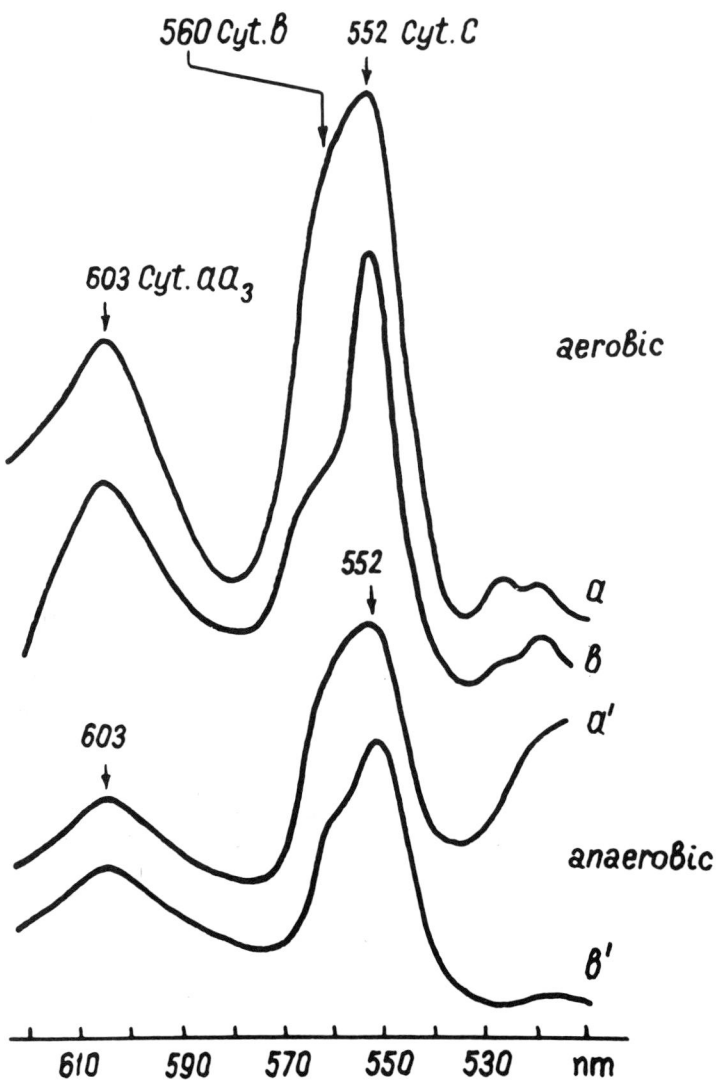

Figure 6. Difference in spectra of cytochromes in mitochondria isolated from 6-day-old rice coleoptiles grown aerobically (a,b) and anaerobically (a′,b′). Curves a and a′ were obtained by cytochrome reduction with dithionite; curves b and b′ by applying ascorbate + N,N,N′,N′-tetramethyl-*p*-phenylendiamine (TMPD). Protein: a - 0.65 mg/ml, a′,b,b′ - 0.76 mg/ml.

cytochromes is 2-2.6 times less than in aerobic mitochondria (Table II).

Histochemical investigations of cytochrome oxidase showed that after transfer of anaerobic seedlings into an ordinary atmosphere, the neoformation of cytochrome oxidase began, the activity of which reaches the level of aerobically grown coleoptiles within 18 hr.[21]

A detailed study of cytochromes of anaerobic coleoptiles was performed recently by Tsuji et al.[20]

Table II. Amount of Cytochromes in Anaerobic and Aerobic Mitochondria

	nmol/mg Protein		
Object	aa$_3$	total b	total c
Mitochondria from anaerobic coleoptiles	0.03	0.10	0.10
Mitochondria from aerobic coleoptiles	0.09	0.21	0.26

Mitochondria from anaerobic coleoptiles can oxidize substrates of the Krebs cycle and NADH but at a consideraly lower rate than mitochondria from aerobic coleoptiles (Table III). An addition of adenosine diphosphate (ADP) to the mitochondrial suspension stimulates oxygen absorption until the ADP is exhausted, then it decreases, i.e., the phenomenon of respiratory control is observed. This indicates that along with the system of electron transport, the system for energy accumulation is preserved in mitochondria under anaerobic conditions.

Table III. Rate of Oxygen Absorption by Mitochondria Isolated from Anaerobic and Aerobic Coleoptiles

	nmol O_2/min/mg Protein	
Substrate	Aerobic	Anaerobic
Succinate (10 mM)	147.2	41.1[a]
NADH (1 mM)	155.4	33.7
α-ketoglutarate (10 mM)	103.3	30.5
Malate (20 mM)	114.5	16.0[b]

[a] Respiratory control of anaerobic mitochondria was 2.0 with succinate as a substrate.
[b] The rate of oxidation was measured after the addition of NAD.

Oxygen absorption by mitochondria isolated from anaerobic coleoptiles as well as from aerobic ones in every substrate is sensitive to cyanide and antimycine A.

The above data on cytochromes and respiratory activity lead us to conclude that the difference between anaerobic and aerobic mitochondria is merely quantitative but not qualitative. On the other hand, the results show that there is no analogy between the anaerobic mitochondria of rice coleoptiles and promitochondria of facultatively anaerobic yeast, which also arise in the oxygen-free environment but are devoid of some important respiratory enzymes of mitochondria.[24-26]

Aerobic and Anaerobic Development of Mitochondria

The detection of actively functioning mitochondria in 5- and 6-day-old coleoptiles grown in oxygen-free environments raises the question whether synthesis of respiratory enzymes as well as biogenesis of mitochondria themselves takes place under anoxia or whether the respiratory apparatus of it pre-exists even in dry seeds and is retained in the process of anaerobic growth of the coleoptiles.

It seems that some data in this relation can be obtained by measuring respiration of whole coleoptiles during their growth. Though rice coleoptiles grow vigorously under anoxia, as noted earlier, nevertheless, the intensity of coleoptile respiration (calculated per coleoptile) upon return to air remains unchanged during 3-7 days of anaerobic germination (Figure 7b). Bearing in mind that the weight of a 7-day-old coleoptile is at least 10 times the weight of a 3-day-old specimen, the coincidence of the amount of absorbed oxygen by these plants is even more impressive. On the contrary, in aerobically germinated coleoptiles, the intensity of respiration increases until the fifth day of growth and then decreases through the seventh day (Figure 7a). These results favor the hypothesis that in the period of anaerobic growth of coleoptiles (at least, from the third to seventh day) the neoformation of respiratory enzymes does not occur.

Another approach to the problem is to use chloramphenicol (CAP) as a selective inhibitor of protein synthesis at the level of mitochondrial ribosomes. The mitochondrial system of protein synthesis is responsible for the neoformation of peptides that are included in the structure of enzymatic complexes localized on the internal mitochondrial membranes.[26-28]

For test purposes, rice seeds were germinated under anaerobic and aerobic conditions up to six days, with and without chloramphenicol. During the growth of coleoptiles, comparative electron microscope and biochemical studies of mitochondria were carried out.

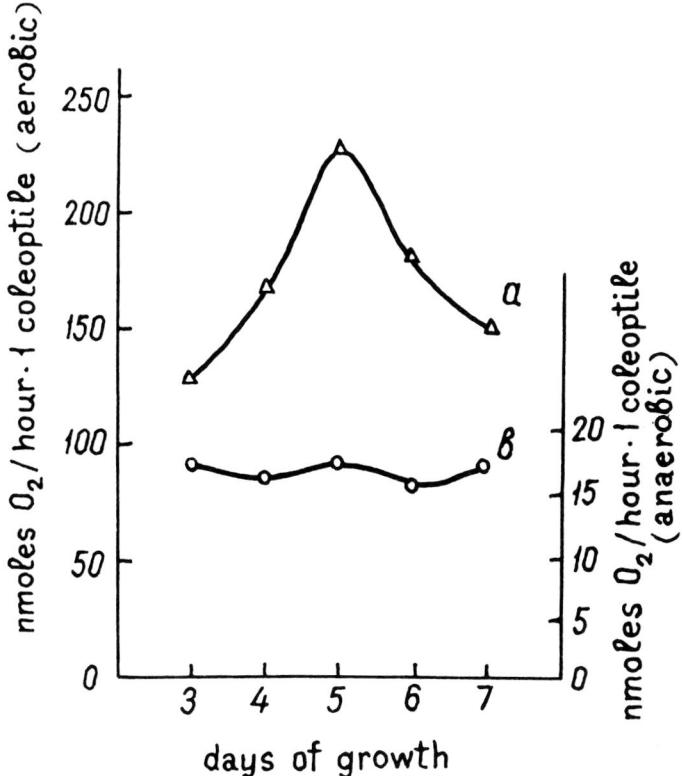

Figure 7. Oxygen absorption by rice coleoptiles of various ages grown (a) aerobically, and (b) anaerobically.

Aerobic Development of Mitochondria in the
Presence of Chloramphenicol

Electron microscope studies of mitochondria during rice seed germination in the presence of oxygen[29,30,40] have shown that the formation of mitochondrial ultrastructure takes place during the first 48 hr of germination regardless of the presence or absence of chloramphenicol in the medium of germination (Figures 8a,b; 9a,b). After 48 hr in water and in CAP the ultrastructure and the volume of mitochondria and the number of cristae were similar (Table IV). These observations indicate that in the early stages of aerobic development, mitochondrial systems of protein synthesis do not play a significant role in development of mitochondrial ultrastructure. These findings substantiate the results of a number of authors.[31,32]

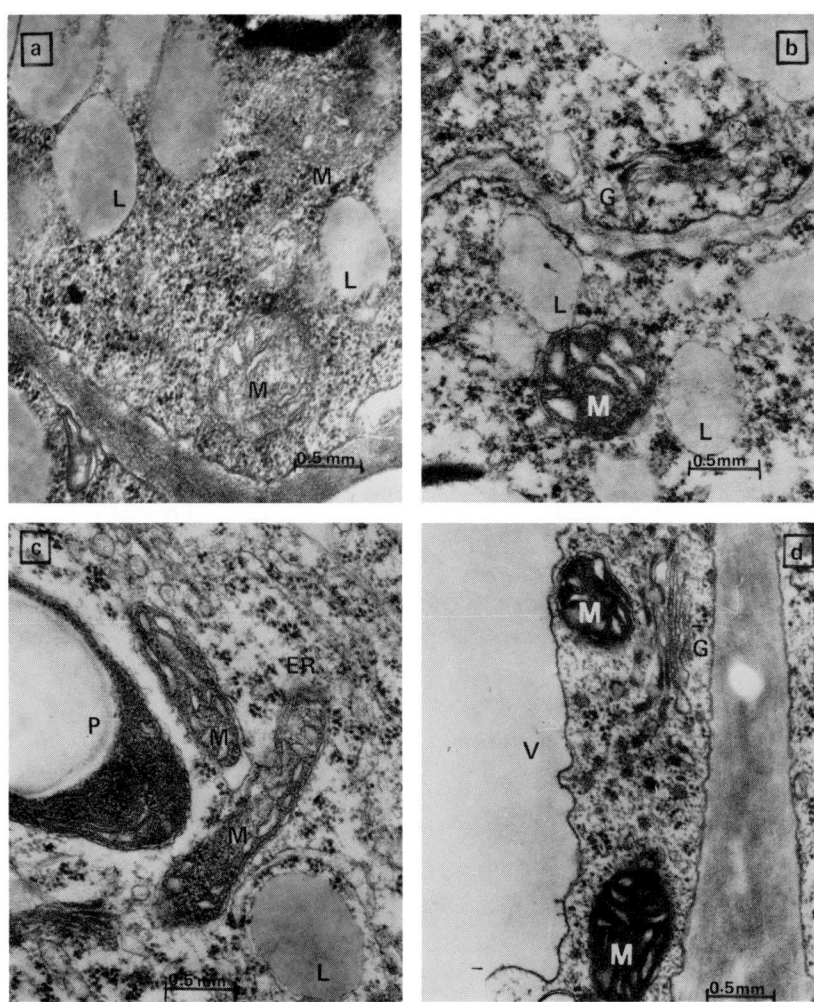

Figure 8. Ultrastructure of mitochondria in rice coleoptiles grown in water under
aerobic conditions: (a) 1 day, (b) 2 days, (c) 3 days, (d) 5 days.
Abbreviations: M - mitochondria; P - plastid; ER - endoplasmic reticulum;
L - lipid droplet; G - dictyosome.

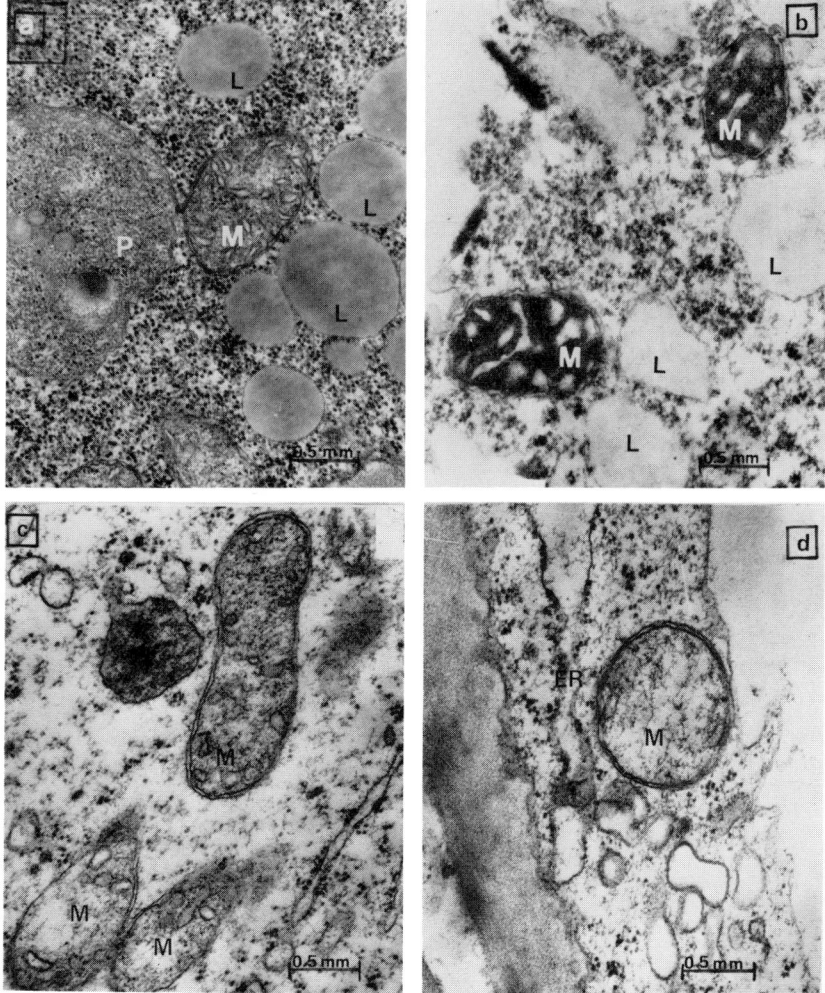

Figure 9. Ultrastructure of mitochondria in rice coleoptiles grown in chloramphenicol solution under aerobic conditions: (a) 1 day, (b), (c) and (d) for 2, 5 and 6 days.   (Abbreviations as in Figure 8.)

Table IV. Changes in Parameters of Coleoptiles' Mitochondrion (M) During Aerobic and Anaerobic Germination of Rice Seeds With and Without Chloramphenicol

Age of Seedlings (days)	Aerobic Germination				Anaerobic Germination			
	In Water		In Chloramphenicol		In Water		In Chloramphenicol	
	Cristae per M	Area per M (μm^2)	Cristae per M	Area per M (μm^2)	Cristae per M	Area per M (μm^2)	Cristae per M	Area per M (μm^2)
1	8	0.232 ± 0.022	8	0.336 ± 0.023	4.5	0.246 ± 0.018	3	0.201 ± 0.016
2	11	0.267 ± 0.018	10	0.210 ± 0.020	10	0.158 ± 0.020	8	0.136 ± 0.012
3	13	0.367 ± 0.018	10	0.264 ± 0.031	9	0.181 ± 0.019	8	0.156 ± 0.012
4	–	–	5	0.230 ± 0.019	9	0.264 ± 0.016	8	0.218 ± 0.017
5	12	0.122 ± 0.001	4	0.261 ± 0.042	–	–	–	–
6	12	–	2	0.305 ± 0.023	8	0.261 ± 0.021	7	0.196 ± 0.017

Experimenting with seeds of pea and haricot, these researchers have shown that the formation of mitochondria in the early stages of seed germination probably takes place by assembling membranes of cristae from proteins that pre-existed in the mitochondria and in the cytoplasm of dry seeds, but not by protein synthesis *de novo*.

From 48-72 hr, substantial differences occurred in mitochondria in water and in CAP. The volume of mitochondria increased in coleoptiles grown in water without chloramphenicol, and their shape became oblong and dumbbell in form (Figure 8c). But the volume of mitochondria in 5-day-old coleoptiles decreased almost 3-fold, and their shape became rounded or oval (Figure 8d). In water, cristae number reached a peak by the third day of germination and remained at that level until the end of the experiment. These findings imply that the division of mitochondria and related *de novo* synthesis of mitochondrial protein takes place between the third and fifth day. After three days of germination in the presence of chloramphenicol, profound changes were observed in the mitochondrial ultrastructure. The mitochondria increased in size, were irregularly shaped, and the mitochondrial matrix was clarified. The most striking change was the progressive loss of mitochondrial cristae: the number of mitochondrial cristae decreased to four after five days of germination (Figure 9c), to two by the sixth day, and were absent in some mitochondria cristae (Figure 9d). The remaining singular cristae were vesiculated. There remained only the inner limiting membrane which was parallel to the outer mitochondrial membrane.

The progressive loss of cristae observed after three days of germination with chloramphenicol is indicative of the absence of mitochondrial ribosome synthesis of some protein components, which are crucial for assembling membranes of mitochondrial cristae. At present, it is known that such components are polypeptides synthesized on mitochondrial ribosomes and included into the composition of three enzymatic complexes localized on the internal mitochondrial membranes: cytochrome oxidase, cytochrome b and oligomycin-sensitive ATPase.[27,33] The progressive loss of cristae has been noted by a number of authors who studied the chloramphenicol action on mitochondria of yeast,[34] paramecium,[35] algae,[36] and animal cells.[37]

Biochemical investigations carried out along with the cytological studies[38-40] demonstrated inhibition of mitochondrial protein synthesis by chloramphenicol but the intensity of coleoptile respiration was not reduced, on the contrary, the stimulation of respiration was observed even after 6 days of germination (Figure 10). By the seventh day, the intensity of respiration per whole coleoptile was 2.5 times more than that of the control.

The enhanced respiration of coleoptiles growing in a chloramphenicol solution is practically insensitive to KCN, an inhibitor of cytochrome

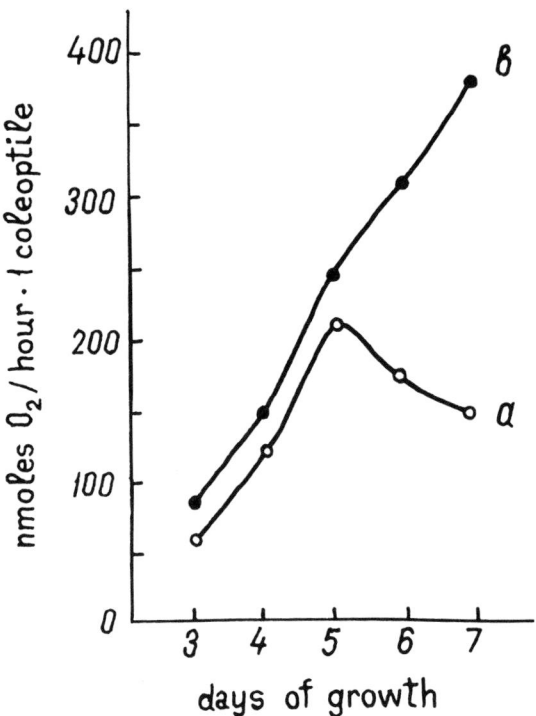

Figure 10.   Oxygen absorption in rice coleoptiles of various ages grown aerobically
(a) in water and (b) in chloramphenicol solution.

oxidase (Figure 11), while in the control plants, cyanide-sensitivity is high
up to the seventh day.

Similar results were obtained with mitochondria isolated from coleoptiles grown in a chloramphenicol solution.[40]   As seen in Figure 12, in
this case KCN inhibits oxygen absorption in the course of succinate oxidation by mitochondria only by 10% (Figure 12b) while in mitochondria
isolated from coleoptiles of plants grown without chloramphenicol, KCN
reduces oxygen absorption by 90%.   However, if salicyl-hydroxamic acid,
a specific inhibitor of the alternate cyanide-insensitive pathway of electron
transport in plant mitochondria,[41] is added to the mitochondrial suspension
together with cyanide, oxygen absorption by mitochondria of plants grown
in a chloramphenicol solution is almost completely inhibited (Figure 12b).
These results show that in mitochondria of plants grown in a chloramphenicol solution, the alternate cyanide-insensitive oxidase is functional.
The nature of cyanide-insensitive respiration, its physiological role as well

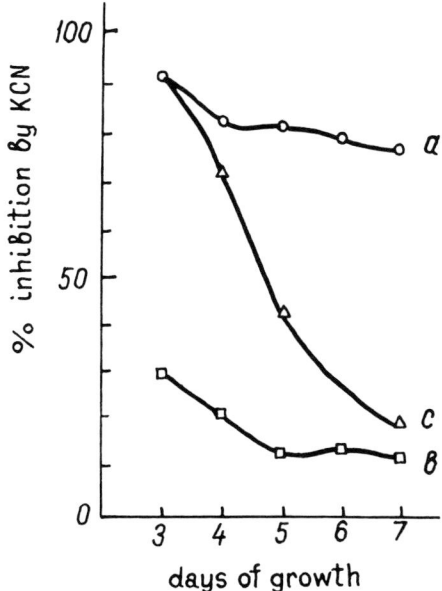

Figure 11. Changes in sensitivity of coleoptiles' respiration to KCN (10^{-3} M), cultivated under aerobic conditions (a) in water, (b) in chloramphenicol solution, (c) 3-day germination in water, then in chloramphenicol solution.

as biogenesis of enzymes of cyanide-insensitive respiration, remains unclear
in spite of efforts in recent years in a number of laboratories.[42-47]

Differential spectra of cytochromes (Figure 13) also indicate a considerable lack of cytochrome oxidase in the mitochondria in such experiments.
Absorption at 603 nm (maximum of cytochrome oxidase absorption) is
practically absent in mitochondria isolated from the plants grown in chloramphenicol solution. But a maximum at 552 nm and a shoulder in the
560 nm range suggest the presence of cytochromes b and c in such mitochondria.

A decrease in cytochrome oxidase activity is connected to the discontinuance of its neoformation due to the chloramphenicol-induced inhibition
of mitochondrial synthesis of polypeptides of this enzyme and simultaneous
breakdown of pre-existing stores of the enzyme.

Under conditions of inhibition of mitochondrial protein synthesis, considerable transformation takes place in the terminal site of the respiratory
chain.

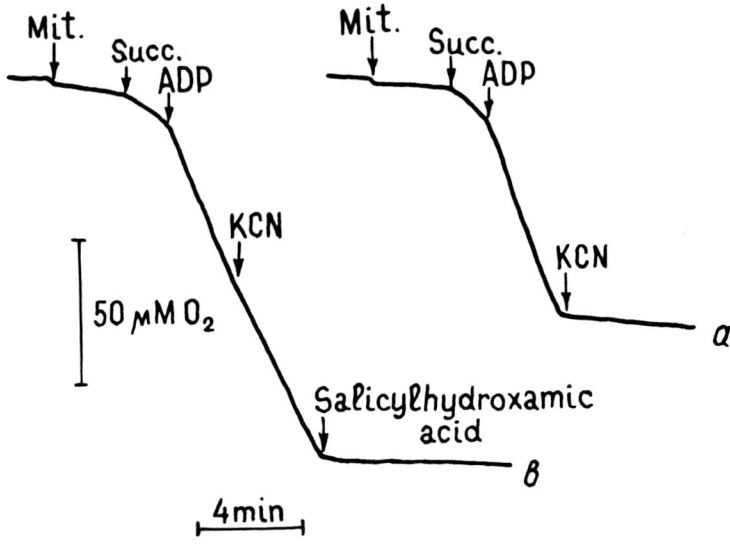

Figure 12. Polarographic recording of O_2 absorption in the course of succinate oxidation by mitochondria isolated from 6-day-old coleoptiles grown aerobically (a) in water, (b) in chloramphenicol solution. Arrows show introducing into the polarographic cell: Mit. - mitochondria; Succ. - succinate, 10 mM; ADP, 0.4 mM; KCN, 1 mM; salicylhydroxamic acid, 4 mM.

The dehydrogenase sites of the respiratory chain do not limit electron transport. Such a conclusion can be drawn both from the similarity of the rates of NADH and Krebs cycle substrates' (except malate) oxidation by mitochondria of the experimental and the control plants (Table V), and from high succinate dehydrogenase activity revealed in plants grown in a chloramphenicol solution.[30]

Table V. Rate of Substrate Oxidation by Mitochondria Isolated from 6-day-old Coleoptiles Grown Aerobically in Water and in Chloramphenicol (CAP) Solution

Substrate	nmol O_2/min/mg Protein	
	Water	CAP Solution
Succinate (10 mM)	147.2	125.7
NADH (1 mM)	155.4	177.7
Malate (20 mM)	114.5	55.0
α-Ketoglutarate (10 mM)	103.3	106.1

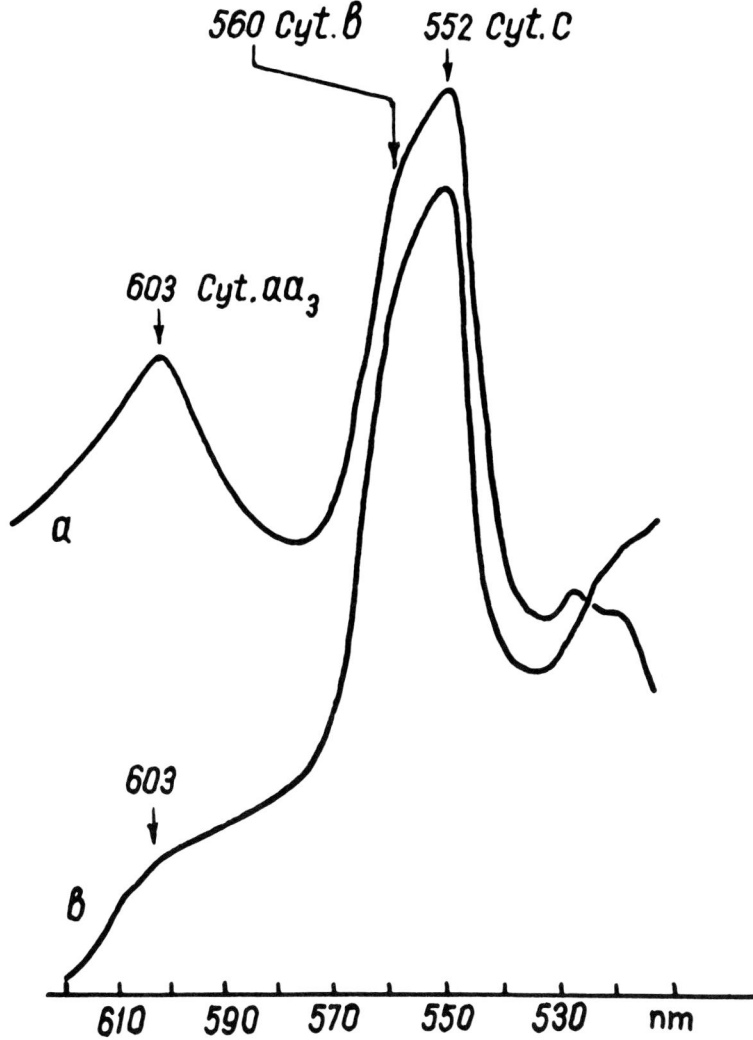

Figure 13. Differential spectra of cytochromes in mitochondria isolated from 6-day-old coleoptiles grown aerobically (a) in water and (b) in chloramphenicol solution. Curves were obtained by cytochrome reduction with dithionite.

Anaerobic Development of Mitochondria in the
Presence of Chloramphenicol

Mitochondria of coleoptiles grown under anaerobic conditions for 24 hr have a less-developed structure than that of mitochondria under aerobic conditions (Figures 8a and 14a). But after 48-hr germination, both in water and in a chloramphenicol solution, mitochondria are well shaped and similar morphologically to mitochondria of aerobic coleoptiles (Figures 14b and 15b). These data suggest that the formation of mitochondrial ultrastructure under anaerobic conditions both in water and in a chloramphenicol solution takes place anaologous to that of aerobic coleoptiles and does not depend on the presence in the environment of either oxygen or an inhibitor of mitochondrial protein synthesis.

After 96-and 144-hr germination in water (Figure 14c,d), an enlargement of mitochondrial volume occurs, and the intracristal space of some cristae are dilated. The latter may be caused by a low energetic supply of the cell organelles due to the absence of a mechanism for aerobic decay of the respiratory substrate. Similar results were obtained earlier[15] in experiments with rice coleoptiles grown anaerobically in water.

In coleoptiles germinated in a chloramphenicol solution, the intracristal space of mitochondria is also gradually dilated and the matrix is partially clarified (Figure 15c,d). However, contrary to similar experiments under aerobic conditions, the presence of chloramphenicol under anoxia does not cause the advanced destruction of mitochondrial cristae.

Another distinction of the anaerobic experiment was that no substantial differences between coleoptiles grown in water and in a chloramphenicol solution were detected regarding the respiratory activity. The intensity of respiration of anaerobic coleoptiles growing in a chloramphenicol solution is similar to the control (Figure 16). Finally, the respiration of anaerobic and aerobic coleoptiles differed qualitatively since cyanide-sensitivity of the respiration in coleoptiles grown under anaerobic conditions was not greatly reduced by the chloramphenicol solution. As in control plants grown anaerobically in water (Figure 17), cyanide-sensitivity of coleoptiles grown anaerobically in the presence of chloramphenicol was 80%.

These observations in coleoptiles were confirmed in the experiments with isolated mitochondria. The results presented in Table VI imply that the anaerobic germination in a chloramphenicol solution does not induce the alternate cyanide-insensitive pathway of respiration in mitochondria as it does in the aerobic experiments.

In general, the above data suggest there is no biosynthesis of polypeptides of respiratory enzymes in mitochondria under anaerobic conditions. Perhaps the respiratory enzymes of anaerobic mitochondria in rice coleoptiles originate, in part, from the pre-existing enzymes of dry seeds and,

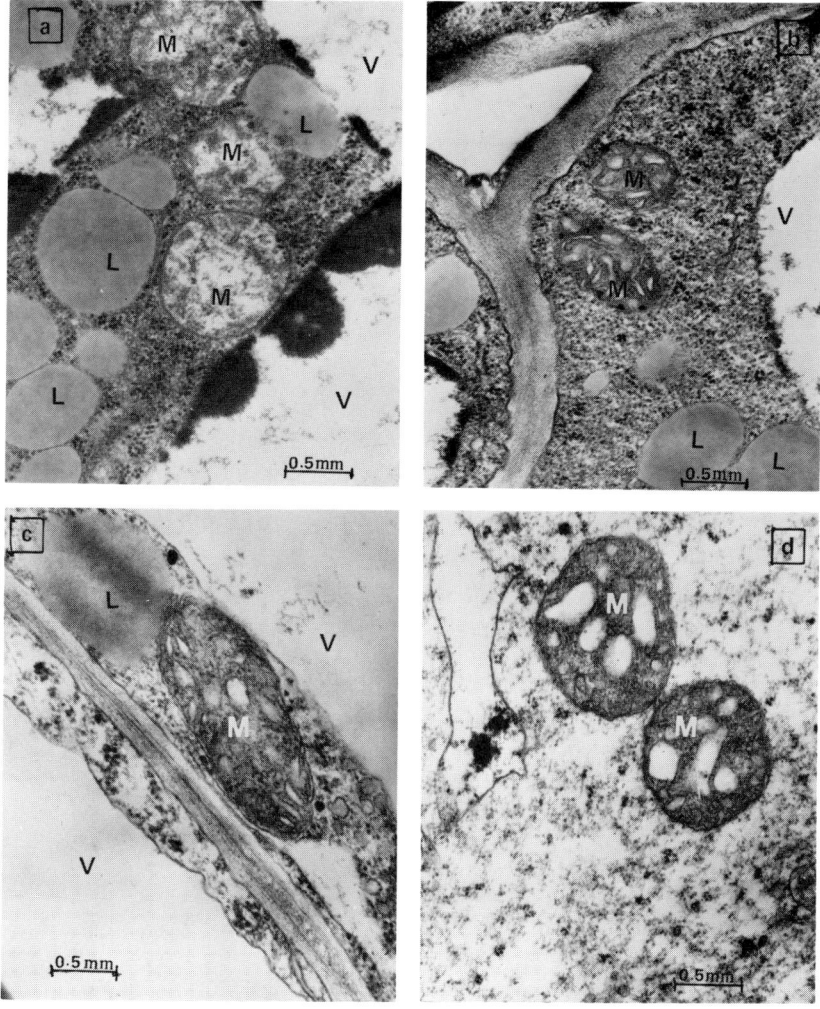

Figure 14. Ultrastructure of mitochondria in rice coleoptiles grown in water under anaerobic conditions. (a) 1 day, (b) 2 days, (c) 4 days, (d) 6 days. Abbreviations: M - mitochondrion; ER - endoplasmic reticulum; V - vacuole; L - lipid droplet.

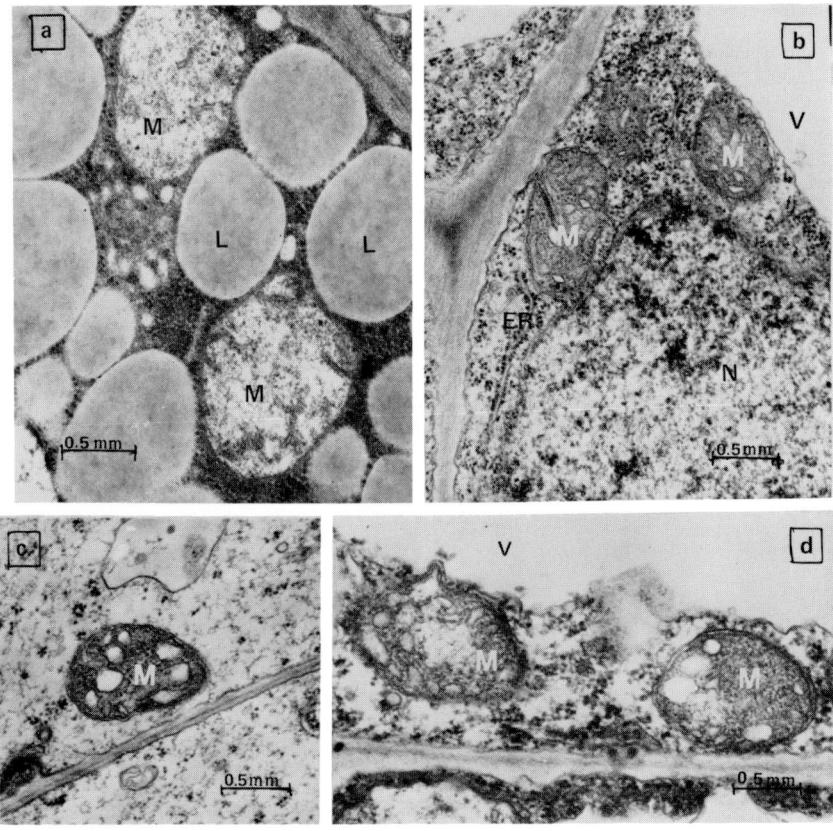

Figure 15.   Ultrastructure of mitochondria in rice coleoptiles grown in chloramphenicol solution under anaerobic conditions.   (a) 1 day, (b) 2 days, (c) 4
days, (d) 6 days.   Abbreviations: M - mitochondria; ER - endoplasmic
reticulum; V - vacuole; L - lipid droplet.

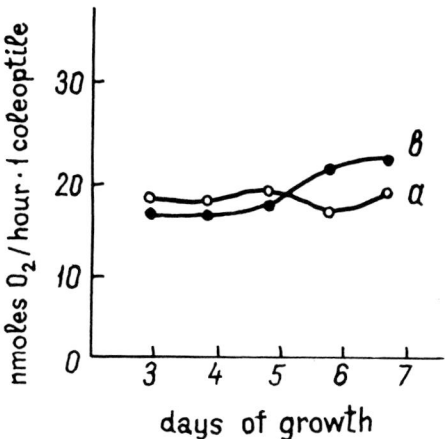

Figure 16. Oxygen absorption by coleoptiles of various ages grown anaerobically (a) in water, (b) in chloramphenicol solution.

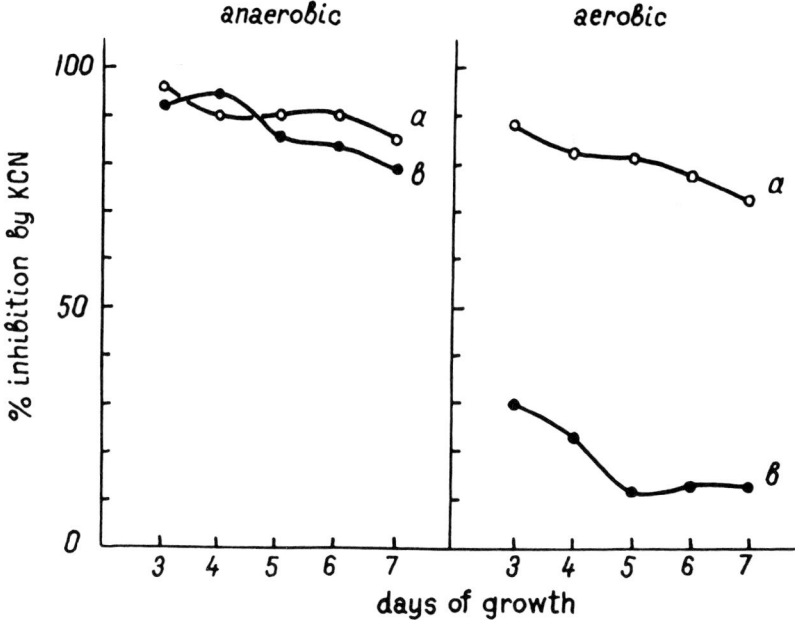

Figure 17. Changes in the sensitivity of the respiration of coleoptiles to KCN (10^{-3} M), cultivated under aerobic and anaerobic conditions (a) in water, (b) in chloramphenicol solution.

Table VI. Cyanide Sensitivity of the Respiration of Mitochondria Isolated from Coleoptiles Grown With and Without Chloramphenicol

Mitochondria Isolated From	Rate of Succinate Oxidation (nmol O_2/min/mg Protein)	Inhibition by KCN (%)
Anaerobic Coleoptiles	38.5	93
Anaerobic Coleoptiles + CAP	31.3	85
Aerobic Coleoptiles	130.0	92
Aerobic Coleoptiles + CAP	125.7	19

in part, are formed in the first days of seed germination by assembling mitochondrial membranes from pre-existing protein precursors. In such mitochondria, synthesis of the alternative oxidase does not proceed in a chloramphenicol medium. All processes of the coleoptile cells are directed to biosynthesis of the enzymes of anaerobic metabolism. Under these conditions, essential transformation of mitochondrial membrane observed in experiments with aerobic plants in chloramphenicol solution does not take place (Table IV).

Thus, on the basis of these results, the development of mitochondria in rice coleoptiles can be divided into two stages. The first lasts from poorly differentiated mitochondria of the seed embryo to the completely formed mitochondria within 48 hr of germination. This stage is characterized by the insensitivity to both an inhibitor of mitochondrial protein synthesis and molecular oxygen, which indicated that mitochondria are probably assembled from a pre-existing material. In this stage, no essential differences in the biogenesis of mitochondria of coleoptiles grown in the presence or absence of oxygen have been detected.

The second stage (2-6 days in our experiment) is characterized in aerobic coleoptiles by an enlargement of mitochondrial volume and the neoformation of components of the mitochondrial membrane. Therefore, the presence of chloramphenicol in the growth medium influences the mitochondria formation. In anaerobic coleoptiles during the second stage, the biosynthesis of respiratory enzymes, at least on mitochondrial ribosomes, does not proceed. That is why mitochondria of such coleoptiles are insensitive to chloramphenicol. As a result, in the second stage there are significant differences in the development of mitochondria of coleoptiles under aerobic and anaerobic conditions.

The Effect of Secondary Anoxia on the
Mitochondrial Ultrastructure

The electron microscope investigations discussed earlier demonstrate the extreme resistance of coleoptile cells' fine structure to anoxia. However, in the experiments mentioned, the rice seedlings were under conditions of anoxia from the very beginning of seed germination until conclusion of the experiment. One could conclude, therefore, that the resistance to anoxia arises and is preserved in coleoptiles under conditions of anoxia, *i.e.*, when the mechanism of oxygen metabolsim of the cell does not function at all.

In view of the above results in anaerobic coleoptiles, we were interested whether the fine structure of rice coleoptile cells would be resistant to prolonged *secondary* anoxia, *i.e.*, anoxia induced after preliminary germination of the seeds in the presence of molecular oxygen. Under these conditions, the enzymatic systems of aerobic metabolism would have begun to function actively in the coleoptile cells, and the coleoptiles would be absorbing molecular oxygen from the environment.[10,11,48]

We also tested the tolerance to anoxia of ultrastructure of the root and leaf of aerobically grown rice seedlings. Unlike coleoptiles, these organs require oxygen in the environment for their growth.

Intact Seedlings

Seedlings were grown for 6 days under conditions of normal aeration and then the whole seedlings with attached seeds were transferred to conditions of prolonged anoxia.

The data show that the transfer of aerobically grown seedlings to conditions of strict anoxia does not induce any degradation of mitochondria and other organelles of coleoptile cells even if anoxia lasts 3 or 5 days.[49,50] Nevertheless, under these conditions, mitochondrial ultrastructure changes (Figure 18). Initially, in an oxygen-free environment, most mitochondrial cristae increase in number and are arranged parallel 8-10 or more cristae in a stack. Usually, the stacks lie transversely to the long axis of the organelle, but they can lie along its axis. As the period of anoxia is prolonged from 1 to 3-5 days, the mitochondrial cristae become more organized. A part of the mitochondria remains unchanged in size and shape (Figure 18b), while part transformed into large organelles with many layers of cristae (Figure 18c). These mitochondria are often irregularly shaped and larger in size (1.2-1.5 μm long, whereas normal mitochondria are 0.5-0.6 μm long). The entire inner compartment of such mitochondria are filled with densely packed parallel cristae.

Such mitochondria have not been described previously in plant cells as far as we know. They are found in animal cells and perform especially

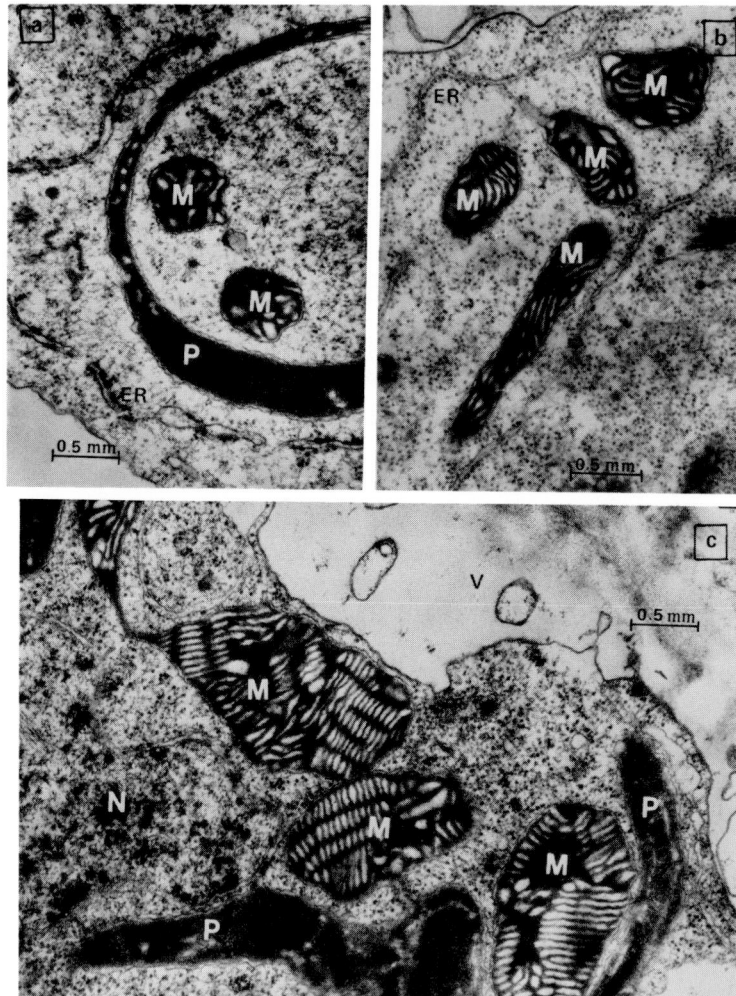

Figure 18. Ultrastructure of mitochondria in rice coleoptile cells (intact seedlings) under secondary anoxia (a) 6-day germination under aerobic conditions, (b,c) the same and then 5-day anoxia in water. Abbreviations: M - mitochondria, P - plastid, ER - endoplasmic reticulum, N - nucleus.

active work, *e.g.*, in cells of flight muscles in insects and cells of myo-cardium in mammals.[51,52]

The appearance of such unusual mitochondria in rice coleoptile cells under conditions of total exclusion of oxygen can hardly be attributed to the beginning of organelle degradation. Rather, it points to hypertrophy of the mitochondrial apparatus, arising in rice coleoptile cells because of a shortage in oxygen supply.

Similar changes of mitochondrial ultrastructure with an insignificant enlargement of mitochondrial volume can be observed in the cells of aerobically grown leaves after they are transferred to conditions of anoxia (Figure 19b). In this case, the degree of order of the cristae is also gradually increased, and after 5 days of anoxia, practically all leaf mito-chondria have stacks of parallel arranged cristae, which extend through the inner mitochondrial compartment.

In mitochondria, from root cells of seedlings under the same conditions, parallel cristae appear as early as one day after anoxia. Within 2 days of anoxia (Figure 19d), mitochondria of roots also increase in size and con-tain orderly arranged cristae (stacks of 10 and more cristae). However, after 3 days of anaerobiosis, mitochondria begin to degrade. Similar destructive changes in the root cells of adult rice and pumpkin plants are encountered.[53]

Thus, the tolerance of the ultrastructure of cells of coleoptiles, primary leaf and root of aerobically grown rice seedlings to anoxia, is different. The cells of coleoptile and primary leaf are the most resistant as they retain the undamaged ultrastructure even after 5 days of anoxia, while clear signs of degradation of ultrastructure of the root cells were observed within 3 days.

One reason for the high resistance of the coleoptile cell ultrastructure of aerobically grown rice seedlings to anoxia is probably due to their ability to develop vigorous glycolytic processes in the absence of oxygen.[54,55] Under these conditions, glycolytic processes proceed so intensively that they provide sufficient energy to maintain the processes of synthesis in the coleoptile and leaf tissue and promote the preservation of the ultra-structure of cell organelles.

The following explanations may account for the notably lower tolerance of the ultrastructure of root cells of seedlings. Either the glycolytic processes in the root cells are less vigorous and cannot satisfy the energetic requirements of the cell or, on the contrary, the glycolytic processes in the root cells are too intensive, thereby bringing about a rapid exhaustion of freely utilized organic substances in the cells. Finally, the possibility of a greater sensitivity of the root cells to toxic products of anaerobic metab-olism cannot be excluded.

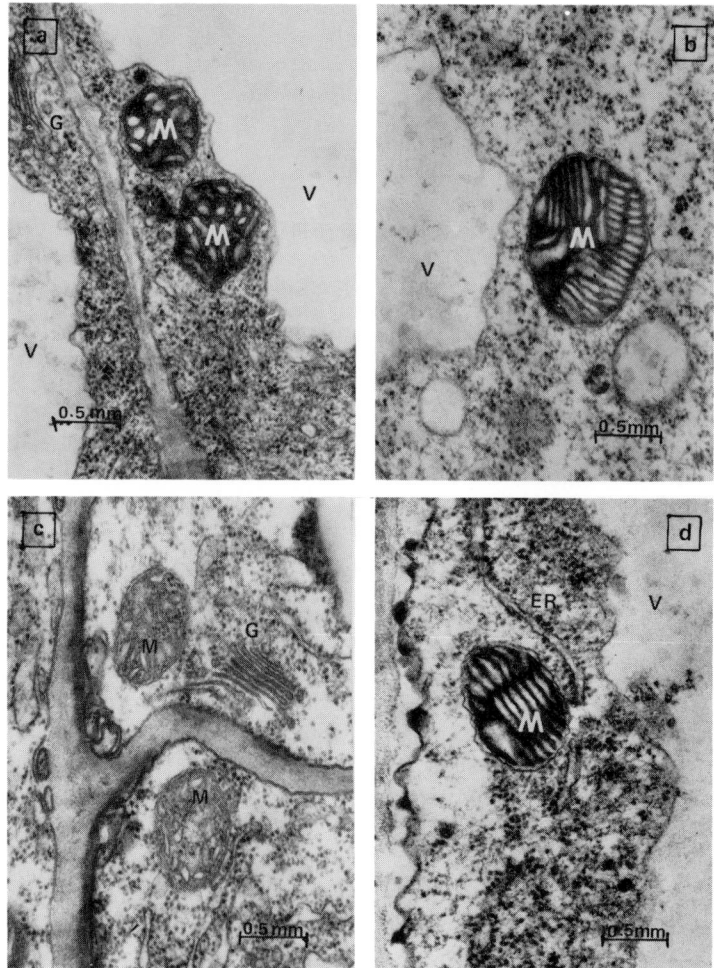

Figure 19.   Ultrastructure of mitochondria in leaves' (a,b) and roots' (c,d) cells of intact rice seedlings under secondary anoxia: (a) 5 days of germination under aerobic conditions; (b) the same and then 5 days of anoxia in water; (c) 5 days of germination under aerobic conditions; (d) the same and then 2 days of anoxia in water.   Abbreviations: M - mitochondrion; ER - endoplasmic reticulum; G - dictyosomes; V - vacuole.

The high resistance of the ultrastructure of coleoptile cells to anaerobiosis[10],[14-16] is not limited by the seedlings grown from the very beginning of seed germination under conditions of strict anoxia. It is preserved to a considerable extent also in aerobically grown rice coleoptiles in which the enzymatic apparatus of the cell oxygen metabolism has been normally formed and has functioned before the induction of anoxia.

In the following experiments, we tried to elucidate the causes of the high resistance of mitochondria and other cell organelles of rice seedlings to anoxia. Is the resistance the feature of mitochondria themselves or is it explained by the physiological peculiarities of the whole plant?

For this purpose, we studied the changes of mitochondrial ultrastructure of coleoptiles, leaves and roots after their detachment from the seed and placement under conditions of strict anoxia.

Detached Organs

The action of secondary anoxia on detached parts of the seedlings is significantly different from the effect on intact seedlings. As early as 1 day after anoxia, mitochondria of detached coleoptiles and leaves are irregularly shaped and cristae are strongly dilated (Figure 20a). This is the initial stage of mitochondria disintegration, which is complete after 2 days of anoxia (Figure 20b,c). Similar destructive changes of mitochondria take place in excised roots as early as one day after anoxia (Figure 20d). Meanwhile, in detached coleoptiles, roots and leaves placed in water under aerobic conditions for 2 days, pathological changes of mitochondria do not occur. Under such conditions, the mitochondrial ultrastructure is similar to that of mitochondria of intact aerobically grown seedlings.

These data led us to hypothesize that the cells of detached coleoptiles, leaves and roots of rice seedlings do not contain sufficient amounts of storage substances to sustain the normal activity of glycolytic processes under conditions of long-term anaerobiosis. Consequently, carbon starvation of the cells soon brings about degradation of the cell ultrastructures.

To test this hypothesis as well as to elucidate the possible role of the transport of organic substances from the seed as an induction of the tolerance of these plants to anoxia, changes of the mitochondrial ultrastructure were studied in detached organs of seedlings supplied with exogenous glucose.

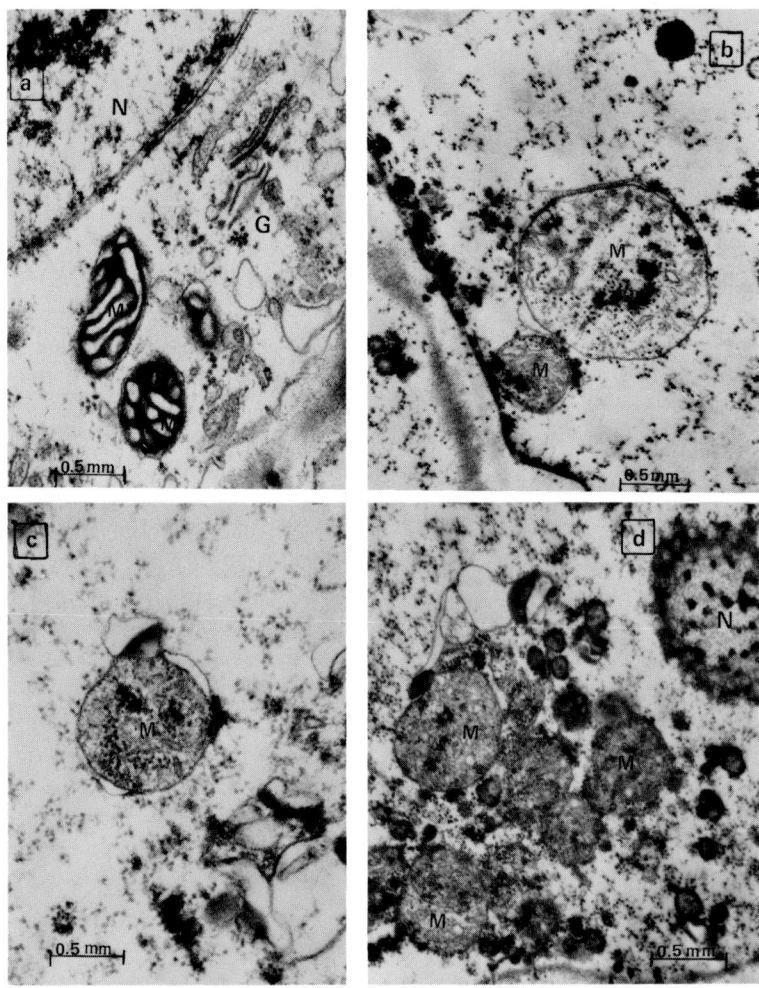

Figure 20.   Ultrastructure of mitochondria of excised organs of rice seedlings under secondary anoxia. Pretreatment: 5 days of germination under aerobic conditions. (a) coleoptiles, 1 day of anoxia in water; (b) coleoptiles, 2 days of anoxia in water; (c) leaves, 2 days of anoxia in water; (d) root, 1 day of anoxia in water.

Detached Organs Fed with Glucose

No destructive changes of mitochondria or other cell organelles were observed when the detached coleoptiles and primary leaves were placed in a 0.5%-1% glucose solution, even after 5 days of anoxia.

On the contrary, under these conditions a rearrangement of the mitochondrial ultrastructure similar to that described in Figure 18 for coleoptile mitochondria of the whole seedlings under anaerobic conditions is noted. Also, the longer the coleoptiles and leaves were kept under anoxia, the higher was the degree of order of the cristae (Figure 21a,b).

In excised roots placed in a 0.5% glucose solution, all mitochondria have a parallel arrangement of the cristae after 1 and, especially 2 days of anaerobiosis (Figure 21c). Only after 3 days of anoxia of the roots, the pattern of destruction of the mitochondria and other cell organelles was similar to that described earlier in roots of the whole seedlings under anaerobic conditions (Figure 21d). Thus, the placement of detached roots in a glucose solution promotes the preservation of ultrastructure of the cell organelles for 2-3 days, *i.e.*, for the same time as in roots on intact seedlings.

The root cells proved to be less resistant to anoxia than the cells of coleoptiles and leaves. This might be due to a higher activity of the anaerobic metabolism in the root so that the organic substances supplied to the root cells either from the seed or from a 0.5% glucose solution may have been insufficient to maintain the sugar content at a desirable level. However, additional investigations have shown that increasing the glucose concentration up to 3% does not induce a significant enhancement of the resistance of root cells. Under these conditions, destructive changes in the root cells are manifested after 96 hr of anaerobiosis.

Further studies are needed to explain the cause of the low resistance of the root cells to anoxia. Even on the basis of the present knowledge, it appears reasonable to assume that the glycolytic processes in the oxygen-free environment cannot provide the root cells with sufficient energy and substances to maintain the processes of biosynthesis for long periods of time. At present, the hypothesis of a greater sensitivity of the root cells to toxic products of anaerobiosis cannot yet be discarded.

Thus, the rearrangement of mitochondrial ultrastructure under anoxia, manifested in the enlargement of the area of the inner mitochondrial membrane and in parallel stacking of cristae, is peculiar to all organs of the seedling (coleoptile, leaf and root). It is observed both in the whole seedlings under anaerobic adaptation and in the detached organs of the seedling when the process of adaptation proceeds in a medium enriched by glucose. It should be emphasized that the noted changes of mitochondrial ultrastructure take place only when an active rearrangement of

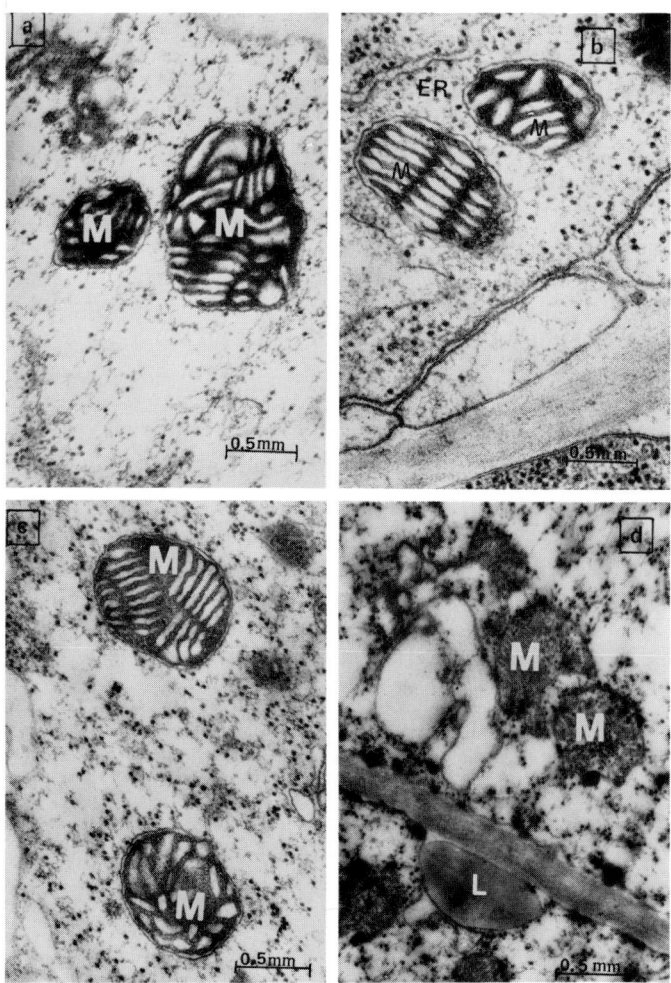

Figure 21. Ultrastructure of mitochondria of excised organs of rice seedlings under secondary anoxia and glucose feeding. Pretreatment: 5 days of germination under aerobic conditions: (a) coleoptiles, 5 days of anoxia in 0.5% glucose solution; (b) leaf, 3 days of anoxia in 1% glucose solution; (c) root, 2 days of anoxia in 0.5% glucose solution; (d) the same, 3 days of anoxia in 0.5% glucose solution.

metabolism occurs in the cell and the life activity of the cell is retained for a long time under anaerobic conditions. If the cell died quickly (as in the case with detached organs of the seedling without glucose feeding), the above changes of mitochondrial ultrastructure were not observed but, on the contrary, a growing disintegration of mitochondria and other cell organelles occurred.

In addition to the organs of young rice plants,[49,50] we have observed an analogous rearrangement of mitochondrial ultrastructure in the cells of excised roots of adult plants of rice and pumpkin in a glucose solution under anaerobic conditions (Figures 37, 38).

Morisset[57] has reported the appearance of the characteristic parallel arrangement of cristae in mitochondria of isolated tomato roots after 72-hr anaerobiosis.

Consequently, the described rearrangement of mitochondrial ultrastructure is peculiar not only to rice seedlings, but seems to be more universal.

The protective role of glucose in the preservation of the fine structure of cells under anoxia, revealed in the experiments on additional feeding with exogenous glucose, should be particularly stressed. On the one hand, these experiments show that life of the cell in the oxygen-free environment can be substantially prolonged by carbon feeding. On the other hand, the experiments demonstrate that a more rapid degradation of the fine structure of organelles and death of the cells of detached organs of rice seedlings under conditions of anoxia is not a result of asphyxia or poisoning by toxic products of anaerobic metabolism but primarily caused by a carbon starvation.

In general, the data obtained led us to conclude that the remarkable capacity of rice coleoptiles to grow under conditions of a total exclusion of molecular oxygen and to preserve undamaged ultrastructure and functions of mitochondria and other cell organelles should be explained not by the tolerance of the subcellular structures themselves to anoxia but, first of all, by the ability of rice seedlings, unlike other higher plants, to transport, easily, organic substances over relatively long distances from the grain to growing parts of the seedling, even under conditions of strict anoxia. Of course, such a conclusion does not exclude but, on the contrary, supposes the presence of the ability in rice coleoptile cells to mobilize organic substances provided by the seed in the absence of oxygen. Pradet and Prat[140] recently have shown that during anaerobic germination, rice coleoptiles are able to keep high levels of energy change due to glycolysis. Naturally, during promptly developing glycolytic processes, organic substances are consumed earlier than under conditions of free access to molecular oxygen. In intact plants, the intensive wastage of storage substances by the cells under anoxia is compensated for by the active inflow of assimilates from the seed. As to detached organs of the seedling deprived of the capacity to receive organic substances from the seed or from the environment, carbon starvation arises and is soon followed by a disorganization of metabolism and degradation of the cell ultrastructure.

ROOT ANAEROBIOSIS AND OXYGEN TRANSPORT FROM SHOOTS TO ROOTS

The physiological role of leaves in root aeration and particularly in oxygen supply under conditions of restricted gas exchange of the rhizosphere has long been a matter of interest to researchers.[58-64]

Since the 1960s, the interest in the problem of oxygen transport from shoots to roots has greatly increased. On the one hand, it was probably associated with the appearance of sensitive new methods for experimental study of this phenomenon. On the other hand, it was related to the practical importance of the problem of establishing scientific bases of plant cultivation (including agricultural plants) under the conditions of overmoistening, irrigation, hydroponics and in connection with mass loss of winter crops under a continuous ice crust. The problem arises when the root system experiences an acute lack of molecular oxygen supply. In this regard, investigators began to use widely in their experiments not only representatives of wild flora related to both hydrophytes and mesophytes, but many species of agricultural plants.[65-96]

By the time the results of our first investigations in this field were published,[75] reports from several laboratories had appeared in which authors revealed notable oxygen transport from shoots to roots using different methods in a number of plants, including mesophytes.[97-103]

During the last 12 years, we used three methodical approaches for the experimental study of the problem—polarographic, electron microscope and chemoluminescent techniques—along with the method of radioactive indicators as well as inhibitors of mitochondrial respiration. The main results of these studies on cotton plants, pumpkin and rice are described below.

Polarographic Study of Oxygen Transport

The polarographic technique permits oxygen concentrations to be registered in solutions and even directly in tissues of live organisms, by means of special detectors of molecular oxygen, i.e., "oxygen electrodes."[104-112] Due to the attributes of the oxygen electrode, it is possible to observe continuously, dynamics of oxygen absorption by the organism. That explains why, in our earlier studies of oxygen transport on plants, we selected this method, assuming that with the aid of an oxygen electrode, we could directly record and quantitatively measure the translocation of molecular oxygen from aerial parts to roots of plants.[75,113]

Method

Hard electrodes, platinum wire or plate (oxygen electrode), are widely used in polarographic investigations as cathodes for measuring dissolved oxygen. Saturated calomel or silver electrode are the most frequently used as anodes. The platinum electrode can be both open or membrane-coated. In the latter case, both cathode and an auxiliary electrode (anode) are in a definite volume of electrolyte solution separated by a semipermeable membrane from the medium studies.

In our work, both open[53,75] and membrane-coated oxygen electrodes[80,81] were used. Preliminary tests proved that the electrodes used were rather sensitive to the presence of oxygen in the medium studied and provided reliable and reproducible results.

In experiments on elucidation of the physiological role of aerial parts for molecular oxygen supply of roots, the roots were placed in darkened thermostatic vessels with a nutrient solution, while shoots were left in air (Figure 22). During the experiments the solution was either vigorously stirred with a magnet stirrer, or was not stirred at all. To rule out the possibility of oxygen penetration from the external atmosphere to the rhizosphere directly through the surface of the nutrient solution, the solution was covered with a 5- to 6-mm layer of Vaseline® oil at the point of root attachment to the shoot. Control determinations showed that no significant amount of oxygen diffused from the ambient atmosphere into the rhizosphere through the Vaseline oil layer, even with vigorous agitation of the nutrient solution.

Thus, in the present experiments we tried to judge the amount of oxygen translocation from the shoots to the roots and rhizosphere by measuring oxygen concentrations in the nutrient solution with an oxygen electrode.

Results and Discussion

Figure 23 presents the results of an experiment with a 14-day-old cotton plant, from which it can be seen that after terminating aeration of the nutrient solution, a continuous decrease of oxygen concentration in the solution due to root respiration up to its complete depletion, is noted. Oxygen content in the solution was not measurable after the roots had been in the nutrient solution for 8 hr. Repeated bubbling of air through the nutrient solution was required to increase the oxygen concentration in it.

Consequently, in this and other similar experiments we failed to detect any diffusion of molecular oxygen from the roots of cotton plants into the surrounding solution.

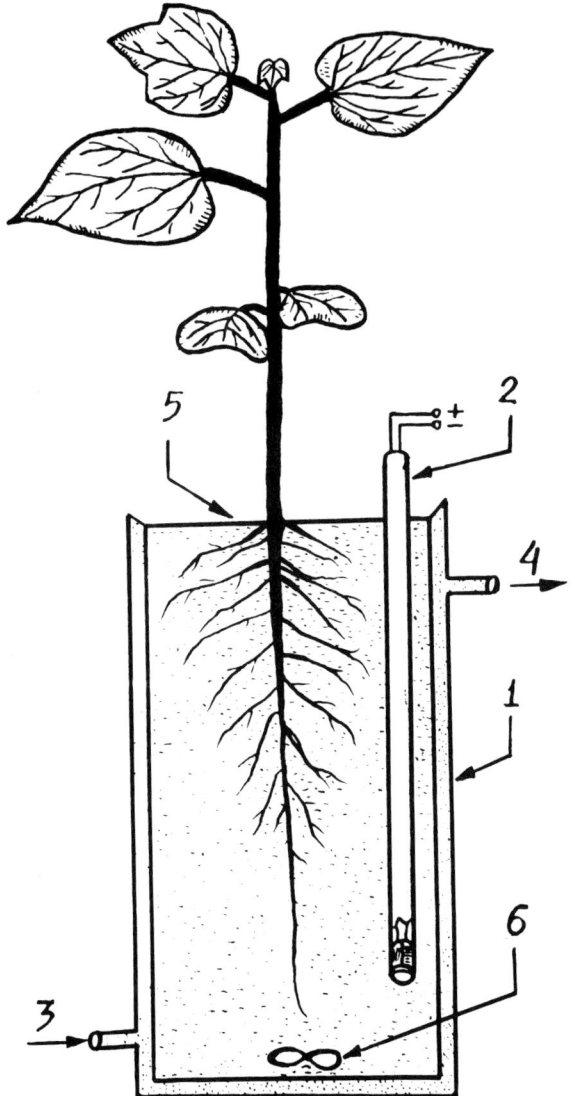

Figure 22.   Scheme of the experiment: 1. incubation vessel; 2. oxygen electrode; 3. inlet for the thermally controlled water; 4. outlet for the thermally controlled water; 5. Vaseline oil layer; 6. rotating magnet.

Nevertheless, it may be supposed that translocated oxygen, although not diffused from roots into the ambient nutrient solution, is transferred in noticeable quantities into roots and is completely utilized in the process of root respiration. To test the validity of this assumption, the experiments

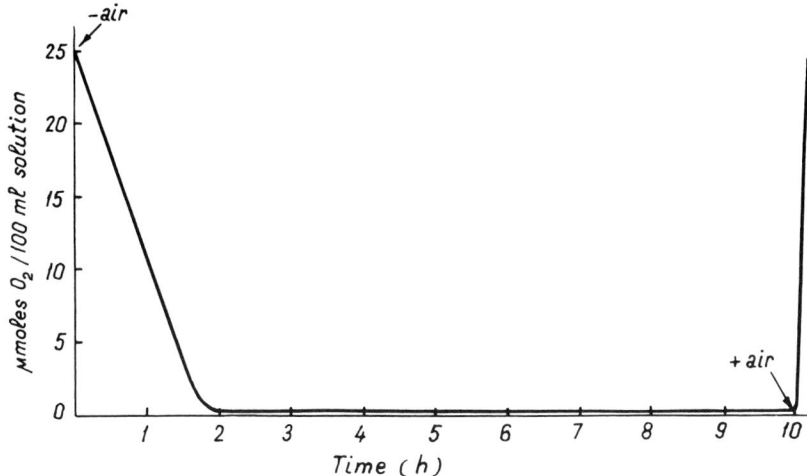

Figure 23. Oxygen uptake by roots of cotton plant. Volume of the nutrient solution, 100 ml; temperature of the nutrient solution, 27°C; fresh weight of roots, 1.46 g. Legend: +air: bubbling by air; −air: without bubbling by air.

were expanded to compare the dynamics of oxygen uptake from the nutrient solution by roots of the same plant before and after consecutive removal of the aerial parts. After removal of the aerial parts, the root has, naturally, only one source of oxygen, the nutrient solution, while the roots of an intact plant could receive oxygen also from the aerial parts if oxygen transport occurs from shoots to roots of the plants. In this case, an additional inflow of oxygen from the aerial parts should lessen oxygen uptake from the nutrient solution, while the excised root having only one source of oxygen should exhaust more rapidly the oxygen stores in the nutrient solution.

Figure 24 shows the results of one such experiment with a 25-day-old cotton plant. The data show that removal of leaves and even all the aerial parts of the plant did not notably influence the dynamics of oxygen uptake by roots from the nutrient solution. Similar findings were obtained with cotton plants of different ages.

Since the duration of the experiments was comparatively short, the procedure of removal of aerial parts itself should not strongly influence the intensity of oxygen uptake from the nutrient solution by roots. Stores of respiratory substrates in the cells of excised roots are sufficient to ensure root respiration on a more or less identical level for many hours.[114] This is substantiated also by the results of the experiment represented in Figure 24.

Thus, the data obtained strongly suggest that in the overall oxygen balance of cotton roots, the fraction of oxygen translocated from shoots was not significant.

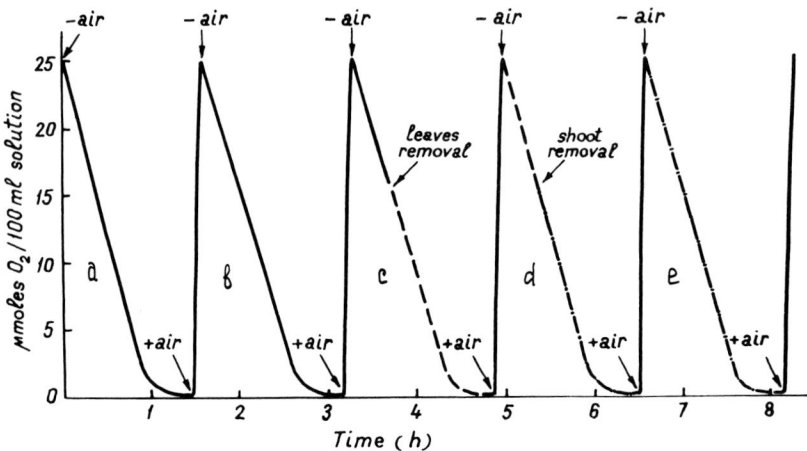

Figure 24. Oxygen uptake by roots of cotton plants before and after detachment of leaves and shoot. Volume of nutrient solution, 100 ml; temperature of the nutrient solution, 27°C; fresh weight of roots, 2.76 g. See Figure 23 and text for other details.

Similar results were found in previous experiments with pumpkin (Mozaleevskaya variety), in which oxygen content in root tissues and the rhizosphere was measured by means of open platinum electrodes.[75,77,115] Results of a typical experiment with a 30-day-old pumpkin plant are reproduced in Figure 25. In this experiment, roots of the plant were placed in a 3-liter pot with Knop nutrient solution. Oxygen content in the solution was measured by a platinum electrode placed in the rhizosphere. To place the oxygen electrode as close as possible to the routes of oxygen transport, another platinum electrode, 0.02 mm in diameter, was inserted directly into the root tissue. The solution was not agitated.

As seen from these findings, after placing the root in the nutrient solution, the oxygen electrodes registered a continuous decrease of oxygen concentration. After 8 hr, all the measurable dissolved oxygen was utilized. Even after rather long-term staying of roots in the solution, no diffusion of oxygen from the roots into the surrounding nutrient solution was detected. The subsequent bubbling of air through the solution caused a rapid rise in oxygen concentration, which meant that the previous fall of current in the polarographic circuit was indeed caused by oxygen uptake in the process of root respiration.

Thus, unlike Brown,[62] we failed to reveal notable oxygen transport from shoots to roots of pumpkin and subsequent diffusion of oxygen from roots to rhizosphere, nor could we detect this phenomenon in cotton plants.

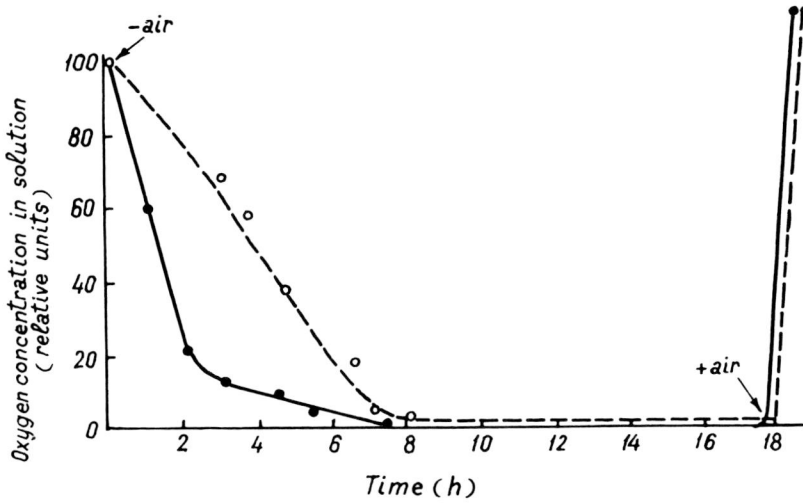

Figure 25.   Oxygen uptake by roots of pumpkin from nonagitated solution. Volume of the nutrient solution, 3 liter; temperature of the nutrient solution, $27°C$; fresh weight of roots, 20 g. Electrode readings: solid line inside the root, dotted line in the solution. Legend: See Figure 23.

Of practical interest are the results of observations, presented in Figure 26, which shows the "topography" of dissolved oxygen in the nutrient solution.[53]   In this experiment, roots of a 30-day-old pumpkin plant were placed in a 3-liter pot with Knop nutrient solution.   Three platinum electrodes for oxygen registration were placed at different points in the solution: one electrode was placed directly into the mass of roots, while another was fixed at the outer surface of root mass and, therefore, registered oxygen content in the solution in the immediate vicinity of the root system.   Finally, the third electrode was placed in the nutrient solution some distance from the roots.   In this experiment, as in the experiment shown in Figure 25, the solution was not agitated.

Records of the platinum electrode placed into the mass of roots are noteworthy.   As early as 30 min after the aeration was stopped, practically all oxygen inside the root mass was used up, while the records of the third electrode indicated the presence in the solution of about 80% of the initial oxygen content.   Only after 3 hr was all the oxygen used up in the stationary solution.

Results of these experiments demonstrate that after 30-40 min, even in a very intensively aerated nutrient solution, a portion of the roots begins to experience oxygen deficiency, if oxygen transport from the green parts

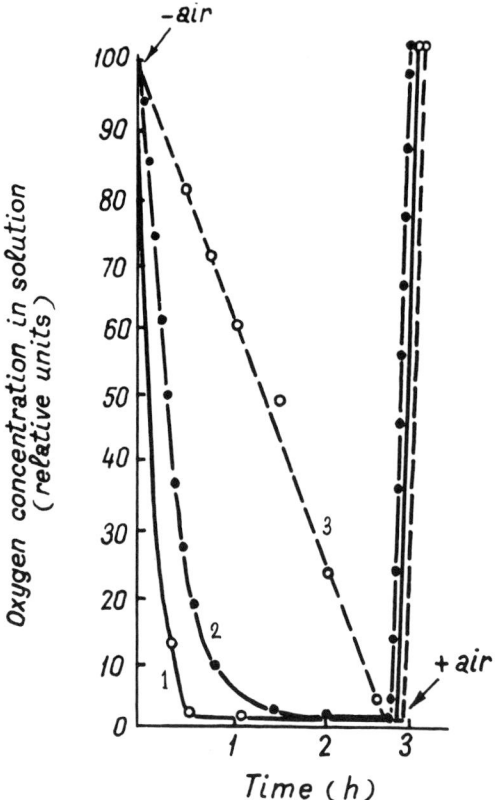

Figure 26. Distributional pattern of oxygen in the rhizosphere of *Cucurbita Pepo* L. in nonagitated solution. Volume of the nutrient solution, 3 liter; temperature of the nutrient solution, 27°C; fresh weight of roots, 100 g. Legend: See Figure 23. 1. readings of the electrode placed inside the root mass; 2. readings of the electrode placed in the solution in immediate proximity of roots; 3. readings of the electrode placed in the solution some distance away from roots.

fails to compensate for the lack of oxygen in the solution. This circumstance should be borne in mind in developing optimal oxygen regime of roots of plants grown without soil (in hydroponic nutrient solutions).

Unlike mesophytes such as pumpkin and cotton plants, hydrophytes such as rice transport oxygen from shoots to roots and oxygen is even translocated from roots into the surrounding medium. Figure 27 represents the results of the study of oxygen transport in rice carried out with the same polarographic method. It is seen that molecular oxygen partially diffuses into the rhizosphere where it can be readily registered by an oxygen electrode.

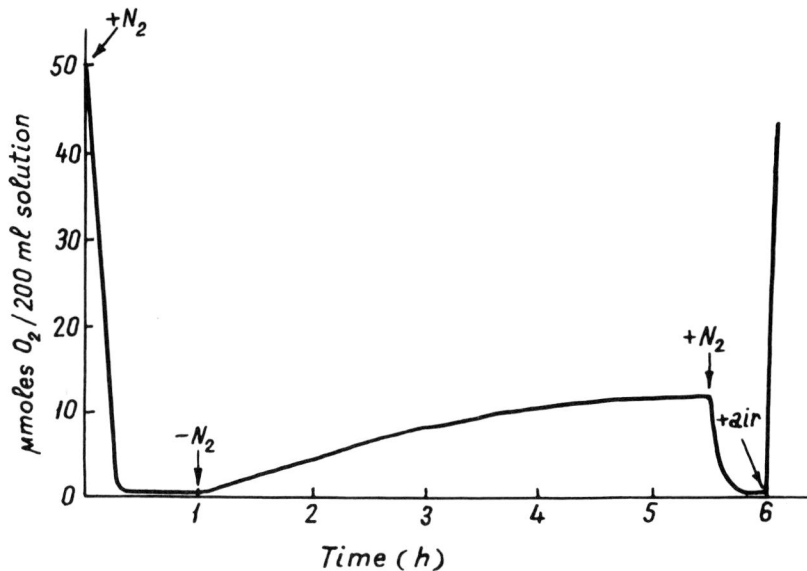

Figure 27.   Oxygen diffusion from rice roots into the solution. Volume of the nutrient solution, 200 ml; temperature of the nutrient solution, $27°C$; fresh weight of roots, 6.4 g. Legend: $+N_2$: bubbling by nitrogen; $-N_2$: without bubbling by nitrogen; $+air$: bubbling by air.

Thus, the overall results of the above experiments with pumpkin and cotton plants show that molecular oxygen transported from the aerial parts of these plants (unlike rice plants) is not a significant portion of the normal oxygen needs of roots.

Study of Oxygen Transport by Measuring Chemiluminescence of Roots

Along with the polarographic technique for the study of oxygen transport in plants, we used also the chemiluminescent method based on the phenomenon of ultraweak glowing.[78,79]

Ultraweak glowing was discovered by a group of Italian scientists[116] and thoroughly examined by Tarusov and co-workers.[117]

The main point of the phenomenon of ultraweak glowing is that every live cell emits ultraweak glowing, which can be measured only with the aid of very sensitive equipment. The phenomenon of an ultraweak glow takes place only in the presence of oxygen in the medium. Under conditions of anoxia, the intensity of glowing begins to drop and, finally, the glow emitted by cells and organisms is completely "extinguished." This circumstance was used in our experiments for studying the role of the aerial parts of plants in supplying molecular oxygen to roots.

Method

If oxygen is transported from shoots to roots and induces aerobic conditions there, an ultraweak glow of root cells should be detectable even in those experiments in which oxygen has been removed from the rhizosphere. Under these conditions, the glowing of roots would be maintained by oxygen translocated from the shoots exposed to the air. On the contrary, if under conditions of root anaerobiosis the phenomenon of chemiluminescence of the roots is absent, it would be reasonable to assume that the amount of oxygen translocated from the aerial parts is insufficient to induce normal aerobic conditions in the root cells.

Figure 28 shows a schematic drawing of the setup for measuring ultraweak glowing in roots used in our experiments. Roots of 5-day-old pumpkin seedlings were submerged in a thermostatic glass (A) with 80 ml of Knop nutrient solution at 22°C. The glass was placed in a dark chamber (B). To avoid possible oxygen diffusion from the atmosphere into the radical zone, the solution's surface was covered with a Vaseline oil layer. The overground parts of the seedlings remained in air. The glass bottom was above the photomultiplier's aperture (C).

The intensity of the root's ultraweak glowing at 450-650 nm with a maximum of 530 nm was measured mainly from the middle and lower part of the roots by means of a quantometer apparatus equipped with a photomultiplier (C). To reduce noise, the latter was cooled down to −5°C in a refrigerator. High voltage was supplied by a high-voltage stabilizer (D). Impulses from the photomultiplier were amplified by a wide-beam amplifier (E) and then summed up on a self-recording potentiometer (G).

Results and Discussion

A preliminary experiment with excised pumpkin roots submerged into the solution, through which currents of air or gaseous nitrogen were passed, showed that under conditions of normal aeration excised roots revealed a distinctly pronounced effect of ultraweak glowing (Figure 29). During the early stages of air bubbling through the solution, a "flash" of ultraweak glowing of roots was observed, the intensity of which began to drop, reaching a stationary level in about 1.5-2 hr. After 3 hr, the air was replaced by bubbling gaseous nitrogen through the solution. As seen from Figure 29, the intensity of the root's glowing began to decrease sharply. In 2 hr, the intensity approached the background level, *i.e.*, the root's

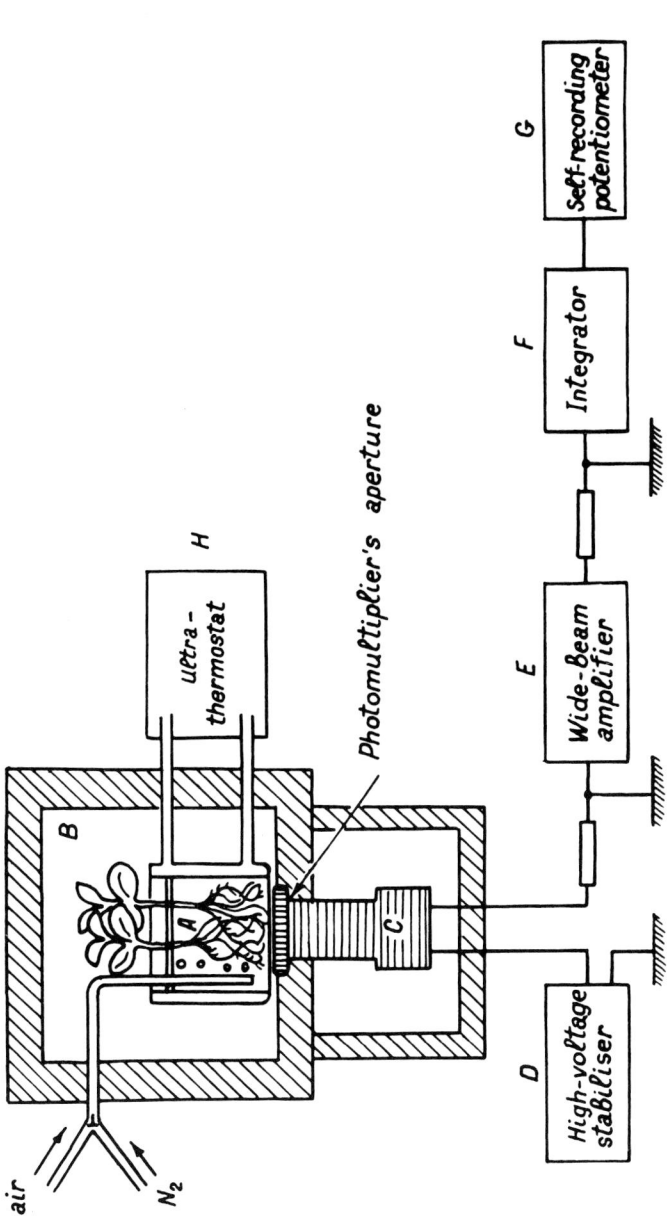

Figure 28. Scheme of the setting-up for the estimation of ultraweak glowing in plant roots. Legend: A, thermostatized glass; B, dark chamber; C, photomultiplier; D, high-voltage stabilizer; E, wide-beam amplifier; F, integrator; G, self-recording potentiometer; H, ultrathermostat.

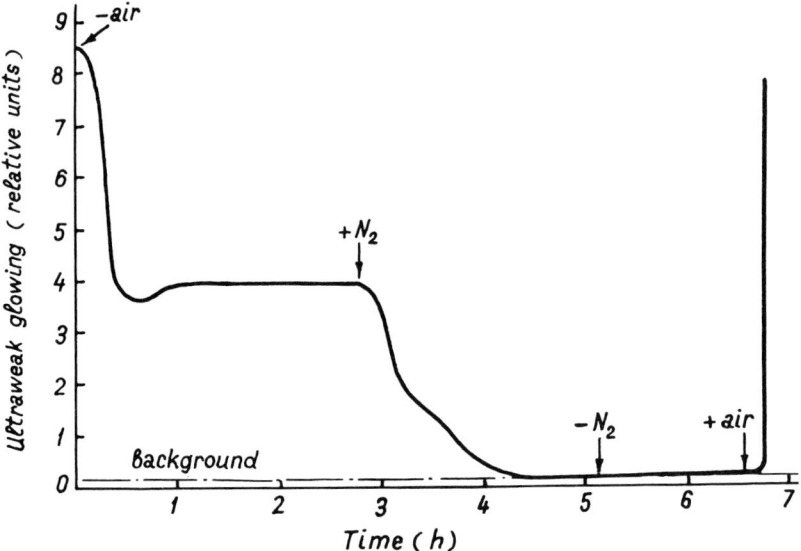

Figure 29. Ultraweak glowing of excised pumpkin roots under different conditions of rhizosphere aeration. Legend: +air, +N_2: bubbling by air or nitrogen; −N_2: without bubbling by air or nitrogen.

glowing practically ceased. A subsequent discontinuance of nitrogen bubbling failed to change the picture. Only a repeated bubbling of air through the solution with roots led to a rapid increase of the root's glowing.

Figure 30 demonstrates the results of an experiment with intact pumpkin plants in which overground parts were in air while the roots were in the solution. The surface of the solution was covered with Vaseline oil to isolate it from the ambient atmosphere. In this case, although the aerial parts of the plants were profusely supplied with oxygen, the roots' glowing in intact plants failed to differ from the glowing of the excised roots. As in the case of excised roots, the removal of oxygen from the radical zone by bubbling through a flow of gaseous nitrogen led to a total "extinction" of the roots' glowing and failed to glow for a long time following the discontinuation of nitrogen bubbling through the solution.

The results of experiments on intact plants favors the view that for pumpkin plants, oxygen transport from the aerial parts cannot provide the roots with an adequate amount of molecular oxygen to permit the process of peroxide oxidation of membrane lipids, which causes the chemiluminescence of the root cells. These data coincide very well with the results of our earlier studies using the polarographic technique (see page 48).

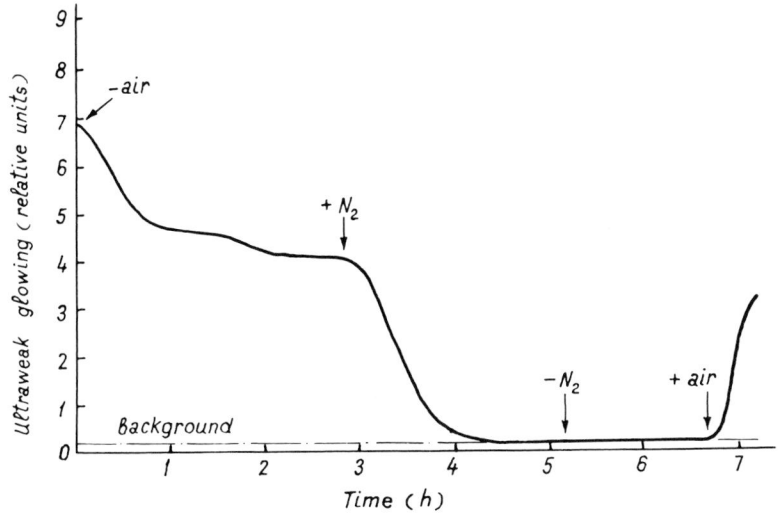

Figure 30.   Ultraweak glowing of roots of intact pumpkin plants under different conditions of rhizosphere aeration. Legend: See Figure 29.

Quantitative Assay of Oxygen Transport in Plants

Although the aforementioned experiments using the polarographic and chemiluminescent techniques have shown that the portion of oxygen translocated from the aerial parts to roots of pumpkin and cotton plants were not sizable, the studies gave no quantitative measure of the amount of oxygen transported.

Subsequently, we attempted to develop a method that would allow measurement of the proportion of transported oxygen as related to the total oxygen consumed by roots of intact plants.[80]

Method

The recent discovery of a specific inhibitor of alternative cyanide-resistant oxidase of mitochondria, a derivative of hydroxamic acid[41] has made it possible to develop the new approach to the problem. With the combined action of the above-mentioned inhibitor and an inhibitor of cytochrome oxidase (KCN), mitochondrial respiration can be almost completely suppressed.

By administering *m*-iodobenzhydroxamic acid with KCN in root cells the entire oxygen uptake from the surrounding medium can be blocked, while the oxygen transport from the aerial parts to the roots is not altered. Using excised roots, we measured the amount of oxygen absorbed by roots in the absence and presence of inhibitors. With these data we were able to calculate the amount of oxygen transported from shoots to roots, which equals oxygen diffused from the roots of an intact plant in the medium $+O_2$ utilized by the excised roots when inhibitors are administrated.

Results and Discussion

Figure 31 illustrates the results of an experiment with excised roots. It is seen that KCN inhibits only 65% of respiration of cotton roots. However, after introducing *m*-iodobenzhydroxamic acid into the nutrient solution, oxygen uptake of roots from the ambient medium was inhibited up to 95%. With these data we made a quantitative assessment of the amount of oxygen transported from shoots to roots.

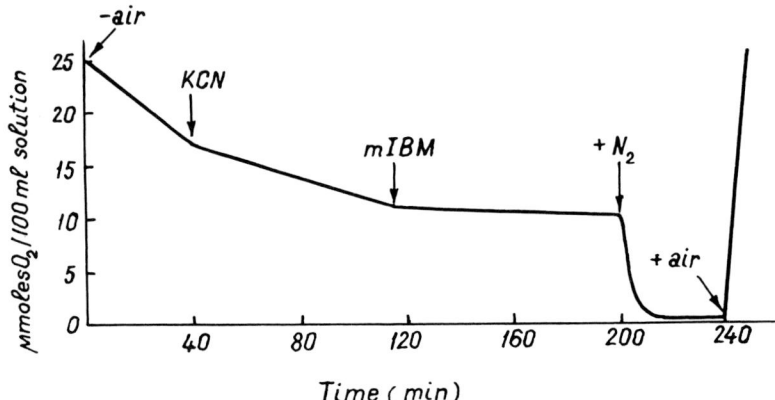

Figure 31.   Oxygen uptake by excised roots of cotton plant before and after administration of KCN, 10^{-3} *M*, and *m*-iodobenzhydroxamic acid, $5 \cdot 10^{-5}$ *M*, (mIBM) to the solution. Volume of the nutrient solution, 100 ml; temperature of the nutrient solution, 27°C; fresh weight of roots, 1.15 g. Legend: $^+$air, $^+$N$_2$: bubbling by air or nitrogen; −air: without bubbling by air.

If the intensity of oxygen uptake by roots in the presence of the combined inhibitors is equal to or exceeds the amount of oxygen supplied from aboveground parts under anaerobic conditions in the rhizosphere, no oxygen will diffuse from the roots into the surrounding medium. Under these conditions, all the transported oxygen will be consumed by respiration of root cells. On the contrary, if the amount of oxygen translocated from the shoots exceeds the needs of residual root respiration, then the surplus of oxygen transported to the roots will diffuse into the surrounding nutrient solution and can be measured quantitatively by means of an oxygen electrode.

Figure 32 presents the results of a typical experiment with intact plants in which we used a 12-day-old cotton plant at the stage of development of one true leaf. During the experiment, roots of the plant were placed into a thermostatically controlled vessel at 27°C; the vessel contained a nutrient solution, whereas aboveground parts remained in air. Prior to the experiment, the nutrient solution was aerated and during the experiment it was continuously agitated. The rhizosphere was isolated from the ambient atmosphere by Vaseline oil, which covered the surface of the solution. Changes in oxygen concentration of the nutrient solution were measured with an oxygen electrode (see Figure 22).

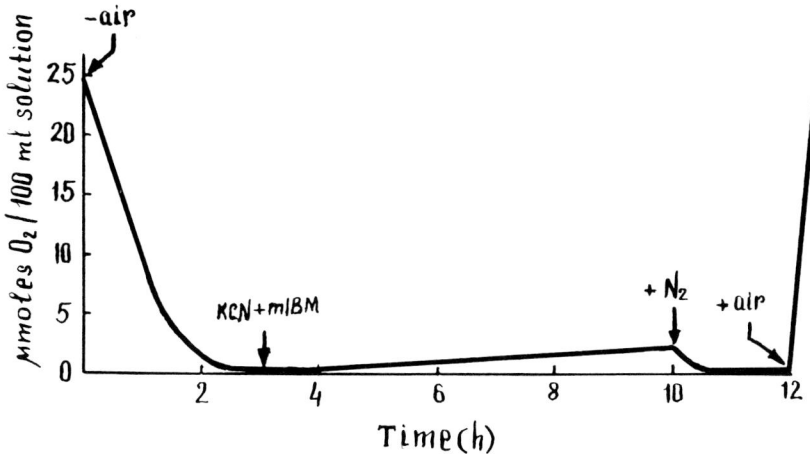

Figure 32.   Oxygen diffusion from roots of intact cotton plant to the solution after administration of inhibitors. Volume of the nutrient solution, 100 ml; temperature of the nutrient solution, 27°C; fresh weight of roots, 1.28 g. Legend: see Figure 33.

As seen in Figure 32, after inhibition of root respiration by cyanide and *m*-iodobenzhydroxamic acid, a weak diffusion of oxygen from roots was noted and, as a result, the presence of oxygen was clearly detected in the solution. Subsequent bubbling of gaseous nitrogen through the nutrient solution removed all oxygen from the roots.

We assessed quantitatively the portion of oxygen transported from shoots to the roots knowing both the amount of oxygen consumed by roots before and after the administration of the inhibitors on the basis of experiments with excised roots, and the amount of oxygen diffused from roots of an intact plant.

According to our calculations in the experiment with a cotton plant shown in Figure 32, the fraction was equal to 8.3%

$$X = \frac{b+c}{a} \cdot 100 = \frac{20.8 + 6.9}{333.9} \cdot 100 = 8.3\%$$

where X is the portion of oxygen transported from the shoots to the roots as a percentage of the total oxygen consumed by roots under normal aerobic conditions; a is the amount of oxygen consumed by excised roots before inhibition of respiration ($\mu g\ O_2$/g fresh weight/hr); b is the quantity of oxygen consumed by excised roots after inhibition of their respiration by cyanide and m-iodobenzhydroxamic acid ($\mu g\ O_2$/g fresh weight/hr); c is the amount of oxygen diffused from the roots of an intact plant into the rhizosphere after inhibition of their respiration by KCN and m-iodo-benzhydroxamic acid ($\mu g\ O_2$/g fresh weight/hr); b + c is the amount of transported oxygen ($\mu g\ O_2$/g fresh weight/hr).

The portion of transported oxygen obtained in our experiments is much lower than the corresponding values predicted from an anlysis of steady-state oxygen diffusion in the experiments with maize root models.[90,92,93] This may be attributed to the differences both in the methods employed and in the plant species used as test objects.

Thus, the results of present studies show that the portion of oxygen transported from the aerial parts to the roots of cotton plants is very small. It is obvious that in the absence of direct aeration of the rhizosphere region the amount of oxygen transported was not enough to maintain the normal aerobic metabolism in root cells.

Electron Microscope Study of the Role of Oxygen Transported

With the results of polarographic measuring of oxygen transport, the data on ultraweak glowing of root cells under anaerobic conditions, and the experiments on quantitative assessment of the amount of transported oxygen, we can now consider the physiological role of transported oxygen for vital activity of roots.

At extreme conditions of living when any direct access of atmospheric oxygen into the rhizosphere is restrained or excluded for some reason or other, transport of even a small quantity of oxygen to roots may play a significant physiological role.

One approach to elucidate this problem can be achieved by use of an electron microscope to study the fine structure of root cells under conditions when any direct access of molecular oxygen into the radical zone is excluded. Results of such experiments are described below.

Method

The results of our previous studies on cell ultrastructure under anoxia were the basis for using the change in mitochondrial fine structure as a criterion for evaluation of cell oxygen supply and, in particular, the role of transported oxygen. We have shown that among cell organelles, mitochondria are especially sensitive to the absence of oxygen. On one hand, in an oxygen-free medium, mitochondria showed characteristic changes increasing stepwise the longer they were subjected to conditions of anoxia up to their degradation. On the other hand, there were adaptational types of changes in mitochondrial ultrastructure which did not result in their destruction. The stages of changes can be followed under experimental conditions.

Ultrastructural Changes in the Initial Period of Anoxia. Changes in mitochondrial cristae were observed as follows: the intracristal space was expanded, sometimes a hypertrophied enlargement of a cavity of one or two cristae took place (Figure 33b), but sizes and shape of mitochondria as well as electron density of the matrix were not altered.

The Development of Pathological Changes. In this case, mitochondria and the intermembrane space were considerably increased in size, vesiculation and fragmentation of cristae which occurred along with a marked rise of vesicles. The mitochondrial matrix was clarified and many lipid droplets appeared in the inner mitochondrial compartment and in the intermembrane space (Figure 33c). Often in the same cell, another type of mitochondrial change occurred simultaneously: prior to degradation they decreased in size, being considerably condensed, and all of the inner mitochondrial compartment was filled by dense granules and isolated vesicles. In some places, the outer membrane of the mitochondria moved away from the inner membrane (Figure 33d).

Organelle Destruction. In this stage, the mitochondrial matrix disappears, the number of vesicles (remnants of destroyed cristae) decreases, and ruptures of the limiting membrane are observed (Figure 33e).

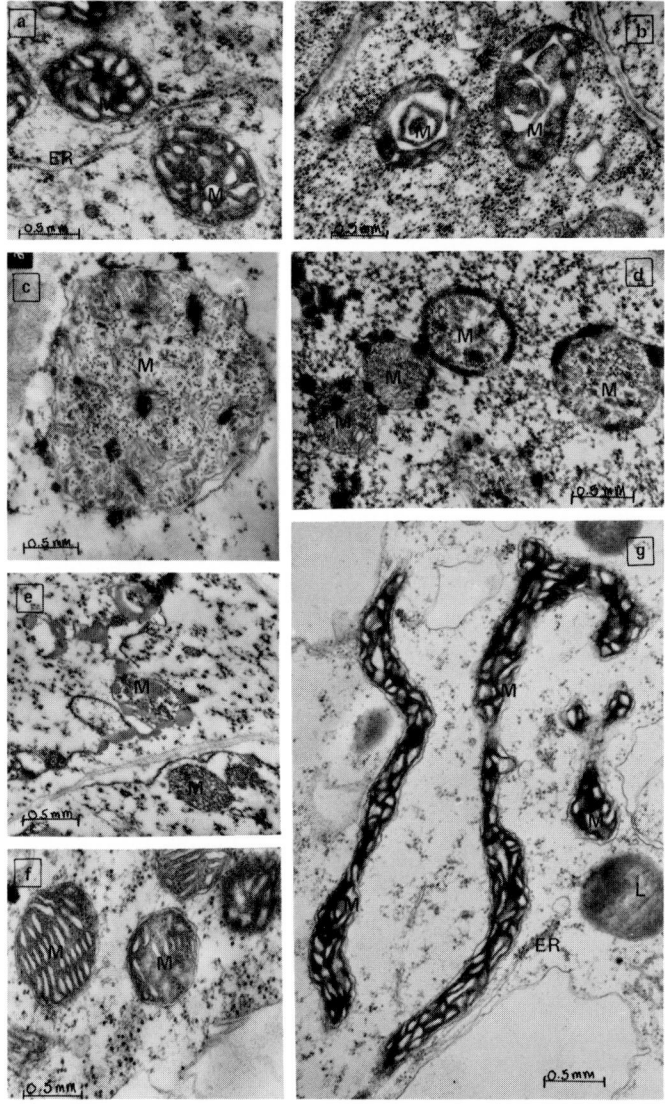

Figure 33. Ultrastructural changes of mitochondria in plant cells under anaerobic conditions. Legend: (a) aerobic conditions (control); (b) initial period of mitochondrial changes under anoxia; (c,d) the development of pathological changes of mitochondria under anoxia; (e) mitochondrial destruction under anoxia; (f,g) adaptational types of mitochondrial changes under anoxia. Abbreviations: M - mitochondrion; ER - endoplasmic reticulum; L - lipid droplet.

Similar mitochondrial changes under conditions of anoxia have been reported by Mölbert and Guerritore,[118] Rouiller[119] in animal cells, and Wrischer[120] and Geiger and Christy[121] in plant cells.

Adaptational Types of Changes. According to our studies, an additional type of mitochondrial alteration is characteristic for cells subjected to an acute oxygen deficiency but which are highly adapted to these conditions.[49,50,122,123]

Cristae of the organelles increase in number and take a parallel arrangement, unusual for plant mitochondria (Figure 33f). Meanwhile, mitochondria sharply enlarge in size in some species but remain almost unaltered in others. Another type of adaptational reorganization of mitochondria consists of the appearance of long and branched organelles, which often form a mitochondrial network, sometimes of a great length (up to 55 μm) and piercing through the cell cytoplasm (Figure 33g).

The above-described changes in mitochondrial ultrastructure were utilized to evaluate the role of oxygen transported from the aerial parts to roots under conditions of root anaerobiosis.

For this purpose, roots of cotton, pumpkin and rice plants were immersed into nutrient solutions that had been deoxygenated. To rule out the possibility of the roots being poisoned with carbon dioxide under prolonged exposure, the nutrient solution was either continuously bubbled with gaseous nitrogen or was regularly renewed with preliminary removal of oxygen. Under these conditions with the aboveground parts in air, oxygen transport from the aerial green parts of plants was the single source of oxygen for the roots. Electron microscope studies of root cell ultrastructure were carried out during the experiments. In control experiments, the nutrient solution was continuously aerated. The experiments were performed in a phytotron chamber at 25°C, 35°C and 42°C; the temperatures of both the air and the nutrient solution were the same in all cases. The experiments lasted for 48 or 72 hr. After the corresponding exposure, the roots were fixed for electron microscope examination.

Results and Discussion

Intact Plants. Ultrastructural investigations of root cells of 8- to 10-day-old and 24- to 30-day-old cotton plants showed that 48 and 72 hr of root anerobiosis at 25°C induced no pathological changes in mitochondria or other cell organelles. However, a majority of the mitochondria in the cells of the meristem and the zone of extension had a parallel arrangement of cristae (Figure 34b,c), which could be indicative of the cells suffering from molecular oxygen deficiency and adapting to the deficiency. Such orderly arrangement of cristae was mostly observed in the zone of meristem where

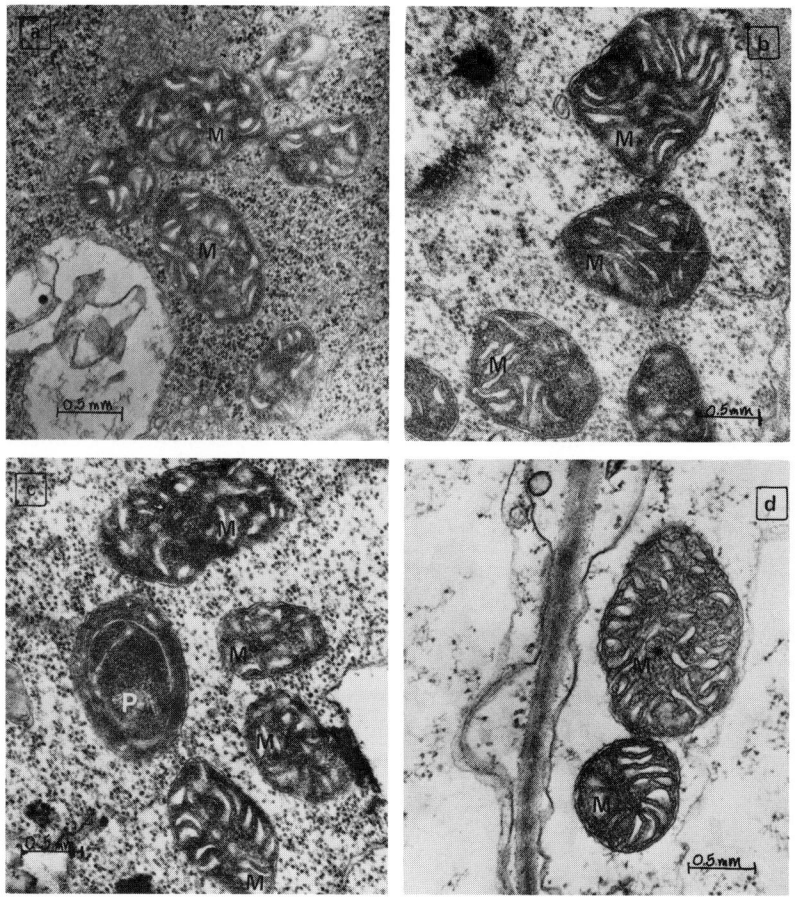

Figure 34.   Ultrastructure of mitochondria in root cells of intact cotton plants under root anaerobiosis (25°C). Legend: (a) aerobic conditions in rhizosphere (control), 9-day-old plants, meristem cells; (b) 3-day root anaerobiosis (9-day-old plants, meristem cells); (c) 3-day root anaerobiosis (26-day-old plants, extension zone); (d) 3-day root anaerobiosis (9-day-old plants, differential zone). Abbreviations: M - mitochondrion; P - plastid.

the oxygen supply was probably more impeded because of a reduction of intercellular spaces (Figure 34b). In the zone of differentiation (1.5-2 cm from root tips), the orderly arrangement of cristae peculiar to anaerobiosis occurred more rarely (Figure 34d). Similar experiments performed at 35° C gave like results: no mitochondrial destruction was found in the root cells of the cotton plant, but changes in the fine structure of mitochondria strongly suggested that the root cells were suffering from oxygen deficiency.

At 42°C, the experimental plants suffered from anaerobiosis as early as 1 hr after initiation of treatment. After 3 hr under anaerobic conditions, all root cell organelles were destroyed, whereas no destructive changes of mitochondria were observed under conditions of aeration at the same temperature, even after 12 hr.

Although the studies did not reveal pathological changes in the fine structure of root cells after 48- and 72-hr exposure to root anaerobiosis at 25° and 35°C, they provided evidence that under these conditions the root cells experienced an oxygen starvation to a certain degree. On the other hand, the experiments imply that oxygen transport from shoots to roots of cotton plants (which, as noted earlier, is not a sizable value as compared with normal oxygen demands of roots) seemed to maintain the mitochondrial ultrastructure of root cells undamaged for a relatively long time. Adaptational processes taking place in the cell under conditions of anoxia seem to contribute to its stability.

In the earlier experiments with pumpkin[53,77] carried out in conditions close to those of the aforementioned experiments with cotton plants, after 50-70 hr anaerobiosis, pathological changes in mitochondrial ultrastructure of root cells in the zones of meristem and extension were observed. The following studies were designed to elucidate the causes of the differences in reaction to root anaerobiosis in experiments between cotton and pumpkin plants.

Unlike the typical mesophytes, cotton and pumpkin, a hydrophyte, rice, showed no significant changes in the fine structure of mitochondria or other cell organelles after 72 hr of root anaerobiosis. The mitochondrial ultrastructure of root cells of the experimental plants under conditions of root anaerobiosis was not different from the control plants (Figure 35a,b).

High tolerance of the fine structure of rice root cells to anaerobiosis may have at least two reasons. It may be associated with metabolic peculiarities of roots of the plants adapted to life under conditions of acute oxygen deficiency in the environment. Crawford drew this conclusion[124-127] based on studies on the tolerance of wild plants to anaerobiosis. He developed the metabolic concept of plant tolerance to soil

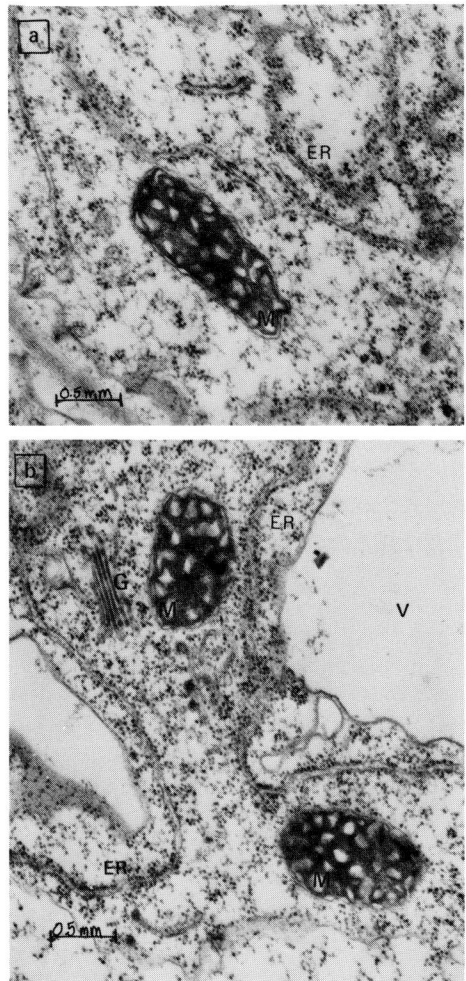

Figure 35.   Ultrastructure of mitochondria in cells of intact rice plants under root anaerobiosis. Legend: (a) aerobic conditions in rhizosphere (control); (b) 3-day root anaerobiosis. Abbreviations: M - mitochondrion; ER - endoplasmic reticulum; G - dictyosome; V - vacuole.

anaerobiosis based on the results of studies of root metabolism of plants tolerant and intolerant to flooding.

Crawford's ideas were further confirmed and developed in works of other researchers.[128-131]

An alternative concept regarding rice[53] and other hydrophytes[132] was suggested by the authors on the basis of a comparative study of tolerance to anoxia of the cell ultrastructure in excised roots of these plants. According to our concept, root cells of adult plants of hydrophytes should not have a higher tolerance to anaerobiosis than those of mesophytes. The ability of the plants to grow on anaerobic soils should be primarily ascribed to the capacity of hydrophytes to aerate their roots due to gas transport within the plant rather than peculiar metabolism of roots itself. The results of electron microscope examinations of rice roots supporting this point of view are described below.

Excised Roots. In experiments, detached roots of 14-day-old rice plants were placed in flasks filled with water and constantly bubbled with gaseous nitrogen. As a control, the excised roots of a mesophyte (pumpkin) were placed simultaneously in the flasks. The experiments were carried out at $20°C$ and $32°C$. After 1, 3, 5 and 7 hr exposure to the oxygen-free environment, the root cells were examined under the electron microscope.

As in Figure 36c, the cell ultrastructure of rice roots had no enhanced tolerance to anoxia as compared with pumpkin root cells: pathological changes in the fine structure of mitochondria were noted as early as 7 hr after exposure of the rice roots to anaerobic conditions at $20°C$.

Increasing the ambient temperature to $32°C$, accelerates the appearance of destructive changes of mitochondria (Figure 36e). As early as 1 hr after exposure to anoxia, an initial stage of mitochondrial changes was observed; in 3 hr, all mitochondria were pathologically changed.

Similar results have been obtained with pumpkin roots (Figure 36d). In this case, after 5–hr anoxia at $20°C$, an initial stage of mitochondrial damage was observed; after 7–hr exposure to anaerobic conditions, pronounced changes in mitochondria and other organelles were observed. At $32°C$ as little as 3 hr of exposure caused pathological changes of mitochondria and the organelle was essentially destroyed after 5 hr (Figure 36f).

The findings substantiate the theory that root cells of rice plants in the natural habitat in soils practically devoid of oxygen are not more tolerant to anaerobiosis as compared with root cells of plants cultivated under conditions of considerably more aerated soils and they may be even inferior to them in this respect. Their tolerance to root anaerobiosis is explained mainly by their receipt of enough oxygen by translocation from the aerial parts.

However, in the above experiments (Figure 36), the ultrastructural destruction of root cells in the oxygen-free environment can be explained not only by lack of oxygen but also by carbohydrate starvation. The detached roots had no opportunity to obtain assimilates from the green parts of plants.

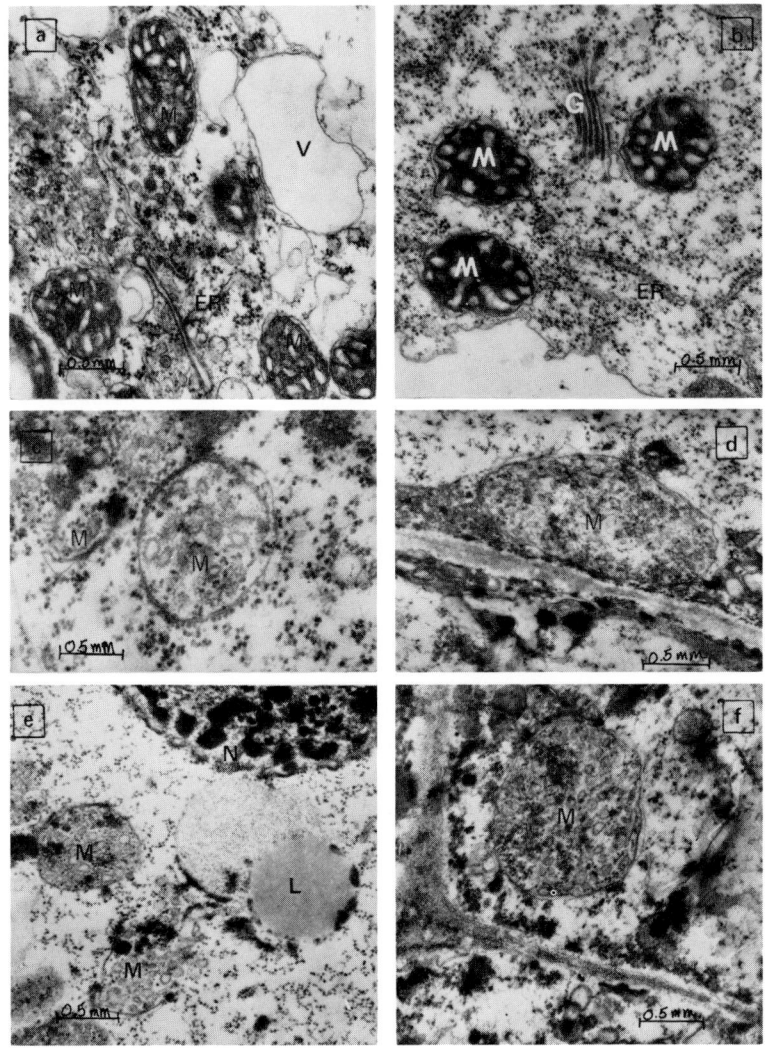

Figure 36. Ultrastructure of mitochondria in cells of excised rice and pumpkin roots under anaerobiosis. Legend: (a, c, e) rice roots (cortical parenchyma, extension zone); (a) 24 hr in aerated water, 20° C (control); (c) 7-hr exposure to anaerobiosis in water, 20° C; (e) 3-hr exposure to anaerobiosis in water, 32° C; (b, d, f) pumpkin roots (cortical parenchyma, extension zone); (b) 24 hr in aerated water, 20° C (control); (d) 7-hr exposure to anaerobiosis in water, 20° C; (f) 5-hr exposure to anaerobiosis in water, 32° C. Abbreviations: M - mitochondrion; ER - endoplasmic reticulum; V - vacuole; L - lipid droplet; G - dictyosome.

Excised Roots Fed by Glucose. To test the validity of carbohydrate starvation, experiments with excised roots of rice and pumpkin were supplied with exogenous glucose under anaerobic conditions to imitate an influx of assimilates from the green parts. The results of the experiments were in agreement with the assumption given above.[123] From Figures 37 and 38 it can be seen, even after 48-72 hr exposure, that the detached roots show no destructive changes in mitochondrial ultrastructure, if they were supplied by exogenous glucose. On the contrary, an essential rearrangement of ultrastructure of the mitochondrial apparatus was even found. The rearrangement proceeded differently in rice and pumpkin. In rice, an increase in number of cristae and an insignificant change in size and shape of mitochondria were observed (Figure 37b,c). The cristae took an orderly parallel arrangement which was most pronounced after 72-hr anoxia (Figure 37c). The parallel stacks of cristae often crossed the whole inner compartment of the organelles. Similar rearrangement was observed earlier in the cells of whole rice seedlings as well as in excised tomato roots under anoxia.[49,50,56,57,122]

After a 96 hr exposure to anaerobiosis, destructive changes appeared in the mitochondria of rice roots, which was indicative of rising cell damage (Figure 37d).

In pumpkin, together with the appearance of mitochondria with parallel stacks of cristae, a transformation of the organelles' size and shape occurred (Figure 38). In the cells of cortical parenchyma, very long mitochondria (up to 7-9 μm long) appeared in close contact with each other. After 24-72 hr anoxia, the long mitochondria in the cells increase in number; in many cases the long mitochondria seemed to fuse, forming a mitochondrial network. Sometimes the overall length of the network averages 55 μm (Figure 38c). A characteristic arrangement of profiles of the endoplasmic reticulum in the cells is of interest. Long and shortened profiles of the endoplasmic reticulum carrying a large number of ribosomes occur along the extended mitochondria or among them and in close contact with mitochondria in the sites of their contacts. In fact, near every mitochondrion is a profile of the endoplasmic reticulum.

After 96-hr exposure to anoxia, destructive changes occur in mitochondria and other organelles of root cells of pumpkin as well as rice (Figures 37d, 38d).

It should be noted that both in experiments without glucose supply and in those with glucose feeding, destructive changes in mitochondria and other subcellular structures of pumpkin and rice roots developed under anoxia almost simultaneously. It means that ultrastructure of roots of the hydrophyte, rice, does not have higher tolerance to anaerobiosis than the roots of the typical mesophyte, pumpkin.

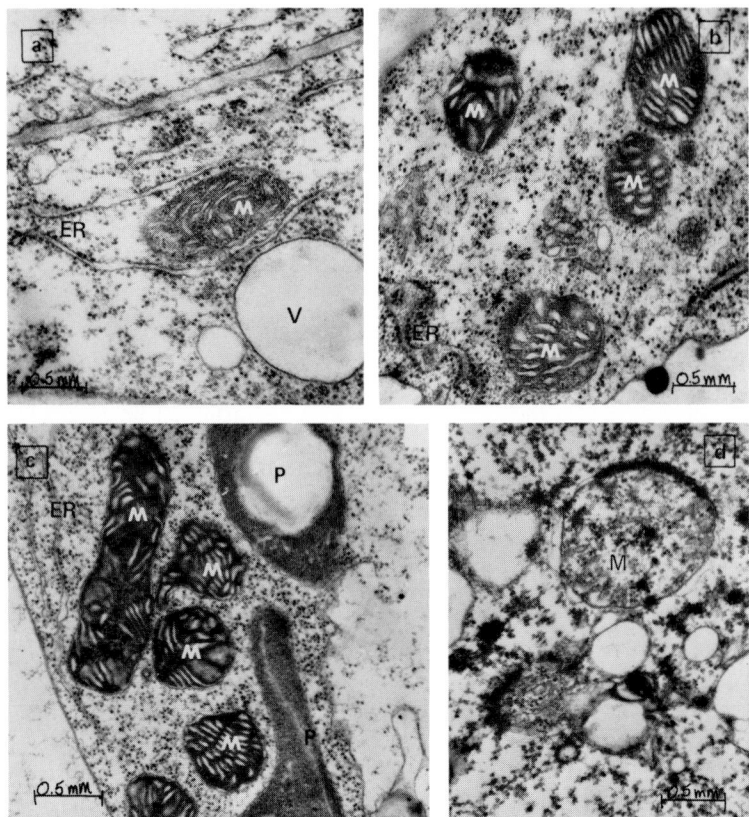

Figure 37. Ultrastructure of mitochondria in cortical parenchyma cells of excised rice roots under conditions of anoxia and glucose feeding. (a) 48–hr in aerated water (control); (b) 48-hr exposure to anoxia in 3% glucose solution; (c) 72–hr exposure to anoxia in 3% glucose solution; (d) 96–hr exposure to anoxia in 3% glucose solution. Abbreviations: M - mitochondrion; ER - endoplasmic reticulum; V - vacuole; P - plastid.

Thus, these investigations confirm the ability of root cells of an adult rice plant (as well as other hydrophytes) to maintain normal metabolism and undamaged ultrastructure when the direct access of oxygen from the ambient atmosphere into the rhizosphere is excluded; it is explained in that the root cells of these plants are abundantly supplied by molecular oxygen transported from the aerial parts of the plants and, therefore, weakly depend upon oxygen content in the environment, rather than their tolerance to oxygen deficiency.

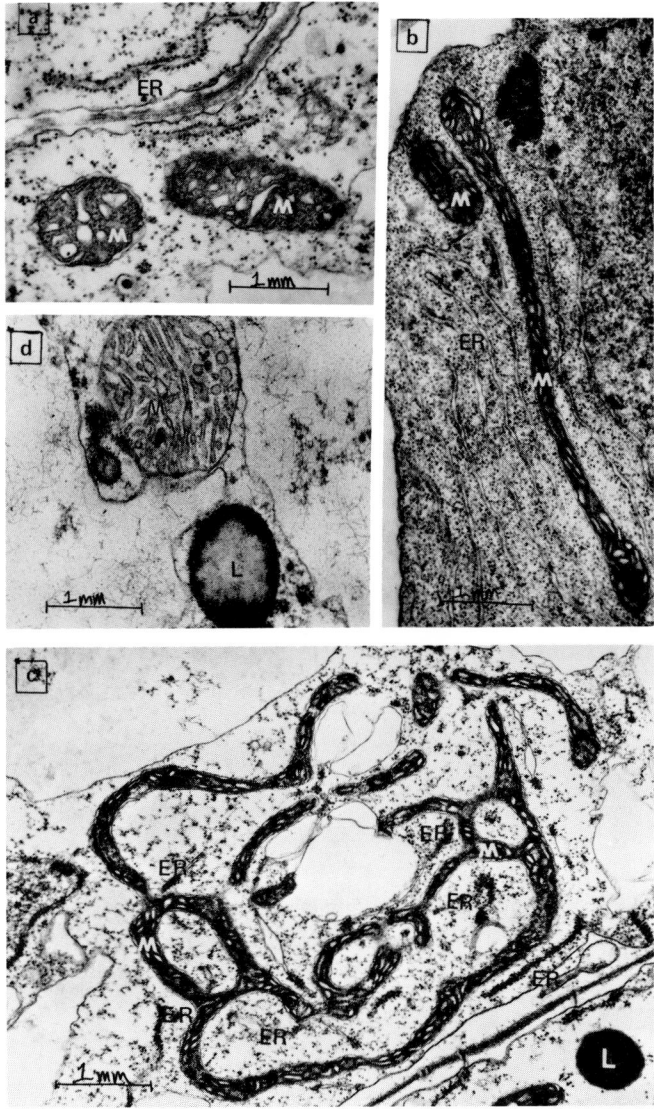

Figure 38.   Ultrastructure of mitochondria in cortical parenchyma cells of excised
pumpkin roots under conditions of anoxia and glucose feeding:
(a) 48-hr in aerated water (control); (b) 24-hr exposure to anoxia
in 3% glucose solution; (c) 72-hr exposure to anoxia in 3% glucose
solution; (d) 96-hr exposure to anoxia in 3% glucose solution.
Abbreviations: M - mitochondrion; ER - endoplasmic reticulum;
G - dictyosome.

The results of the experiments on exogenous glucose feeding are also interesting from two other points of view. First, the observed active rearrangement of mitochondria under anaerobic conditions in the presence of sufficient amounts of freely utilized sugars in the cell appears to substantiate that in the oxygen-free environment, mitochondria play a vital role in the physiology of the cell. In particular, these observations are interesting in view of the concept of cell bioenergetics (chemiosmotic theory of energy transformation in biological systems) advanced by Mitchell[133] and experimentally validated by Racker[134] and Skulachev.[135,136] According to this theory, reciprocal transformation of chemical (ATP) and physical (potential) forms of energy readily occurs on membranes of a living cell. In comparison with chemical energy, the physical one is advantageous in that it can be easily and rapidly transported along the membrane over a relatively long distance where it can then be transferred back to the universal form of chemical energy (ATP) utilized by the cell. It is postulated that the potential may be transferred from one membrane to another.

In our experiments, pumpkin root cells deprived of the aerobic source of energy (generated in abundance by mitochondria) in an oxygen-free environment were obliged to utilize the energy that was released in the nonstructuralized cytoplasm during glycolysis. Under these conditions, the branched network of mitochondria had many advantages because it increased the surface area available for contact between the mitochondrial membrane with both the cytoplasm where glycolysis proceeds and the endoplasmic reticulum which has a great number of ribosomes attached to it. This may provide rapid transport of energy (potential) along the membrane to the sites of protein synthesis where the physical potential energy may be transferred back to the chemical form of energy (ATP) and be readily utilized by the cell.

Thus, in the oxygen-free environments, mitochondrial membranes, which lose the capacity to generate energy using molecular oxygen, may function as collectors of energy released during glycolysis, to transform chemical and physical forms of energy and, finally, to transport the energy over relatively long distance, to the sites of biosynthesis of proteins and other cell substances.

If these concepts can be demonstrated by electron microscopy and direct biochemical and biophysical investigations, then pumpkin roots may be a good illustration of the adaptation of a higher plant to oxygen deficiency manifested at the subcellular level.

Second, the results of the above experiments on the effect of glucose feeding on the resistance to anoxia of the mitochondrial ultrastructure are also useful for understanding the nature of mesophyte roots suffering under conditions of root anaerobiosis. In the case of cotton and pumpkin, we found their capacity for root aeration by oxygen transport was very limited. As a result, under long-term root anaerobiosis in these plants, one can observe changes in the mitochondrial ultrastructure of roots characteristic of cells experiencing an oxygen deficiency.

One unfavorable effect of root anaerobiosis is that the organic compounds of the root cells cannot be oxidized by the normal aerobic metabolism. Such explanation implies that oxygen starvation-induced asphyxia should be the immediate cause of damage and destruction of mesophyte roots under long-term root anaerobiosis. However, the results of the present experiments on glucose feeding of roots show that the presence of sufficient stores of organic substances in the cell is essential to retain the cell ultrastructure undamaged in the oxygen-free environment. Lack of oxygen in the rhizosphere not only alters considerably the pathways of utilization of respiratory substrate, but also intensifies markedly the consumption of organic substances in root cells (Pasteur effect). An elevated rate of glycolysis in the anaerobic environment in turn brings about a rapid depletion of the resources of utilizable organic substances in the root cells.

This situation is further aggravated because transport of organic compounds from leaves to roots under conditions of root anaerobiosis is reduced considerably as demonstrated recently in experiments with C^{14}-sucrose translocation.[137] This is seen clearly from Figure 39, which presents the results of several experiments with 52-day-old cotton plants whose roots were under anaerobic conditions for 6, 24, 48 and 120 hr. A solution of radioactive sucrose was then introduced into their leaves for 3 hr, while the roots remained in the anaerobic medium. Even short-term anaerobiosis (9 hr) had an influence upon transport of assimilates from leaves to roots of cotton plants. After 27 and 51 hr in the oxygen-free environment, the inflow of radioactive sucrose to the roots was considerably decreased, and after 123 hr of root anaerobiosis, transport of radioactive assimilates to roots were scarcely observable. If we consider that along with a depression of assimilate influx to roots, and sharply accelerated expenditure of organic substances, in the absence of oxygen it will be evident that under conditions of root anaerobiosis the root cells should experience not only an oxygen but also a carbon deficiency. Under long-term anaerobiosis, it should in turn bring about an energetic starvation of roots and, as a result, an inhibition not only of metabolism and functions of the root but also of its growth.

Together with inhibition of the descendant flow of assimilates, the absence of oxygen in the rhizosphere depresses considerably the ascent of water transport from roots as was shown in our recent experiments on tritiated water transport from roots to aerial parts of 12-day-old maize plants.[138]

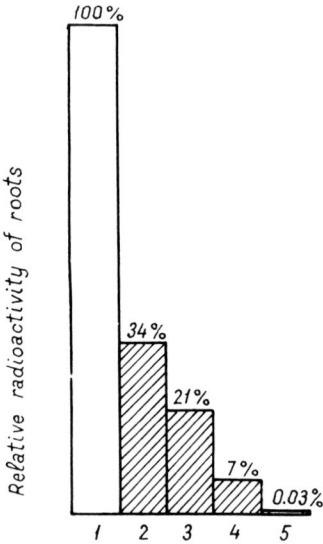

Figure 39. Translocation of C^{14}-sucrose from the leaves to the roots of cotton plants under root anaerobiosis: 1. control, normal aeration; 2. 9-hr root anaerobiosis; 3. 27-hr root anaerobiosis; 4. 51-hr root anaerobiosis; 5. 123-hr flooding. In the experiments, 2 through 4 anaerobiosis in the rhizosphere was induced by bubbling of N_2 through the nutrient solution, but in experiment 5 it was induced by flooding of plants cultivated on sandy soil.

Thus, the depressing action of long-term anaerobiosis on mesophyte roots is not due only to inhibition of aerobic utilization of respiratory substrate, but to a more rapid exhaustion of the stores of organic substances in the root cells not compensated for an influx of assimilates from leaves.

In the case of rice and other hydrophytes, flow of oxygen from green parts of plants facilitates the possibility of aerobic metabolism in roots and, consequently, more economic expenditure of organic compounds. Aerobic conditions in roots in turn also facilitate the influx of assimilates from leaves.

CONCLUSIONS

The electron microscope and biochemical studies of rice coleoptiles have mainly demonstrated the possibility for some higher plants to form and preserve for a long time all subcellular structures typical for cells of aerobically grown plants, including even functionally active organelles of oxygen metabolism under conditions of a total exclusion of molecular oxygen from the environment. Mitochondria of anaerobically grown rice coleoptiles not only retain their ultrastructure undamaged but are capable of transporting electrons of a respiratory substrate to molecular oxygen as well as to accumulate energy in macroergic bonds of ATP (coupled phosphorylation).

Comparative investigations carried out using an inhibitor of mitochondrial protein synthesis (chloramphenicol) to study biogenesis of mitochondria in rice coleoptiles in the presence of molecular oxygen as well as in an oxygen-free environment demonstrated, along with some similarities, substantial differences in the formation of mitochondria. Inhibitors of protein synthesis on mitochondrial ribosomes under conditions of strict anoxia cause the formation of mitochondria to differ both in ultrastructure and in respiratory function from mitochondria of cells of coleoptiles grown in the presence of the inhibitor but under conditions of normal aeration.

Marked differences also have been revealed in composition of soluble protein of aerobically and anaerobically grown coleoptiles.[139]

As found through the cytological examinations of the cells of intact rice seedlings as well as their detached organs with and without exogenous glucose feeding, the high tolerance of cell ultrastructure of rice coleoptiles to anaerobiosis is the property of the whole seedling rather than of the subcellular structures themselves. Here, the authors would like to emphasize particularly the ability of rice seedlings to transport readily organic compounds from the seed into growing coleoptile under conditions of strict anoxia. Owing to this transport and the ability of coleoptiles to intensify sharply glycolytic processes in the oxygen-free environment, the cells of this organ are sufficiently supplied both with energy and plastic substances under anoxia.

On the other hand, the results of the experiments on glucose feeding and its protective role under anoxia clearly show the importance of storage of organic substances in the cell for the maintenance of the integrity of its metabolism, ultrastructure and functions under oxygen-free conditions when a considerably more intensive wastage of organic compounds takes place.

Consequently, the more rapid degradation of fine structure of the cells and their destruction in excised organs under conditions of anoxia is first of all a result of carbon starvation rather than asphyxia or poisoning by toxic products of anaerobic metabolism.

On the whole, the data from the above experiments led us to conclude that the remarkable tolerance of ultrastructure of rice coleoptile cells to anoxia and the ability of coleoptiles to grow vigorously in oxygen-free environment is a result of an active metabolic adaptation. The latter is realized through substantial rearrangement of protein metabolism and energetics of the cell and especially through the unusual ability of rice seedlings to easily transport organic substances over a relatively long distance under conditions of strict anoxia.

Thus, rice seedlings constitute a marvelous biological system for the study of molecular, subcellular and cellular mechanisms of the adaptation of higher plants to oxygen deficiency. The authors propose to call such tolerance *true* physiological *tolerance* of plants to *anaerobiosis*, as it is manifested under conditions of a total exclusion of molecular oxygen from the environment and realized by the organism through a cardinal rearrangement of the cell metabolism.

The nature of the tolerance to anaerobiosis in adult hydrophytes and, in particular, in adult rice plants, is different. Although these plants, like young rice seedlings (coleoptiles), readily grow on flooded soils with extremely restricted gas exchange, the fine structure of their root cells has no enhanced tolerance to anoxia. The reason such plants are able to grow in flooded soils is the ability of hydrophytes to develop anatomical and morphological features of structure that facilitate the transport of molecular oxygen from the green organs to roots. The plant induces aerobic conditions not only in the root cells but even in the rhizosphere. Therefore, the ability of the plant to thrive in flooded soil is due to its ability to oxidize its rhizosphere, rather than by peculiarities of metabolism of the root cells themselves.

It should be stressed that by providing sufficient oxygen to the roots, hydrophytes simultaneously can provide easy transport of assimilates from the green parts to the roots. These considerations permit us to refer hydrophytes into the category of plants with *apparent tolerance to anoxia*, so unlike young rice seedlings, their resistance to soil anaerobiosis is based chiefly on possibility to receive molecular oxygen from other parts of the plant rather than on peculiarities of root cell metabolism.

Mesophytes include most of the important agricultural plants of the world. One should refer them to the category of plants labeled *intolerant to anaerobiosis*.

It can be seen from the data in this chapter, based on the experiments with such mesophytes as pumpkin and cotton plants, that the role of oxygen transported from aboveground parts to the roots of such plants was overestimated in a number of previous publications. The study of oxygen transport from the green parts to roots of pumpkin and cotton plants using a number of methods—polarographic, chemiluminescent and electron microscope techniques—as well as the results of the experiments with C^{14}-sucrose transport, support the hypothesis that oxygen translocated from the aboveground parts cannot provide normal aerobic metabolism in the root cells when a direct access of atmospheric oxygen into the rhizosphere is restricted or excluded.

Calculations undertaken in the experiments with cotton plants imply that the fraction of oxygen entering the roots from the green parts of the plant is about 8% of the total root respiration need for oxygen under conditions of normal aeration. Of course, the values found for cotton are not considered to be absolute for plants in general. Most likely it will fluctuate within some range, depending on species, age of plants, environmental conditions, etc. However, it is evident that if this value were two or even three times more than had been found under conditions of our experiment with cotton plants, it would be far from providing optimal conditions for aerobic respiration.

Nevertheless, oxygen translocated from the aerial parts appears to play some physiological role in the vital activity of roots under conditions of root anaerobiosis. The results of electron microscope studies of roots of intact plants and detached roots exposed to anoxia favor this assumption. The fine structure of the cells of detached roots having no connection with the aboveground parts proves to be far more vulnerable under conditions of anaerobiosis than roots of intact plants whose aboveground parts are in an ordinary atmosphere.

However, these observations can be also interpreted from another point of view. The study of cell ultrastructure of detached roots with and without exogenous glucose feeding as well as the results of the experiments on C^{14}-sucrose transport from leaves to roots dramatizes the situation that arises in the root cells of these plants under long-term anaerobiosis with regard to their supply of organic compound. Accelerated wastage of stores of organic substances in root cells in the oxygen-free environment is not compensated by the corresponding influx of assimilates from aboveground organs as far as transport of organic compounds from leaves to roots under conditions of root anaerobiosis is markedly inhibited under conditions of root anaerobiosis in mesophytes.

As a result, oxygen starvation of roots under conditions of long-term root anaerobiosis is aggravated by carbon starvation. It is the authors' opinion that carbon starvation is the immediate cause of the destruction of root cells and subsequent death of the root itself if root anaerobiosis lasts too long.

The findings that have been elaborated in this chapter have direct application to the agrotechnics of cultivating plants, especially on soils with restricted gas exchange and in hydroponics. It is obvious that the problem of oxygen supply of roots and aeration of the rhizosphere deserves as much attention as that of soil fertilization, since the effective utilization of the nutrient elements by plants is possible only under conditions of aerobic metabolism of the root cells.

REFERENCES

1. Wada, S. "Growth Patterns of Rice Coleoptiles Grown on Water and Under Water," *Sci. Rep. Tôhoku Univ.* 27:Ser. 4, 199 (1961).
2. Tsuji, H. "Effect of Anaerobic Condition on the Dry Weight Decrease of Germinating Rice Seeds," *Bot. Mag. (Tokyo)* 81(959): 233 (1968).
3. Kordan, H. A. "Rice Seedlings Germinated in Water with Normal and Impeded Environmental Gas Exchange," *J. Appl. Ecol.* 9(2):527 (1972).
4. Kordan, H. A. "Patterns of Shoot and Root Growth in Rice Seedlings Germinating under Water," *J. Appl. Ecol.* 11(2):685 (1974).
5. Morinaga, T. "The Favorable Effect of Reduced Oxygen Supply upon the Germination of Certain Seeds," *Am. J. Bot.* 13(2):159 (1926).
6. Durhan, V. M. and P. S. Wellington. "Studies on the Germination of Wheat Grains during Maturation," *Ann. Bot.* 25(98) (1961).
7. Siegel, S. and L. A. Rosen. "Effects of Reduced Oxygen Tension on Germination and Seedling Growth," *Physiol. Plant.* 15(3):435 (1962).
8. Siegel, S. M., L. A. Rosen and C. Guimarro. "Plants at Subatmospheric Oxygen Levels," *Nature* 198(4887):1288 (1963).
9. Heichel, G. H. and P. R. Day. "Dark Germination and Seedling Growth in Monocots and Dicots of Different Photosynthetic Efficiencies in 2% and 20.9% O_2," *Plant Physiol.* 49(2):280 (1972).
10. Vartapetian, B. B., I. P. Maslova and I. N. Andreeva. "Mitochondria of Coleoptiles of Rice (*Oryza sativa*) Grown under Anaerobic Conditions," *Fiziol. Rast.* 19(1):106 (1972) (In Russian).
11. Tsuji, H. "Respiratory Activity in Rice Seedlings Germinated under Strictly Anaerobic Conditions," *Bot. Mag.* (Tokyo) 85(999):207 (1972).
12. Costes, C. and B. B. Vartapetian. "Plant Grown under Vacuum: Ultrastructure and Functions of Mitochondria," (In press).
13. Aleshin, E. P., B. V. Yakovlev, V. J. Gulak and G. P. Kolesnikov. "The Ways of Energy Transfer in Rice Coleoptiles," *Dokl. Vaskhnil* 4(4):23 (1971) (In Russian).
14. Vartapetian, B. B., I. N. Andreeva and I. P. Maslova. "Ultrastructure of Rice Coleoptile Cells under Anaerobic and Aerobic Conditions," *Dokl. USSR Acad. Sci.* 196(5):1231 (1971) (In Russian).
15. Ueda, K. and H. Tsuji. "Ultrastructural Changes of Organelles in Coleoptile Cells during Anaerobic Germination of Rice Seeds," *Protoplasma* 73(2):203 (1971).
16. Opik, H. "Effect of Anaerobiosis on Respiratory Rate, Cytochrome Oxidase Activity and Mitochondrial Structures in Coleoptiles of Rice (*Oryza sativa* L.)," *J. Cell Sci.* 12(3):725 (1973).
17. Tsuji, H. "Growth and Metabolism in Plants under Anaerobic Conditions," *Environ. Control Biol.* 11(2):79 (1973) (In Japanese).
18. Vartapetian, B. B. and A. J. Maslov. "Respiration of Coleoptiles of Anaerobically Grown Rice Seedlings," *Fiziol. Rast.* 21(4):807 (1974) (In Russian).

19. Vartapetian, B. B., A. I. Maslov, I. N. Andreeva and G. I. Kozlova. "Cytochromes of Mitochondria of Rice Coleoptiles Grown under Conditions of Strict Anoxia," *Fiziol. Rast.* 20(6):1279 (1973) (In Russian).

20. Tsuji, H., T. Katoh and K. Ueda. "Growth and Metabolism in Plants under Anaerobic Conditions," in *Abstracts of the Twelfth International Botanical Congress* (Leningrad: "Nauka," 1975), p. 373.

21. Vartapetian, B. B., A. I. Maslov and I. N. Andreeva. "Cytochromes and Respiratory Activity of Mitochondria in Anaerobically Grown Rice Coleoptiles," *Plant Sci. Letters* 4(1):1 (1975a).

22. Karnovsky, M. J. "A Formaldehyde-Glutaraldehyde Fixative of High Osmolality for Use in Electron Microscopy," *J. Cell Biol.* 27(2):137 A (1965).

23. Wiebel, E. R. "Stereological Principles for Morphometry in Electron Microscopic Cytology," in *International Review of Cytology* Vol. 26, G. H. Bourne and J. F. Danielli, Eds. (New York: Academic Press, 1969), p. 235.

24. Linnane, A. W., J. M. Haslam, H. B. Lukins and P. Nagley. "The Biogenesis of Mitochondria in Microorganisms," *Ann. Rev. Microbiol.* 26:163 (1972).

25. Criddle, R. S. and G. Schatz. "Promitochondria of Anaerobically Grown Yeast. I. Isolation and Biochemical Properties," *Biochemistry* 8(1):322 (1969).

26. Schatz, G. in *Membranes of Mitochondria and Chloroplasts,* E. Racker, Ed. (New York: Van Nostrand-Reinhold, 1970), p. 257.

27. Tzagoloff, A. M., M. S. Rubin and M. F. Sierra. "Biosynthesis of Mitochondrial Enzymes," *Biochim. Biophys. Acta* 301(1):71 (1973).

28. Wilson, S. B. and A. L. Moore. "The Effect of Protein Synthesis Inhibitors on Oxidative Phosphorylation by Plant Mitochondria," *Biochim. Biophys. Acta* 292(3):603 (1973).

29. Andreeva, I. N. and B. B. Vartapetian. "Ultrastructure of Rice Coleoptile Mitochondria during Biogenesis," in *Abstracts of the Twelfth International Botanical Congress* (Leningrad: "Nauka," 1975), p. 347.

30. Andreeva, I. N., G. I. Kozlova and B. B. Vartapetian. "Aspects of the Formation of Mitochondria in Rice Coleoptiles during Germination under Aerobic and Anaerobic Conditions," *Fiziol. Rast.* 23(1): 111 (1976) (In Russian).

31. Nawa, Y. and T. Asahi. "Effect of Cycloheximide on Development of Mitochondria in Germinating Pea Cotyledons," *Agric. Biol. Chem.* 37(4):937 (1973).

32. Malhotra, S. S., T. Solomos and M. Spencer. "Effect of Cycloheximide, d-threo-chloramphenicol, Erythomycin and Actinomycin D on *de novo* Synthesis in the Cotyledons of Germinating Pea Seeds," *Planta (Berlin)* 114(2):169 (1973).

33. Mahler, H. R. in *Genetic Functions of Cytoplasmic Organoids, Genetics of Mitochondria and Their Interrelationships with Viruses,* Abstr. Symposium (Leningrad: "Nauka," 1974), p. 72.

34. Clark-Walker, G. D. and A. W. Linnane. "The Biogenesis of Mitochondria of *Saccharomyces cerevisiae.* A Comparison between

Cytoplasmic Respiratory Deficient Mutant Yeast and Chloramphenicol Inhibited Wild Type Cells," *J. Cell. Biol.* 34(1):1 (1967).

35. Adautte, A. M., M. Balmefrezol, J. Beisson and J. André. "The Effects of Erythromycin and Chloramphenicol on the Ultrastructure of Mitochondria in Sensitive and Resistant Strains of Paramecium," *J. Cell. Biol.* 54(1):8 (1972).

36. Smith-Johannsen, H. and S. P. Gibbs. "Effects of Chloramphenicol on Chloroplast and Mitochondrial Ultrastructure in *Ochromonas danica*," *J. Cell Biol.* 52(3):598 (1972).

37. King, M. E., G. C. Godman and D. W. King. "Respiratory Enzymes and Mitochondrial Morphology of Hela and L Cells Treated with Chloramphenicol and Ethidium Bromide," *J. Cell Biol.* 53(1):127 (1972).

38. Maslov, A. I. and B. B. Vartapetian. "The Induction of the Cyanide-Insensitive Respiration in Rice Seedlings," *Dokl. USSR Acad. Sci.* 216(2):455 (1974) (In Russian).

39. Vartapetian, B. B. and A. I. Maslov. "Respiratory Activity of Mitochondria in Rice Coleoptiles Grown Anaerobically," in *Abstracts of the Twelfth International Botanical Congress* (Leningrad: "Nauka," 1975), p. 374.

40. Vartapetian, B. B., A. I. Maslov and I. N. Andreeva. "Biochemical and Cytological Studies of Mitochondrial Development in Chloramphenicol-Induced Rice Coleoptiles," *Biochim. Biophys. Acta* 411(3):357 (1975).

41. Schonbaum, G. R., W. D. Bonner, B. T. Storey and J. T. Bahr. "Specific Inhibition of the Cyanide-Insensitive Respiratory Pathway in Plant Mitochondria by Hydroxamic Acids," *Plant Physiol.* 47(1): 124 (1974).

42. Storey, B. T. and J. T. Bahr. "The Respiratory Chain in Plant Mitochondria. 1. Electron Transport Between Succinate and Oxygen in Skunk Cabbage Mitochondria," *Plant Physiol.* 44(1):115 (1969).

43. Bendall, D. S. and W. D. Bonner. "Cyanide-Insensitive Respiration in Plant Mitochondria," *Plant Physiol.* 47(2):236 (1971).

44. Bonner, W. D., E. L. Christensen and J. T. Bahr. in *Biochemistry and Biophysics of Mitochondrial Membranes*, G. F. Azzone, E. Carafoli, A. L. Lehninger, E. Quagliarello and N. Silprandi, Eds. (New York: Academic Press, 1972), p. 113.

45. Lambowitz, A. M., E. W. Smith and C. W. Slayman. "Oxidative Phosphorylation in Mitochondria: Studies on Wild Type, Poky and Chloramphenicol-Induced Wild Type," *J. Biol. Chem.* 247(15):4859 (1972).

46. Lance, C. "Respiratory Control and Oxidative Phosphorylation in Arum Maculatum Mitochondria," *Plant Sci. Letters* 2(2):165 (1974).

47. Bahr, J. T. "Control of Cyanide-Insensitive Respiration," in *Abstracts of the Twelfth International Botanical Congress* (Leningrad: "Nauka," 1975), p. 348.

48. Tsuji, H. "Rice in Oxygen Uptake during Air-Adaptation of Anaerobically Treated Rice Seedlings," *Bot. Mag. (Tokyo)* 82(972):226 (1969).

49. Kursanov, A. L., B. B. Vartapetian, I. N. Andreeva and G. I. Kozlova. "Tolerance to Anoxia of Cell-Fine Organization of Rice Seedlings Grown under Aerobic Conditions," *Fiziol. Rast.* 20(3):564 (1973) (In Russian).

50. Vartapetian, B. B., I. N. Andreeva and A. L. Kursanov. "Appearance of Unusual Mitochondria in Rice Coleoptiles at Conditions of Secondary Anoxia," *Nature* 248(5445):258 (1974). (See also Erratum 250, No. 5461, 84 (1974).

51. Novikoff, A. B. in *The Cell: Biochemistry, Physiology, Morphology,* Vol. 2, J. Brachet and A. E. Mirsky, Eds. (New York: Academic Press, 1961), p. 299.

52. Munn, E. A. *The Structure of Mitochondria* (New York: Academic Press, 1974).

53. Vartapetian, B. B., I. N. Andreeva, I. P. Maslova and N. G. Davtian. "The Oxygen and Ultrastructure of Root Cells," *Agrochim.* 15(1):1 (1970).

54. Taylor, D. L. "Influence of Oxygen Tension on Respiration, Fermentation and Growth in Wheat and Rice," *Am. J. Bot.* 29(9):721 (1942).

55. Phillips, J. W. "Studies on Fermentation in Rice and Barley," *Am. J. Bot.* 34(2):62 (1947).

56. Morisset, C. "Compartment des racines isolées de *Lycopersicum esculentum* (solanacées), cultivees *in vitro* et somises á l'anoxia," *C. R. Acad. Sci. (Paris)* 276, Ser. D, 311 (1973).

57. Morisset, C. "Ultrastructural Changes of Mitochondria and Oxygen Take-Up in Excised Roots of *Lycopersicum esculentum,* Cultivated *in vitro* and Submitted to Different Treatments: Prolonged Anoxia, Uncoupling Agent, Plasmolysis," in *Abstracts of the Twelfth International Botanical Congress* (Leningrad: "Nauka," 1975), p. 366.

58. Zimmerman, P. W., A. E. Hitchcock and W. Crocker. "The Movement of Gases into and through Plants," *Contrib. Boyce Thompson Inst.* 3(3):313 (1931).

59. Conway, V. M. "Aeration and Plant Growth in Wet Soils," *Bot. Rev.* 6(4):149 (1940).

60. van Raale, M. H. "On the Oxygen Supply of Rice Roots," *Ann. Bot. Garden Buitenzorg* 51(3):43 (1941).

61. Glasstone, V. F. "Passage of Air through Plants and its Relation to Measurement of Respiration and Assimilation," *Am. J. Bot.* 29(2): 156 (1942).

62. Brown, R. "The Gaseous Exchange between the Root and the Shoot of the Seedling *Cucurbita Pepo*," *Ann. Bot.* 11(44):417 (1947).

63. Vallance, K. B. and D. A. Coult. "Observation on the Gaseous Exchange which Takes Place between *Menyanthes trifoliata* and Environment," *J. Exp. Bot.* 2(5):212 (1951).

64. Scholander, P. F., L. Van Dam and S. L. Scholander. "Gas Exchange in the Roots of Mangroves," *Am. J. Bot.* 42(1):124 (1955).

65. Jensen, C. R., J. Letey and L. H. Stolzy. "Labelled Oxygen: Transport through Growing Corn Roots," *Science* 144(3618):550 (1964).

66. Jensen, C. R., L. H. Stolzy and J. Letey. "Tracer Studies of Oxygen Diffusion through Roots of Barley, Corn and Rice," *Soil Sci.* 103(1):23 (1967).
67. Armstrong, W. "Oxygen Diffusion from the Roots of Some British Bog Plants," *Nature* 204(4960):801 (1964).
68. Armstrong, W. "The Use of Polarography in the Assay of Oxygen Diffusing from Roots in Anaerobic Media," *Physiol. Plant* 20(3):540 (1967).
69. Armstrong, W. "Oxygen Diffusion from the Roots of Woody Species," *Physiol. Plant* 21(3):539 (1968).
70. Armstrong, W. "Rhizosphere Oxidation in Rice: an Analysis of Intervarietal Differences in Oxygen Flux from the Roots," *Physiol. Plant* 22(2):296 (1969).
71. Armstrong, W. "Oxygen Diffusion from the Roots of Rice Grown under Nonwaterlogged Conditions," *Physiol. Plant* 24(2):242 (1971).
72. Armstrong, W. "The Wetland Condition and the Internal Aeration of Plants," in *Abstracts of the Twelfth International Botanical Congress* (Leningrad: "Nauka," 1975), p. 347.
73. Healy, M. T. and W. Armstrong. "The Effectiveness of Internal Oxygen Transport in Mesophyte *(Pisum sativum L.),*" *Planta* 103(4): 302 (1972).
74. Armstrong, W. and E. J. Wright. "Radial Oxygen Loss from Roots: The Theoretical Basis for the Manipulation of Flux Data Obtained by the Cylindrical Platinum Electrode Technique," *Physiol. Plant* 35(1):21 (1975).
75. Vartapetian, B. B. "A Polarographic Investigation of Oxygen Transport in Plants," *Fiziol. Rast.* 11(5):774 (1964) (In Russian).
76. Vartapetian, B. B. "Molecular Oxygen and Water in Cellular Metabolism," (Moscow: "Nauka," 1970), p. 119 (In Russian).
77. Vartapetian, B. B., I. N. Andreeva, N. G. Davtian and I. P. Maslova. "Ultrastructure of *Cucurbita Pepo* Root Cells in Connection with Oxygen Transport," *Fiziol. Rast.* 15(1):19 (1968) (In Russian).
78. Vartapetian, B. B., L. P. Agapova, A. A. Averianov and V. A. Veselovsky. "New Approach to Study of Oxygen Transport in Plants Using Chemiluminescent Method," *Nature* 249(5454):269 (1974).
79. Vartapetian, B. B., L. P. Agapova, A. A. Averianov and V. A. Veselovsky. "Study of Oxygen Translocation from Shoots to Roots of *Cucurbita Pepo* by Measuring Ultraweak Glowing," *Agrochim.* 19(2):173 (1975).
80. Vartapetian, B. B. and N. Nuritdinov. "Molecular Oxygen Transport in Plants," *Naturwissenschaften* 63(5):246 (1976).
81. Nuritdinov, N. and B. B. Vartapetian. "Oxygen Transport from the Shoots of Cotton Plants into the Roots," *Fiziol. Rast.* 23(3): 622 (1976) (In Russian).
82. Greenwood, D. J. "Studies on the Transport of Oxygen through the Stems and Roots of Vegetable Seedlings," *New Phytol.* 66(3): 337 (1967).
83. Greenwood, D. J. "Studies on Oxygen Transport through Mustard Seedlings," *New Phytol.* **66**(3):597 (1967).

84. Greenwood, D. J. "Studies on the Distribution of Oxygen around the Roots of Mustard Seedlings *(Sinapsis Alba L.)*," *New Phytol.* 70(1):97 (1971).

85. Greenwood, D. J. and D. Goodman. "Studies on the Supply of Oxygen to the Roots of Mustard Seedlings *(Sinapsis Alba L.),*" *New Phytol.* 70(1):85 (1967).

86. Chirkova, T. V. and S. V. Soldatenkov. "Paths of Movement of Oxygen from Leaves to Roots Kept under Anaerobic Conditions," *Fiziol. Rast.* 12(2):216 (1965) (In Russian).

87. Chirkova, T. V. "Peculiarities of Oxygen Supply to Woody Plant Roots Subjected to Anaerobic Conditions," *Fiziol. Rast.* 15(3):565 (1968) (In Russian).

88. Luxmoore, R. J., L. H. Stolzy and J. Letey. "Oxygen Diffusion in the Soil-Plant System. I. A Model," *Agron. J.* 62(3):317 (1970).

89. Luxmoore, R. J., L. H. Stolzy and J. Letey. "Oxygen Diffusion in the Soil-Plant System. II. Respiration Rate, Permeability and Porosity of Consecutive Excised Segments of Maize and Rice Roots," *Agron. J.* 62(3):322 (1970).

90. Luxmoore, R. J., L. H. Stolzy and J. Letey. "Oxygen Diffusion in the Soil-Plant System. III. Oxygen Concentration Profiles, Respiration Rates, and the Significance of Plant Aeration Predicted for Maize Roots," *Agron. J.* 62(3):325 (1970).

91. Luxmoore, R. J., L. H. Stolzy and J. Letey. "Oxygen Diffusion in the Soil-Plant System. IV. Oxygen Concentration Profiles, Respiration Rates, and Radical Oxygen Losses Predicted for Rice Roots," *Agron. J.* 62(3):329 (1970).

92. Luxmoore, R. J. and L. H. Stolzy. "Oxygen Diffusion in the Soil-Plant System. V. Oxygen Concentration and Temperature Effects of Oxygen Relations Predicted for Maize Roots," *Agron. J.* 64(6): 720 (1972).

93. Luxmoore, R. J. and L. H. Stolzy. "Oxygen Diffusion in the Soil-Plant System. VI. A Synopsis with Commentary," *Agron. J.* 64(6):725 (1972).

94. Hook, D. D., C. L. Brown and P. P. Kormanik. "Inductive Flood Tolerance in Swamp Tupelo *[Nyssa sylvatica var. biflora (Walt.) Sarg.]*," *J. Exp. Bot.* 22(70):78 (1971).

95. Hook, D. D., C. L. Brown and R. H. Wetmore. "Aeration in Trees," *Bot. Gaz.* 133(4):443 (1972).

96. Hook, D. D. and C. L. Brown. "Root Adaptations and Relative Flood Tolerance of Five Hardwood Species," *Forest Sci.* 19(3):225 (1973).

97. Evans, N. T. S. and M. Ebert. "Radioactive Oxygen in the Study of Gas Transport down the Root of *Vicia faba*," *J. Exp. Bot.* 11(32):246 (1960).

98. Evans, N. T. S. and M. Ebert. "The Effect of Metabolism of the Transport of O^{15}-Labelled Oxygen through *Vicia faba* Roots," *Internat. J. Radiation Biol.* 3(6):627 (1961).

99. Barber, D. A., M. Ebert and N. T. S. Evans. "The Movement of O^{15} through Barley and Rice Plants," *J. Exp. Bot.* 13(39):397 (1962).

100. Soldatenkov, S. V. and Chzhao Sian-Duan. "Role of Bean and Maize Leaves in the Respiration of Roots Deprived of Oxygen," *Fiziol. Rast.* 8(4):385 (1961) (In Russian).
101. Soldatenkov, S. V. and T. V. Chirkova. "Role of Leaves in Respiration of Roots Deprived of Oxygen," *Fiziol. Rast.* 10(5):535 (1963) (In Russian).
102. Jensen, C. R. and D. Kirkham. "Labelled Oxygen: Increased Diffusion Rate through Soil Containing Growing Corn Roots," *Science* 141(3582):735 (1963).
103. van Heid, H., B. M. M. Boer-Bolt and M. H. van Raalte. "The Effect of a Low Oxygen Content on the Medium on the Roots of Barley Seedlings," *Acta Bot. Neerl.* 12(2):231 (1963).
104. Morgan, E. H. and G. Nahas. "Study of Relationship of Arterial Oxygen Tension to Alveolar Oxygen Pressure in Man, Utilizing a Polarometric Method for Whole Blood," *Am. J. Physiol.* 163(3):736 (1950).
105. Montgomery, H. and O. Horowitz. "Oxygen Tension of Tissues by Polarographic Method," *J. Clin. Invest.* 29(9):1120 (1950).
106. Lemon, E. R. and A. E. Erickson. "The Measurement of Oxygen Diffusion in the Soil with a Platinum Microelectrode," *Soil Sci. Soc. Am. Proc.* 16(2):160 (1952).
107. Clark, L., R. Walf, D. Granger and Z. Taylor. "Continuous Recording of Blood Oxygen Tensions by Polarography," *J. Appl. Physiol.* 6(3):189 (1953).
108. Charlton, G. "A Microelectrode for Determination of Dissolved Oxygen in the Tissue," *J. Appl. Physiol.* 16(4):729 (1961).
109. Kovalenko, E. A. "Changes of Oxygen Tension in Dog's Brain Tissue in Rarified Atmosphere," *Patol. Fiziol. Exper. Tezap.* 6(18) (1962) (In Russian).
110. Macdonald, J. K. and G. G. Laties. "Oxygen Electrode Measurements of Potato Slice Respiration at $0°C$," *J. Exp. Bot.* 13(39):435 (1962).
111. Ostrovsky, D. N. and M. S. Gelman. "Estimation of Oxygen Concentration in Biological Fluids by Applying Polarographic Method with Immobile Solid Electrode," *Biochemistry (USSR)* 27(3):532 (1962) (In Russian).
112. Bronk, J. B. and D. S. Parsons. "The Polarographic Determination of the Respiration of the Small Intestine in the Rat," *Biochim. Biophys. Acta* 107(3):397 (1965).
113. Kursanov, A. L. and B. B. Vartapetian. "Plants and Oxygen," *Mediterranea* 27:726 (1968).
114. James, W. O. *Plant Respiration* (Oxford: Clarendon Press, 1953).
115. Vartapetian, B. B. "Aeration of Roots in Relation to Molecular Oxygen Transport in Plants," *Proceedings of the Uppsala Symposium, Plant Response to Climatic Factors,* UNESCO, 1973.
116. Colli, L., U. Facchini, G. Guidotti, R. Dygnani-Lonati, M. Orsenigo and O. Sommariva. "Further Measurements on the Bioluminescence on the Seedlings," *Experientia* 11(12):479 (1955).
117. Tarusov, B. N., J. M. Ivanov and Y. M. Petrusevich. "Ultraweak Glowing in Biological Systems," Moscow State University Publisher (1967) (In Russian).

118. Mölbert, E. and D. Querritore. "Electronenmicroskopische Untersuchungen am Leberparenchym bei akuter Hypoxie," *Beitr. Pathol. Anat. Allgem. Pathol.* 117(1):32 (1957).
119. Rouiller, C. "Physiological and Pathological Changes in Mitochondrial Morphology," *Internat. Rev. Cytol.* 9:227 (1962).
120. Wrischer, M. "Electronenmikroskopische Untersuchungen der Zellnekrobiose," *Protoplasma* 60(4):355 (1965).
121. Geiger, D. R. and A. L. Christy. "Effect of Sink Region Anoxia on Translocation Rate," *Plant Physiol.* 47(2):172 (1971).
122. Vartapetian, B. B., I. N. Andreeva and G. I. Kozlova. "The Resistance to Anoxia and the Mitochondrial Fine Structure of Rice Seedlings," *Protoplasma* 88(2):215 (1976).
123. Vartapetian, B. B., I. N. Andreeva, G. I. Kozlova and L. P. Agapova. "Mitochondrial Ultrastructure in Roots of Mesophyte and Hydrophyte at Anoxia and After Glucose Feeding," *Protoplasma* 91(3): 243 (1977).
124. Crawford, R. M. M. "The Physiological Basis of Flooding Tolerance," *Ber. Dtsch. Bot. Gas.* 82(1/2):111 (1969).
125. Crawford, R. M. M. "Metabolic Adaptations to Anoxia in Plants and Animals," in *Abstracts of the Twelfth International Botanical Congress* (Leningrad: "Nauka," 1975), p. 353.
126. Crawford, R. M. M. and P. D. Tyler. "Organic Acid Metabolism in Relation to Flooding Tolerance in Roots," *J. Ecol.* 57(1):235 (1969).
127. McManmon, M. and R. M. M. Crawford. "A Metabolic Theory of Flooding Tolerance: The Significance of Enzyme Distribution and Behaviour," *New Phytol.* 70(2):299 (1971).
128. Chirkova, T. V., I. V. Khazova and T. P. Astafurova. "On Metabolic Regulation of Plant Adaptation to Temporal Anaerobiosis," *Fiziol. Rast.* 21(1):102 (1974) (In Russian).
129. Chirkova, T. V. "Some Regulatory Mechanisms of Plant Adaptations to Anaerobiosis," in *Abstracts of the Twelfth International Botanical Congress* (Leningrad: "Nauka," 1975), p. 394 (In Russian).
130. Marshall, D. R., P. Broue and A. J. Pryor. "Adaptive Significance of Alcohol Dehydrogenase Isozymes in Maize," *Nature New Biol.* 244(131):16 (1973).
131. Lonhart, Y. B. and J. Baker. "Intra-Population Differentiation of Physiological Response to Flooding in a Population of *Veronica peregrina L.*," *Nature* 242(5416):275 (1973).
132. Andreeva, I. N., L. P. Agapova, G. I. Kozlova and B. B. Vartapetian. "Effect of Anoxia on Mitochondrial Ultrastructure in the Roots of Hydrophytes," *Fiziol. Rast.* 22(1):77 (1975).
133. Mitchell, P. "Chemiosmotic Coupling in Oxidative and Photosynthetic Phosphorylation," *Biol. Rev.* 41(3):445 (1966).
134. Racker, E. "Resolution and Reconstitution of the Inner Mitochondrial Membrane," *Fed. Proc.* 26(5):1335 (1967).
135. Skulachev, V. P. *Energy Transformation in Biomembranes* (Moscow: "Nauka," 1972) (In Russian).
136. Skulachev, V. P. "Mechanism of Action of Molecular Biological Electric Generators," in *Abstracts of the Twelfth International Botanical Congress* (Leningrad: "Nauka," 1975), p. 371.

137. Nuritdinov, N. and B. B. Vartapetian. "Transport of C^{14}-Sucrose in Cotton Plants under Conditions of Root Oxygen Deficiency," *Dokl. USSR Acad. Sci.* 228(2):149 (1976) (In Russian).

138. Ferron, F. B., B. Vartapetian and C. Costes. "Effects de L'anoxie sur la circulation de l'eau à l'obscurité chex jeunes plants de Maïs," *C. R. Acad. Sci. (Paris)* 282 Ser. D, 361 (1976).

139. Maslova, I. P., I. F. Tchernyadeva and B. B. Vartapetian. "Soluble Proteins and Alcohol Dehydrogenase of Rice Seedlings in Anoxia," in *Abstracts of the Twelfth International Botanical Congress* (Leningrad: "Nauka," 1975), p. 365.

140. Pradet, A. and C. Prat. "Métabolisme Energétique au cours de la Germination du Riz en Anoxie," in *Etudes de Biologie Végétale*, R. Facques, Ed. (Paris: Gif-sur-Yvette, C.N.R.S., 1976), p. 56.

3

ENERGY METABOLISM IN PLANTS
UNDER HYPOXIA AND ANOXIA

A. Pradet

Institut National de la Recherche Agronomique
Laboratoire de Physiologie Végétale
Centre de Recherche de Bordeaux
33140 Pont De La Maye, France

J. L. Bomsel

Laboratoire de Physiologie Végétale Appliquée
Université Paris VI
75230 Paris Cedex 05, France

INTRODUCTION: ENERGY METABOLISM AND EVOLUTION

The phrase "life without oxygen" is ambiguous for higher plants at this stage in evolution when the atmosphere contains 21% oxygen. Most of this is known to originate from plants through photosynthesis.[1] Consequently when green cells are placed in an anoxic environment, they synthesize oxygen provided that the atmosphere or water contains CO_2. However, in regard to green plants grown in the dark, the nonchlorophyllous parts of such plants, and nonchlorophyllous plants, exogenous oxygen is essential for respiratory metabolism which results in the production of energy and a variety of substrates from molecules synthesized by green cells.

However, under numerous ecological conditions specified in many chapters of this book, respiration of the plant may either be limited or completely suppressed because oxygen is lacking in the surroundings.

It is now generally accepted that the first living cells were built from organic molecules whose synthesis was favored by the chemical and physical conditions prevailing at that time in the sea. Such cells lived without oxygen. The first metabolic pathways to appear were fermentative and used as substrate amino acids such as glycine or alanine. These were followed by glycolysis and the pentose cycle, the energy yield from all these anaerobic mechanisms of oxidation being relatively low. Later, these mechanisms were completed by more complex metabolic pathways— the tricarboxylic acid cycle and the respiratory chain—which greatly improved energy yield through their effect on cellular oxidation. Life at this stage of evolution was still anaerobic, however, and inorganic molecules such as nitrate received the electrons generated by the respiratory pathways.[2,3]

Thus it appears that the principal metabolic pathways characteristic of aerobic life are derived from mechanisms that already existed when life was anaerobic.[4] This probably facilitated the maintenance of anaerobic metabolic pathways in modern plants that encounter temporary anoxic conditions as, for example, when the soil is flooded. The coefficient of diffusion of oxygen is very high in air but very low when the gas is dissolved in water. Consequently in the cells of eukaryotes there is a concentration gradient between the outside of the cell and the intracellular location of the respiratory system.[5,6] Multicellular organization made oxygen diffusion much more difficult. Early in evolution (Cambrian), animals developed a circulatory system the efficiency of which was further improved by the subsequent development of oxygen carriers. However, muscular metabolism in mammals is partially anaerobic, aerobic mechanisms being unable to supply sufficient energy. In higher plants the problem of diffusion in bulky organs has received considerable attention; diffusion in the extracellular spaces is not a limiting factor as it is in intracellular spaces.[7,8] However, for those plant organs (such as seeds and roots) living in flooded soils the factor limiting respiration became the diffusion of oxygen in the soil. The roots of some plants are able to obtain oxygen from the aerial parts by diffusion through the intercellular spaces. This phenomenon is present in many hydrophytes, and oxygen transport has also been demonstrated in the roots of cultivated plants such as pea (*Pisum sativum* L.), corn (*Lea mays* L.), barley (*Hordeum vulgare*) and cotton (*Gossypium hirsutum* L.). However, in spite of much research it is difficult to evaluate the efficiency of this oxygen transport which might make aerobic metabolism possible in roots placed in anoxic conditions.[9-13]

Thus the seeds and roots of many plants are able to maintain either aerobiosis or hypoxia or anoxia. Their cells contain the metabolic mechanisms permitting adaptation to these conditions. It should be mentioned that adaptation to anoxia is always limited in nature and only permits temporary existence in the absence of respiratory metabolism.

The metabolic pathways permitting the generation of energy and essential cell metabolites from organic molecules are few. Potentiality to synthesize the enzymes permitting the functioning of these metabolic pathways is present in the genomes of most living cells. In contrast, considerable variability is apparent in the way in which living things have been able to regulate the direction and activity of these different metabolic pathways.[14] During the past few years the anoxic survival of aquatic animals has received much attention.[15,16]

Metabolic adaptation to the transition Anaerobiosis $\leftrightarrows$ Hypoxia $\leftrightarrows$ Anoxia has become possible through wide diversity in the modulation of enzyme activities and synthesis and development of different kinds of metabolism in different species.

GENERAL FEATURES OF ANAEROBIC METABOLISM

It has been known since the time of Pasteur that when oxygen is lacking higher plants accumulate ethanol. Later it was shown that some plants placed under the same conditions may synthesize lactate. For a long time, the studies devoted to anaerobic metabolism in plants were mainly attempts to understand these phenomena. The relationships between quantities of carbon dioxide, ethanol and lactate synthesized when different plant tissues in various physiological states were subjected to anoxia were analyzed in numerous papers.[17-19] The aim of such work was mainly to evaluate the activity and efficiency of anoxic, as compared with aerobic, energy metabolism.

Early during this period the ideas of biological energy produced by oxidation and end products derived from the oxidation of sugars were postulated. These concepts inspired further work on the yields of anaerobic metabolism.

The steps in glycolysis involving the oxidation of glucose and synthesis of ethanol and lactate will not be described in detail. Instead a simple outline of the relationships with three essential phenomena is given. These are the production of energy-rich bonds that permit biological utilization of energy for synthesis, the maintenance of equilibrium in the redox balance of the cell, and the production of substrates.

Glycolysis in Anoxia

The energy yield obtained through this mechanism is not high. Transformation of 1 mole of glucose to ethanol or lactate produces only 2 moles of ATP from ADP, giving about 15 kcal of energy for biological use. During this time the glucose loses about 10% of its free energy and this yield is very poor. Many plants are able to increase their rate of glucose oxidation, however, when they are transferred from aerobic to anaerobic conditions. This, termed the Pasteur effect by Warburg,[20] permits an increased production of biologically usable energy at the expense of considerable glucose consumption. The redox balance of the cell is not modified by this mechanism, which is neutral for the oxidation and reduction of pyridine coenzyme. It is not, therefore, a limiting factor in glycolysis activity.

The transition from aerobic to anaerobic metabolism causes certain and sometimes highly transient modifications in the levels of the various substrates of glycolysis.[21,22,23] The concentration of these substrates in the cell is always very low when glycolysis is active. Consequently the turnover of these compounds in the cells is very rapid, varying between a few seconds and minutes, with or without oxygen.

When the functioning of the Krebs cycle is limited by low oxygen levels, an increase in the level of pyruvate is prevented by the synthesis of lactate and ethanol. It is a very general biological law that the intracellular concentrations of substrates involved in the fundamental metabolic pathways, and more especially those of metabolic branchpoints, are regulated in a narrow range; these molecules have a rapid turnover. This permits instantaneous adaptation of the cells to variation in environmental factors.

If life under anaerobic conditions is considered to be a phenomenon only occasionally encountered by higher plants, and then only for a short time, it is possible to envisage a mechanism of survival which relies on glycolysis and the formation of lactate and ethanol. As these compounds are metabolized by the plant when aerobic metabolism is restored, there is no energy waste.

However it is known that some plant tissues or organs, such as apples (*Pyrus*), potato tubers (*Solanum*), rice (*Oryza sativa*) and lettuce (*Lactuca sativa*) seeds, and the roots of some trees[24] are able to sustain anoxia for several weeks. When rice grains are imbibed under anoxic conditions, although no growth of roots or leaves occurs, growth of the coleoptile is possible and one week after germination this organ may measure as much as 40 mm.[25-27] In consequence, two additional problems have to be considered.

Toxicity of Glycolysis Products

Lactate on acidification and ethanol are highly toxic to the cells. Yeasts synthesize large quantities of ethanol but this compound is then excreted. Similarly, the metabolism of actively working muscle produces lactate but part of this is flushed out into the blood in which exist very efficient buffer systems. Do plant cells possess mechanisms which protect them against the toxicity of these molecules? It is generally accepted that the sensitivity of a plant to anoxia originates mainly from the increased intracellular concentration of toxic compounds. It has been shown for certain plant tissues[28,29] that lactate synthesis is inhibited when the synthesis of ethanol begins (Figure 1). Davies has shown[29] that it is the lowering of the pH that induces an inhibition of lactate dehydrogenase activity and an activation of pyruvate decarboxylase (Figure 2). Such plants are thereby protected against the accumulation of lactate.

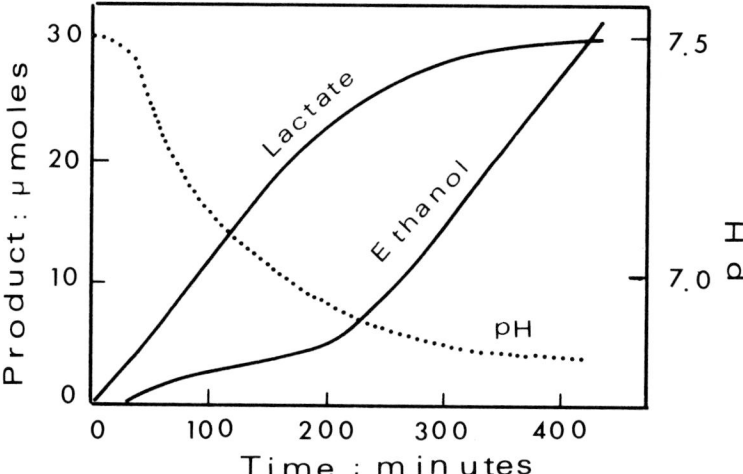

Figure 1. Production of lactate and ethanol from glucose by an extract of pea seeds.

On the other hand, Doireau[30] has shown that the anoxic gaseous medium surrounding seedlings contains large quantities of ethanol and Chirkova and Gutman[31] found ethanol, ethylene and acetaldehyde liberated through lenticels of willow (*Salix alba* L.) and poplar (*Populus petrowskiana* Sch.) when the roots of such trees were submitted to flooding conditions. It is also known that in many plants which are well adapted to anoxia, the rate of ethanol synthesis is considerably reduced under such conditions.[32] By means of these different mechanisms, glycolysis may be able to function for a long time.

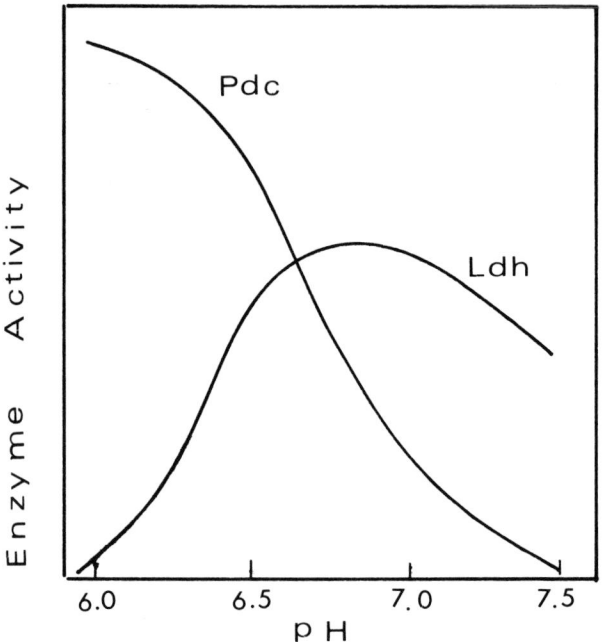

Figure 2. Effect of pH on the activity of pyruvate decarboxylase (Pdc) and lactate dehydrogenase (Ldh) in an extract of parsnip root.

Synthesis in Anoxia

The above interpretations are, however, oversimplified and do not offer suitable explanations for many of the facts. The activity of some enzymatic systems such as alcohol dehydrogenase, malic enzyme or nitrate reductase, is known to increase during anoxia.[33-35] Such an increase in enzyme activity can be explained by an activation of preexisting protein. Recently we have shown (Mocquot, Prat, Pradet[40,41]) that C^{14} leucine is actively incorporated into the proteins of rice seedlings under anoxia, which means that protein synthesis is possible in such material under these conditions. It is not known whether the amino acids used in this synthesis are produced during anoxia or whether they originate from the turnover of proteins. It was stressed by Krebs[36] that although the biosynthesis of some amino acids (aspartate, alanine) is neutral with respect to the redox balance, synthesis of the majority involves more oxidation than reduction, and this in turn implies the production of hydrogen atoms which need to be disposed of. Effer and Ranson[37] have studied *in vivo* the variation in tricarboxylic acid cycle substrates and 18 amino acids in buckwheat (*Fagopyrum*) seedlings. This material does not exhibit any demonstrable Pasteur effect. The authors conclude that the Krebs cycle is able to function under anoxic conditions permitting the synthesis of most organic acids and amino acids. The data obtained by a number of authors on the compounds synthesized during anoxia in higher plants and animals have been reviewed by Crawford.[12] In 1971 this author postulated that metabolic adaptation to anoxia was dependent on a wide diversification of glycolysis products. The same opinion is apparent in the different papers published by the laboratories of De Zwaan and Hochachka since 1972.[15,16] These authors are working mainly on the problems associated with the adaptation of bivalve molluscs to anoxia. It is known that such animals sustain anoxia soon after the closure of the shells and that they are able to survive for several weeks under such conditions. The authors have shown that during anoxia many of the Krebs cycle compounds and amino acids may be synthesized but that the metabolic pathways are modified. The same observation was made by Streeter and Thompson[38,39] who studied the incorporation of different precursors of organic and amino acids in radish leaves subjected to anoxia.

In a recent study,[40] DNA synthesis was shown to be active in rice seedlings growing under anoxic conditions. The level of adenine nucleotides was found to increase twentyfold during coleoptile growth,[41] thus demonstrating that DNA and nucleotide synthesis are possible in anoxia. However it is not known whether the pentose and the bases constituting such nucleotides are synthesized during anoxia or whether they originate from the hydrolysis of nucleic acids. The synthesis of pentose is difficult in anoxia because it heavily modifies the oxido-reduction equilibrium. Nevertheless this pathway has been shown to function under such conditions by Effer and Ranson[37] and Chirkova, Kharova and Astafurova.[42]

Thus it appears that all the amphibolic pathways (those which result in a simultaneous production of substrates and energy) may be active under anoxic conditions. However, it is possible that the direction of these various pathways is modified in comparison to aerobic metabolism. Many authors have postulated that a surplus of energy-rich bonds is synthesized in anoxia when compounds such as succinate or propianate accumulated in the cells.[15,16] Today it is impossible to determine the importance of this phenomenon in higher plants.[15,16] The most important problem arising from anaerobic metabolism of long duration is the maintenance of a redox equilibrium. The NAD/NADH ratio of plants in anoxia has been shown by Effer and Ranson[37] and Yamamoto[43] to be higher than 1. This means that higher plants must possess various mechanisms which enable them to use the protons and regenerate NAD. It is possible that the cells synthesize reduced molecules. Nitrate reduction is another means of restoring the equilibrium between oxidation and reduction. An increase in nitrate reductase activity has been demonstrated in various plants subjected to anoxia.[44-46] It is possible that a sort of "nitrate respiration" functions in anaerobiosis *in vivo* but insufficient evidence for this is available at present.

REGULATORY MECHANISMS

The adaptation of living organisms exposed to changing environmental conditions involves two main types of phenomena which are referred to as coarse and fine mechanisms. Both of them act in different ways on enzyme activity (*cf.*, for instance, Hochachka and Somero[47]).

Coarse Mechanisms

The coarse mechanisms result in a lengthy adaptation to given conditions. They are controlled at the gene level and require relatively long periods of time to become effective. They determine the nature and/or the concentration of the enzymes present in the cells.[48,49] They require new synthesis of proteins or activation of inactive precursors (*e.g.*, enzyme induction by substrate).

A good example of such a regulation is provided by alcohol dehydrogenase. Two genes control the synthesis of this enzyme *Adh1* and *Adh2*. The balance of active gene products specified by *Adh1* and *Adh2* varies drastically in different tissues and in the same tissue in response to different environments such as anaerobiosis.[50] The possible significance of this mechanism for adaptation to anoxic conditions was studied by Marshall *et al.*,[51] in maize (*Zea mays* L.) and *Bromus mollis*.[52] However, the mechanisms of enzyme synthesis regulation in higher plants are not known as well as in prokaryotes.

Fine Mechanisms

The fine mechanisms change the reactivity between enzyme and substrate already present in the cell; their action is almost instantaneous. There are many kinds of such mechanisms, one of the most important being the case of allosteric regulation[53] when "modifiers" or "effectors" bind on the allosteric site of the enzyme, inducing modifications in the tertiary or quaternary structure of the enzymatic protein.[54] The saturation curves (enzyme activity against substrate concentration) are then modified, but there is generally no change in Vmax. The effects are mostly modification of Km (S 0.5) or of the Hill coefficient,[55] or transition from an hyperbolic to a sigmoidal curve, the sigmoidal curves representing greater advantages in regulation (*cf.*, Figure 3). The regulatory enzymes are typically located at the starting point or near the branchpoints of the metabolic pathways. Their activity is modulated in an allosteric manner by one or several end products of the given pathway (feedback control described by Umbarger[56] and Yates and Pardee.[57] This phenomenon can reach a great order of complexity (for the many different types of feedback control see reviews in Stadtman[58] and Priess and Kosuge [59]). In consequence, the first reaction in a pathway can be regulated by the last metabolite or sometimes the last reaction by the first metabolite.

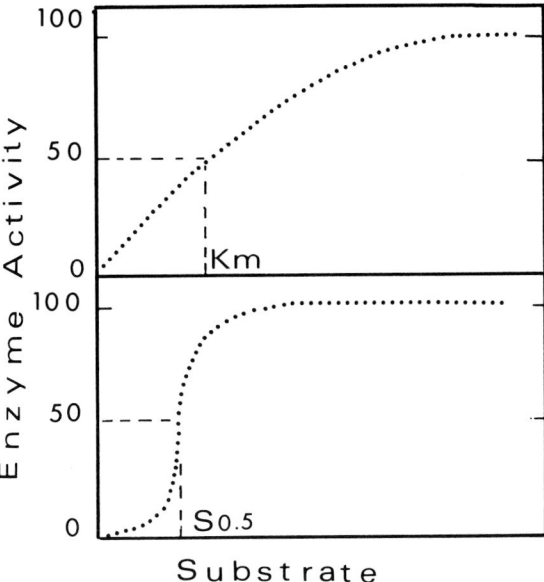

Figure 3. Hyperbolic and sigmoid saturation curves for two kinds of enzymes. Small changes in substrate concentration around Km or S 0.5 or small changes of the apparent E-S affinity produce a larger change in reaction velocity for sigmoid than for hyperbolic curves.

Atkinson *et al.*[60] have pointed out that the substrate concentrations in the cells are generally equal to or slightly below the Km (S 0.5) of the enzymes with which they are involved. Two important exceptions concern the coupling agents of metabolism, *i.e.*, the adenine and pyridine nucleotides: the former are involved in the coupling of metabolic sequences in terms of energy and the latter in terms of electron transport. The enzymes which utilize these nucleotides as a substrate (*e.g.*, kinases or dehydrogenases) have Km (S 0.5) values which are much lower than the physiological concentrations of these nucleotides. The consequences is that their catalytic sites should always be saturated so that the reaction velocity is not determined by the concentration of one single nucleotide but by the ratio between the forms more or less rich in pyrophosphate bonds (*i.e.*, ATP/ADP, ATP/AMP) or by the ratio between the oxidized and reduced forms (*i.e.*, NAD/NADH, NADP/NADPH).

Adenylate Energy Charge: Definition

As mentioned above, the activities of some enzymes (kinases especially) are directly determined by the ATP/ADP or ATP/AMP ratios depending on the affinity constants between the catalytic sites and the different nucleotides.[60] On the other hand, one or several nucleotides serve as allosteric effectors for numerous enzymes. When ATP acts as a positive effector, ADP or AMP are negative and vice versa. When this occurs, enzyme activity is modulated by the ATP/ADP or ATP/AMP ratios in a much more sensitive manner. Atkinson and Walton[61] introduced the new parameter energy charge (E.C.) which integrates these two ratios. These authors pointed out that when a system is equilibrated according to Equation 1,

$$\text{2 ADP} \xrightarrow{\text{Adenylate kinase}} \text{ATP + AMP} \qquad (1)$$

where adenylate kinase has an equilibrium constant close to 1, then the adenylate pool serves as an accumulator of energy-rich bonds. The ratios ATP/ADP and ATP/AMP take precisely defined values (*cf.*, Figure 4) depending on the degree of saturation in energy-rich bonds of the adenylate pool. This latter is expressed by Equation 2:

$$\text{E. C.} = (\text{ATP}) + \tfrac{1}{2}\,(\text{ADP}) \,/\, (\text{ATP}) + (\text{ADP}) + (\text{AMP}) \qquad (2)$$

The curves in Figure 5 show how enzyme activity responds to energy charge (as found in *in vitro* experiments with purified enzymes).[62] Many U- and R-type reactions occur in cells and it has been predicted by Atkinson[62] that the energy charge value would tend to remain stable *in vivo* at a level corresponding to that of the intersection of the U and R curves, *i.e.*, close to 0.8.

It was shown by Bomsel and Pradet[63] that the energy charge values calculated from the data of numerous authors were close to this predicted value (ranging from 0.7 to 0.9) providing that the fixation of the material and the method of analysis employed were suitable (this particular point

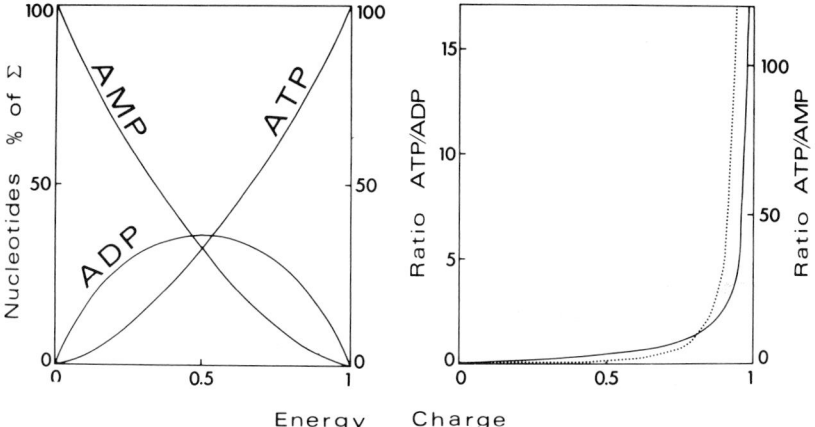

Figure 4. Concentrations of AMP, ADP and ATP (on the left); ratios of ATP/ADP and ATP/AMP (on the right). These values were calculated from a value of 0.8 for the equilibrium of adenylate kinase: $K = (ATP) \times (AMP) / (ADP)^2$. On the left, the dotted line indicates the ATP/AMP ratio.

is discussed elsewhere in this paper). The same result has been obtained by authors who determined energy charge in tissues or cells in their normal physiological state (*cf.*, for instance, tabulation in Bomsel and Pradet,[63] Pradet,[64] and Chapman *et al.*[65]).

In several plant tissues, the adenylate kinase mass action ratio is reached *in vivo* even if variations in energy charge are induced artificially at the same time.[63] Sellami[66] found that the adenylate kinase equilibrium was maintained in different cellular compartments though the value of the apparent equilibrium constant was not the same in the chloroplasts and the nonchloroplastic part of the cell.

Adenylate Energy Charge Regulation

When adenine nucleotides are determined *in vivo* the parameter for which the most constant value for a given tissue under given conditions is to be found is the energy charge. The concentrations of individual nucleotides, together with the size of the adenylate pool, are subject to fluctuations from one experiment to another, whereas the energy charge is not. Several mechanisms may be involved in placing the energy charge at around 0.8. The curves in Figure 5 illustrate how one of these mechanisms comes into operation. For example, when the activity of the biosynthetic pathways exceeds the ATP-regenerating capacity of the catabolic processes the energy charge is lowered. In consequence, the U-type reactions slow down and the R-type reactions are accelerated, thus bringing the energy charge back to its equilibrium value.[62] When the R-type reactions are limited (for instance by lack of oxygen) another mechanism is apparent: the reduction in size of the adenylate pool. This permits the energy charge equilibrium to be achieved with a smaller quantity of energy-rich bonds. This has been observed in animal cells under the action of AMP deaminase; the regulatory properties of this latter have been elucidated by Chapman et al.[67,68] In some microorganisms, the AMP nucleosidase seems to take this role. Some examples demonstrating how this process operates in plants will be discussed later in this paper. The same result (decrease of pool size) can be obtained by compartmentation. It has been shown[66] that the migration of nucleotides from the cytoplasm into chloroplasts, for instance, is responsible for the rapid rise in energy charge in the cytoplasm when leaves are transferred from nitrogen to air.

Adenylate Energy Charge Role

The role of energy charge in the maintenance of cell energy balance has been emphasized above. It should be stressed that even small variations in energy charge around the value of 0.8 will result in considerable variations in the ATP/ADP and ATP/AMP ratios (cf., Figure 4) and consequently in the activity of the enzymes which are regulated by these ratios. But, because of the large number of reactions which are controlled by energy charge, it is difficult to imagine that variation in this parameter alone should be able to direct metabolism towards one pathway rather than another. A significant decrease in energy charge will result mostly in slowing anabolic reactions and facilitating those which regenerate ATP. From a practical point of view, energy charge can be considered an indicator of the state of energy balance of the cell. The role of the adenylate pool size is far from being completely understood (cf., discussion concerns this point in Swedes et al.[69]). Further work on this problem is necessary.

The Redox Charge

This parameter, which now appears in the literature, has been set up to parallel the energy charge in the expression of a cell's reducing power, indicating the degree of saturation in electrons of the pyridine nucleotides. It has been defined (see for instance, Matin and Gottschal,[70] and Vermeersch et al.[71]) as:

$$(NADH) + (NADPH) / (NAD) + (NADH) + (NADP) + (NADPH) \qquad (3)$$

and varies between 0 and 1. Its physiological signification is not very clear since the enzyme specificity for NAD or NADP is rather strict and the ratios NAD/NADH and NADP/NADPH, as far as we know, have

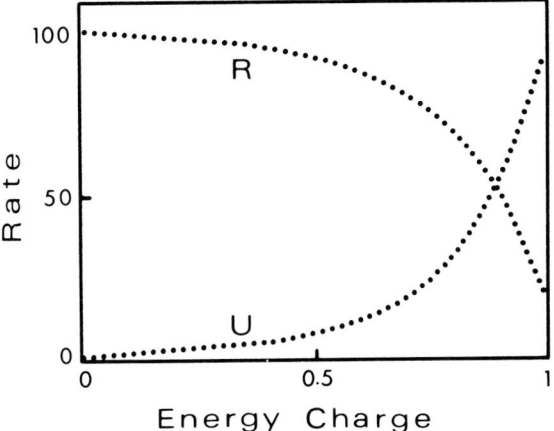

Figure 5. Generalized response to energy charge of enzymes involved in regulation of ATP-regenerating (R) and ATP-utilizing (U) sequences.

always been found to be different. Nevertheless, the regulatory role of pyridine nucleotides has already been underlined (see above, and also a review concerning microorganisms in Sanwal[72]). The lack of secure data on pyridine nucleotide in plant tissues, especially during anoxia, is of general evidence. This is probably due to the difficulty in elaborating a good technology for plant tissues. Here, also, great attention must be given the conditions of fixation. To our knowledge the best works on these problems related to anoxia are those of Effer and Ranson[37] and of Yamamoto's group[43] performed 10 years ago.

Role of Compartmentation

In addition to the above-mentioned example, it should be pointed out that several pathways can be spatially separated in eukaryote cells and this can have great significance in regulation (*cf.*, Atkinson *et al.*[60]). Sanwal[72] has reported regulatory mechanisms which enable microorganisms to compensate for the absence of compartmentation.

Role of pH

Regulation of and by pH has been examined by Davies' group.[14,29] Its role in the regulation of carboxylating and decarboxylating reactions is important. For instance, lactate dehydrogenase and pyruvate decarboxylase, which compete for utilization of pyruvate, exhibit opposite reactions to decrease in pH (Figure 2). At the beginning of anaerobiosis the drop in ATP activates lactate dehydrogenase, which produces lactate. The latter brings about a reduction in cell pH which consequently inhibits the enzyme and activates pyruvate decarboxylase, which is the starting point of the pathway leading to ethanol production.[29] pH also plays a role in the regulation of the Pasteur effect.[23,73,74]

Role of Mineral Ions

Mono- and divalent cations serve as inhibitors or activators of numerous enzymes for which they present various degrees of specificity (*cf.*, review in Stadtman[58]). The binding of divalent cations with nucleotides can interfere with these phenomena. Both ratio and equilibrium of reactions involving nucleotides are markedly dependent on the type and concentration of cations present in the cells.[75-77] The role of inorganic phosphate is also important.[74]

Regulation of Amphibolic Pathways

Atkinson[78] states that, at least in the case of glycolysis, such regulation should be achieved by the joint action of energy charge and of one or several metabolites which act as indicators of the level of the starting materials used for the synthesis. Sanwal,[72] in a well-documented paper (dealing mostly with bacteria but also providing general prospects), stresses that regulation by precursors has been observed up to now only in amphibolic pathways. Turner and Turner in a recent review[74] have collected data dealing with the regulation of glycolysis and Pasteur effect, gluconeogenesis, pentosephosphate pathway and Krebs cycle. Most of the studies relating to these topics also deal with the kinetic properties of isolated enzymes *in vitro*. Transposition of these results *in vivo* remains partly hypothetical, mainly because we do not have sufficient evidence that the kinetic properties of enzymes are the same *in vitro* as they are *in situ*. The assumption that enzymatic proteins could be modified during the isolation processes cannot be definitely ruled out (especially when the starting material is a plant tissue). The catalytic sites of enzymes seem to be much more resistant than the allosteric ones or than the tertiary or quaternary structure of the protein (*cf.*, Yates and Pardee[57]).

The Case of Phospho-Fructo-Kinase (PFK)

This enzyme controls the first step in glycolysis. The regulatory properties of plant enzymes have been extensively studied in a variety of materials (*cf.*, recent review in Turner and Turner[74]). The intriguing point is that plant enzymes, in contrast to those of animals, do not seem to respond to energy charge. They are inhibited by ATP but this inhibition is not relieved by AMP and ADP. The latter seem to serve as inhibitors themselves. Inorganic phosphate, on the other hand, is an activator and able to counteract, at least in part, the inhibition caused by ATP. It is likely that plant PFK are regulated by the ATP/Pi ratio and not by energy charge. In the end the results are similar, however, with maybe a greater sensitivity to activation brought about by a decrease in the adenylate pool size accompanied by a considerable increase in inorganic phosphate.

OXYGEN AND CARBON DIOXIDE UPTAKE AND ENERGY METABOLISM

It is worthy of mention that analysis of the literature relating to the effects of oxygen partial pressure is complicated by the fact that different authors use different units in their expression of oxygen concentration. Plant physiologists tend to speak in terms of percentage oxygen in experiments with gas and molar concentration where aqueous media are involved. Animal physiologists refer to millimeters of mercury or atmospheric percentage. Lake and river ecologists and oceanologists employ a variety of units, the one most often used being the volume of dissolved oxygen. The concentration of impurities in the gas is expressed in parts per million. Nevertheless, it is possible to compare these different units, providing the experimental conditions involved are adequately described (Table I).

Oxygen Partial Pressure

As early as 1902, Friedel[79] showed that reduction of atmospheric pressure to 1/5 its normal value reduced oxygen uptake without affecting CO_2 involvement. Since then this phenomenon has undergone extensive study and has been thoroughly analyzed in a number of reviews.[17,80,81] Many authors have shown that oxygen uptake increases hyperbolically as the oxygen pressure is raised, until a point is reached at which the respiration rate becomes constant (Figure 6). Berry and Norris[82] have termed this point the critical oxygen pressure. This parameter is not easy to appreciate experimentally. Some authors calculate from their data the oxygen partial pressure or concentration that reduces the oxygen uptake to 50% from the value observed at the critical oxygen pressure. This value is sometimes termed Km [81,83] in analogy with enzyme kinetics.

Table I. Dimensional Analysis of Units Expressing Oxygen Concentration in Different Media

Gaseous Medium	Gaseous and Liquid Mediums			Liquid Medium	
ppm = $\mu l/1$	Percentage	Hg (mm)	Atmosphere	Molar Concentration	ppm = $\mu l/1$
2.1×10^5	21	159.6	2.1×10^{-1}	2.5×10^{-4} M	5600
2.1×10^4	2.1	15.96	2.1×10^{-2}	2.5×10^{-5} M	560
1.0×10^4	1	7.6	1.0×10^{-2}	1.2×10^{-5} M	267
10^3	10^{-1}	7.6×10^{-1}	1.0×10^{-3}	1.2×10^{-6} M	26.7
10^2	10^{-2}	7.6×10^{-2}	1.0×10^{-4}	1.2×10^{-7} M	2.67
10	10^{-3}	7.6×10^{-3}	1.0×10^{-5}	1.2×10^{-8} M	0.267
1	10^{-4}	7.6×10^{-4}	1.0×10^{-6}	1.2×10^{-9} M	0.026

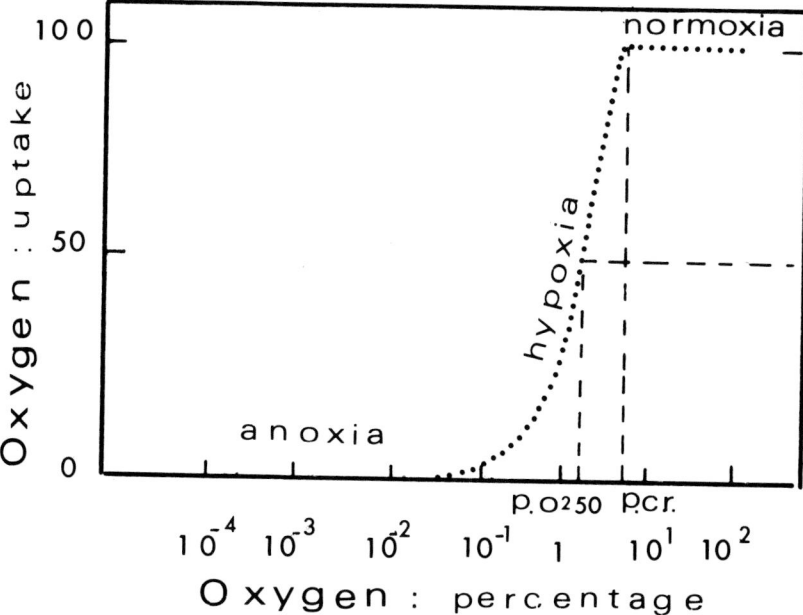

Figure 6. Typical response of oxygen uptake reduction when oxygen partial pressure is lowered under the critical pressure (crp). The partial pressure is expressed as percentage of oxygen in logarithmic scale. According to the tissues, the value of crp, and pO_2 50 (see in the text) may be higher or lower. Consequently normoxia, hypoxia or anoxia are induced by various partial pressures.

There is, however, no simple relationship between the external concentration of substrate and oxidase so that other authors prefer to call this pO_2 50. The susceptibility of cells and tissues subjected to reduced oxygen partial pressure can be compared using these two parameters.

The sensitivity of cells and tissues to oxygen concentration varies considerably depending on the organism involved. pO_2 50 values of about 10^{-7} M were obtained by Longmuir[5,6] for bacteria and yeasts, and in many plant tissues the range is from 10^{-6} M to 10^{-3} M. Interpretation of the results obtained with higher plants has, for the past twenty years, been a source of controversy.

It is now accepted that cytochrome oxidase is the fundamental site of oxygen utilization in living cells. It has long been known that the main factor controlling the observed differences in sensitivity of tissues to oxygen partial pressure is the rate of oxygen diffusion.

Recently it was observed by Armstrong and Gaynard[84] that the parameters controlling sensitivity to oxygen partial pressure in plant tissues have usually been determined using excised organs or blocks of tissues. These authors believe that the intercellular spaces in the tissues may fill up with water after cutting. Yocum and Hackett[85] have shown that pO_2 50 values for slices of *Arum maculation* spadices are higher when the tissues are immersed in well-aerated water than when they are measured in a gaseous medium. Armstrong[84] gives indirect evidence for a similar phenomenon in roots, but it is not certain that this phenomenon is general. Ducet and Chevillotte[7,8] have demonstrated that oxygen diffusion in the cells limits respiratory activity when the partial pressure is lower than the critical oxygen pressure. Moreover, Armstrong's criticisms do not apply to experiments where embryos or seedlings, which exhibit high values of pO_2 50, have been used.[17,64]

Alternate Pathways

Analysis of cellular oxidation in plants has been complicated by the occurrence of various oxidases, such as polyphenol oxidase or ascorbic acid oxidase, and an alternative respiratory pathway that is not inhibited by cyanide and various inhibitors of cytochrome oxidase.

Polyphenol oxidases are very active in most plant cells. In comparison with cytochrome oxidase, the oxygen affinity of these enzymes is very low. They are not inhibited by cyanide. For a long time many people thought that these enzymes were involved in the respiratory metabolism of the cell. The work of Mapson and Burton[86] is representative of this period. Using the methods of enzyme kinetics, these authors analyzed the curves of respiratory activity plotted against oxygen concentration and from their results postulated the existence of two terminal oxidases with very different Km (4×10^{-7} and 4×10^{-4}) in potato tubers.

It is now known that cyanide-resistant respiration is a property of mitochondrial enzymes whereas polyphenol oxidases are soluble enzymes.[87] The mitochondria of many plant tissues—including fruits, *Arum* spadice roots, slices of potato tubers—exhibit such respiration.[88] It is also known to operate *in vivo*. The affinity for oxygen exhibited by this alternate respiratory pathway has not been measured but it is possible that the oxidations involved are not tightly coupled to phosphorylation.[89,90]

Oxygenase

Direct participation of molecular oxygen in metabolic reaction is well known. Numerous oxygenases use oxygen as substrate but the quantity of oxygen employed by the reaction appears low compared to oxidative phosphorylation.

Significance of Oxygen Uptake

After this discussion we might ask what significance oxygen uptake actually has on the metabolic activity observed *in vivo*. During the last twenty years the techniques used to extract mitochondria from plant tissues have been greatly improved. Bonner and his group were pioneers in this field (for a review see Palmer[87]). It is now possible to obtain mitochondria from many tissues that are able to oxidize substrates and produce energy-rich bonds according to theory. The few exceptions include tissues exhibiting cyanide-resistant respiration.

Consequently it is now generally admitted that most cellular oxidation is coupled to phosphorylation. This theory, however, is based on analysis of mitochondrial properties. Very few experiments provide data which permit the *in vivo* P/O ratio to be assessed. In this laboratory we have obtained P/O values of 2 using germinating lettuce seeds *in vivo*,[91] by suppressing oxidative phosphorylation with anaerobiosis. The theoretical P/O ratio is about 3. This is important in understanding hypoxic or anoxic metabolism when it is necessary to measure the respective influence of oxidation and fermentation in the production of ATP. As mentioned above, the respiratory oxidation of glucose produces 20 times more ATP than glycolytic processes, but this is only true when all the oxidations are tightly coupled with phosphorylation.

Definition of Hypoxia and Anoxia

Cellular hypoxia and anoxia can be defined in terms of respiratory metabolism. The cell may be considered to be under hypoxic conditions when the oxygen partial pressure limits the production of energy-rich bonds by the mitochondria. When the production of energy-rich bonds by oxidative phosphorylation becomes negligible in comparison to those generated by fermentation processes, the cell is under anoxia (Figure 7).

The respiratory activity of the cells must be greatly reduced for the anoxic state to be induced. When hypoxia reduces respiration to 5% of its initial value, the oxidative production of ATP is, in the absence of a Pasteur effect, equivalent to that produced by fermentative processes.

Carbon Dioxide Uptake

It is very difficult to appreciate *in vivo* the participation of cellular oxidation in the production of energy. However, oxygen uptake is a very useful criterium which enables the reduction in metabolic activity and energy production in plants under hypoxia to be assessed.

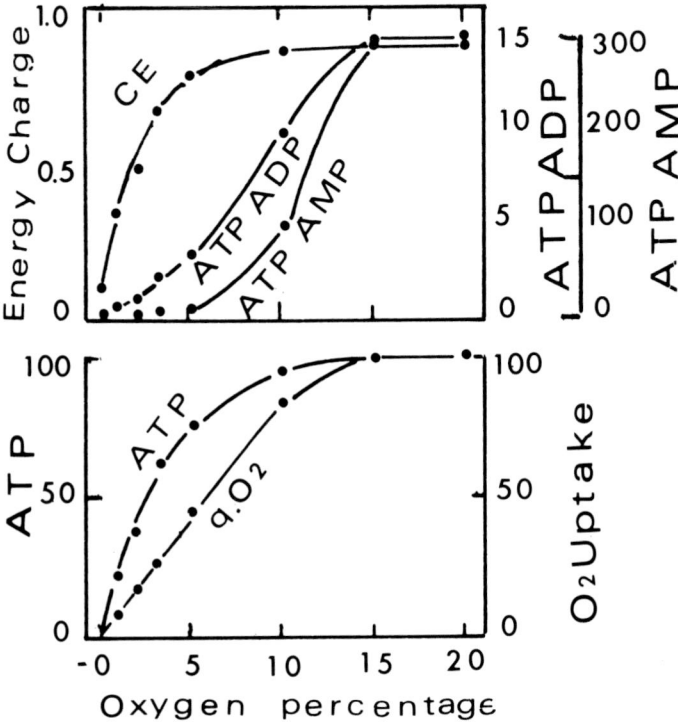

Figure 7. Effect of oxygen partial pressure on O_2 uptake, ATP concentration values of energy charge, and of the ratios ATP/AMP and ATP/ADP in germinating lettuce seeds.[105]

Interpretation of CO_2 data is more complicated. The problem is straightforward if we are studying cells such as yeasts where alcoholic fermentation largely exceeds the other metabolic process. Anaerobic oxidation of 1 mole of glucose to ethanol produces three times less CO_2 than respiration. Very often the production of CO_2 by plant tissues is not lowered by anaerobic conditions. This is the sign that glycolysis activity has been increased at least threefold. When CO_2 production is heavily reduced, glycolysis may still be very active if the cell synthetizes lactic acid. Interpretation of *in vivo* data of CO_2 production becomes even more complicated if the plant produces organic acids by the carboxylation of C_3 sugars. The problem is made even more difficult if the cells oxidize a substrate other than glucose and if the substrates involved in oxidation differ under aerobic and anaerobic conditions.

It is quite impossible to evaluate the activity of anaerobic metabolism in plant tissues from CO_2 data alone.

ADENINE NUCLEOTIDES *IN VIVO*

The estimation of adenine nucleotide concentration requires care. The chromatographic[92,93] and enzymatic[64,94] methods of analysis used today give good results. The main difficulties arise from the halting of enzyme activity.[95] Two kinds of errors usually appear:

1. The turnover of adenine nucleotides is very fast. Enzyme activity must be stopped when the tissues are submitted to experimental conditions. Pradet and Raymond have shown that the level of ATP increased threefold in 5 seconds when etiolated wheat (*Triticum aestivum*) leaves whose energy charge was lowered by nitrogen treatment were transferred to air. Consequently enzyme activity should be suppressed while the tissues are still in the gaseous medium under study. This may be achieved by introducing liquid nitrogen or cold ethyl ether (-80°C) into the anaerobic jars.
2. Another difficulty arises from the inefficiency of many of the methods usually employed to denaturate enzymes. The use of boiling water or buffers and perchloric acid[96] very often gives low values of energy charge as compared to the use of trichloroacetic acid.

These difficulties are not associated specifically with the estimation of adenine nucleotide concentration but constitute a general problem encountered in *in vivo* studies of molecules where the time of turnover involves seconds or minutes.[97]

Energy Charge in Good Conditions of Oxygenation

It is widely accepted now that the distribution of adenine nucleotides in normally metabolizing cells is characterized by high values for the ATP/ADP ratios and higher values for the ratios of ATP/AMP. Consequently the calculation of energy charge gives high values. The values found in animal tissues, bacteria and yeast are always higher than 0.8.[65,98]

In 1967, Bomsel and Pradet[99] calculated energy charge values in aerobic wheat leaves in the dark and in the light. Many values were lower than 0.65. The same results were obtained by Lin and Hanson[100] in aerobic maize roots. From these data it was postulated by various authors[65,100] that energy charge in higher plants was not regulated in the same way as in other living organisms. From the compilation of the most carefully obtained data, Bieleski[101] found in plant tissues an ATP/ADP/AMP ratio of 10/3/1 and calcualted an energy charge of 0.80. Using improved techniques, we detected energy charges higher than 0.8 in wheat leaves[64] and maize roots. During the last few years the measurement of energy charge of many more aerobic plant tissues has been performed in our laboratory. From these more precise experiments, it now appears that the range of energy charge of plant or tissue cells lies between 0.8 and 0.95 if the cells are normally oxygenated (normoxic state). Thus it appears that higher plants do not differ from other organisms in their regulation of the energy charge value *in vivo*.

Energy Charge in Anoxia

When plant tissues or cells are transferred from air to anaerobic conditions, the level of ATP drops immediately as a consequence of oxidative phosphorylation inhibition.[64,91] A new equilibrium between ATP and ADP and AMP is reached after 1 to 15 minutes depending on the cells or tissues involved. During this period the energy-rich bonds originate from fermentative processes. The new energy charge values measured in higher plants during the first hours of anaerobiosis range from 0.1 to 0.7. Low values between 0.1 and 0.2 have been observed in many seeds (lettuce, tomato (*Lycopersicum esculentum*), onion (*Allium cepa*), radish (*Raphanus sativus)* (unpublished data of Pradet and Pradet et al.[91]). Intermediate values (0.5) have been measured in rice seeds and wheat leaves, peas, rice coleoptiles and roots. Higher values (0.7) are found in young maize seedlings.

Subsequent evolution of the energy charge has been studied in different plant tissues or seeds subjected to anaerobic treatment for several hours or days. In lettuce and radish seeds, the energy charge does not vary and remains low. The treatment is not lethal and the energy charge regains its normal level as soon as the seeds are aerated. The effect of anaerobic treatment during germination of radish has been also studied by Moreland et al.[102] They found that the energy charge of seeds maintained under anaerobic treatment did not deviate markedly from the aerated control seeds (0.88). These authors did not detail the methods used, but with our experimental methods described at the beginning of this section we had very different results: energy charge values of 0.9 and 0.2, respectively, for aerobic and anaerobic germinating seeds (Pradet, unpublished data). Rice seeds and wheat leaves exhibit the same initial behavior, but subsequent evolution of the charge is very different. In wheat leaves, Sellami and Bomsel[103] have demonstrated a slow decline in the chrage over 24 hours after which time the cells die. In contrast, in rice embryos and coleoptiles the energy charge, lowered to 0.5 by anaerobic treatment, remains stabilized at this value between 0.75 and

0.85. These values remain stable for many days. During this time RNA, DNA[40] and protein synthesis remain active. When the seeds are aerated, the energy charge rises to 0.90 in about 1 min.

E. coli and baker's yeast are able to grow in anoxia and exhibit high values of energy charge[65,104] and active metabolism. Rice coleoptile is the only plant organ known to be able to grow in anoxia. When the growth of the coleoptile is stopped, the seedling remains viable for several weeks. During this time the value of the energy charge remains high. Germinating lettuce seeds are also able to survive during many days (about two weeks) in anoxia but unlike rice seedlings do not grow; in this material the value of the charge is very low (0.05) and it appears that the metabolic activity is correspondingly reduced (e.g., we were unable to measure CO_2 evolution from this material under anoxia). When materials whose survival in anoxia is poor (e.g., wheat leaves, pea seeds) are submitted to anoxia for days, a slow decline of the charge is observed.

There thus seem to be two strategies for seed survival in anoxia. Rice seedlings are able to grow and survive in anoxia by maintaining high energy charge and corresponding active metabolic activity. Lettuce seeds do not grow but are able to survive in anoxia through low energy charge and reduced metabolic activity.

Energy Charge Under Hypoxia

By studying the effect of oxygen partial pressure, we have shown that it is possible to stabilize the energy charge of a tissue of any level intermediate between that observed in air and that observed in nitrogen (Figure 7). Consequently, for a given tissue with a given pool of enzymes, the level of the charge is a parameter which may be correlated with oxidative phosphorylation activity. At the present time we do not know if the charge level is related to the production of energy by fermentative pathways in vivo.

From these studies it can be observed that the energy charge in vivo is a very well regulated parameter. It can be used to characterize and compare metabolic states of a given tissue with a given pool of enzymes under various conditions. This parameter may be very useful to appreciate in vivo the reduction of metabolic activity under hypoxia. By using this parameter, Raymond, Bruzau and Pradet[13] have developed a new technique to estimate the physiological role of oxygen transported from leaves to roots.

Total Adenine Nucleotides in Anoxia

In microorganisms and animal tissues, many authors have shown that a decline in energy charge is usually accompanied by a decrease in total adenylate. In plant tissues the same phenomenon is observed when anoxic metabolism is able to maintain energy charge at values higher than 0.45 (maize roots, wheat leaves, rice seedlings), but not when energy charge drops to values below 0.3. This phenomenon may be related to the ability to survive under anoxic conditions.

In the study of anaerobic metabolism the analysis of adenine nucleotides *in vivo* complements the data obtained from gaseous exchange experiments.

REFERENCES

1. Tappan, H. "Molecular Oxygen and Evolution" in *Molecular Oxygen in Biology; Topics in Molecular Oxygen Research,* O. Hayashi, Ed. (Amsterdam: North Holland Publishing Co., and New York: American Elsevier Publishing Co., Inc., 1974) p. 80.
2. Hall, J. B. "Evolution of the Prokaryotes" *J. Theor. Biol.* 30(3): 429 (1971).
3. Harvath, R. S. "Evolution of Anaerobic Energy-Yielding Metabolic Pathways of the Prokaryotes" *J. Theor. Biol.* 47(2):361 (1974).
4. Oparin, A. I. *Genesis and Evolutionary Development of Life* (New York: Academic Press, 1968).
5. Longmuir, I. S. "Respiration Rate of Rat Liver Cells at Low Oxygen Concentrations," *Biochem. J.* 65(2):378 (1957).
6. Longmuir, I. S., and A. Bourke. "The Measurement of the Diffusion of Oxygen through Respiring Tissue," *Biochem. J.* 76(2):255 (1960).
7. Chevillotte, P., and G. Ducet. "Respiration du Tubercule de Pomme de Terre. Influence de la Tension d'Oxygne sur des Disques Minces," *Physiol. Veg.* 7(3):305 (1969).
8. Chevillotte, P. "Relation between the Reaction Cytochrome Oxidase—Oxygen and Oxygen Uptake in Cells *in vivo*," *J. Theor. Biol.* 39:277 (1973).
9. Greenwood, D. J. "Studies on the Transport of Oxygen through the Stems and Roots of Vegetable Seedlings," *New Phytol.* 66(3): 337 (1967).
10. Healy, M. T., and W. Armstrong. "The Effectiveness of Internal Oxygen Tranport in a Mesophyte (*Pisum sativum L.*)," *Planta* 103 (3):302 (1972).
11. Vartapetian, B. B., L. P. Agapova, A. A. Averianov and V. A. Veselovsky. "New Approach to Study of Oxygen Transport in Plants Using Chemiluminescent Method," *Nature* 249(5454):269 (1974).

12. Crawford, R. M. M. "Tolerance of Anoxia and the Regulation of Gly-colysis in the Roots," in *Tree Physiology and Yield Improvements*, M. G. R. Cannell and F. T. Last, Eds. (London: Academic Press, 1976).
13. Raymond, P., F. Bruzau and A. Pradet. "Etude du Transport d'Oxygène des Parties Aériennes aux Racines à l'aide d'un Paramètre du Métabolisme: la Charge Energétique," *C. R. Acad. Sci. Paris* (in press).
14. Davies, D. D. "Metabolic Control in Higher Plants," in *Biosynthesis and its Control in Plants*, Vol. 9. Proc. Phytochem. Soc. (London and New York: Academic Press, 1972), p. 1.
15. De Zwaan, A., J. M. F. M. Kluytmans and D. I. Zandee. "Faculta-tive Anaerobiosis in Molluscs," in *Biochemical Adaptation to Envi-ronmental Changes*, Vol. 41. Biochem. Soc. Symp. (London: The Biochemical Society, 1976), p. 133.
16. Hochachka, P. W. "Design of Metabolic and Enzymic Machinery to Fit Life Style to Environment," in *Biochemical Adaptation to Environmental Changes*, Vol. 41. Biochem. Soc. Symp. (London: The Biochemical Society, 1976), p. 3.
17. James, W. O. *Plant Respiration* (Oxford: The Clarendon Press, 1953).
18. Thomas, M. "The Quotient of the Respiratory Gas Exchange," in *Handbuch der Pflanzenphysiologie*, Vol. 12, Part 2, W. Ruhland, Ed. (Berlin: Springer-Verlag, 1960), p. 1.
19. Turner, J. S. "Fermentation in Higher Plants; its Relation to Res-piration; the Pasteur Effect," in *Handbuch der Pflanzenphysiologie*, Vol. 12, Part 2, W. Ruhland, Ed. (Berlin: Springer-Verlag, 1960), p. 42.
20. Warburg, O. "Über die Wirkung von Blausäureäthylester auf die Pasteurische Reaktion," *Biochem. Z.* 172(3):432 (1926).
21. Barker, J., M. A. A. Khan and T. Solomos. "Studies in the Res-piratory and Carbohydrate Metabolism of Plant Tissues. XXI. The Mechanism of the Pasteur Effect in Peas," *New Phytol.* 66(5):577 (1967).
22. Effer, W. R., and S. L. Ranson. "Respiratory Metabolism in Buck-wheat Seedlings," *Plant Physiol.* 42(8):1042 (1967).
23. Faiz-ur-rahman, A. T. M., A. J. Trewavas and D. D. Davies. "The Pasteur Effect in Carrot Root Tissue," *Planta* 118(3):195 (1974).
24. Hook, D. D., C. L. Brown and R. H. Wetmore. "Aeration in Trees," *Bot. Gaz.* 133(4):443 (1972).
25. Taylor, D. L. "Effects of Oxygen on Respiration, Fermentation and Growth in Wheat and Rice," *Science* 95(2457):129 (1942).
26. Vartapetian, B. B., I. N. Andreeva and I. P. Maslova. "Ultrastruc-ture of Rice Coleoptile Cells under Aerobic and Anaerobic Condi-tions," *Dokl. Acad. Sci. U. S. S. R.* 196(5):1231 (1971).
27. Tsuji, H. "Respiratory Activity of Rice Seedlings Germinated under Strictly Anaerobic Conditions," *Bot. Mag. Tokyo* 84(2):102 (1971).
28. Barker, J. and A. F. El Saifi. "Studies in the Respiratory and Carbohydrate Mechanisms of Plant Tissues. I. Experimental Studies of the Formation of Carbon Dioxide, Lactic Acid and other Products in Potato Tubers under Anaerobic Conditions," *Proc. Roy. Soc. London, Ser. B* 140(899):362 (1952).

29. Davies, D. D., S. Grego and P. Kenworth. "The Control of the Production of Lactate and Ethanol in Higher Plants" *Planta* 118(4):297 (1974).

30. Doireau, P. "Exhalation d'Alcool et d'autres Substances Volatiles en rapport avec le Catabolisme Fermentaire, au cours de la Germination du Haricot," *C. R. Acad. Sci. Paris, Ser. D* 273(8):741 (1972).

31. Chirkova, T. V., and T. S. Gutman. "Physiological Role of Branch Lenticels of Willow and Poplar under Conditions of Root Anaerobiosis," *Fiziol. Rast.* 19(2):352 (1972).

32. Crawford, R. M. M. "Alcohol Dehydrogenase Activity in Relation to Flooding Tolerance in Roots," *J. Exp. Bot.* 18(56):458 (1967).

33. App, A. A., and A. N. Meiss, "Effect of Aeration on Rice Alcohol Dehydrogenase," *Arch. Biochem. Biophys.* 77(1):181 (1958).

34. Hageman, R. H., and D. Flesher. "The Effect on Anaerobic Environment on the Activity of Alcohol Dehydrogenase and other Enzymes of Corn Seedlings," *Arch. Biochem. Biophys.* 87(2):203 (1960).

35. Crawford, R. M. M., and M. McManmon. "Inductive Response of Alcohol and Malic Dehydrogenase in Relation to Flooding Tolerance in Roots," *J. Exp. Bot.* 19(60):435 (1968).

36. Krebs, H. A. "The Pasteur Effect and the Relation Between Respiration and Fermentation," in *Essays in Biochemistry,* Vol. 8, P. N. Campbell and F. Dickens, Eds. (New York: Academic Press, 1972), p. 1.

37. Effer, W. R., and S. L. Ranson. "Some Effects of Oxygen Concentration on Levels of Respiratory Intermediates in Buckwheat Seedlings," *Plant Physiol.* 42(8):1053 (1967).

38. Streeter, J. G., and J. F. Thompson. "*In vivo* and *in vitro* Studies on γ-aminobutyric Acid Metabolism with the Radish Plant (*Raphanus sativus L.*)," *Plant Physiol.* 49(4):579 (1972).

39. Streeter, J. G., and J. F. Thompson. "Anaerobic Accumulation of γ-aminobutyric Acid and Alanine in Radish Leaves (*Raphanus sativus L.*)," *Plant Physiol* 49(4):572 (1972).

40. Mocquot, B., A. Pradet and S. Litvak. "DNA synthesis and Anoxia in Rice Coleoptiles," *Plant Sci. Lett.* 9(4):365 (1977).

41. Pradet, A., and C. Prat. "Metabolisme Energetique au cours de la Germination du Riz en Anoxie," in *Etudes de Biologie Vegetale,* R. Jacques, Ed. (Gif-sur-Yvette, Paris:C.N.R.S., 1976), P. 561.

42. Chirkova, T. V., I. V. Kharova and T. P. Astafurova. "Metabolic Regulation of Plant Adaptation to Conditions of Temporary Anaerobiosis," *Fiziol. Rast.* 21(2):102 (1973).

43. Yamamoto, Y. "Variation of Nicotinamide Adenine Dinucleotide Phosphate Level in Bean Hypocotyls in Relation to O_2 Concentration," *Plant Physiol.* 41(3):519 (1966).

44. Garcia-Novo, F., and R. M. M. Crawford, "Soil Aeration, Nitrate Reduction and Flooding Tolerance in Higher Plants," *New Phytol.* 72 (5):1031 (1973).

45. Chirkova, T. V. "Role nitratnovo dichaniya kornei v zhiznediyatelnosti nekotori drevesnich rastenii v usloviach vremnovov anaerobioza," *Vestnik Leningrad Univ.* 21:118 (1971).

46. Lambers, H. "Respiration and NADH Oxidation of the Roots of Flood-tolerant and Flood-intolerant *Senecio* Species as Affected by Anaerobiosis," *Physiol. Plant.* 37(2):117 (1976).

47. Hochachka, P. W., and G. N. Somero. *Strategy of Biochemical Adaptation* (Philadelphia, London and Toronto: W. B. Saunders Co., 1973).

48. Jacob, F., and J. Monod. "Genetic Regulatory Mechanisms in the Synthesis of Proteins," *J. Mol. Biol.* 3(3):318 (1961).

49. Wainwright, S. *Control Mechanisms and Protein Synthesis* (New York: Columbia University Press, 1972).

50. Freeling, M., and D. Schwarz. "Genetic Relationships between the Multiple Alcohol Dehydrogenases of Maize," *Biochem. Genet.* 8(1):28 (1973).

51. Marshall, D. R., P. Broue and A. J. Pryor. "Adaptative Significance of Alcohol Dehydrogenase Isozyme in Maize," *Nature (New Biol.)* 244(131):16 (1973).

52. Brown, A. D. H., D. R. Marshall and J. Munday. "Adaptedness of Variants at an Alcohol Dehydrogenase Locus," *Aust. J. Biol. Sci.* 29(4):389 (1976).

53. Monod, J., J. P. Changeux and F. Jacob. "Allosteric Proteins and Cellular Control Systems," *J. Mol. Biol.* 8(4):306 (1963).

54. Koshland, D. E., Jr. "The Molecular Basis for Enzyme Regulation," in *The Enzymes,* Vol. I, P. D. Boyer, Ed. (New York: Academic Press, 1970), p. 341.

55. Atkinson, D. E. "Regulation of Enzyme Activity," *Annu. Rev. Biochem.* 35(1):85 (1966).

56. Umberger, H. E. "Evidence for a Negative Feedback Mechanism in the Biosynthesis of Isoleucine," *Science* 123(3202):848 (1956).

57. Yates, R. A., and A. B. Pardee. "Control of Pyrimidine Biosynthesis in *Escherichia coli* by a Feedback Mechanism," *J. Biol. Chem.* 221 (2):757 (1956).

58. Stadtman, E. R. "Mechanisms of Enzyme Regulation in Metabolism," in *The Enzymes,* Vol I, P. D. Boyer, Ed. (New York: Academic Press, 1970), p. 397.

59. Preiss, J., and T. Kosuge. "Regulation of Enzyme Activity in Metabolic Pathways," in *Plant Biochemistry,* J. Bonner and J. E. Varner, Eds. (New York: Academic Press, 1976), p. 277.

60. Atkinson, D. E., P. J. Roach and J. S. Schwedes. "Metabolite Concentrations and Concentration Ratios in Metabolic Regulation," *Advan. Enzyme Regul.* 13:393 (1975).

61. Atkinson, D. E., and G. M. Walton. "Adenosine Triphosphate Conservation in Metabolic Regulation. Rat Citrate Cleavage Enzyme," *J. Biol. Chem.* 242(13) 3239 (1967).

62. Atkinson, D. E. "The Energy Charge of the Adenylate Pool as a Regulatory Parameter. Interactions with Feedback Modifiers," *Biochemistry* 7(1):4030 (1968).

63. Bomsel, J. L., and A. Pradet. "Study of Adenosine 5'-Mono-, Di-, and Triphosphates in Plant Tissues. IV. Regulation of the Level of Nucleotide *in vivo* by Adenylate Kinase: Theoretical and Experimental Study," *Biochem. Biophys. Acta* 162(2) 230 (1968).

64. Pradet, A. "Contribution à l'Etude du Rôle joué par les Nucléotides Libres *in vivo* des Semences de Laitue et des Feuilles de Blé," *These de Doctorat d'Etat es-Sciences,* Faculte des Sciences de Paris, (Paris. C.N.R.S., No. A.O. 3605).

65. Chapman, A.G., L. Fall and D.E. Atkinson. "Adenylate Energy Charge in *Escherichia coli* Growth and Starvation," *J. Bacteriol.* 103 (3):1072 (1971).

66. Sellami, A. "Evolution des Adenosine Phosphates et de la Charge Energétique dans les Compartiments Chloroplastiques et non-Chloroplastiques des Feuilles de Ble," *Biochem. Biophys. Acta,* 423(3):524 (1976).

67. Chapman, A. G., and D. E. Atkinson. "Stabilization of Adenylate Energy Charge by the Adenylate Deaminase Reaction," *J. Biol. Chem.* 248 (23):8309 (1973).

68. Chapman, A. G., A. L. Miller and D. E. Atkinson. "Role of the Adenylate Deaminase Reaction in Regulation of Adenine Nucleotide Metabolism in Ehrlich Ascites Tumor Cells," *Cancer Res.* 36(3):1144 (1976).

69. Swedes, J. S., R. J. Sero and D. E. Atkinson. "Relation of Growth and Protein Synthesis to the Adenylate Energy Charge in an Adenine Requiring Mutant of *Escherichia coli," J. Biol. Chem.* 20(17):6930 (1975).

70. Matin, A. and J. C. Gottschal. "Effect of Growth Rate on NADP and NADPH Contents and Ratios under Carbon or Ammonium Limitation," *J. Appl. Chem. Biotechnol.* 26(6):323 (1976).

71. Vermeersch, J., D. Lechevalier and R. Moneger. "Sur les Nucleotides Pyridiniques du Blé et de la Spirodèle cultivés dans diverses Conditions (Obscurité, Lumierè rouge de faible Puissance, Dichlorophényldiméthylurée)," *C. R. Acad. Sci. Paris, Ser. D* 284(18):1785 (1977).

72. Sanwal, B. D. "Allosteric Control of Amphibolic Pathways in Bacteria," *Bacteriol. Rev.* 34(1):20 (1970).

73. Faiz-ur-rahman, A. T. M., A. J. Trewavas and D. D. Davies. "The Control of Glycolysis in Aged Slices of Carrot Root Tissues," *Planta* 118(3):211 (1974).

74. Turner, J. F. and D. H. Turner. "The Regulation of Carbohydrate Metabolism," *Annu. Rev. Plant Physiol.* 26:159 (1975).

75. Larsson-Raznikiewicz, M. and B. Schierbeck. "Activation and Inhibition of the Phosphoglycerate Kinase Reaction by ATP^4," *Biochem. Biophys. Acta* 481(2):283 (1977).

76. Purich, D. L., and H. J. Fromm. "Studies on Factors Influencing Enzyme Response to Adenylate Energy Charge," *J. Biol. Chem.* 247 (1):249 (1972).

77. Blair, J. McD. "Magnesium Potassium, and the Adenylate Kinase Equilibrium. Magnesium as a Feedback Signal of the Adenine Nucleotide Pool," *Eur. J. Biochem.* 13(2):384 (1970).

78. Atkinson, D. E. "Enzymes as Control Elements in Metabolic Regulation," in *The Enzymes,* Vol. 1, P. D. Boyer, Ed. (New York: Academic Press, 1970), p. 461.

79. Friedel, J. "L'assimilation Chlórophylienne aux Pressions Interieures a la Pression Atmosphérique," *Rev. Gen. Bot.* 14(8):337 and 14(9):369 (1902).

80. Beevers, H. *Respiratory Metabolism in Plants* (New York: Row Peterson and Co., 1960).

81. Ducet, G., and A. J. Rosenberg. "Leaf Respiration," *Annu. Rev. Plant Physiol.* 13:171 (1962).

82. Berry, L. J., and W. E. Norris. "Studies of Onion Root Respiration. I. Velocity of Oxygen Consumption in Different Segments of Root at Different Temperatures as a Function of Partial Pressure of Oxygen," *Biochem. Biophys. Acta* 3(4):593 (1949).

83. Fischer, R. B. *The Mutability of Km in Oxygen in the Animal Organism,* F. Dickens and E. Neil, Eds., I.U.B. Symposium (London: Pergamon Press, 1964), p. 339.

84. Armstrong, W., and R. J. Gaynard. "The Critical Oxygen Pressures for Respiration in Intact Plants," *Physiol. Plant.* 37(3):200 (1976).

85. Yocum, C. S., and D. P. Hackett. "Participation of Cytochromes in the Respiration of the Aroid Spadix," *Plant Physiol.* 32(3):186 (1957).

86. Mapson, L. W., and W. G. Burton. "The Terminal Oxidases of the Potato Tuber," *Biochem. J.* 82(1):19 (1962).

87. Palmer, J. M. "The Organization and Regulation of Electron Transport in Plant Mitochondria," *Annu. Rev. Plant. Physiol.* 27:133 (1976).

88. Solomos, T. "Cyanide-Resistant Respiration in Higher Plants," *Annu. Rev. Plant Physiol.* 28:279 (1977).

89. Tomlinson, P. F., and D. E. Moreland. "Cyanide-Resistant Respiration of Sweet Potato Mitochondria," *Plant Physiol.* 55(2):365 (1975).

90. Lance, C. "Respiratory Control and Oxidative Phosphorylation in *Arum maculatum* mitochondria," *Plant Sci. Lett.* 2(3):165 (1974).

91. Pradet, A., A. Narayanan and J. Vermeersch. "Etude des Adénosine-5'-mono, di et tri-phosphates dans les Tissus Végétaux. III. Métabolisme Energétique au cours des Premiers Stades de la Germination des Semences de Laitue," *Bull. Soc. Fr. Physiol. Veg.* 14(1):107 (1968).

92. Bieleski, R. L. "Levels of Phosphate Esters in *Spirodela,*" *Plant Physiol.* 43(8):1297 (1968).

93. Cole, C. V., and C. Ross. "Extraction, Separation, and Quantitative Estimation of Soluble Nucleotides and Sugar Phosphates in Plant Tissues," *Anal. Biochem.* 17(3):526 (1966).

94. Pradet, A. "Etude des Adénosine-5'-mono, di et tri-phosphates dans les Tissus Végétaux. I. Dosage Enzymatique," *Physiol. Veg.* 5(3):209 (1967).

95. Bieleski, R. L. "The Problem of Halting Enzyme Action when Extracting Plant Tissues," *Anal. Biochem.* 9(4):431 (1964).

96. Ikuma, H., and R. M. Tetley. "Possible Interference by an Acid-Stable during the Extraction of Nucleoside Di- and Triphosphate from Higher Plant Tissues," *Plant Physiol.* 58(3):320 (1976).

97. Kennedy, R. A., and L. E. Williams. "Effect of Different Killing Techniques on Early Labeled Photosynthetic Products in C_4 Plants," *Plant Physiol.* 59(2):207 (1977).

98. Atkinson, D. E. "Adaptations of Enzymes for Regulation of Catalytic Function," in *Biochemical Adaptation to Environmental Change,* Vol. 41. Biochem. Soc. Symp. (London: The Biochemical Society, 1976), p. 205.

99. Bomsel, J. L. and A. Pradet. "Etude des Adénosine-5'-mono, di et triphosphates dans les Tissus Végétaux. II. Evolution *in vivo* de l'ATP, l'ADP et l'AMP dans les feuilles de Blé en Fonction de Différentes Conditions de Milieu," *Physiol. Veg.* 5(3):223 (1967).

100. Lin, W., and J. B. Hanson. "Phosphate Absorption Rates and Adenosine 5'-Triphosphate Concentrations in Corn Root Tissue," *Plant Physiol.* 54(3):250 (1974).

101. Bieleski, R. L. "Phosphate Pools, Phosphate Transport, and Phosphate Availability," *Annu. Rev. Plant Physiol.* 24:225 (1973).

102. Moreland, D. E., G. G. Hussey, C. R. Shriner and F. S. Farmer. "Adenosine Phosphates in Germinating Radish (*Raphanus sativus* L.) Seeds," *Plant Physiol.* 54(4):560 (1974).

103. Sellami, A., and J. L. Bomsel. "Evolution de la Charge Energétique du Pool Adenylique des Feuilles de Blé au cours de l'Anoxie. Etude de la Réversibilité des Phénomènes Observés," *Physiol. Veg.* 13(3): 611 (1975).

104. Ball, W. J. Jr., and D. E. Atkinson. "Adenylate Energy Charge in *Saccharomyces cerevisiae* during Starvation," *J. Bacteriol.* 121(3): 975 (1975).

4

METABOLIC ADAPTATIONS TO ANOXIA

R. M. M. Crawford

Department of Botany
The University
St. Andrews, Scotland

INTRODUCTION

The many similarities in the major metabolic pathways of plants and animals suggest that for the majority of biochemical processes there is one optimal solution. These similarities do not exclude the existence of variants in the metabolic pathways of different organisms. However, when variants occur it would be expected that they will be molded by the nature of the environmental stress imposed on the organisms and not by phylogenetic relationships. From this argument it should follow that conditions of similar ecological adversity should call forth similar metabolic adaptations in tolerant species regardless of their taxonomic relationships. This hypothesis of ecological similarity in metabolic adaptation is of particular interest when we compare the adaptations of plants and animals to conditions of anoxia. Most species of animals and plants are clearly divided on whether they can withstand prolonged periods of partial or complete anoxia. In plants, the study of metabolic adaptation is confined largely to roots and seeds, whereas in animals there is a wealth of information covering the adaptations of parasitic species of protozoa, platyhelminths and nematodes, mud-inhabiting annelids, the larval stages of chironimid insects, intertidal molluscs, insects at diapause, and diving and hibernating mammals and reptiles. In the metabolism of these animals, four features stand out as common to all species capable of surviving anoxia:

1. control of metabolic rate;
2. diversification of the end products of glycolysis;
3. provision of adequate carbohydrate supplies;
4. coupling of metabolic pathways to facilitate proton disposal and provide additional ATP.

If we examine each of these features in turn, it is possible to demonstrate a remarkable similarity between plants and animals in many aspects of their metabolic adaptation to anaerobiosis.

METABOLIC ADAPTATION

Control of Metabolic Rate

Probably the best-known effect of oxygen depletion on metabolic rate is the acceleration of glycolysis that takes place when tissues are deprived of an adequate supply of oxygen (the Pasteur effect). Due to the low Km values of cytochrome oxidase, which range from 4×10^{-6} to $2.4 \times 10^{-8} M$,[1] the concentration of oxygen in the plant tissue must be less than 1% before availability alone proves a limitation to its utilization.[2] This concentration is readily reached in germinating seeds and in certain tissues of the root. In the meristematic zone of roots the supply of oxygen can be shown to be limiting even under well-aerated conditions.[3] Admittedly, this limitation only applies to the apical 3-5 mm of the root, but nevertheless indicates that any reduction in the supply of oxygen to the root will increase the state of anaerobiosis that already exists. In flood-intolerant species, a reduction in soil aeration has been shown to be accompanied by a rapid increase in glycolysis (the Pasteur effect) as well as by a dramatic induction of alcohol dehydrogenase (ADH) activity. Flood-intolerant species that behave in this way show themselves to be distinct from flood-tolerant plants where no such induction takes place (Figures 1,2).[4-6]

Ethanol accumulation resulting from increased glycolysis can readily lead to membrane destruction by lipid solubilization and this will inactivate mitochondrial enzyme activity and further increase the preponderance of glycolytic activity. Leakage of cell contents will quickly follow, and the exudation of soluble sugars facilitates the growth of soil pathogens which then invade the roots and eventually cause death.

Flood-tolerant plants differ from intolerant species in not exhibiting this acceleration of glycolysis and show little or no induction or change in the kinetic properties of ADH when subjected to partial anoxia. This distinction between flood-tolerant and intolerant species has now been confirmed by several authors and extended to the behaviour of flood-tolerant and intolerant varieties of a single species. Thus electrophoretic

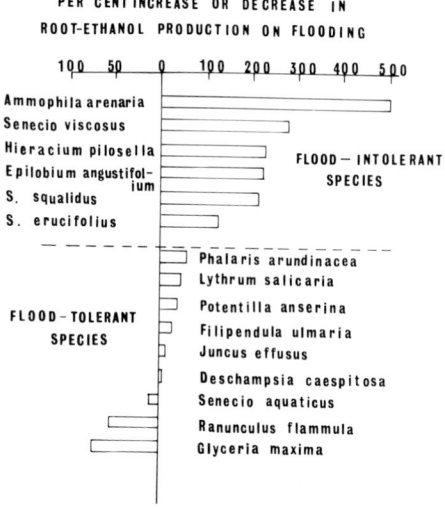

Figure 1. Changes in rate of ethanol production in roots of 15 species of higher plants of varying flooding tolerance after flooding in sand culture for one month as compared with unflooded controls. Ethanol production was determined in excised roots incubated with 2.5% glucose under nitrogen.[5]

studies of *Zea mays* [7] have shown that alternate forms of ADH affect the fitness of individual plants to withstand flooding. In strains of maize possessing a fast migrating, active and heat-labile form of ADH, there is a greater reduction in growth on flooding than in strains possessing the less-active, slower migrating form of the enzyme. This pattern of behavior has also been observed in *Trifolium subterraneum* where it has been suggested that the differential behavior of ADH is sufficiently regular, that provided standard portions of roots are taken, the activity of this one enzyme might be used in selecting flood-tolerant provenances of this species.[8] In *Lupinus angustifolius*, another species with varying tolerance of flooding, the flood-tolerant strain contains only one band of ADH isoenzymes, whereas the increase in ADH activity that is found on flooding intolerant plants is associated with the possession of an additional and distinct ADH isoenzyme band.[9]

The seeds of higher plants also differ in their ability to withstand anoxia after imbibition. Species in which the seeds do not germinate after such a treatment are said to suffer from soaking injury. Many crop species are prone to this form of injury, which causes serious losses in years

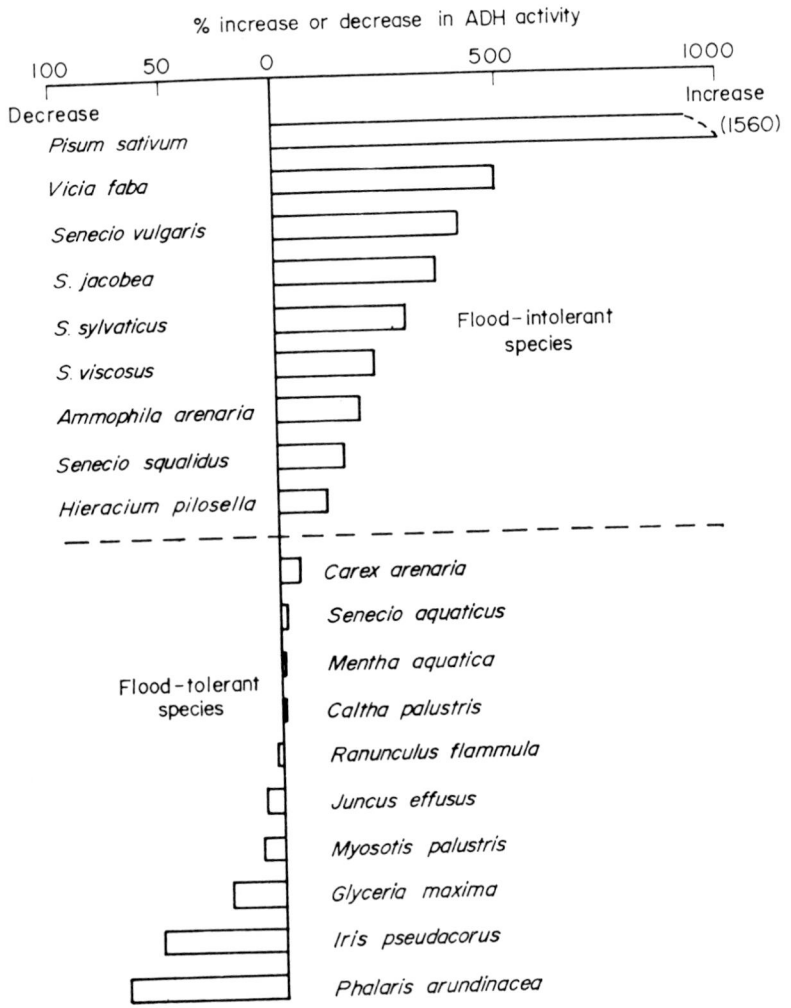

Figure 2. Changes in alcohol dehydrogenase activity (ADH) of 19 species after flooding in sand culture for one month as compared with unflooded controls. Enzyme activity was calculated on a soluble protein basis.[6]

of cold, wet springs. By contrast, many of our wild species and certain selected forms of grasses commonly used in agriculture are resistant to soaking injury. These tolerant species distinguish themselves from the sensitive species by their much lower rates of respiration, a minimum Pasteur effect and a partial replacement of ethanol by lactate as the end product of glycolysis (Figures 3,4).[10] This low temperature effect on the ability to withstand soaking injury must not be confused with low-temperature aggravation of the effects of anoxia reported in peas.[11]

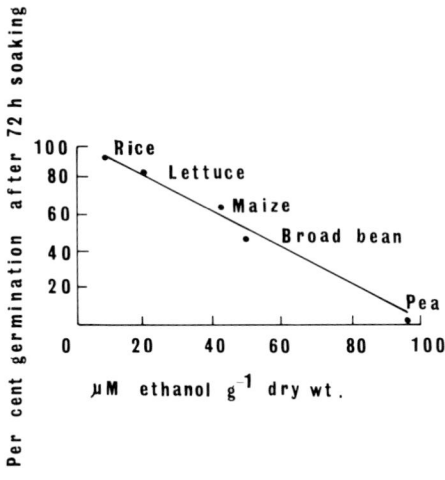

Figure 3. Relationship between ethanol content of seeds after 72 hr soaking, with viability.[10]

Cases in which low temperature aggravates the effects of soaking injury are all reported in seeds that are very sensitive to soaking injury and succumb after a few days immersion. Such species belong to the quick-germinating and rapidly growing plants of agriculture and horticulture. However, in wild species and certain cultivated grasses with a greater tolerance of anoxia in the imbibed condition, lowering the temperature during soaking increases the tolerance of the plants to anoxia (Figure 5).[10]

This behavior in relation to metabolic rate, temperature and tolerance of anoxia is also found in certain diving animals. Paul Bert (1870) was the first to show that in diving ducks the metabolic rate decreased with the duration of the dive.[12] Similar behavior has also been found in seals, alligators and turtles.[13 - 15] The latter is extremely tolerant of anoxia and can survive under conditions of total anaerobiosis. During prolonged dives, the rate of glycolysis decreases as well as the rate of the consumption of stored oxygen. The diving turtle (*Pseudemys scripta elegans*) is not only capable of surviving under conditions of complete anaerobiosis, but when the temperature of the environment is lowered, the capacity of the animal to remain submerged is increased.[16]

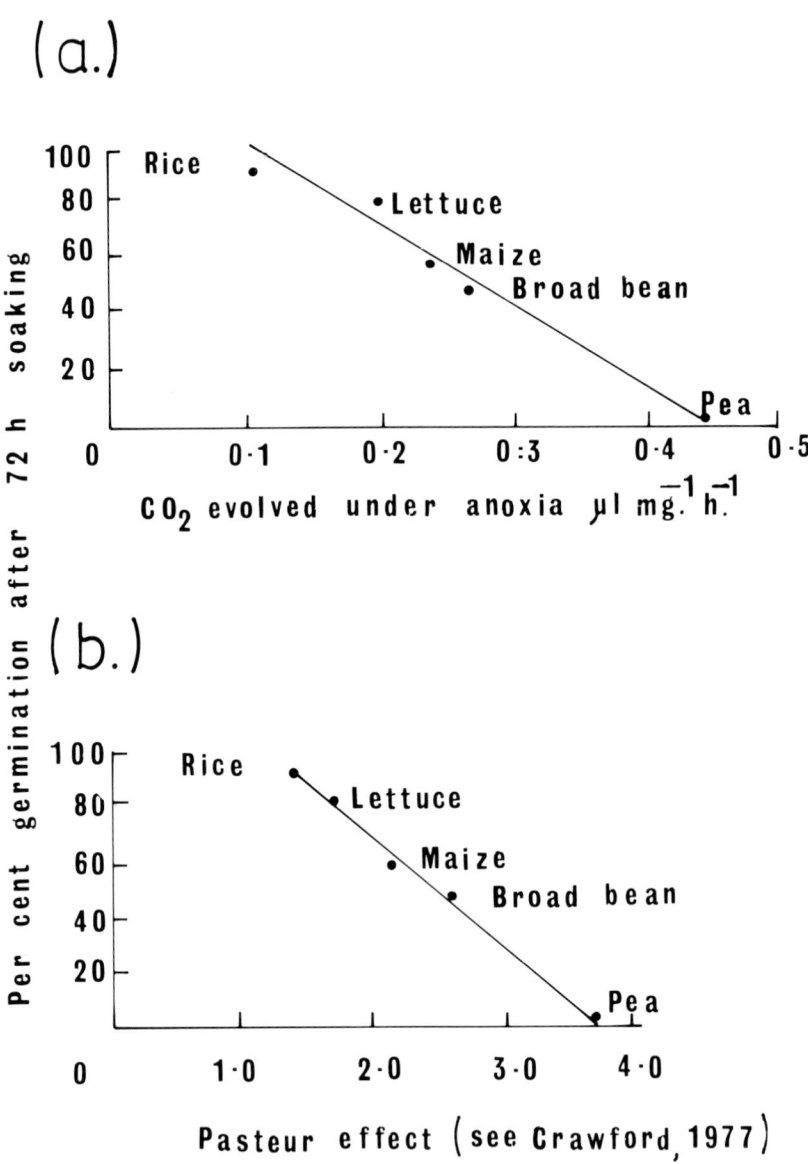

Figure 4. Relationship between viability after 72 hr soaking and (a) anaerobic respiration rate $QCO_2 N_2$ (μl $CO_2 mg^{-1}$ dry wt.h^{-1}) and (b) Pasteur effect as determined from carbon dioxide evolution.[10]

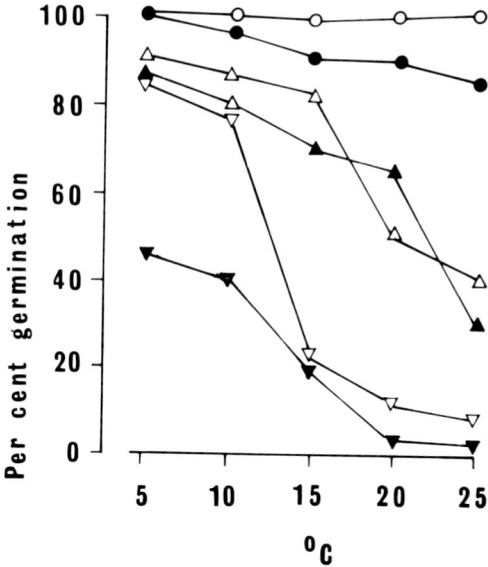

Figure 5. Deterioration of seed viability of six graminaceous species during storage in imbibed condition for six weeks under complete anoxia in anaerobe jars: ○, *Oryza sativa*; ●, *Festuca rubra*; △, *Lolium perenne*; ▲, *Dactylis glomerata*; ▽, *Agrostis tenuis*; ▼, *Zea mays*.[10]

Diversification of the End Products of Glycolysis

In animal species tolerant of anoxia, lactate is not the only metabolite produced as a result of an accumulating oxygen debt. In insects, alanine is commonly reported as accumulating during anaerobiosis, and in some cases glycerol and alanine accumulate simultaneously.[17] In molluscs, the simultaneous accumulation of malic and propionic acids, together with alanine, is a common feature of anaerobiosis.[18] A comparative list of the range of metabolites produced in plants and animals under conditions of complete or partial anoxia is given in Table I. This list includes the production of lactate by seeds, malate in a number of marsh species, shikimic acid in the tuberous roots of a number of aquatic plants, and glycerol in *Alnus incana*. In plants, amino acid accumulation as a result of flooding has been noted in certain flood-tolerant trees.[36,37] In seeds, the greatest lactate accumulation is found during the initial stages of germination. The seed is a dense structure with a minimal amount of vacuolation in its cells, and it is therefore probable that tissue acidification limits the extent of lactate accumulation. In this connection, Davies[38] has

Table I. Substances Reported to Accumulate Under Anaerobic Conditions
in Higher Plants and Animals[a]

Substance	Animal Occurrence	Plant Occurrence
Lactic Acid	Vertebrate skeletal muscle[19]	Germinating seeds,[20] tubers[21]
Pyruvic Acid	Vertebrate skeletal muscle[19]	Willow roots[22]
Formic Acid	Parasitic helminths[23]	----
Acetic Acid	Bivalve mulluscs, cestodes[24]	----
Acetoin	Nematodes[23]	----
Propionic Acid	Molluscs, cestodes[23]	----
Butyric Acid	Parasitic protozoa[23]	----
Succinic Acid	Bivalve mulluscs[24]	Seeds[20]
Malic Acid	----	Roots of marsh plants[26,27]
Shikimic Acid	----	Iris and water lily roots[28]
Glycolic Acid	----	Willow roots[22]
Ethanol	Parasitic protozoa,[23] Helminths[23]	Flood-intolerant roots and seeds[3,4,10]
Sorbitol	Insects[17]	----
Glycerol	Insects[17]	Alder roots[28]
Alanine	Sea turtles,[25] molluscs[29]	Flood-tolerant roots[22]
Aspartic Acid	Marine annelids[30]	Flood-tolerant roots[22]
Glutamic Acid	Marine annelids[30]	Flood-tolerant roots[22]
Serine	----	Flood-tolerant roots[22]
Proline	----	Flood-tolerant roots[22]
Octopine	Cephalopods[31]	----
γ amino butyric acid	----	Tomato roots[32]
	----	Radish leaves[33]
Methyl Butyrate	Parasitic nematodes[34]	----
Methyl Valerate	Parasitic nematodes[34]	----
Glycerophosphate	Insects[17]	----
Hydrogen	Parasitic protozoa[23]	----
Ethylene	----	Roots and fruits[35]

[a]References are intended only as a source of further information and do not indicate priority of discovery or complete range of occurrence.

discussed the limits imposed on organic acid accumulation by tissue acidification and pointed out that the differing pH optima of carboxylating and decarboxylating enzymes will act as a pH stat and limit the accumulation of organic acids in the plant cytoplasm. In the growing plant, however, with its large capacity for vacuole storage of metabolites, organic acid accumulation will not be limited by pH reduction if the organic acids, malate, shikimate etc., are stored in the vacuole.

The distinction between flood-tolerant and intolerant plants in their ability to accumulate malate under conditions of partial anoxia[26] has now been extended to a study of flood-tolerant and intolerant varieties of the same species. In an elegant study of races of *Veronica peregrina* growing around an ephemeral pool in California, Linhart and Baker[27] showed that plants raised from parents that germinated underwater in the center of the pond and grew for several weeks in a flooded soil, accumulated malate when subjected to experimental flooding, whereas populations collected from the edge of the pond did not exhibit any accumulation of malate when flooded.

In the ascending xylem-sap of birch trees (*Betula pubescens*) growing in wet sites, there can be noted a considerably higher concentration of malate than that found in sap of trees growing on drier soils.[28] When growth begins in the early spring, the deeper anchoring roots of trees are frequently deprived of oxygen as the winter rise of the water table has not yet subsided to its summer level. These roots have large supplies of starch stored in their xylem parenchyma. This carbohydrate storage can be likened to the storage of glycogen that takes place in the liver of the human infant before parturition, which facilitates the tolerance of up to 30 minutes of anoxia at birth. Similar glycogen stores are found in the muscle of the oyster and are again related to the ability of the tissues to endure anoxia. It is possible to argue, therefore, that the carbohydrate stores of the root allow this organ to respire anaerobically in spring and that the oxygen debt of the root is transferred to the shoot where it is repaid with the assimilation of malate into carbohydrate via pyruvate, with no loss in fixed carbon to the plant as a whole. In the study of birch trees referred to above, the molecular ratio of malate to ethanol was in the region of 100:1, and it would appear that malate was therefore the principal substance transferring the oxygen debt from the root to the shoot.

Provision of Adequate Carbohydrate Supplies

The anaerobic metabolism of subterranean food reserves is inefficient from a purely energetic point of view. However, it appears that there are many species of plant and animals which endure this energetic inefficiency as the price that must be paid for the privilege of living in a flooded or aquatic environment, free from the competition of purely dryland species. Thus, many of the higher plants that have evolved a preference for aquatic or marsh habitats are frequently characterized by the possession of large tubers and well-developed rhizomes, which lie submerged in anaerobic mud throughout the winter resting period. The carbohydrate reserves of these well-developed submerged organs are used by anaerobiosis in the early stages of the growing season when growth is first resumed and shoots are renewed. Before shoot renewal there will be no means of submerged tissue aeration, and the early growth of these aquatic plants must be supported by anaerobic respiration.

In an earlier account of flooding tolerance,[39] it was suggested that the accumulation of malate might be made possible by the deletion of malic enzyme in the roots of flood-tolerant species. The original scheme has been modified by Chirkova *et al.*,[40] who found that malic enzyme was present in *Glyceria aquatica*, a flood-tolerant plant, but had its activity reduced when flooded and increased by aeration. By contrast, in species intolerant of flooding, exposure to anaerobiosis caused an increase in malic enzyme. Thus, some form of metabolic control of malic enzyme rather than deletion appears to be associated with the ability of the plants to withstand flooding. This modification of our original theory (Figure 6) is also necessitated by the findings of Davies *et al.*,[41] who showed that malic enzyme was present in a large number of both flood-tolerant and intolerant species. Davies and co-workers detected malic enzyme activity in flooded roots of flood-tolerant plants but did not examine the

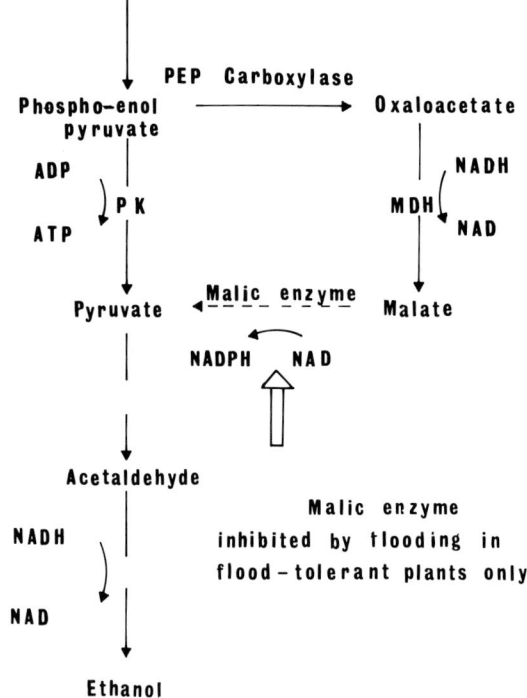

Figure 6. A possible metabolic differentiation between flood-tolerant and nonflood-tolerant plants. In intolerant species there is an acceleration of glycolysis; malic enzyme is not inhibited and ethanol is the exclusive end product of glycolysis. In flood-tolerant species there is no acceleration of glycolysis and malic enzyme is inhibited, allowing malate to accumulate a proportion of the oxygen debt.

effect of flooding on the kinetics of malic enzyme activity or of aeration on the level of enzyme activity in relation to flooding tolerance. In experiments with potato, it has further been shown by Davies and co-workers that oxalic acid is a powerful inhibitor of malic enzyme, and this may play a role in the differential response of flood-tolerant and nonflood-tolerant plants to periods of anaerobiosis.

Coupling of Metabolic Pathways

Plants tolerant of anoxia characteristically produce a number of metabolites when subjected to low-oxygen stress. This behavior is analagous to the mixed fermentations discussed by Krebs.[42] Most biochemical syntheses involve an excess of oxidations over reductions; therefore, if a tissue is to perform any anabolic activities under anaerobiosis, additional forms of proton disposal are necessary. In addition, the coupling of metabolic pathways can yield significant increases in ATP production. Thus, in facultative anaerobic animals such as the oyster, there is a simultaneous production of succinate, alanine and proprionate[18] and in this case, coupling of the pathways of carbohydrate and amino acid metabolism can increase the yield of ATP formation to 8 mols/mol of glucose. In plants, the production of shikimic acid is achieved by the coupling of the pentose phosphate pathway with glycolysis, which allows the cycling of $NADP$-$NADPH_2$ as dehydroshikimic acid utilizes $NADPH_2$ when reduced to shikimic acid (Figure 7). In eutrophic plants, flooding increases nitrate reductase activity, suggesting that the uptake of nitrate and subsequent amino acid synthesis is proving to be a very efficient method of proton disposal.[36,37]

The yield of ATP that may be expected in plants as a result of anaerobic respiration is problematic and no conclusions can be presented with any certainty. If oxaloacetic acid is carboxylated by phosphoenolpyruvate carboxylase, which is likely, as this enzyme is highly active in most plant tissues, then the accumulation of malate will not result in ATP production. It will however deflect pyruvate from being reduced to ethanol, which may have a survival value for this plant. If, however, phosphoenolpyruvate carboxykinase is involved in malic acid synthesis, then the normal ATP production of 2 mols/mol glucose will result. It is intriguing to speculate whether any plant is capable of metabolizing malate to acetate as postulated, but not yet proven in goldfish and carp.[18] An examination of some of our native and less-studied plants tolerant of anoxic conditions may yet yield more examples of metabolic advantage from coupling of metabolic pathways.

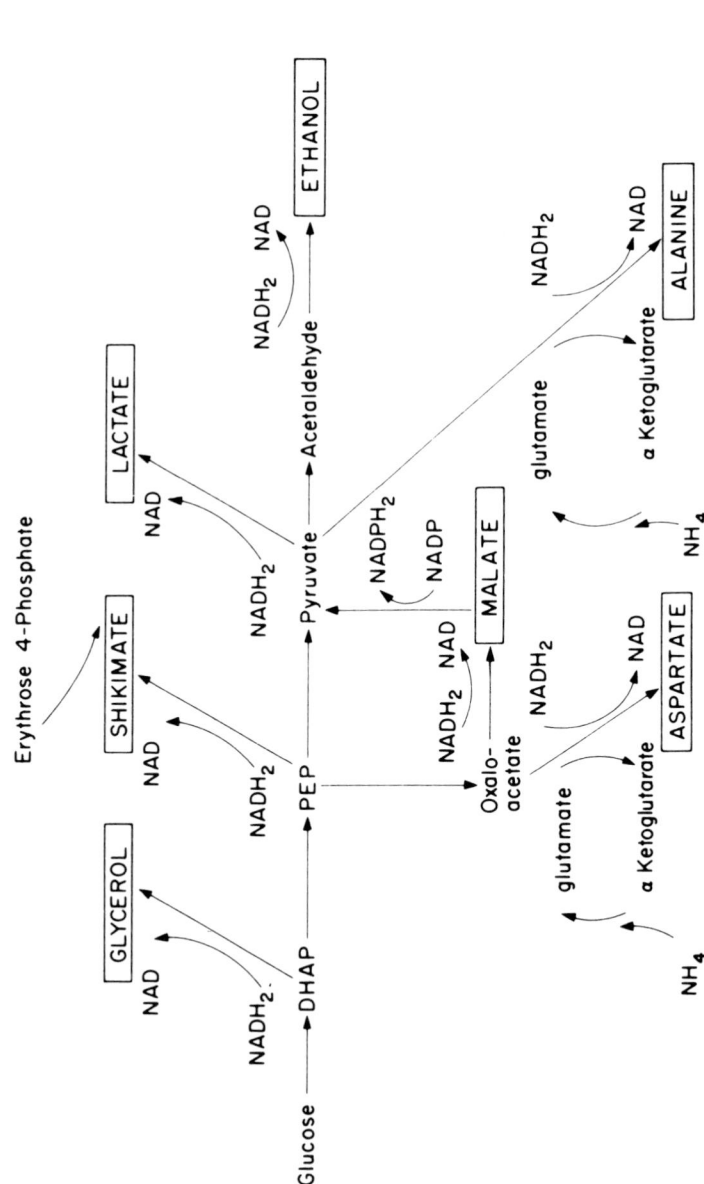

Figure 7. Summary diagram illustrating the various means of proton disposal and the range of end products found in plants capable of enduring prolonged periods of total or partial anoxia. (See Table I for references).

EXCEPTIONAL CAUSES OF DEATH AND MECHANISMS OF ADAPTATION TO FLOODING

It would be both presumptious and erroneous to extend the present comparison to cover the reactions of all plants to flooding. In many species, particularly in fruit trees, the bark is rich in cyanogenic glycosides, and flooding of these plants can cause death within 24 hours.[43] In these species, death is so rapid that no comparison can be made of ethanol production in tolerant and intolerant varieties. In plants with adventitious roots in direct contact with free water, as in rice, more than 95% of the ethanol produced in the roots will diffuse quickly into the surrounding medium. In these cases, an acceleration of glycolysis may not necessarily be harmful to the growing plant,[44] whereas in the germinating rice seed before rupture of the testa, the limitation of ethanol production was associated with tolerance of anoxia. Similarly, in swamp trees (*Nyssa sylvatica* var. *biflora*), there may be a ready diffusion of ethanol out of the roots on flooding, so that increased glycolytic rates do not produce toxic quantities of ethanol in the plants tissues.[45] However, in the majority of perennial aquatic plants with thickened roots, rooted in anaerobic muds, or in the perennating tubers and rhizomes of marsh plants as well as in germinating seeds, ethanol is not readily lost by diffusion, and in these species, just as with animals, the control of glycolysis is regularly associated with tolerance of anoxia.

DETOXIFICATION OF ANAEROBIC SOILS

Higher plants with their specialization of nutrient uptake by roots encounter further difficulties in returning to the aquatic or wetland habitat, as many of the reduced soil mineral ions are phytotoxic. In mineral-rich lakes submerged plants can absorb ions through their leaves, but in nutrient-poor waters the root still serves as the major mineral-capturing site of the higher plants. It is a curious feature of many wetland species that despite abundance of water in their soils, they exhibit many of the xeromorphic features characteristic of drought-adapted plants. The reduction of leaves in the genus *Juncus* and the hairy folded leaves of the cross-leaved heath (*Erica tetralix*) are two typical examples. Reduction of the water requirements of wetland plants will reduce the in-flow of water to the root and with it, the arrival at the root surface of reduced phytotoxins. The detoxification rate necessary to protect the root will therefore be reduced, and it appears that it is the downward diffusing oxygen from the shoot to the root that is the principal means of oxidizing reduced ions. The action of this downward diffusing oxygen in a root of

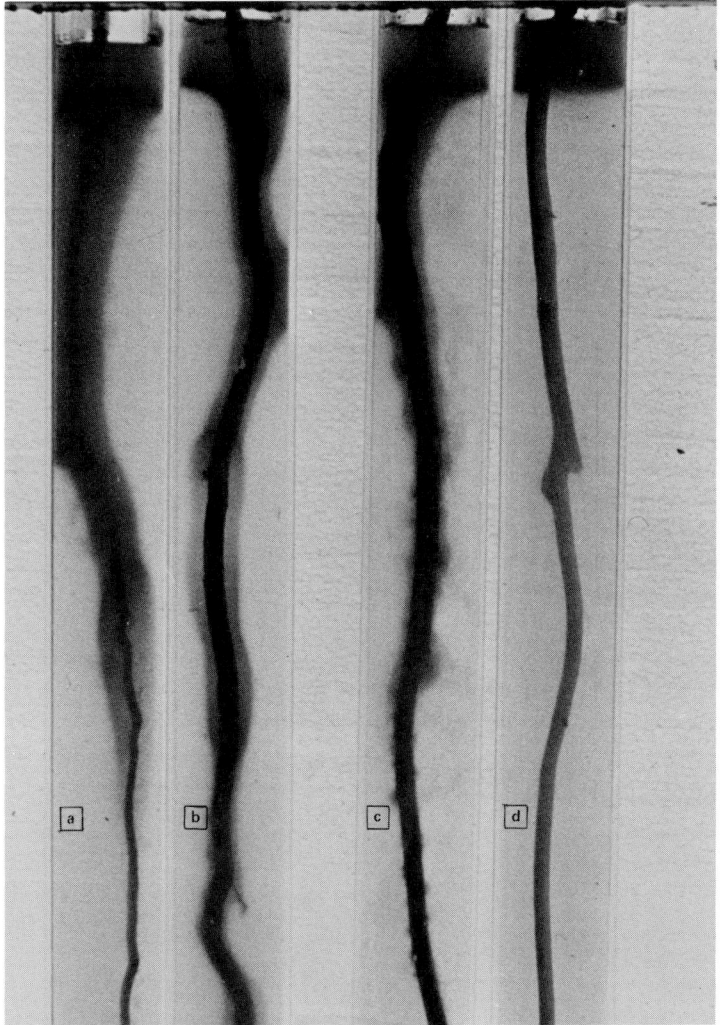

Figure 8. Oxygen transport in tree roots of Sitka spruce and lodgepole pine demonstrated by the oxidation of reduced indigo carmine solution to a colored form by oxygen diffusing from the roots. From left to right the roots are (a) lodgepole pine with tissues external to the xylem removed; (b) lodgepole pine, bark intact; (c) Sitka spruce, bark intact; (d) Sitka spruce, tissues external to the xylem removed. The roots are submerged to a depth of 40 cm in the dye. Photograph of unpublished research supplied by courtesy of M. P. Coutts and J. J. Philipson, Forestry Commission, Northern Research Station, Roslin, Scotland.

lodgepole pine is simulated by the oxidation of a reduced and invisible dye by roots of *Pinus contorta* growing below the water table (Figure 8). The importance of control of water uptake for the survival of bog plants has been demonstrated with the use of anti-transpirants.[46] Spraying the leaves of *Erica cinerea* (a flood-sensitive heath) not only increases the ability of this species to survive flooding but reduces its iron content. The experiments illustrate the underlying cause of the effect well known in farming, that rushes not only grow in wet places, but make them still wetter. The reduced transpiration of the rush will allow the water table to rise and favor the competitive power of the rushes against the original dry-land vegetation.

CONCLUSIONS

Coevolution has long been recognized as a general morphological rule in areas of environmental stress. The desert plants of the old and new worlds resemble each other in form even though taxonomically distinct. The adaptations that occur in plants and animals in relation to anoxia give a striking example of coevolution at the molecular level. The remarkable fact that molecular evolution has selected the same mechanisms in such widely different groups as plants and animals, shows that there can be only a few efficient metabolic solutions to any particular environmental stress. It must be expected that similar situations exist in other areas of plant and animal metabolism. Further biochemical investigations into the success or failure of organisms in relation to environmental stress will bring a new dimension into ecology and allow practical field studies to benefit from recent advances in molecular biology.

REFERENCES

1. Hayaishi, O. "Biological Oxidations," *Ann. Rev. Biochem.* 31:25 (1962).
2. Greenwood, D. J. "Effect of Oxygen Distribution in the Soil on Plant Growth," *Proc. 15th Easter School Agric, Sci. Univ. Nottingham*, 202 (1968).
3. Crawford, R. M. M. "Tolerance of Anoxia and the Regulation of Glycolysis in Tree Roots," in *Tree Physiology and Yield Improvement*, M.G.R. Cannell and F.T. Last, Eds. (New York: Academic Press, 1976).
4. Crawford, R. M. M. "The Control of Anaerobic Respiration as a Determining Factor in the Distribution of the Genus *Senecio, J. Ecol.* 54:413 (1966).

5. Crawford, R. M. M. Alcohol Dehydrogenase Activity in Relation to Flooding Tolerance in Roots, *J. Exp. Bot.* 18:458 (1967).

6. Crawford, R. M. M., and M. McManmon. "Inductive Responses of Alcohol and Malic Dehydrogenases in Relation to Flooding Tolerance in Roots, *J. Exp. Bot.* 19:435 (1968).

7. Marshall, D. R., P. Broué, and A. J. Pryor. "Adaptive Significance of Alcohol Dehydrogenase Isoenzymes in Maize," *Nature New Biol.* 244:16 (1973).

8. Francis, C. M., A. C. Devitt, and P. Steele. "Influence of Flooding on the Alcohol Dehydrogenase Activity of Roots of *Trifolium subterraneum* L.," *Aust. J. Plant Physiol.* 1:9 (1974).

9. Marshall, D. R., P. Broué, and R. N. Oram. "Genetic Control of Alcohol Dehydrogenase in Narrow Leaved Lupins," *J. Hered.* 65: 198, (1974).

10. Crawford, R. M. M. "Tolerance of Anoxia and Ethanol Metabolism in Germinating Seeds," *New Phytol.* 79:511 (1977).

11. Perry, D. A. and J. G. Harrison. "The Deleterious Effect of Water and Low Temperature on Germination of Pea Seed," *J. Exp. Bot.* 21:504 (1970).

12. Bert, P. *Lecons sur la physiologie comparée de la respiration,* Paris (1870).

13. Scholander, P. F., L. Irving, and S. W. Grinnell. "On the Temperature and Metabolism of the Seal During Diving," *J. Cell. Comp. Physiol.* 19:67 (1942).

14. Anderson, H. T. "Physiological Adjustments to Prolonged Diving in the American Alligator," *Acta Physiol. Scand.* 53:23 (1961).

15. Belkin, D. A. "Anoxia: Tolerance in Reptiles," *Science* 139:492, (1963).

16. Jackson, D. C. "Metabolic Depression and Oxygen Depletion in the Diving Turtle," *J. Appl. Physiol.* 24:503 (1968).

17. Gilmour, D. *The Metabolism of Insects,* Edinburgh (1965).

18. Hochachka, P. W. and G. N. Somero. *Strategies of Biochemical adaptation,* Philadelphia (1973).

19. Hochachka, P. W. and K. B. Storey. "Metabolic Consequences of Diving in Animals and Man," *Science* 187:613 (1975).

20. Wager, H. G. The Effect of Anaerobiosis on Acids of the Tricarboxylic Cycle in Peas," *J. Exp. Bot.* 12:34 (1961).

21. Davies, D. D., S. Grego, and P. Kenworthy. The Control of the Production of Lactate and Ethanol by Higher Plants. *Planta* 118: 297 (1974).

22. Dubinina, I. M. "Metabolism of Roots under Various Levels of Aeration," *Soviet Plant Physiol.* 8:314 (1961).

23. Von Brand, T. *Biochemistry of Parasites* (New York: 1966).

24. Gade. G., H. Wilps, J. H. Kluytmans, and A. de Zwaan. "Glycogen Degradation and End Products of Anaerobic Metabolism in the Fresh Water Bivalve *(Anodonta cygnea)*," *J. Comp. Physiol.* 104:79 (1975).

25. Hochachka, P. W., T. G. Owen, J. F. Allen, and G. G. Whittow. "Multiple End Products of Anaerobic Metabolism in Diving Vertebrates," *Comp. Biochem. Physiol.* 50 B:17 (1975).

26. Crawford, R. M. M. and P. D. Tyler. "Organic Acid Metabolism in Relation to Flooding Tolerance in Roots," *J. Ecol.* 57:237 (1969).

27. Linhart, Y. B. and I. Baker. "Intra-Population Differentiation of Physiological Response to Flooding in a Population of *Veronica peregrina* L." *Nature (London)* 242:275 (1973).

28. Crawford, R. M. M. "Physiologische Ökologie:ein Vergleich der Anpassung von Pflanzen und Tieren an sauerstoffarme Umgebung," *Flora* 161:209 (1972).

29. de Zwaan, A. and T. C. M. Wijsman. "Anaerobic Metabolism in Bivalva (Mollusca); Characteristics of Anaerobic Metabolism," *Comp. Biochem. Physiol.* 54 B:313 (1976).

30. Zebe, E. *In-Vivo* Untersuchungen uber den Glucose-Abbau bei *Arenicola marina* (Annelida, Polychaeta)," *J. Comp. Physiol.* 101:133 (1975).

31. Hochachka, P. W., P. H. Hartline, H. H. A. Fields. "Octopine as an End Product of Anaerobic Glycolysis in the Chambered Nautilus," *Science* 195:72 (1977).

32. Fulton, J. M., A. E. Erickson, N. E. Tolbert, "Distribution of C^{14} Among Metabolites of Flooded Aerobically Grown Tomato Plants," *Agron. J.* 56:527 (1964).

33. Streeter, J. G. and J. F. Thompson. Anaerobic Accumulation of y Aminobutyric Acid in Radish Leaves (*Raphanus sativus* L)," *Plant Physiol.* 49:572 (1972).

34. Bryant, C. *The Biology of Respiration* (London: 1961).

35. Pratt, H. K. and J. D. Goeschl. "Physiological Roles of Ethylene in Plants," *Ann. Rev. Plant Physiol.* 20:541 (1969).

36. Chirkova, T. V. "Role Nitratnovo Dichaniya Kornei v Zhiznediyatelnosti Nekotori Drevensnich Rastenii v Usloviach Vremnovov Anaerobioza," *Vestnik Leningr. gos. Univ.* 21:118 (1971).

37. Garcia-Novo, F. and R. M. M. Crawford. "Soil Aeration, Nitrate Reduction and Flooding Tolerance in Higher Plants," *New Phytol.* 72:1031 (1973).

38. Davies, D. D. "Control of and by pH," *Soc. Exp. Biol. Symp.* 27:513 (1973).

39. McManmon, M. and R. M. M. Crawford. "A Metabolic Theory of Flooding Tolerance: the Significance of Enzyme Distribution and Behaviour," *New Phytol.* 70:299 (1971).

40. Chirkova, T. V., I. V. Khazova, and T. P. Astafurova. "Metabolic Regulation of Plant Adaptation to Conditions of Temporary Anaerobiosis," *Soviet Plant Physiol.* 21:82 (1974).

41. Davies, D. D., K. H. Nascimiento, K. D. Patil. "The Distribution and Properties of Malic Enzyme in Flowering Plants," *Phytochemistry* 13:2417 (1974).

42. Krebs, H. "The Pasteur Effect and the Relations Between Fermentation and Respiration," in *Essays in Biochemistry,* P. N. Campbell and F. Dickens, Eds, Vol. 8., London, pp. 1-34.

43. Rowe, R. N. and D. V. Beardsell. "Waterlogging of Fruit Trees," *Hort. Abstr.* 43:534 (1973).

44. John C. D. and H. Greenway. "Alcoholic Fermentation and Activity of Some Enzymes under Anaerobiosis," *Aust. J. Plant Physiol.* 3:325 (1976).

45. Hook, D. D. C. L. Brown, and P. P. Kormanik. "Inductive Flood-Tolerance in Swamp Tupelo (*Nyssa sylvatica* var *biflora* (Walt. Sarg.)," *J. Exp. Bot.* 20:78 (1971).
46. Jones, H. E. "Comparative Studies of Plant Growth and Distribution in Relation to Waterlogging. II An Experimental Study of the Relationship Between Transpiration and the Uptake of Iron in *Erica cinerea* L. and *E. tetralix*," *J. Ecol.* 59, 167 (1971).

5

SOME REGULATORY MECHANISMS OF
PLANT ADAPTATION TO TEMPORAL ANAEROBIOSIS

T. V. Chirkova

Department of Physiology and Biochemistry
of Plants
Leningrad State University
Leningrad, USSR

INTRODUCTION

Among the adaptations that enable some plant species to be resistant to oxygen deficiency, physiological-biochemical adaptations representing a wide set of metabolic reactions regulated by the plant are significant. Some help the root system avoid anaerobiosis either by transporting atmospheric oxygen through leaves and branch lenticels[1,2] or by utilizing its inner sources—the nitrates and peroxides.[3−8]

Adaptations of another type ensure function of the organism under decreased partial oxygen pressure. In this case, some decrease of the metabolic activity may occur, (compensatory alterations in the metabolism) which render toxic products harmless, or remove them. The reversal of the terminal part of glycolysis and the Krebs cycle, related to enhancing reductive processes and the increase of NADH, offers an example of the regulatory mechanism induced by hypoxia. This chapter will present some general ideas resulting from our studies on this problem.

METABOLIC CONTROL MECHANISMS

The rearrangement of exchange processes to be observed in case of oxygen deficiency may be caused by some modifications in the ratio of the pathway for glucose oxidation and, in particular, in the increase of the glycolytic portion.[9-11] In plants susceptible to hypoxia, a brief increase in the glycolitic activity does occur. But their rapid destruction under anaerobic conditions does not allow the phenomenon to be regarded as adaptive for plants of this type. Poisoning of tissues by toxic products of anaerobic respiration (lactate, ethanol), of an amount that considerably increases in case of anoxia, must be one reason for the destruction. In plants resistant to oxygen deficiency, the increase in the amount of glycolytic turnovers, which leads to the energizing of glycolysis, is accompanied by the fall of ethanol and lactic acid content as the N_2 exposition in the atmosphere lengthens.[12] This fact, when established, led us to consider the fate of both products of anaerobic respiration and the mechanisms of the final steps of glycolysis.

Experimental

Plants differing in resistance to oxygen deficiency were used as objects of investigation: resistant ones were willow (*Salix alba* L.), mannagrass (*Glyceria aquatica* Wahib.) and 7-8 day shoots of rice (*Oryza sativa* L.); susceptible ones were poplar (*Populus Petrowskiana* Schraed.), bean (*Phaseolus vulgaris* L.) and 7-8 day shoots of wheat (*Triticum vulgare* L.).

Exudation of Ethanol

In plants resistant to oxygen deficiency, the ability to remove the fermented toxic products was observed. By applying gas chromatography and nuclear magnetic resonance, it was shown that in willow, by contrast with poplar in the case of root anaerobiosis, both a rise of ethylene, ethanol and acetaldehyde in the aboveground part into the leaves together with the transpiration current, and a partial removal of these substances through the branch lenticels and the leaves' stomata[13] may occur. That ethyl alcohol is likely to enter the leaves from anaerobic roots is indicated by a number of authors.[14-16]

It is known that ethanol exudes into the solution surrounding the roots.[17,18] In our experiments, also, ethanol exuded into the solution near the roots, but varied a great deal depending on the ecological nature of the plant (Figure 1). The estimation of the ethyl alcohol content in the solution from under the roots of wheat shoots showed that the removal of ethanol by the plants increased with the duration of the time the plants were kept in the nitrogen atmosphere. In this case, the amount of ethanol in the roots also increased; that is, the accumulation of the anaerobic respiration product proceeded at a much greater rate than its detoxification by being removed from the tissues. In rice, the opposite was observed. Both the exudation into the solution and the ethanol content in the roots dropped with the length of the anaerostatic period. This may be interpreted as the change of the fashion by which ethanol was rendered harmless in the roots of the plant resistant to hypoxia. With the increase of exposure in N_2, ethanol exudation into the solution surrounding the roots was reduced, presumably by alterations in metabolism.

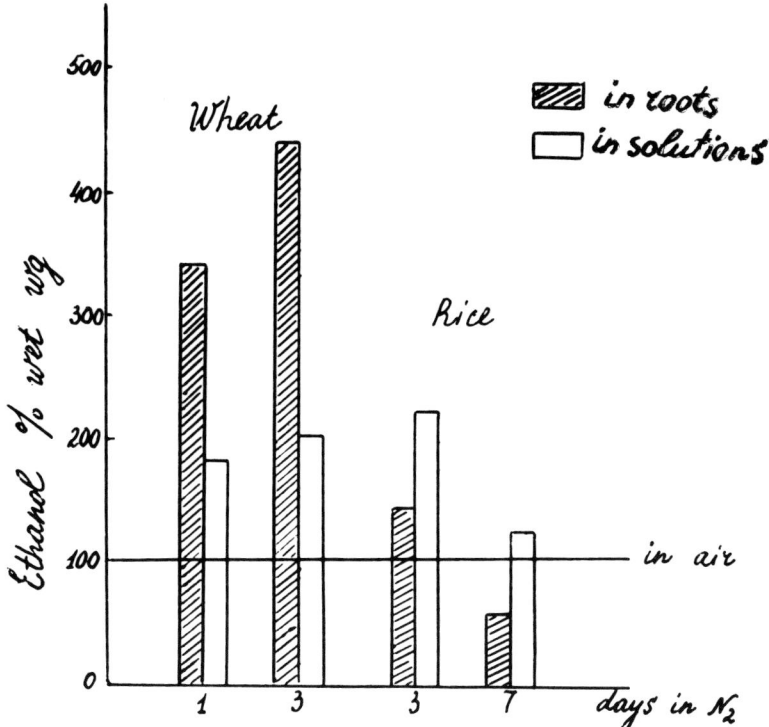

Figure 1. The influence of anaerobic conditions on the ethanol content in roots and in solutions from under the roots of wheat and rice.

Resistance of Roots to Lactate and Ethanol

Before proceeding to the question of utilizing both ethanol and lactate in metabolic reactions, the root resistance of experimental plants to various concentrations of these substances was estimated. It turned out that the superficially sterilized roots of rice and willow remained viable in 1 M solutions of ethanol and lactate prepared with sterile distilled water, for a considerably longer time than the roots of wheat and poplar. The roots of willow were killed in 1 M solution in three days, and in 10^{-1} M in seven days; while in poplar, the growth points of roots died off in the same concentrations in two and four days, respectively. It means that the resistance of plants to oxygen deficiency correlates with their resistance to higher concentrations of the glycolytic products. The resistance increased still more if the plant was kept under anaerobic conditions. Oxiphyl plants, on the other hand, lost their viability in this case much earlier than under aerobic conditions. Concentrations of 10^{-2}-10^{-3} M appear to be the optimum for ethanol and lactate as no signs of injury were detected in the plants during the three experimental weeks.

Oxygen Consumption of Roots in Solutions of Lactate and Ethanol

By applying ethanol and lactate as respiration substances, oxygen consumption by the roots and leaves of willow and poplar was estimated. As seen from Figure 2, the respiration activity of willow tissues, by contrast with poplar, proceeded at a much greater rate in the solutions of the above substances, in comparison with the aqueous control. The increase was particularly pronounced when the willow plants were first kept anaerobically. As to the susceptible plants, prior anaerobiosis caused a decrease in the consumption of oxygen by its roots and leaves. Thus, as willow was more resistant to the increased concentrations of the glycolytic products, it showed much more ability to utilize them than was the case with poplar. In similar experiments with shoots of wheat and rice, it was shown that roots of rice after anaerobiosis preferred to utilize ethanol and lactate, even to glucose, as respiration substrates.

Metabolism of Ethanol-^{14}C under Anaerobic Conditions

These studies indicated the possibility of ethyl alcohol and lactic acid being metabolized both under aeration and after the preliminary effect of anaerobiosis. To determine the behavior of organic substances in plant tissues during anaerobiosis, ethanol -1-^{14}C was added into the solution surrounding the roots. The ^{14}C content was estimated in the fractions of carbohydrates and organic and amino acids.

As shown in Table I, ethanol-^{14}C under aeration entered the root tissues of plants, thereby actively participating in the metabolism. A part of the alcohol or its metabolites products was transported to the aboveground organs. However, there were essential differences in the utilization of ethyl alcohol between willow and poplar. Radioactive carbon in roots and leaves of willow was detected in all fractions under investigation, being distributed among them rather uniformly. In poplar

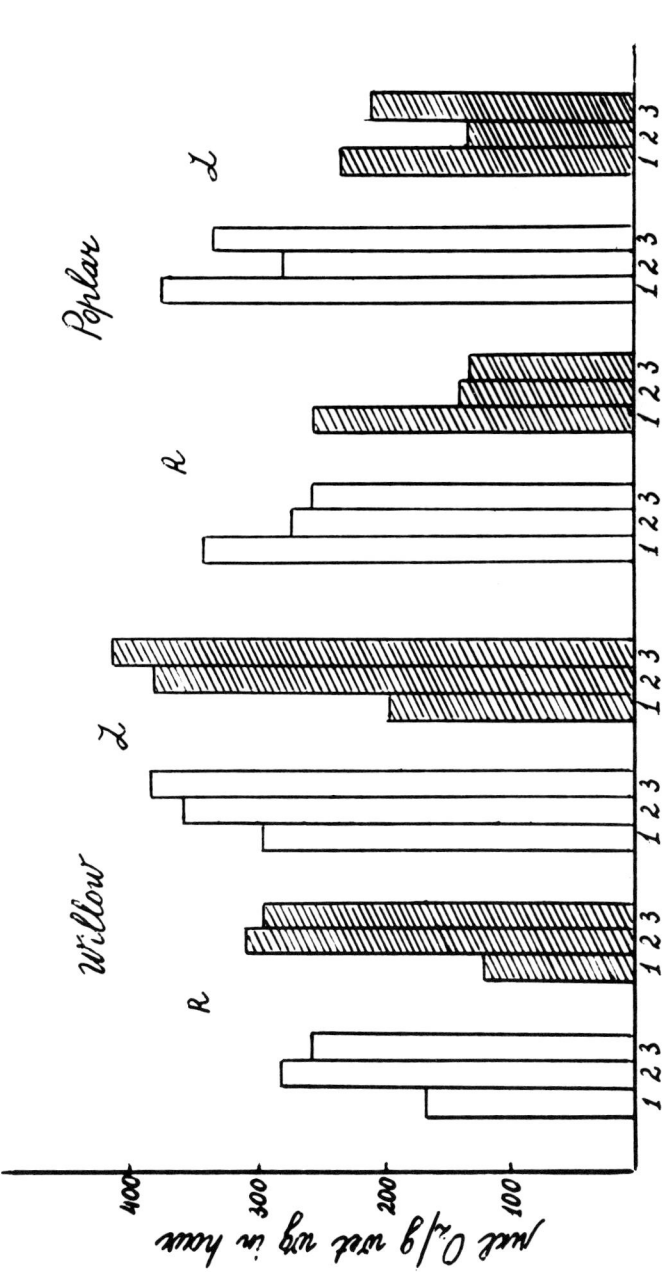

Figure 2. The influence of ethanol and lactate on the consumption of oxygen by roots (R) and leaves (L) of willow and poplar after different aeration conditions. Aeration—unshaded columns; anaerobiosis/24 hr for poplar, 3 days for willow—shaded columns; 1. control (H_2O); 2. ethanol (10^{-2} M); 3. lactate (10^{-2} M).

Table I. The Influence of Different Conditions of Plant Aeration on ^{14}C Distribution Among the Fractions of the Metabolites of Roots and Leaves in Willow and Poplar

Variant:	Roots				Leaves			
	Extract	Sugars	Organic Acids	Amino Acids	Extract	Sugars	Organic Acids	Amino Acids
				Willow				
			Radioactivity imp/min/g wet weight · 10^3					
Aeration	22.1	16.9	7.1	6.9	18.5	4.7	3.9	8.7
Anaero-biosis 3 days	18.9	3.0	6.6	6.9	18.0	4.6	5.2	6.3
				Poplar				
Aeration	23.6	0	5.6	15.2	35.3	1.9	15.5	15.9
Anaero-biosis 24 hr	14.7	0	4.2	9.6	21.6	2.3	14.8	3.7

roots, the greatest radioactivity was found in the amino acid fraction, a lesser part in the organic acids, and none in the sugars.

This method also showed that in tissues of the plant resistant to hypoxia, ethanol under aeration was utilized more efficiently in metabolism than it was in the susceptible plants. At the same time, the absence of the label in the carbohydrate fraction taken from the roots and leaves of poplar was interpreted as evidence that in case of secondary utilization of alcohol, no sugar production occurs. This may explain the decrease of the tissue respiration activity of poplar in the solution of ethanol, particularly after anaerobic conditions, as the content of carbohydrates in the tissues are known to decline sharply.[11]

Under anaerobiosis, the utilization of alcohol takes place in tissues of both plants. The total radioactivity of the extracts in the willow after 3 days of anoxia was the same as the aerated control. It implies that ethanol metabolism by tissues of a resistant plant proceeds actively not only after anaerobic action in an aerated environment, but even in cases where oxygen is completely excluded. The data to be found in literature only refer to the utilization of ethyl alcohol after anaerobic conditions, when the plants were transferred to the open air.[19-21]

While comparing the data under examination with those obtained in estimating the consumption of oxygen by willow tissues in the ethanol solution, the following was observed. The rise in the respiration activity of willow roots and leaves in the ethanol solution after anoxia may be explained by increased consumption of ethanol under aeration. Under anaerobiosis, utilization of this substrate remained the same as the control variant—that is, before exposure to N_2. In this case, oxygen of such compounds as, for example, the nitrates, is likely to play, at least in part, the role of the terminal hydrogen acceptor. The presence of nitrate respiration in willow, by contrast with its absence in poplar, has been shown by the author.[5]

In poplar, ethanol metabolism under anaerobic conditions was, after remaining in the nitrogen atmosphere for 24 hr, decreased by 40% of ^{14}C content in the extract, mainly at the expense of the amino acid fraction. The results obtained show that ethanol is well utilized by willow plants in comparison with poplar.

The study of the utilization of ethanol-^{14}C by wheat and rice roots led us to the same conclusions. In this case, however, it should be particularly noted that ^{14}C content in the carbohydrate fraction in rice roots, under anaerobic conditions, increased as much as 2.5 times. This shows once more that during glyconeogenesis, ethanol is likely to be utilized in the tissues of plants resistant to hypoxia. This is also proved by the fact that under anaerobic conditions, the carbohydrate content in adapted plants is considerably higher than in nonadapted ones.[11]

The qualitative analysis of the organic acid fraction in which, in the absence of oxygen, radioactivity markedly increased, revealed the following information (Table II).

The radioactivity of both the DTC and the monocarbon acid fraction is higher in roots of rice than wheat. [14]C accumulation in the latter takes place chiefly in the monocarbon acids, of which the prevailing one is lactic acid. Under aeration, the metabolism of the lactic acid in wheat is hampered. Thus, even if some secondary ethanol transformation under anaerobic conditions is possible in plants of the wheat type, a considerable part of ethanol is transformed into another toxic product—lactate. But in adapted plants under anaerobic conditions, a considerable amount of DTC is formed, and is easily metabolized under aeration. Some increase of lactic acid in rice roots, which occurs here, is not ruinous because, the adapted plants are more resistant to high lactate concentrations and lactates enter the respiratory metabolism more intensively after anoxia, as shown above and confirmed by experimental data.

Table II. The Influence of Different Conditions of Aeration on ^{14}C Distribution from Ethanol-^{14}C in Organic Acids of Wheat and Rice Roots

Variant	DTC[1]	Citric	Malic	Succinic	Fumaric	Monocarbonic
			Rice			
			Radioactivity imp/min/g wet weight · 10^3			
Aeration	1086	265	500	174	144	447
Anaerobiosis 3 days	1917	430	896	372	217	1046
Anaerobiosis + 8 hr aeration	1169	271	584	157	157	526
			Wheat			
Aeration	79	30	49	0	0	84
Anaerobiosis 24 hr	117	50	67	0	0	390
Anaerobiosis + 8 hr aeration	62	62	0	0	0	272

[1]Di- and tricarbon acids.

Alcohol Dehydrogenase Activity and Resistance to Hypoxia

As the main enzyme associated with the formation or oxidation of ethanol is alcohol dehydrogenase (ADH), a study was also undertaken to determine the activity of the enzyme relative to the concentration of its substrates—ethyl alcohol and acetaldehyde (Figure 3).[22] Under aeration, the greatest activity of ADH in roots of willow and rice was observed at the 10^{-2} M concentration of ethanol. After 3 days anaerobiosis, the oxidation rate of alcohol increased sharply, and the peak of the reaction in willow shifted to the 10^{-3} M concentration. This may indicate an increase of the enzyme affinity for the substrate. The activity of ADH with acetaldehyde as the substrate in the roots of the same plants was very low and was reduced still more after anoxia. This indicates that the reaction of the reduction of acetaldehyde to ethanol was inhibited.

In poplar, bean and wheat, the greatest activity was observed at higher concentrations of ethanol than in resistant plants (10^{-1} M), but rate decreased after 24 hr of anaerobiosis. The reduction of acetaldehyde,

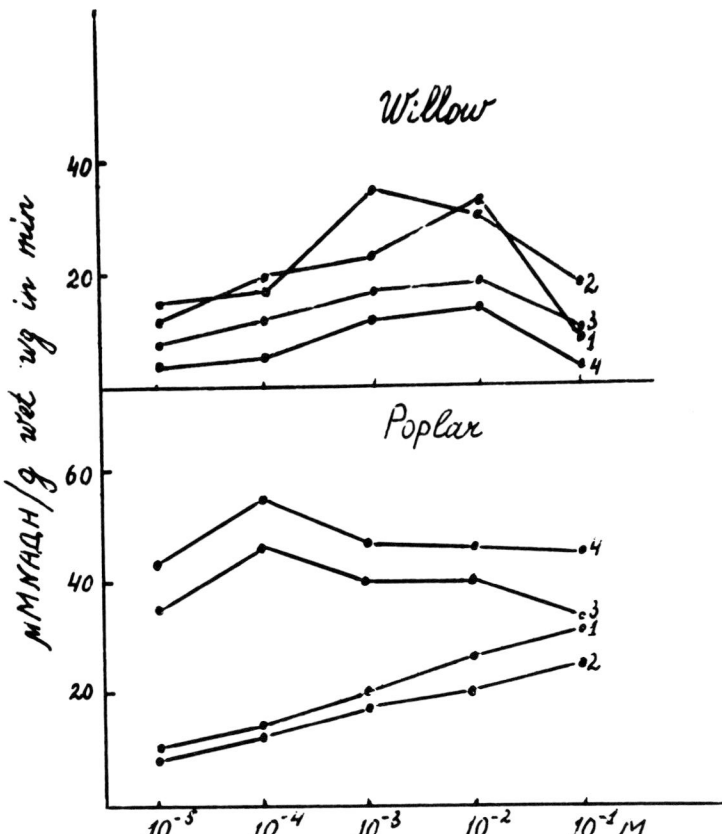

Figure 3. The influence of various ethanol concentrations (1,2) and acetaldehyde (3,4) on ADH activity in roots of willow and popular after different plant aeration conditions. (1,3) aeration; (2,4) anaerobiosis.

however, proceeded at a greater rate than in resistant plants and continued to increase markedly after exposure to N_2. In this case, the lower the acetaldehyde concentration (10^{-4} M), the higher the rate of the enzyme action. In all probability, this indicates a great affinity of ADH roots of susceptible plants for acetaldehyde. It may be said that some shift of alcohol dehydrogenase reaction toward oxidation of ethanol was characteristic for roots of resistant plants after anaerobic effect, and towards reduction of acetaldehyde for nonresistant plants. It may be due to these shifts that accumulation of ethanol occurs in susceptible plants, while in resistant ones it undergoes a radical change.

The results obtained agree with the data available in literature on the decrease of K_M ADH, with acetaldehyde as substrate in susceptible plants, after continued submerging.[23]   There are also some data on pea shoots showing that they possess a higher value of this index for ethanol than for acetaldehyde.   The opposite is the case with rice shoots.[24]

Thus, there appears to be a direct relationship between the ADH affinity for one or another substrate and the degree of plant adaptation to oxygen deficiency.   That is, the ecological type to which a plant belongs is manifested in the mechanisms of regulation of enzymatic systems.   The characteristic feature of the mechanism acting in plants resistant to hypoxia is the ability for secondary utilization of the glycolytic products.

Pyridine Nucleotides and Adaptation to Oxygen Deficiency

Both the mentioned shifts in the final stages of glycolysis and the increase of its activity must lead to the increase of the reduced pyridine nucleotide (PN) content in the tissues of resistant plants.   The reduced PN forms located in the mitochondria, in the microsomes and in the glycolytic systems of cytoplasm, were estimated *in vivo* by fluorescent microscopy.   The amount of /NAD/F/H in the plant roots (the elongation zone) was judged by the fluorescent activity of their /FPN/ which was measured by means of a luminescent microscope ML-4 at 471 nm after excitation at 365 nm.   The level of the reduced PN in the roots of plants resistant to hypoxia was shown to be 1.5 - 2.5 times higher than in the nonresistant ones (Figure 4).[25]   The increase of the reduction ability in the respiratory chain is regarded as the basis of compensatory reactions.[26,27]

As seen from Figure 4, the reduced PN content in the roots of adapted plants remains for a long time at the level close to the initial PN content.   This presumably indicates the possibility of retaining a stable reducing and oxidizing balance of the system.   The decreasing amount of reduced PN with prolonged anaerobiosis points to the disturbance of congruence in the interaction of separate components in the given system.   This decrease proceeds much slower in roots of resistant plants than in those susceptible to oxygen deficiency.   Hence, the former, as contrasted with the latter, seem to possess some regulatory mechanisms that provide long-term functioning of the respiratory metabolism under hypoxia or anoxia.   These mechanisms are to provide both the highest possible ATP formation and reoxidation of NADH.   As is known, it is ATP and NAD/P/ that play the main part in the regulation of $ADP+P_n$-NAD/P/H cell metabolism.[28,29]

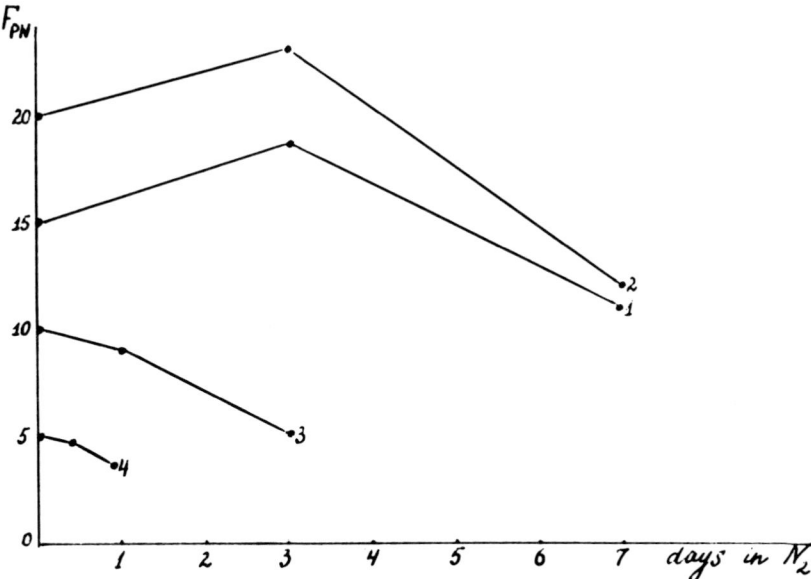

Figure 4. The influence of anaerobic conditions on the root F_{PN} of plants with differing resistance to oxygen deficiency. 1. rice; 2. mannagrass; 3. wheat; 4. bean.

Besides the possibility of transferring $\bar{e}$ and H^+ to the oxygen of nitrates and peroxides, as well as to the NADH oxidation in the glyconeogenesis reactions, it was possible for the tissues of resistant plants to regenerate oxidized cofactors in some reducing reactions. This is more advantageous for the plants than the production of ethanol and lactate.

Changes in the Krebs Cycle

Crawford hypothesizes about the possible preferential production of malate in the tissues of plants resistant to oxygen deficiency instead of the production of ethanol and lactate, which accumulate in nonresistant forms.[23,30] As supposed, this can be achieved by transformation of the second half of the Krebs cycle, the transformation being induced by the rise of the reduced cofactors content.

The pyruvate produced from anaerobic decomposition of carbohydrates gives, with the aid of phosphoenolpyruvate dikinase and ATP, phosphoenol pyruvate (PEP), which carboxylates to yield oxalacetic acid (OAA). It seems that under anaerobic conditions, in case of ATP deficiency, a direct incorporation of PEP in the reaction of OAA production by PEP-carbo-xylase is also possible. Though we did not undertake any studies of PEP-carboxylase activity, it is quite reasonable to suppose that PEP-carboxy-kinase is also likely to take part in the process. As shown on animal tissues under hypoxia, the OAA formed with the help of PEP-carboxykinase is coupled with the formation of 2 mol of high-energy compounds.[31] As to the plant objects, the same assumption is advanced by Davies.[32] OAA is reduced by NAD-specific malate dehydrogenase (MDH) to give malate. Then, depending upon the induction of NADF-specific MDH - malic en-zyme-, malate either accumulates or decarboxylates to give pyruvate.

Malic Enzyme Activity

In this study, the estimation of malic-enzyme activity showed that in susceptible plants (bean, wheat) anaerobiosis causes induction of NADF-dependent MDH, while in resistant plants (mannagrass, rice), activity de-creased under the same conditions (Figure 5). This conforms with the above data on differences in accumulation of pyruvate, ethanol and lac-tate in plants of various ecological groups in the case of lack or deficien-cy of oxygen. In bean, for example, the increase of the rate of malate decarboxylation with the help of malic enzyme must lead to the forma-tion of pyruvate which, under the influence of ADH and LDH, is con-verted into ethanol and lactic acid. Rapid accumulation of fermenting products leads to inhibition of the anaerobic dehydrogenase's activity ac-cording to the principle of negative feedback. This accumulation retards the utilization of pyruvate as well. In mannagrass, formation of pyruvate from malic acid is halted under the same conditions, and the amount formed goes to malic acid synthesis. The anaerobic dehydrogenase acti-vation does produce some accumulation of ethanol and lactate at the onset of anaerobic exposure, but due to the gradual removal of pyruvate, the reaction becomes directed towards malate.

Our data about the malic enzyme decrease after anaerobiosis in the roots of resistant plants and its increase in nonresistant forms support the metabolic scheme of Crawford. At the same time, the data raise some doubts concerning the complete absence of malic enzyme action in the roots of the plants resistant to flood, as it is reported by Craw-ford.[23] Similar doubt is also expressed by Davies, who estimated the activity of the enzyme in 38 kinds of plants, including those that toler-ate flooding.[32]

Metabolism of ^{14}C-Glucose in the Absence of Oxygen

The possibility of the above biochemical shifts is also supported by our data on the metabolism of ^{14}C-glucose in the roots of wheat and

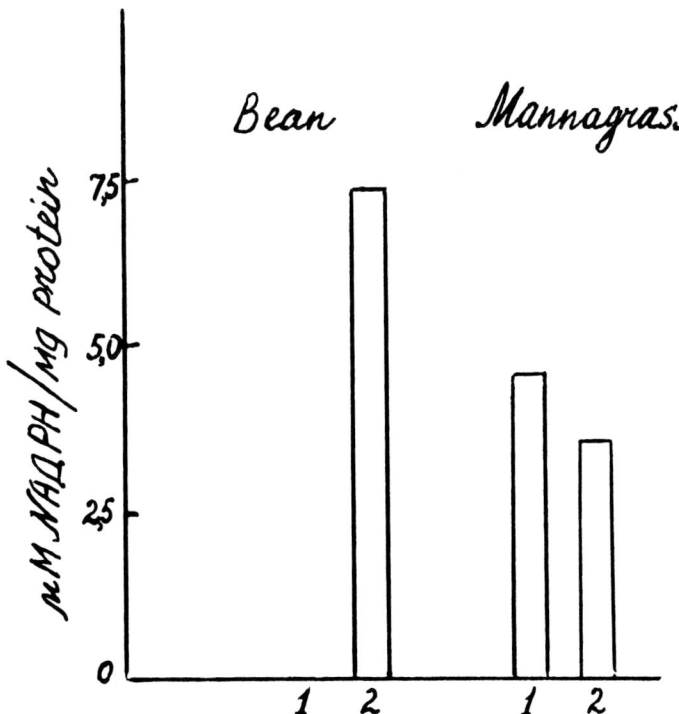

Figure 5. The influence of anoxia on the malic-enzyme activity in the roots of
bean and mannagrass. 1. aerated control; 2. exposure to N_2 /20 hours
for bean, 3 days for mannagrass.

rice.[33]   Figure 6 shows that the ^{14}C content in organic acids increased
markedly, although total radioactivity decreased in the extracts of roots
in the absence of oxygen.   In this case, plants of various ecological types
have some characteristic differences.   In wheat, the amount of labeled
carbon in the fraction mentioned increased with prolongation of anaerobiosis.   In rice, the accumulation of label in the organic acids was highest
after three days of anoxia.   After seven days of exposure, the radioactivity level decreased.

A more detailed analysis of the organic acid fraction showed (Figure
7) that in wheat after one and, in particular, three days of anaerobiosis,
the main rise was observed in monocarbonic acids, among which lactic
acid predominates.   In di- and tricarbonic acids (DTC), a negligible accumulation of label occurred only in case of a brief anoxia followed by
the fall of DTC radioactivity.   In the roots of rice, on the contrary, the

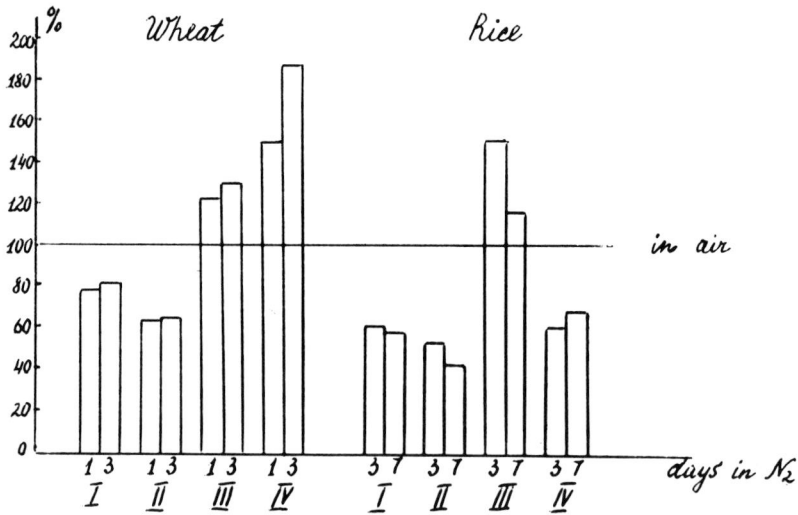

Figure 6. The influence of different plant aeration conditions on the distribution of radioactivity from ^{14}C-glucose among fractions of organic substances of roots of rice and wheat. I. aqueous-alcohol extract; II. carbohydrates; III. organic acids; IV. amino acids.

highest concentration of ^{14}C in DTC occurred after three days of anaerobiosis. After 7 days in the N_2 atmosphere, the amount of ^{14}C slightly decreased, exceeding the level of aeration. The changes that took place in monocarbonic acids were of a similar type, though the total amount of label in this fraction of rice is considerably smaller.

Of all DTC in the roots of wheat and rice, malic acid had the greatest share of radioactivity. This was valid in the case of anaerobic conditions as well and especially pronounced in the roots of rice. While in wheat some increase of ^{14}C amount in malic and, to a lesser degree, succinic acids was observed, this was only after 24 hr of anoxia. In rice, the increase was considerably greater and was detected over a much longer period of time under anaerobic conditions (7 days). Of interest also were the interconversions of malic and succinic acids which occurred in roots of rice with the increase of the anaerostatic time. After three days of anoxia, malic acid possessed the greatest radioactivity and succinic acid somewhat less. A more prolonged effect (7 days) altered the ratio in the content of labeled carbon between these acids; its share decreased in malate and increased in succinate. Thus, some similarity is observed in the ethanol and glucose transformation products under anaerobic conditions, which probably testifies to the identity of their metabolic pathways.

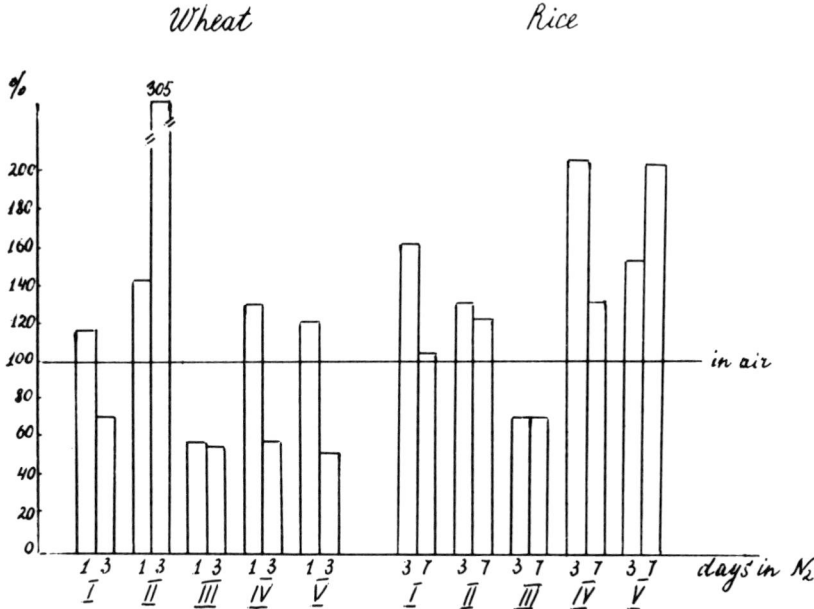

Figure 7. The influence of anoxia on ^{14}C labeling into organic acids from ^{14}C-glucose in roots of rice and wheat. I. DTC; II. monocarbonic; III. citric; IV. malic; V. succinic.

At the same time, the data once more emphasize that ethyl alcohol may be used by the cells of the resistant plant's roots as a normal respiratory substrate.

That the enhanced content of succinate under anoxia is associated with the utilization of NADH accumulated under these conditions, has been shown in a number of studies on animals by Kondrashova.[34,35] The accumulation of succinate is functionally significant because its oxidation is achieved with the aid of succinate-dehydrogenase, which functions faster and remains in an oxidized state under oxygen deficiency for a longer period of time than the NAD-dependent enzymes. Moreover, as was shown in the case of helminths, for example, when fumarate is reduced to give succinate fumaratereductase, there takes place an ATP formation associated with flavins under anaerobic conditions.[36] It is possible to suggest, therefore, that accumulation of succinic acid under hypoxia plays the role of an additional mechanism for ATP synthesis, and is related to the deoxidation of one more NADH molecules.

Such a change in the Krebs cycle reaction proceeding not only to malate but to succinate as well is quite likely to occur in plants resistant to anaerobiosis, in particular under the restricted conditions of gas exchange. The increase of succinic acid content in case of prolonged root anaerobiosis in pumpkin and tomatoes was observed by Dubinina.[37] The contribution of the glyoxalate cycle to the process of malate and succinate production is not to be excluded either.[38,39]

The above changes in enzymic reactions of the DTA cycle are also possible, in part, in the roots of nonresistant plants. In these, however, the changes do not seem to be adaptive because an increase in the length of anaerobic exposure causes the accumulation of labeled malate and succinate to decrease and the lactic acid content to increase. Thus, if the accumulation of intermediate products in nonresistant plants is due to the disturbance of metabolism in adapted forms, it is the result of regulatory rearrangements. These are necessary for the organism to find some possibilities for ATP formation and oxidation of reduced PN.

CONCLUSIONS

Plants resistant to oxygen deficiency are adapted to utilize the energy of glycolysis and, at the same time, to render its toxic products harmless. Detoxification of these plants can be achieved, for example, by exudation into the solution surrounding the roots, transport to the aboveground part, discarding into the air, and enhancing its ability for secondary metabolism. Some increase in the resistance to high concentrations of these substances is also possible. In addition, in the plants mentioned there are some regulatory systems that prevent production of ethanol and lactate by a reversal of the final step of glycolysis and the Krebs cycle, to produce some intermediate products of exchange harmless for the plants. These are malate and succinate, which are readily metabolized under the normalization of gas conditions.

REFERENCES

1. Soldatenkov, S.V., and T.V. Chirkova. "On the Role of Leaves in Respiration of Roots Deprived of Oxygen," *Fiziol. Rast.* 10(5):535 (1963). (In Russian.)
2. Chirkova, T.V. "Peculiarities of Oxygen Supply to Woody Plant Roots Subjected to Anaerobic Conditions," *Fiziol. Rast.* 15(3):565 (1968). (In Russian.)
3. Dubinina, I.M. "On the Nitrate Respiration of Plant Roots Under Oxygen Deficiency in the Nutrient Medium," *Fiziol. Rast.* 12(6): 980 (1965). (In Russian.)
4. Grineva, G.M., and F.G. Karimova. "Exudation of Substances and Nitrate Reduction by Maize Roots Under Root Anaerobiosis Conditions," *Fiziol. Rast.* 14(2):247 (1967). (In Russian.)

5. Chirkova, T.V. "The Role of Nitrate Respiration of Roots in the Life of Some Woody Plants Under Temporal Anaerobiosis Conditions," *Vest. Leningrad. Univ.* (21):118 (1971). (In Russian.)

6. Garcia-Novo, F., and R.M.M. Crawford. "Soil Aeration, Nitrate Reduction and Flooding Tolerance in Higher Plants," *New Phytol.* 72 (5):1031 (1973).

7. Ullrich, "Zur Sauerstoffabhängigkeit des Transportes in den Siebröhren," *Planta* 57(4):402 (1961).

8. Chirkova, T.V., E.L. Sokolovskaya, and I.V. Khazova. "Effect of Temporal Anaerobiosis on the Activity and Isoenzyme Composition of Peroxidase in Plant Roots," *Fiziol. Rast.* 20(6):1236 (1973). (In Russian.)

9. Taylor, D.L. "Influence of Oxygen Tension on Respiration, Fermentation and Growth in Wheat and Rice," *Am. J. Bot.* 29(9):721 (1961).

10. Grineva, G.M. "On the Formation and Exudation of Alcohol by Plant Roots Under Aerobic Conditions," *Fiziol. Rast.* 10(2):432 (1963). (In Russian.)

11. Chirkova, T.V. "The Role of Anaerobic Respiration in the Adaptation of Some Woody Plants to Temporal Anaerobiosis," *Vest. Leningrad. Univ.* (3):88 (1973).(In Russian.)

12. Chirkova, T.V., I.V. Khazova, and T.P. Astafurova. "On Metabolic Regulation of Plant Adaptation to Temporal Anaerobiosis," *Fiziol. Rast.* 21(1):102 (1974). (In Russian.)

13. Chirkova T.V., and T.C. Gutman. "On Physiological Role of Branch Lenticels of Willow and Poplar Under Conditions of Root Anaerobiosis," *Fiziol. Rast.* 19(2):352 (1972). (In Russian.)

14. Kenefick, D.J. "Formation and Elimination of Ethanol in Sugar Beet Roots," *Plant Physiol.* 37(3):434 (1962).

15. Fulton, J.M., and A.E. Erickson. "Relation Between Soil Aeration and Ethyl Alcohol Accumulation in Xylem Exudate of Tomatoes," *Soil Sco. Am. Proc.* 28(5):610 (1964).

16. Leblova, S., J. Zimakova, D. Safrova, and F. Barthova. "Occurrence of Ethanol in Pea Plants in the Course of Growth Under Normal and Anaerobic Conditions," *Biol Plant Acad. Sci. Bohemosl.* 11(6):417 (1969). (In Czechoslovakian.)

17. Grineva, G.M., I.N. Andreeva, V.A. Lipasova. "Respiratory Metabolism and the Ultrastructure of Mitochondria of Maize Root Cells After Short- and Long-Duration Anaerobiosis," *Dok. AN SSSR* 181 (1):248 (1968). (In Russian.)

18. Bolton, E.F., and A.E. Erickson. "Ethanol Concentration in Tomato Plants During Soil Flooding," *Agron. J.* 62(2):220 (1970).

19. Cossins, E.A. "Utilization of Alcohol in Germination Pea Seedlings," *Nature* 194 (4833):1095 (1962).

20. Cameron, D.C. and E.A. Cossins. "Studies of Intermediary Metabolism in Germinating Pea Cotyledons. The Pathway of Ethanol Metabolism and the Role of Tricarboxylic Acid Cycle," *Biochem. J.* 105(1):323 (1967).

21. Sherwin T., and E.W. Simon. "The Appearance of Lactic Acid in Phaseolus Seeds Germinating Under Wet Conditions," *J. Exp. Bot.* 20(65):776 (1969).

22. Chirkova, T.V. "Metabolism of Ethanol and Lactate in the Tissues of Trees with Different Resistance to Oxygen Deficiency," *Fiziol. Rast.* 22(5):952 (1975). (In Russian.)

23. McManmon, M., and R.M.M. Crawford. "A Metabolic Theory of Flooding Tolerance: The Significance of Enzyme Distribution and Behaviour," *New Phytol.* 70(2):299 (1971).

24. Leblova, S., and J. Hlochova. "Rice and Pea Alcohol Dehydrogenase," *Collection Czechosl. Chem. Com.* 40(10):3220 (1974).(In Czechoslovakian.)

25. Chirkova, T.V., E.V. Dragunova, and M.P. Burgova. "Redox Reactions of Flavoproteins and Pyridine Nucleotides from Roots of Plants Different in Their Resistance to Oxygen Deficiency Studied *In Vivo*," *Fiziol. Rast.* 24(1):126 (1977). (In Russian.)

26. Kondrashova, M.N. "Regulation of Mitochondria Respiration Under Enhanced Attacks on the Cell," *Biofizika* 15(2):312 (1970).(In Russian.)

27. Labori, A. *Regulation of Exchange Processes* (Moscow: Medicina, 1970). (In Russian.)

28. Atkinson, D.E., and G.M. Walton. "Adenosine Triphosphate Conservation in Metabolic Regulation. Rat Liver Citrate Cleavage Enzyme," *J. Biol. Chem.* 242(13):3239 (1967).

29. Veech, R.L., D. Raijman, and H.A. Krebs. "Equilibrium Relations Between the Cytoplasmic Adenine–Nucleotide System and Nicotinamide–Adenine Nucleotide System in Rat Liver," *Biochem. J.* 117 (3):499 (1970).

30. Crawford, R.M.M., and P.D. Tyler. "Organic Acid Metabolism in Relation to Flooding Tolerance in Roots," *J. Ecol.* 57(1):235 (1969).

31. Hochachka. P.W., and G.N. Somero. *Strategies of Biochemical Adaptation* (Moscow: Mir, 1977). (In Russian.)

32. Davies, D.D., K.H. Nascimento, and K.D. Patil. "The Distribution and Properties of NADP Malic Enzyme in Flowering Plants," *Phytochem.* 13(11):2417 (1974).

33. Chirkova T.V., and G.E. Nastinova. "Metabolism of ^{14}C-glucose in the Roots of Wheat and Rice Seedlings Under Various Conditions of Aeration," *Fiziol. Rast.* 24(2):291 (1977). (In Russian.)

34. Kondrashova, M.N. "Accumulation and Utilization of Succinic Acid in Mitochondria," in col. *Mitochondria. Molekulyarnye mekhanizmy fermental'nykh reaktsyj* (Moscow: "Nauka", 1972), p. 151.(In Russian.)

35. Kondrashova, M.N., E.I. Maevsky, G.B. Babyan, I.R. Saakyan, and R.N. Akhmetov. "Adaptation to Hypoxia Through Switching Over Metabolism to the Conversion of Succinic Acid," in col. *Mitochondria. Biokhimiya i ultrastruktura* (Moscow: "Nauka", 1973), p. 112.

36. Benediktov, I.I. "Transport of Electrones in Mitochondria of the Trematode Fasciola Hepatica," *Trudy Vsesoyuznogo Inst. gel'mintol.* 17:57 (1971). (In Russian.)

37. Dubinina, I.M. "Root Metabolism Under Various Levels of Aeration," *Fiziol. Rast.* 8(4):395 (1961). (In Russian.)

38. Zemlianukhin, A.A., and A.M. Makeev. "Metabolism of Succinic Acid a-2,3-[14]C Introduced into Maize Leaves," *Fiziol. Rast.* 16(2):344 (1969). (In Russian.)

39. Kondrashova, M.N., and M.A. Rodionova. "The Realization of the Glyoxylic Cycle in the Mitochondria of Animal Tissue," *Dok. AN SSSR* 196(5):1225 (1971). (In Russian.)

6

PYRUVATE CONVERSIONS IN HIGHER PLANTS DURING NATURAL ANAEROBIOSIS

Sylva Leblová

Department of Biochemistry
Charles University
Prague, Czechoslovakia

INTRODUCTION

Animals under anaerobiosis reoxidize reduced nicotinamide adeninedinucleotide, formed during carbohydrate degradation at the stage of oxidation of phosphoglyceraldehyde to phosphoglycerate, along the only path available, by transfer of hydrogen to pyruvate with formation of lactate. This process results in a shift of the cell pH towards the acidic region and in rigor mortis.

Plants resist a lack of oxygen substantially better. An example of life under anaerobic conditions is seed germination, called natural anaerobiosis. We would like to focus on this stage in plant growth from the point of view of metabolism.

Carbohydrates are degraded in plants by the same mechanism as in animals. Therefore, under anaerobiosis, plants can also reoxidize NADH in a reaction catalyzed by lactate dehydrogenase, involving transfer of hydrogen to pyruvate with the formation of lactate.[1-5] The problem of acidosis connected with this process has already been mentioned. However, plants can also reoxidize NADH in a reaction catalyzed by alcohol dehydrogenase, involving transfer of hydrogen to acetaldehyde with formation of ethanol,[6-17] as they contain active pyruvate decarboxylase catalyzing conversion of pyruvate into acetaldehyde. Another possibility would be metabolism of the glucose molecule to lactate and ethanol, as occurs under anaerobiosis, *e.g.* in heterofermentive bacteria *Leuconostoc mesenteroides.*[18]

Finally, it would be possible that malate would be accumulated primarily in germinating seeds during anaerobiosis, as occurs in a number of roots:[19-24]

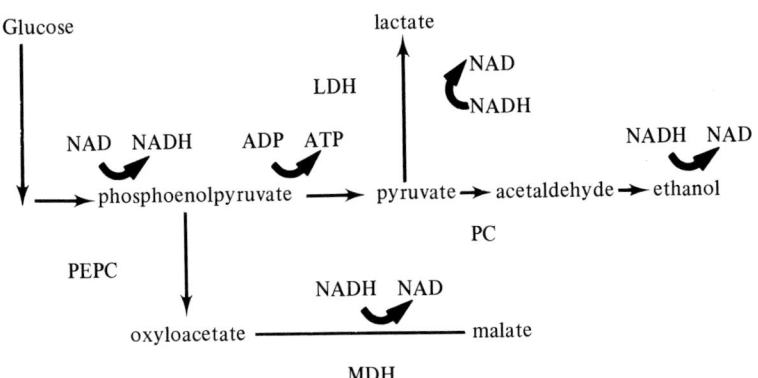

MDH	-	malate dehydrogenase
LDH	-	lactate dehydrogenase
ADH	-	alcohol dehydrogenase
PC	-	pyruvate dehydrogenase
PEPC	-	phosphoenolpyruvate carboxylase

OCCURRENCE OF ETHANOL

Ethanol has been found in a number of germinating seeds, such as pea, bean, broad bean, soybean, maize, rice, barley and wheat,[6-16] and in seedlings of buckwheat,[17] in castor bean on transfer into a nitrogen atmosphere[25] and in seedlings of bean[26] and maize in humid air.[27] Ethanol has also been found in roots,[19-24,28-32] in potato tubers,[33-36] in ripening seeds[37] and in fruit.[38-41]

We found that many mono- and dicotyledonous plants with proteins, glycides and fats as reserve substances form ethanol. During swelling of seeds of bean, lentil, broad bean, pea, lupine, soybean, barley, maize, wheat, oats, rye, cucumber, melon, flax, sunflower and rape, the maximum ethanol concentration was attained characteristically for each kind of plant, generally between 30 and 60 hr of swelling when specified cultivation conditions (the amount of water in the medium, volume of the vessel, number of seeds, temperature) were maintained. The ethanol content then decreased. The amount of ethanol formed and the rate and time of its increase and decrease were significant for each kind of plant.

Because of justified comments on the nonselectivity of certain methods for the determination of ethanol, we used three methods in parallel, namely:

1. an oxidimetric method with potassium dichromate as the oxidant, in a titration or photometric modification; [42]
2. enzyme determination with tests from Boehringer Company, where the increase in the absorbance at 340 nm is proportional to the amount of ethanol; and
3. gas chromatography with a Porapak Q column, permitting analysis of aqueous extracts of plant tissues.[13]

During swelling of seeds* of broad bean (*Vicia faba*), lentil (*Lens esculenta*), bean (*Phaseolus vulgaris* L.), pea (*Pisum sativum*), lupine (*Lupinus albus*), soybean (*Glycine max.*), barley (*Hordeum distichon* L.), maize (*Zea mays*, L.), wheat (*Triticum aestivum* L.s.s.), oat (*Avena sativa* L.), rye (*Secale cereale* L.), cucumber (*Cucumis sativus* L.), melon (*Cucumis melo* L.), flax (*Linum usitatissimum* L.), sunflower (*Helianthus annuus* L.) and rape (*Brassica napus* L., var. *arvensis* [Lam]), the maximum ethanol concentration was attained characteristically for each kind of plant, generally between 30 and 60 hr after swelling, when specified cultivation conditions (the amount of water in the medium, volume of the experimental vessel, number of seeds, temperature) were maintained. The ethanol content then decreased. The amount of ethanol formed and the rate and time of its increase and decrease were significant for a particular kind of plant.[11-13,43] The results obtained are illustrated in Figure 1.

When the seedlings were incubated in nitrogen, the ethanol concentration increased constantly and ethanol content decreased after transfer to the air. The results obtained with germinating peas are depicted in Figure 2. The ethanol concentration increased after 24-hr incubation in nitrogen, *e.g.*, 4 times for bean and maize, 9 times for wheat and 12 times for sunflower as compared with the control plants.

OCCURRENCE OF LACTATE

Lactate was detected, for example, in germinating seeds of buckwheat, mangrove, pea, broad bean, bean, castor bean, barley, mustard and sunflower.[1-5,10,13,25,26,44-46] The amount of lactate in plant

---

*Seeds, the variety of which is not stated, were conventional commercial types not designated in more detail.

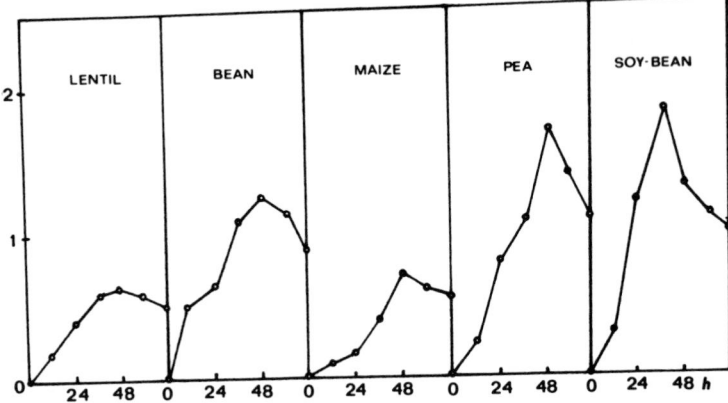

Figure 1. Ethanol content in the experimental plants during the first three days of germination. Abscissa: time in hr; ordinate: mg of ethanol/g of fresh weight of germinating seeds.

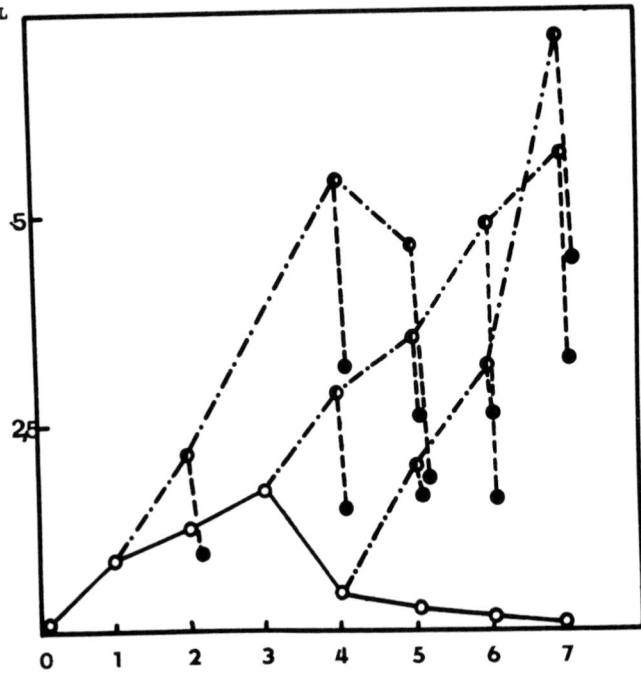

Figure 2. Ethanol content in 1-7-day-old pea plants grown for various periods in nitrogen, and decrease in the ethanol content after transferring the plants from nitrogen to the air: --o-- ethanol content in germinating plants; •-•-1-day-old, 3-day-old, 4-day-old plants transferred from nitrogen to the air; and ----decrease in the ethanol content after transferring plants from nitrogen to air.

tissues of both germinating and ripening seeds, and also in leaves, stalks, roots or bulbs is considerably affected by the external conditions, such as variations in the oxygen pressure, nitrogen atmosphere, excess of water, different salt contents in the medium, etc.[1,17,25,26,33-36,47-53]

In our experiments, the photometric method according to Barker and Summerson,[54] modified by Hullin and Noble,[55] and the enzymatic method using the tests of the Boehringer Company were used in parallel for the determination of the lactate content in plant tissues. The lactate contents in several seedlings among those studied is shown in Figure 3. The conditions mentioned for ethanol, i.e., the amount of water in the medium, the number of seeds in the cultivation vessel[43] and the composition of the atmosphere, also exerted an effect on the lactate content. If germinating seeds were incubated in nitrogen, the lactate content increased constantly. After 24 hr, its level increased 10 times in soybean, pea, broad bean and cucumber, and as much as 30 times in bean. The lactate was rapidly degraded on transfer from nitrogen into the air.[56]

Finally, both these anaerobic metabolites are formed in germinating seeds: lactate is accumulated in ripening seeds and its concentration increases to a maximum during 12-14 hr swelling; then the lactate concentration decreases more or less rapidly. At the end of a three-day period the lactate content usually increases again. The ethanol concentration maximum was attained later than that of lactate in all 16 plants studied. The concentration of ethanol at the maximum was 10 times higher than the lactate concentration. The formation first of lactate and then of ethanol was described by Barker et al. in a model experiment with potato tubers after incubation in nitrogen, by Kobr and Beevers[25] with castor beans grown in nitrogen atmosphere and by Davies et al.[57] using in vitro cell-free extracts prepared from pea seeds and parsnip roots.

The successive formation of lactate and then ethanol is typical of germinating seeds and could be explained using data on the optimum pH values for carboxylating and decarboxylating enzymes. While the former exhibit an optimum pH in the alkaline region, that of pyruvate dehydrogenases is at pH 6.[58] We assume that the production of lactate in intact germinating seeds may lead to a decrease in the pH in the cell (the concentration of lactate in our experiments is comparable with the amount of lactate rising in cell-free extracts in Davies' experiments) and thus pyruvate decarboxylase with a pH optimum in the weakly acidic region is activated. Consequently, pyruvate ceases to be reduced and is decarboxylated to give acetaldehyde, which becomes a hydrogen acceptor forming ethanol in a reaction catalyzed by alcohol dehydrogenase. The results of experiments with four-day-old pea seedlings incubated in nitrogen also support this assumption (Figure 4).

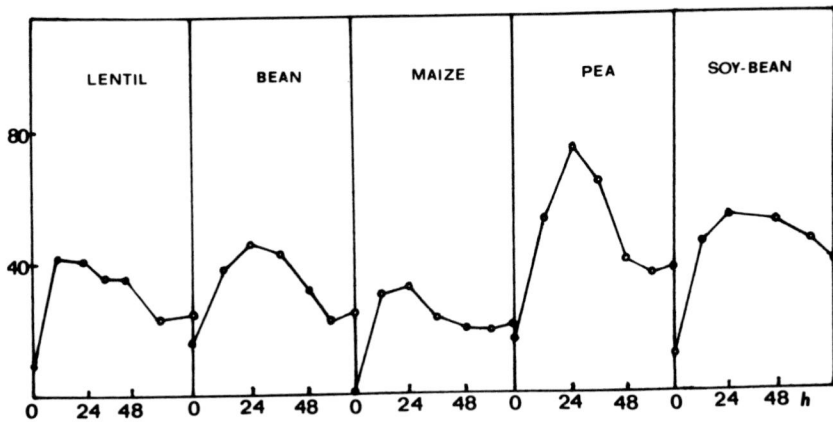

Figure 3. Lactate content in the experimental plants during the first three days of germination. Abscissa: time in hours; ordinate: μg of lactate/g of fresh weight of germinating seeds.

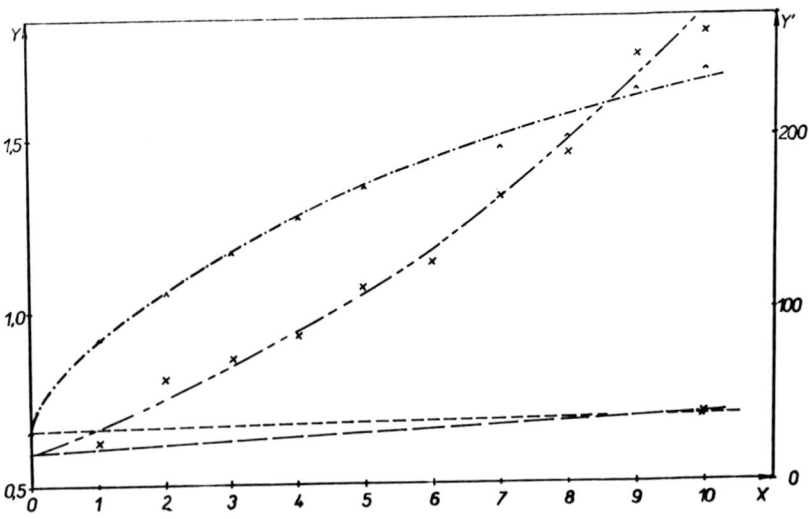

Figure 4. Ethanol and lactate contents in 4-day-old pea seedlings transferred into nitrogen. Abscissa: time in hr; ordinate, left: mg ethanol/g of fresh weight; ordinate, right: lactate content in μg/g of fresh weight: ▬•▬•▬• lactate content in N_2; ▬▬▬▬ lactate content in air; ▬ ▬ ethanol content in air; and ▬•▬• ethanol content in N_2.

OCCURRENCE OF ALCOHOL DEHYDROGENASE

Ethanol is oxidized by alcohol dehydrogenase which, as follows from Figure 5, was present in all the plants studied. The course of the activity exhibits a maximum which was reached similarly in all six plants, later than the highest ethanol concentration.

It has been assumed that alcohol dehydrogenase is induced by ethanol or by acetaldehyde.[22,59,60] Other authors explain this activity decrease as a consequence of inhibition by substances formed in the seed during germination.[61] We have also studied this problem. That we were able to bring about an increase in the alcohol dehydrogenase activity by anaerobiosis as well as by application of substrate to the medium speaks in favor of induction of enzyme synthesis by the substrate (Figure 6). However, on the other hand, if ADH was isolated from germinating seeds after one, two or three days of swelling, an enzyme with identical specific activity was obtained after application of ammonium sulfate and chromatographic separation, although the initial ADH activities in the sodium phosphate extracts were different. This finding suggests that an ADH activity inhibitor was present in the extracts, which was removed during the purification process.

Gibberellic acid, kinetin, and gibberellic acid with kinetin increased the ADH activity in bean hypocotyls by 34, 42 and 64%, respectively. This can be prevented by addition of chloramphenicol.

OCCURRENCE OF LACTATE DEHYDROGENASE

Lactate dehydrogenase is already present in a rather high concentration in the dry seed. Values of about 1000 activity units were found in bean, pea and soybean. During germination, the activity immediately decreased (with the exception of broad bean); the amount of decrease depended on the increase of the lactate concentration. The activity reached a minimum after 12-24 hr germination and then increased. The second maximum appeared after 36-48 hr germination. In peas, this maximum was followed by a rapid decrease, but in maize, lentils and soybeans the activity was reduced slowly during the following growth period (Figure 7).

It can be stated that the decrease in the enzyme activity is caused by a decrease in the pH value of the cell (changes in pH optimum), the rapid increase in the lactate content participating in this process. It may also be admitted that accumulation of lactate induces the enzyme and, therefore, enzyme induction by the substrate may be involved. This explanation is given without experimental justification by Sherwin in a study on the pea enzyme.[26] It is interesting to consider the behavior

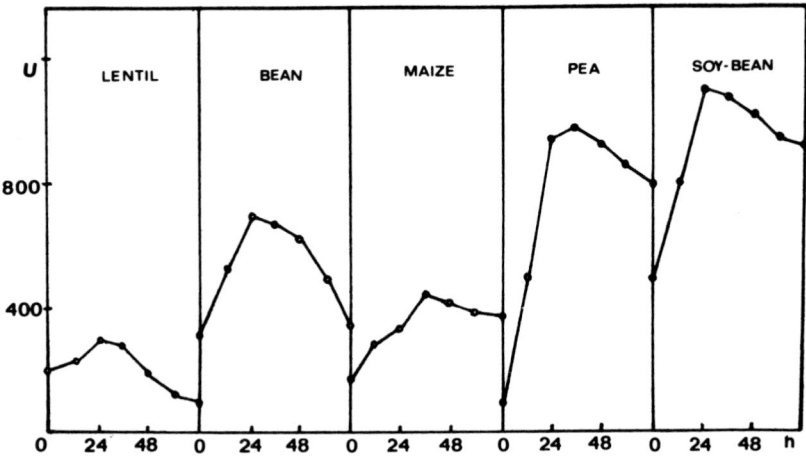

Figure 5. Alcohol dehydrogenase activity in germinating seeds during the first three days. Abscissa: time in hours; ordinate: alcohol dehydrogenase activity in units/g of fresh weight of germinating seeds.

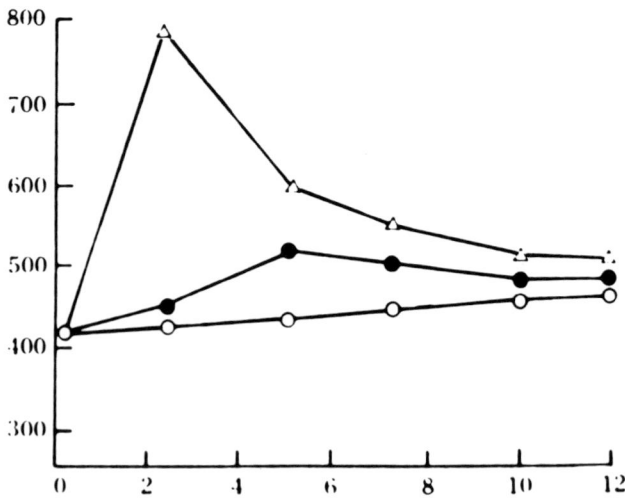

Figure 6. The effect of acetaldehyde on the activity of broad bean alcohol dehydrogenase. Abscissa: incubation time in acetaldehyde in hr; ordinate: enzyme activity in units/mg of protein: ooooo control seeds; ••••• plants incubated in 0.05 M acetaldehyde solution; and△△△△plants incubated in 0.25 M acetaldehyde solution.

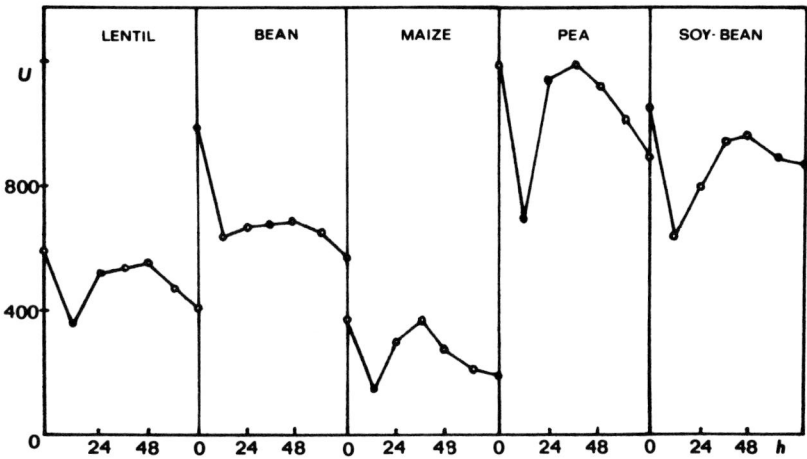

Figure 7. Lactate dehydrogenase activity during the first three days of germination. Abscissa: time in hr; ordinate: lactate dehydrogenase activity in units/g of fresh weight of plant tissues.

of lactate dehydrogenase and of lactate during the entire observation period: the relationship between the enzyme activity and the substrate concentration was one of indirect proportion. This was true also of seeds of germinating broad bean, the behavior of which was quite different from that of the other plant species studied.

CONCLUSIONS

It can be stated in conclusion that regardless of the type of reserve substance present in the seeds of the individual species, a certain amount of lactate is already formed during ripening (10-40 μg/g of fresh weight). With the exception of broad bean, the amount of lactate increased under the above given experimental conditions during the first day of germination alone. The increase was well defined and was approximately fourfold.

Ethanol was formed during germination in a concentration by one order higher than that of lactate. The maximum ethanol concentration was attained in all species studied about 20 hr later than that of lactate. It appeared that this effect can be explained on the basis of knowledge of the carboxylating and decarboxylating enzymes. While the first category, *e.g.*, phosphoenolpyruvate carboxylase or ribulose diphosphate carboxylase, exhibited a pH optimum at about 7.9, the optimum for decarboxylating enzymes, *e.g.*, wheat pyruvate decarboxylase, lies at pH 6.0.[58] The decrease in the lactate concentration, and the change in the pH value of the cell related to it, may activate (induce) pyruvate decarboxylase as the first enzyme of pyruvate metabolism on the pathway via acetaldehyde to ethanol. According to our results, it was precisely acetaldehyde which induced the formation of alcohol dehydrogenase (Figure 6).

A decrease in the pH due to the formation of lactate in plant cells can be explained not only by the activation of pyruvate decarboxylase,[58] but also by inactivation of the lactate dehydrogenase that catalyzes the reduction of pyruvate into lactate and whose pH optimum lies in a slightly alkaline region.[62,63] The optimum pH for alcohol dehydrogenase as an enzyme catalyzing the reduction of acetaldehyde to ethanol lies around pH 7, with small fluctuations towards the acidic or alkaline region, depending on the species of plant.

On the other hand, the enzymes connected with the oxidation of the two anaerobic products, i.e., alcohol and lactate dehydrogenase, catalyze the reaction in the oxidation direction in a pH region of 8.3-9.3.

The dependence of the activity of soybean lactate dehydrogenase on the substrate concentration was sigmoid. This allosteric enzyme had Krebs cycle intermediates, such as malate, succinate, pyruvate and acetate, as modulators.[62,63] The activity of the other enzyme connected with the anaerobic metabolism, i.e., alcohol dehydrogenase, was also affected by malate, succinate, pyruvate and also lactate. As the concentration of a number of acids varies during anaerobiosis,[20,21,64-66] it is possible that these acids play a regulating role in the anaerobic metabolism of plant cells. Both enzymes, LDH and ADH as SH enzymes, are also affected by SH poisons and activated, for example, by cystein or mercaptoethanol. Alcohol dehydrogenase as a metalloprotein is also inhibited by chelating agents. Among natural metabolites, fatty acids undoubtedly also participate in the ADH regulation by competing for a bonding site with ethanol: this effect increases with increasing chain length. The effect of chloride anions and metal cations is also interesting. The ADH activity is strongly inhibited by heterocycles such as pyrazole, imidazole and pyridine, and oximes and amides. Both dehydrogenases are also regulated by ATP, ADP and AMP, which decrease the reaction rate in dependence on the pH: with increasing pH the inhibition effect decreases.

REFERENCES

1. Brown, J. M., H. A. Dufred, and C. F. Hill. "Respiratory Metabolism in Mangrove Seedlings," *Plant Physiol.* 44(2):287 (1969).
2. Barthová, J., J. Matoušková and S. Leblová. "Anaerobní metabolismus klíčících rostlin," Ústav vědeckotechnických informací, *Rostlinná výroba Praha* 19(46):1221 (1973).

3. Cossins, E. A. "Formation and Metabolism of Lactic Acid during Germination of Pea Seedlings," *Nature* 203(4948):989 (1964).
4. Schneider, A. "Untersuchungen über das Auftreten der Milchsäure in höheren Pflanzen," *Planta* 29(4):747 (1939).
5. Schneider, A. "The Formation of Lactic Acid in Higher Plants Especially during Germination," *Planta* 32(3):234 (1940).
6. Cossins, E. A. "Utilization of Ethanol-2-C^{14} by Pea Slices," *Nature* 194(4833):1095 (1962).
7. Cossins, E. A. and E. R. Turner. "Losses of Alcohol and Alcohol Dehydrogenase Activity in Germinating Seeds," *Ann. Bot.* 26(104): 591 (1962).
8. Cossins, E. A. and E. R. Turner. "The Metabolism of Ethanol in Germinating Pea Seedlings," *J. Exp. Bot.* 14(41):290 (1963).
9. Goksöyr, J., E. Boeri and R. K. Bonnichsen. "The Variation on Alcohol Dehydrogenase and Catalase during the Germination of Green Pea (*Pisum sativum*)," *Acta. Chem. Scand.* 7(4):652 (1953).
10. James, W. O. "Reaction Paths in the Respiration of the Higher Plants," *Advan. Enzymol.* 18:281 (1957), reprint from China.
11. Leblová, S., I. Zimáková, D. Sofrová and J. Barthová. "Occurrence of Ethanol in Pea Plants in the Course of Growth under Normal and Anaerobic Conditions," *Biol. Plant.* 11(6):417 (1969).
12. Leblová, S., D. Ehlichová and J. Barthová. "Výskyt ethanolu a alkoholdehydrogenazy v klíčních rostlinach, Ústav Vědeckotechnických informací," *Rostlinna vyroba Praha* 19(46):1209 (1973).
13. Leblová, S., E. Sinecká and V. Vaníčková. "Pyruvate Metabolism in Germinating Seeds during Natural Anaerobiosis," *Biol. Plant.* 16(6): 406 (1974).
14. Peterson, C. A. and E. A. Cossins. "Participation of the Glyoxalate Cycle in the Metabolism of Ethanol by Castor Bean Endosperm Tissues," *Can. J. Biochem.* 44(4):423 (1966).
15. Philips, J. W. "Studies on Fermentation in Rice and Barley," *Amer. J. Bot.* 34(2):62 (1947).
16. Stafford, H. E. and B. Vennesland. "Alcohol Dehydrogenase of Wheat Germ," *Arch. Biochem. Biophys.* 44(2):404 (1953).
17. Effer, W. R. and S. L. Ranson. "Respiratory Metabolism in Buckwheat Seedlings," *Plant Physiol.* 42(8):1042 (1967).
18. Rauschenbach, P. and H. Simon. "Untersuchungen des Kohlenhydratstoffwechsels mit Wasserstoff Markerung. VI. Tritium-Fixierung in Athanol und Lactat nach Fermentation von Glukose in H_2O/HOT oder von T-markierten Glukose durch Leuconostoc mesentoroides," *Hoppe-Seyler's Z. Physiol. Chem.* 349(10):1330 (1968).
19. Crawford, R. M. M. "The Control of Anaerobic Respiration as a Determining Factor in the Distribution of the Genus Senecio," *J. Ecol.* 54(2):403 (1966).
20. Crawford, R. M. M. "Alcohol Dehydrogenase in Relation to Flooding Tolerance in Roots," *J. Exp. Bot.* 18:458 (1967).
21. Crawford, R. M. M. "The Physiological Basis of Flooding Tolerance," *Ber. Dtsch. Bot. Ges.* 82(1/2):111 (1969).
22. Crawford, R. M. M. and M. McManmon. "Inductive Responses of Alcohol and Malic Dehydrogenase," *J. Exp. Bot.* 19(60):435 (1968).

23. Crawford, R. M. M. and P. D. Tyler. "Organic Acid Metabolism in Relation to Flooding Tolerance in Roots," *J. Ecol.* 57(1):235 (1969).

24. McManmon, M. and R. M. M. Crawford. "A Metabolic Theory of Flooding Tolerances. The Significance of Enzyme Distribution and Behavior," *New Phytol.* 70(2):299 (1971).

25. Kobr, M. J. and H. Beevers. "Glukoneogenesis in the Castor Bean Endosperm. I. Changes in Glycolytic Intermediates," *Plant Physiol.* 47(1):48 (1971).

26. Sherwin, T. and E. W. Simon. "The Appearance of Lactic Acid in Phaseolus Seeds Germinating under Wet Conditions," *J. Exp. Bot.* 20(65):776 (1969).

27. Grinĕva, G. M., I. N. Andreeva and E. A. Stupišina. "Effect of Flooding on the Growth Respiration and Concentration of Oxygen in the Tissues of Rudimentary and Stem Roots of Maize Plants," *Fiziol. Rast.* 17(4):655 (1970). (In Russian).

28. Aubertin, G. M., R. W. Rickman and J. Letey. "Plant Ethanol Content as an Index of the Soil Oxygen Status," *Agron. J.* 58(3):305 (1966).

29. James, W. O. and A. F. Ritchie. "The Anaerobic Respiration of Carrot Tissue," *Proc. Roy. Soc. (London) B* 143:302 (1954).

30. Lowe, C. and W. O. James. "Carrot Tissue and Ethanol," *New Phytol.* 59(3):288 (1960).

31. Betz, A. "Der Athanolumsatz in meristematischen Wurzelspitzen von Pisum sativum," *Flora* 146(4):532 (1958).

32. Kenefick, D. G. "Formation and Elimination of Ethanol in Sugar Beet Roots," *Plant Physiol.* 37(3):434 (1962).

33. Barker, J. and A. F. El-Saifi. "Studies in the Respiratory and Carbohydrate Metabolism in Plant Tissues. II. Interrelationship between the Rates of Production of CO_2, Lactic Acid and of Alcohol in Potato Tubers under Anaerobic Conditions," *Proc. Roy. Soc. (London) B* 140:385 (1952).

34. Barker, J. and A. F. El-Saifi. "Studies in the Respiratory and Carbohydrate Metabolism in Plant Tissues. I. Experimental Studies of the Formation of CO_2, Lactic Acid and other Products in Potato Tubers under Anaerobic Conditions," *Proc. Roy. Soc. (London) B* 140:362 (1952).

35. Barker, J. and A. F. El-Saifi. "Studies in the Respiratory and Carbohydrate Metabolism in Plant Tissues. III. Experimental Studies of the Formation of CO_2 and Changes in Lactic Acid and other Products in Potato Tubers in Air following Anaerobic Conditions," *Proc. Roy. Soc. (London) B* 140:508 (1953).

36. Barker, J. and A. F. El-Saifi. "Studies in the Respiratory and Carbohydrate Metabolism in Plant Tissues. IV. The Relation between the Rate of CO_2 Production in Potato Tubers in Air Following Anaerobic Conditions, and the Accompanying Changes in Lactic Acid and Sugar Concentration," *Proc. Roy. Soc. (London) B* 140:523 (1953).

37. Wagner, H. G. "The Effect of Artificial Wilting on the Production of the Ethanol by Ripening Pea Seeds," *New Phytol.* 58(1):68 (1959).

38. Fidler, J. G. "A Comparison of Anaerobic and Aerobic Respiration of Apples," *J. Exp. Bot.* 2:41 (1951).

39. Masquelier, J. and G. Vitte. "Aerobe Bildung von Alkohol und verschiedenen ählichen flüchtingen Verbindungen durch Pflanzen," *C. R. Acad. Sci. Paris, Ser. D* 265:1924 (1967).
40. Masquelier, J. and G. Vitte. "Sur d'importance du taux d'ethanol dans les jus fruits," *Bull. Soc. Pharm.* 108:49 (1967).
41. Masquelier, J., L. Queylouis and G. Vitte. "Sur la presence d'ethanol dans les jus de fruits," *Bull. Soc. Pharm.* 106:35 (1967).
42. Widmark, E. M. P. "Eine Mikromethode zue Bestimmung von Aethylalcohol in Blut," *Biochem. Z.* 131:471 (1922).
43. Sinceka, E. "Conversion of Pyruvate in Higher Plants," Unpublished dissertation, Charles University, Prague, 1974.
44. Abdul-Baki, A. A. "Metabolism of Barley Seed during Early Hours of Germination," *Plant Physiol.* 44(5):733 (1969).
45. King, J. "The Isolation, Properties and Physiological Role of Lactic Dehydrogenase from Soybean Cotyledons," *Can. J. Bot.* 48(3):533 (1970).
46. Missen, A. W. and A. T. Wilson. "The Metabolism of *Sinapis Alba* Seeds in Water under Anaerobic Condition," *Phytochem.* 9(7):1473 (1970).
47. Barber, D. A. "Lactic Acid Formation and Carbon Dioxide Fixation," *Nature* 180(4594):1053 (1957).
48. Barron, E. S., G. K. K. Link, R. M. Klein and B. E. Michel. "The Metabolism of Potato Slices," *Arch. Biochem. Biophys.* 28(3):377 (1950).
49. Barker, J. and L. M. Mapson. "Studies in the Respiratory and Carbohydrate Metabolism in Plant Tissues. V. Experimental Studies of the Formation of CO_2 and of the Changes in Lactic Acid, Sucrose and Certain Fractions of Ketoacids in Potato Tubers in Air Following Anaerobic Conditions," *Proc. Roy. Soc. (London) B* 141(904):321 (1953).
50. Barker, J. and L. M. Mapson. "Studies in the Respiratory and Carbohydrate Metabolism in Plant Tissues. VI. Analysis of the Interrelationships Between the Rate of CO_2 Production and the Changes in the Contents of Lactic Acid, Sucrose and of Certain Fractions of Ketoacids in Potato Tubers in Air Following Anaerobic Conditions," *Proc. Roy. Soc. (London) B* 141(904):338 (1953).
51. Barker, J. and L. M. Mapson. "Studies in the Respiratory and Carbohydrate Metabolism in Plant Tissues. XII. Further Studies of the Formation of CO_2 and the Changes in Lactate, Alcohol, Sucrose, Pyruvate and α-Ketoglutarate in Potato Tubers in Nitrogen and in Air Following Anaerobic Conditions," *Proc. Roy. Soc. (London) B* 157(968):383 (1963).
52. Čirkova, T. A. "Metabolizm etanola i laktata v tkanjach drevjesnych rastjenij, razligajuščichsja po ustojčivosti i nedostatku kisloroda," *Fiziol. Rast.* 22:952 (1975).
53. Hasson-Porath, E. and A. Poljakoff-Mayber. "Lactic Acid Content and Formation in Pea Roots Exposed to Salinity," *Plant Cell Physiol.* 11(6):891 (1970).
54. Barker, S. B. and W. L. Summerson. "The Colorimetric Determination of Lactic Acid in Biological Material," *J. Biol. Chem.* 138(2):535 (1941).

55. Hullin, R. P. and R. L. Noble. "Determination of Lactic Acid in Microgram Quantities," *Biochem. J.* 55:289 (1953).
56. Matoušková, J. "Metabolism of Lactate," Unpublished dissertation, Charles University, Prague, 1974.
57. Davies, D. D., S. Greco and P. Kenworthy. "The Control of the Production of Lactate and Ethanol by Higher Plants," *Planta* 118(4): 297 (1974).
58. Davies, D. D. "Metabolic Control in Higher Plants," in *Biosynthesis and its Control in Plants,* B. V. Milborrow, Ed. (London: Academic Press, 1973), pp. 1-20.
59. App, A. A. and A. N. Meiss. "Effect of Aeration on Rice Alcohol Dehydrogenase," *Arch. Biochem. Biophys.* 77(1):181 (1958).
60. Hageman, R. H. and D. Flesher. "The Effect of Anaerobic Environment on the Activity of Alcohol Dehydrogenase," *Arch. Biochem. Biophys.* 87(2):203 (1960).
61. Suzuki, Y. and K. Hyuwa. "Activation and Inactivation of Alcohol Dehydrogenase in Germinating Pea Cotyledons," *Physiol. Plant* 27(2):121 (1972).
62. Barthová, J., P. Hrbas and S. Leblová. "Isolation and Properties of Plant Lactate Dehydrogenase," *Coll. Czech. Chem. Commun.* 38(7): 2174 (1973).
63. Barthová, J., P. Hrbas and S. Leblová. "Some Structural and Kinetic Characteristics of Lactate Dehydrogenase from Soybean Seeds (*Glycine max.* L.)," *Coll. Czech. Chem. Commun.* 39(11):3383 (1974).
64. Mazelis, M. and D. Vennesland. "CO_2 Fixation into Oxaloacetate in Higher Plants," *Plant Physiol.* 32(6):591 (1957).
65. Streeter, J. G. and J. F. Thompson. "Anaerobic Accumulation of Gamma-aminobutyric Acid and Alanine in Radish Leaves," *Plant Physiol.* 49(4):572 (1972).
66. Streeter, J. G. and J. F. Thompson. "*In Vivo* and *In Vitro* Studies on Gamma-aminobutyric Acid Metabolism with the Radish Plant," *Plant Physiol.* 49(4):579 (1972).

7

ETHANOL METABOLISM IN PLANTS

Edwin A. Cossins

Department of Botany
University of Alberta
Edmonton, Alberta, Canada T6G 2E9

INTRODUCTION

The importance of ethanol in plant metabolism has been the subject of much speculation. Following the demonstration that various fruits produce alcohol,[1-3], it was proposed by Pfeffer[4] that ethanol was a normal respiratory intermediate in plants. However, later workers were unable to demonstrate rates of ethanol consumption which approached respiratory rates,[5,6] so these early concepts were modified to recognize pyruvate as a key intermediate in both aerobic and anaerobic respiration. In this respect, there is now excellent evidence[7] that the Embden-Meyerhof-Parnas pathway is the major route for ethanol and lactate production in higher plants. There are also many reports that part of this ethanol and lactate is metabolized when aerobic conditions are reestablished. This chapter reviews this evidence and considers the metabolic pathways by which ethanol is utilized. Although reference is made to related work on animal tissues and microorganisms, emphasis is given mainly to those plant species that commonly experience periods of anaerobiosis under environmental conditions.

EVIDENCE FOR ETHANOL UTILIZATION IN PLANTS

A literature review shows that a wide variety of plant tissues can metabolize ethanol. In most reported cases, ethanol accumulated during periods of natural or experimentally imposed anaerobiosis is consumed when the tissues are placed in air. In other instances, plant tissues have been examined for ability to utilize ethanol supplied under controlled conditions. In early work, changes in ethanol content were commonly monitored by colorimetric and titrimetric assays using potassium dichromate as an oxidizing agent.[8,9] More sensitive analyses have been made by enzymatic assays of ethanol using purified alcohol dehydrogenases.[10] Gas chromatography has also been used successfully for estimating ethanol and volatile products of its metabolism.[11] Research on the biochemical nature of ethanol utilization has also been greatly aided by use of ethanol solutions labeled with C^{14} and H^3.[12-15]

Germinating Seedlings

More than 70 years ago, the Japanese researcher Sawa[16] showed that growth of rice seedlings (*Oryza sativa*) was stimulated by dilute ethanol solutions. Similar data were reported by Nagai.[17] Taylor[18] examined this effect in more detail and found that adding increasing quantities of ethanol, besides inhibiting shoot and root growth, tended to depress the respiratory quotient (RQ). For example, when the seedlings received 8% ethanol, the RQ fell to 0.8. As the theoretical RQ for complete oxidation of ethanol is 0.67, Taylor concluded that some oxidation of ethanol took place under these conditions. Some support for this conclusion comes from the work of Phillips[19] in which colorimetric analysis of ethanol contents in young rice seedlings showed a depletion in air after a 6-hr period in nitrogen. By contrast, seedlings in the first leaf stage continued to accumulate alcohol throughout the subsequent aerobic period and thus did not appear to utilize endogenous ethanol. In parallel work on barley (*Hordeum vulgare*) seedlings, data for ethanol were more variable but indicated that the alcohol formed in a 5-hr period of anaerobiosis was not consumed after 5 hr in air. Cossins and Turner[20] found that 100 g fresh weight of barley seeds accumulated 33 mg of ethanol when soaked aseptically in water for 24 hr. When germinated in moving air for 120 hr, the ethanol content of the seeds dropped to 4 mg/100 g fresh weight, but some 45 mg ethanol were collected from the air stream, thus confirming that barley seedlings have little ability to metabolize ethanol.

Other cereals, however, are able to oxidize ethanol. In wheat (*Triticum vulgare*), Cossins and Turner[20] found small losses of alcohol, which could not be attributed to volatility. Corn (*Zea mays*) coleoptiles (Table I), root tips, shoots and leaves (Table II) clearly formed a variety of labeled products when incubated with ethanol-1-C^{14}.[12]

In early Russian work, Kostychev[5] showed that ethanol is normally formed during germination of peas (*Pisum sativum*). This natural anaerobiosis has since been confirmed by several other workers.[21-24] The major reason for this phenomenon in peas appears related to poor penetration of oxygen through the intact seed coat. Once the testa is ruptured by the emerging radicle, a marked depletion of ethanol occurs.[21,22,24] Such large losses of alcohol could not be attributed to volatility, and they were accelerated by cutting or removing the seed coat.[23] It follows that pea cotyledons metabolize a large portion of the ethanol accumulated during the early stages of germination. This was subsequently confirmed in ethanol-C^{14} feeding experiments.[12,23,25] Data from the latter work

Table I. Metabolism of Ethanol-1-C^{14} by Corn Coleoptiles[a]

Fraction	cpm x 10^{-3}	% Incorporated C^{14}
Ethanol-Solubles	769	37.0
Organic acids	259	12.0
Acidic amino acids	115	6.0
Neutral and basic amino acids	123	6.0
Lipids	156	8.0
Sugars	15	0.7
Volatile on acidification	46	2.0
Residue	551	27.0
CO_2	733	35.0
Total C^{14} Incorporated	2053	

[a]Used with permission.[12] Incubated in air at $25°C$ for 4 hr.

are given in Table III and show that 1- and 3-day-old pea cotyledons utilize ethanol carbon for a wide variety of syntheses. Only relatively small amounts of C^{14} were incorporated into CO_2, and incubation of the 3-day-old tissues in oxygen, which accelerated ethanol losses *in vivo*,[25] caused an even more extensive distribution of the label.

In a survey of several other seedlings, Cossins and Turner[20] concluded that germination is commonly accompanied by depletion of ethanol from the storage organs. Losses of ethanol in excess of that collected in air passing over the seedlings were found in *Phaseolus multiflorus* (French bean), *P. vulgaris* (Runner bean), *Vicia faba* (Broad bean) and *Ricinus communis* (Castor bean). Endosperm slices of the latter species were

Table II. Metabolism of Ethanol-1-C^{14} by Corn Seedlings[a]

Fraction	Root Tips		Shoots		Leaves	
	cpm x 10^{-3}	% Incorporated C^{14}	cpm x 10^{-3}	% Incorporated C^{14}	cpm x 10^{-3}	% Incorporated C^{14}
Ethanol-Solubles	178.5	75	190.0	30	206.0	48
Organic acids	36.1	15	28.0	4	40.5	9
Acidic amino acids	28.0	12	18.0	3	62.5	14
Neutral and basic amino acids	23.0	10	7.7	1	16.0	4
Lipids	10.6	4	55.5	9	48.0	11
Sugars	8.0	3	10.7	2	21.0	5
Volatile on acidification	67.0	28	37.0	6	–	–
Residue	39.0	16	381.0	61	130.0	32
CO_2	22.1	9	57.5	9	96.0	22
Total C^{14} Incorporated	240.0		629.0		432.0	

[a]Used with permission.[12] Five-day old corn seedlings dissected into root tips, shoots, and leaves, incubated in air at 25°C for 4 hr.

Table III. Metabolism of Ethanol-1-C^{14} by Pea Cotyledon Slices[a]

Fraction	1-Day-Old Cotyledons[b]		3-Day-Old Cotyledons[c]	
	cpm x 10^{-3}	% Incorporated C^{14}	cpm x 10^{-3}	% Incorporated C^{14}
Ethanol-Solubles	145.0	69	377.0	93.0
Organic acids	72.0	34	76.1	18.0
Acidic amino acids	45.0	27	25.7	6.0
Neutral and basic amino acids	4.50	2	1.9	0.4
Lipids	6.1	3	268.0	66.0
Sugars	3.7	2	5.2	1.0
Residue	62.0	29	21.0	5.0
CO_2	2.4	1	5.0	1.0
Total C^{14} Incorporated	209.0		403.0	

[a]Used with permission.[12]
[b]Incubated in air at 25°C for 4 hr.
[c]Incubated in O_2 at 25°C for 1 hr.

supplied ethanol-C^{14} by Cossins and Beevers.[12] As summarized in
Table IV, this tissue has an active ethanol metabolism. Considerable
radioactivity was lost from the ethanol-soluble fraction on acidification
with acetic acid. This volatile material may include acetate, as fats are
extensively hydrolyzed in this tissue during germination.[26] The similarity
between the percentages for incorporation in sugars and CO_2 suggested
that ethanol was utilized by the glyoxylate cycle and subsequently incorporated into sugars. This possibility was later confirmed by Peterson and
Cossins.[27]

Table IV. Metabolism of Ethanol-1-C^{14} by Castor Bean Endosperm Slices[a]

Fraction	cpm x 10^{-3}	% Incorporated C^{14}
Ethanol-Solubles	165.0	64
Organic acids	7.0	3
Acidic amino acids	3.0	1
Neutral and basic amino acids	8.0	3
Lipids	6.0	2
Sugars	43.5	17
Volatile on acidification	70.0	27
Residue	57.0	22
CO_2	33.0	13
Total C^{14} Incorporated	255.0	

[a]Used with permission.[12] Slices of endosperm from 5-day-old castor beans were
incubated in air at $25°C$ for 4 hr.

Effer and Ranson[28] carried out extensive work on the respiratory
metabolism of buckwheat (*Fagopyrum esculentum*) seedlings. Under
anaerobic conditions, CO_2, ethanol and lactate production did not account
for all the carbohydrate consumed, but on return to air, CO_2 output
increased and the contents of ethanol and lactate declined. The rate of
ethanol depletion in surface-sterilized seedlings was such that 75% of the
1500 μmol/100 g dry weight accumulated during the 6.5-hr anaerobic
period was consumed within 6.5 hr in air. Very little ethanol was present
after 12 hr in air. During this loss, temporary increases occurred in
acetaldehyde. The authors concluded that at least part of the ethanol
was metabolized via the tricarboxylic acid cycle which, from their analyses

of cycle intermediates, appeared to increase in activity on return of the seedlings to air.

Root and Stem Tissues

A form of natural anaerobiosis or aerobic fermentation is known to occur in the meristematic regions of roots.[29-32] In detailed studies of this process in pea root tips, Betz[31] analyzed ethanol contents of successively older root segments using a sensitive enzyme assay. He found that ethanol accumulations were greatest in the 0-2 mm tip segment even when the roots were kept in an aerobic environment. In oxygen, however, ethanol contents were depleted by amounts approximating the observed increases in oxygen uptake. It is clear that root tips must have ability to metabolize ethanol, but the exact reasons for its formation under aerobic conditions are so far not fully explained.

Direct evidence for ethanol metabolism in corn root tips was obtained by Cossins and Beevers[12] (Table II). Similar data have been reported by Liu et al.[13] for pea root tips. In this latter species, however, a major product of ethanol-C^{14} metabolism was ethyl β-glucoside. Ability to form this product was slightly higher for the second 7-mm root segment (9.8 cpm/mg)tissue) than for the apical 7-mm segment (7.4 cpm/mg tissue). Liu et al.[13] therefore concluded that conjugation of ethanol to form the β-glucoside may be one reaction responsible for the lower ethanol contents of older root segments. Paper chromatography of other labeled products revealed that pea root tips also formed a variety of amino and organic acids from ethanol carbon.

The formation and oxidation of ethanol by the nodulated roots of several leguminous species have been reported by Ludwig et al.[33] Chemical analyses showed that ethanol was present in small amounts even under aerobic conditions. Increasing amounts of ethanol added to detached nodules resulted in a progressive decrease in the RQ and, in the presence of 80% oxygen, this ratio approached 0.67. These manometric determinations were complemented by analyses of ethanol after the incubation period. In experiments with cow pea nodules (*Vigna sinensis*), for example, rates of ethanol oxidation of 0.79-1.99 mg/g dry wt/hr were observed. Roots of this species and of red clover (*Trifolium pratense* L.) also oxidized ethanol at rates of 0.25 mg/g dry wt/hr and 0.78 mg/g dry wt/hr.

Although the pathway for ethanol utilization in root nodules has still not been established, it is clear from recent work by van Straten and Schmidt[11] that oxidation to acetaldehyde is probably an initial reaction.

Using gas chromatography, these workers found that acetaldehyde was formed as ethanol was consumed. With longer incubations (2-6 hr), the nodules also produced acetone. When acetaldehyde was supplied, ethanol and acetone were formed, with the latter product assuming importance as incubations were extended to 6 hr. The production of these two carbonyl compounds from ethanol-C^{14} had earlier been reported by Cossins and Turner[23] for pea cotyledons. Thus, the initial stages of ethanol oxidation in plants probably occurs by a pathway common to several species.

The use of C^{14} and H^3-labeled ethanol solutions has shown that stem tissues also oxidize this compound. Besides formation of CO_2, ethanol carbon readily enters into the synthesis of several important cellular constituents. For example, subapical segments of etiolated 8-day-old pea seedlings incubated in 0.1 mM auxin converted ethanol-1-C^{14} and ethanol-2-C^{14} mostly into nonvolatile products which were soluble in 80% ethanol.[34] Both carbons of ethanol also labeled a lipid fraction but C-1 contributed more to respiratory CO_2 than C-2. Light-grown pea seedlings also incorporated ethanol-1-C^{14} into ethanol-soluble compounds in the work of Cossins and Beevers.[12] In this case (Table V), the components of this fraction were fractionated by ion-exchange chromatography. Despite the relatively long period of ethanol metabolism examined in this experiment, only 16% of the label was evolved as CO_2. This emphasizes that ethanol is not completely oxidized in pea shoots but is incorporated into a large number of cellular components.

An interesting paper by Ragland and Hackett[14] reports on the metabolic fates of ethanol labeled in the C-1 position with C^{14} and H^3. The radioactive substrate was supplied to 1-cm stem and hypocotyl sections prepared from etiolated seedlings of pea and mungbean (*Phaseolus aureus* Roxb.), respectively. The main purpose of the experiments was the generation, *in situ*, of tritiated NADH by active alcohol dehydrogenases present in both tissues. It was assumed that this reduced nucleotide would be reoxidized by molecular oxygen to yield tritiated water, whereas H^3-labeled NADPH, generated from glucose-1-C^{14}, H^3, might be more important in biological reductions such as lipid biosynthesis. In both cases, however, the presence of C^{14} allowed analysis of carbon flow into a variety of other metabolic products.

In both tissues, ethanol-1-C^{14}, H^3 was not incorporated to any great extent into the tissue fractions, but a rapid conversion to $C^{14}O_2$ and tritiated H_2O was evident. This strongly suggests that ethanol is converted to acetate (or acetyl CoA), which is then oxidized to CO_2 and H_2O, conceivably by way of the tricarboxylic acid cycle. These data are, therefore, markedly different from those summarized in Table V. However,

Table V. Metabolism of Ethanol-1-C^{14} by 14-Day-Old Pea Seedlings[a]

Fraction	Shoots		Leaves	
	cpm x 10^{-3}	% Incorporated C^{14}	cpm x 10^{-3}	% Incorporated C^{14}
Ethanol-Solubles	450	57	308.0	75
Organic acids	89	11	26.0	6
Acidic amino acids	80	10	26.0	6
Neutral and basic amino acids	26	3	8.4	2
Lipids	100	13	151.0	37
Sugars	57	7	42.0	10
Volatile on acidification	56	7	42.0	10
Residue	214	27	70.0	17
CO_2	125	16	33.0	8
Total C^{14} Incorporated	800		411.0	

[a]Used with permission.[12] Sections of shoots and leaves incubated in darkness at $25°C$ in air for 4 hr.

considering the different physiological condition of the tissues used (etio-lated versus green), these differences are perhaps not surprising.

Leaves

It has been known for some time that leaf tissues readily form ethanol when exposed to anaerobic conditions.[35] In detailed studies of detached rhododendron (*Rhododendron ponticum*) leaves, Bourne and Ranson[36] showed that all of the ethanol accumulated in N_2 was depleted after 220 hr in air. Although some ethanol was collected from the air passing over the leaves, this only represented 60 mg of the 240 mg which were lost by 100 g fresh weight of leaves. In all experiments, rates of ethanol depletion in air were never greater than 40% of the rates of ethanol accumulation in N_2. The authors conclude that ethanol may be oxidatively consumed, and in support of this, quote data of Clarke[37] who supplied ethanol-C^{14} to rhododendron leaf disks in air and found some labeling of intermediates of the tricarboxylic acid cycle.

The data in Table V for pea leaves show that C-1 of ethanol entered into synthesis of several ethanol-soluble compounds. Of these, the lipids were most important, although C^{14} also entered the sugars, organic and amino acids. Only a relatively small percentage of the ethanol utilized was converted into CO_2 even though the tissues were incubated in darkness. As the leaves were obtained from green seedlings grown for 14 days in air, it follows that for this species at least, ability to metabolize ethanol was not related to prior exposure to anaerobic conditions.

Fruits

Much research has been carried out on the aerobic and anaerobic metabolism of apple and other fruits. For example, Thomas[6] kept mature apples under anaerobic conditions followed by long periods in air. Analyses showed no losses of accumulated ethanol or acetaldehyde over a 14-day period. In a detailed comparison of aerobic and anaerobic respiration of apples, Fidler[38] showed that ethanol accumulated during storage in air. It appears, therefore, that apples do not normally consume ethanol at appreciable rates, if at all. This was confirmed by Cossins and Beevers,[12] who showed that very small incorporation of C^{14} occurred when apple slices were incubated with 45 x 10^4 cpm of ethanol-1-C^{14} for 4 hr in air (Table VI).

At the respiratory climacteric, ripening fruits commonly produce ethylene in relatively large amounts.[39] The metabolic origins of this olefin are complex and may vary in different species. In tomatoes, at least, there is very good evidence that ethanol is a precursor of ethylene. In

Table VI. Metabolism of Ethanol-1-C^{14} by Apple Slices[a]

Fraction	cpm x 10^{-3}	% Incorporated C^{14}
Ethanol-Solubles	8.50	95
Organic acids	0.34	4
Acidic amino acids	0.32	4
Neutral and basic amino acids	1.30	15
Lipids	0.38	4
Sugars	0.41	5
Volatile on acidification	5.00	48
Residue	Not active	
CO_2	0.42	5
Total C^{14} Incorporated	8.90	

[a]Used with permission.[12] Slices of apple tissue incubated at $25°C$ in air for 4 hr.

detailed studies of ethylene biosynthesis *in vitro*, Meheriuk and Spencer[40] used a subcellular fraction prepared from ripening tomatoes. Although several compounds supported ethylene formation, the amounts formed when ethanol was added were second only to those collected when propionate was supplied. It is likely that ethanol carbon contributed directly to ethylene production, as Gibson[41] detected labeling of this hydrocarbon when ethanol-2-C^{14} was supplied to *Penicillium digitatum.* There is also good evidence that various intermediates of the tricarboxylic acid cycle act as precursors of ethylene.[39] Thus, Meheriuk and Spencer[40] suggested that ethylene may arise by metabolism of ethanol to acetyl CoA followed by entry into tricarboxylic acid cycle. They also point out that a more direct pathway may also be used. In this regard, Shimokawa and Kasai[42] showed that acetaldehyde-1,2-C^{14} was an effective precursor of ethylene in apple tissue slices and in later *in vitro* studies[43] this conversion was shown to be dependent on flavin mononucleotide.

Storage Tissue Disks

Wetzel[44] reported a rapid oxidation of ethanol solutions added to slices of carrot root. In one experiment, 160 mg of ethanol were consumed when 100 g of slices were placed in 5% ethanol at $37°C$ for 1 hr. He further claimed to have demonstrated an enzymic conversion of ethanol → acetaldehyde → acetic acid, but unfortunately, no exact details of his experimental procedure were given. Some doubts that carrot slices have ability to oxidize ethanol arose when Lowe and James[45] published the results of ethanol-C^{14} feeding experiments. In this work, labeled ethanol was excessively diluted with unlabeled ethanol and supplied to the slices for periods of 16-22 hr. Large amounts of ethanol uptake occurred during this period, but release of $C^{14}O_2$ was insignificant, and paper chromatography failed to detect radioactivity in the organic acids. Although it can be argued that high rates of ethanol metabolism should have been detected in these feeding experiments, it follows that the low specific radioactivity of the ethanol supplied could easily have masked carbon flow into a variety of products.

Cossins and Beevers[12] repeated these ethanol-1-C^{14} feeding experiments but supplied substrate of high specific radioactivity in only micromolar amounts. Their results, given in Tables VII and VIII, showed that the supplied ethanol was rapidly consumed; some was converted to alcohol-soluble material and some to CO_2. Most of the radioactivity in the former materials resided in the organic acids, amino acids and lipids. Table VIII shows that intermediates of the tricarboxylic acid cycle, related amino acids and long-chain fatty acids all derived carbon from ethanol in these experiments.

Barker and el-Saifi[47] showed that potato tubers accumulated large amounts of lactic acid and smaller amounts of ethanol when held under anaerobic conditions. When placed in air after an anaerobic period, the endogenous lactate content declined rapidly and small decreases in ethanol were noted. Work with ethanol-C^{14} has confirmed that this tissue slowly metabolizes ethanol.[12,34] As in other plant tissues, ethanol carbon is widely distributed in such experiments (Table IX), indicating that the initial products are precursors of secondary pathways.

From this survey, it is clear that all tissues examined so far have some ability to metabolize ethanol-C^{14}. Although quantitatively large amounts of ethanol are commonly oxidized when tissues are returned to air after a period of anaerobiosis, this is not a prerequisite for alcohol utilization. Cossins and Beevers[12] compared different tissues for ability to metabolize

Table VII. Metabolism of Ethanol-1-C^{14} by Carrot Slices[a]

Fraction	cpm x 10^{-3}	% Incorporated C^{14}
Ethanol-Solubles	850.5	
Organic acids	176.0	16.0
Amino acids	315.0	30.0
Lipids	56.7	5.0
Sugars	4.6	0.5
Volatile on acidification	297.0	29.0
Residue	117.0	11.0
CO_2	76.0	7.0
Total C^{14} Incorporated	1043.0	

[a]Used with permission.[12] Incubated at 25°C for 4 hr.

Table VIII. Incorporation of C^{14} from Ethanol-1-C^{14} by Carrot Tissue Slices,
Organic, Amino Acid and Lipid Fractions

Fraction	cpm x 10^{-3}	% Incorporated C^{14}
Organic Acids		
Malic	81.00	10.00
Citric	42.00	5.00
Glyoxylic	0.62	0.07
Glycollic	8.60	1.00
Succinic	20.00	3.00
Unidentified peak	4.40	0.50
Lipid		
Nonsaponifiable	3.10	0.40
Saponifiable material	53.50	6.00
Amino Acids		
Glutamic and glutamine	173.10	21.00
Aspartic and asparagine	38.00	4.50
Neutral and basic	25.10	3.00

[a]Used with permission.[12] Incubated at 25°C for 4 hr. Organic acids were separated by ion exchange chromatography and identified by cochromatography with authentic organic acids using n-propanol:ammonia 60:40 v/v; phenol:water 8:3 v/v; n-butanol: acetic acid:water 4:1:5 v/v/v. Amino acids were separated by ion-exchange chromatography. Lipids were saponified by the methods of Newcomb and Stumpf.[46]

ethanol under fairly comparable conditions (Table X). In each case the tissues were sliced to facilitate penetration of the labeled substrate. The slices (0.5 g fresh wt) were then incubated with 15 μmol of ethanol-1-C^{14} (radioactivities are given in Table X) and incubated in 125-ml stoppered Warburg flasks in the dark at 25°C. The only liquid in the flask was the 0.1 ml of ethanol-1-C^{14} solution delivered by microsyringe onto the tissue slices. After killing and extraction, the alcohol-soluble extracts were dried *in vacuo* to remove all residual traces of labeled substrate. Some tissues utilized nearly all the ethanol-1-C^{14} supplied; in corn coleoptiles and pea leaves, 93% of the C^{14} was present in metabolic products. In agreement with Wetzel's earlier data,[44] carrot root tissue also metabolized most of the ethanol supplied. The small amount of ethanol utilized by apple tissue has been noted above, but in this case, additional amounts of ethanol carbon could have been lost as ethylene.

PATHWAYS OF ETHANOL METABOLISM IN PLANTS

Data from ethanol-C^{14} feeding experiments are generally consistent with an initial oxidation to acetaldehyde mediated by alcohol dehydrogenase. The fate of this aldehyde depends, however, on the nature of

Table IX. Metabolism of Ethanol-1-C^{14} by Potato Tuber Slices[a]

Fraction	cpm x 10^{-3}	% Incorporated C^{14}
Ethanol-Solubles	158.0	83.0
Organic acids	34.0	18.0
Acidic amino acids	51.0	27.0
Neutral and basic amino acids	8.4	4.0
Lipids	7.8	4.0
Sugars	1.6	0.8
Volatile on acidification	32.0	16.0
Residue	27.0	14.0
CO_2	4.4	2.0
Total C^{14} Incorporated	189.0	

[a]Used with permission.[12] Incubated at $25°C$ for 4 hr.

intermediary metabolism in the tissue under study. As noted earlier, there is good evidence that apple fruits utilize acetaldehyde as a precursor of ethylene and some details of the reactions involved have been elucidated.[42,43] Aldehyde dehydrogenases are also present in plants and have properties[47] that implicate them in ethanol metabolism. Purified acetaldehyde dehydrogenase from peanut cotyledons[48] has allosteric properties, is NAD-dependent, and forms acetate-C^{14} when labeled substrate is supplied. Oppenheim and Castelfranco[48] did not obtain evidence for acetyl CoA as an intermediate in this oxidation although the enzyme was inhibited by this compound. This latter property might have significance in controlling rates of ethanol oxidation as it follows that acetyl CoA is a logical precursor of other compounds formed from ethanol carbon.[12]

Evidence for acetaldehyde formation from ethanol *in vivo* has also been obtained.[11,23,49] In these cases, however, acetone was also formed, and accumulations of this ketone occurred as the period of ethanol utilization was extended. The mechanism of acetone production in plants remains obscure but could involve an intermediary formation of acetoacetate as in mammals (Scheme 1).

This aspect of ethanol metabolism warrants more detailed study as acetone is commonly among the postclimacteric volatiles of ripening fruits.[50] Evidence for its metabolism is documented for some plant species.[51]

Acetylcoenzyme A, arising from ethanol in pea cotyledons, is quickly incorporated into the tricarboxylic acid cycle.[15] Using C^{14}-labeled ethanol, acetaldehyde and acetate, these workers followed the early sequence of incorporation into individual cycle acids and respiratory CO_2. When the rates of incorporation of ethanol-1-C^{14} and -2-C^{14} were compared (Table XI), it was clear that C-1 was more rapidly converted to

Table X. Percentage of Ethanol-1-C[14] Utilized by Higher Plant Tissues[a]

Tissue	Ethanol-1-C[14] Added (cpm x 10^{-3})	Ethanol Utilized	
		(cpm x 10^{-3})	% Added cpm
Storage Tissues			
Carrot[b]	1200	1043.0	86
Potato	450	189.0	42
Germinating Seedlings			
1-day-old pea cotyledon	850	209.0	24
3-day-old pea cotyledon[c]	675	403.0	60
Castor bean endosperm	450	255.0	57
Fruit Tissues			
Apple	450	8.9	2
Coleoptile Tissues			
Corn	2200	2053.0	93
Shoots, Roots and Leaves			
Pea shoots	850	800.0	93
Pea leaves	850	411.0	49
Corn root tips	850	240.0	28
Corn shoots	850	629.0	74
Corn leaves	850	432.0	51

[a]Used with permission.[12]
[b]Incubated at 25°C in O_2 for 4 hr. All other treatments incubated at 25°C in air for 4 hr.
[c]Slices incubated at 25°C for 1 hr in O_2.

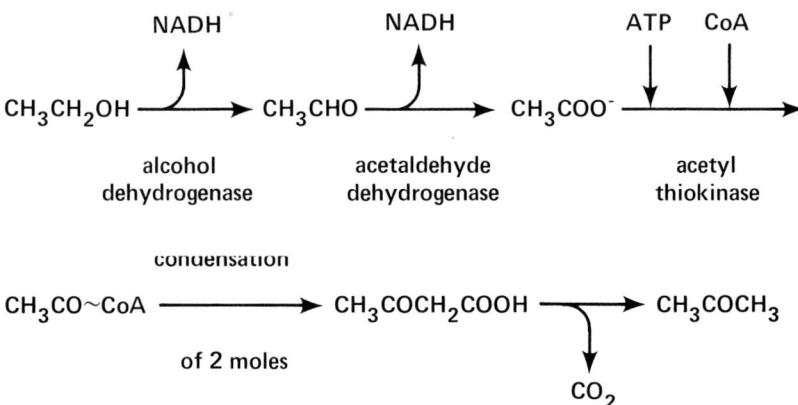

Scheme 1.   Possible reactions for conversion of ethanol to acetyl coenzyme A and acetone.

CO_2 than the methyl carbon.   However, only small amounts of $C^{14}O_2$ were evolved in either case even after 1 hr of ethanol metabolism.   After 5 min, more than 50% of the incorporated label was present in the organic acid fraction and this mainly resided in citrate and isocitrate (Figure 1). As the period of ethanol feeding was extended, the percentage of C^{14} in the organic acids declined and that in the acidic amino acids and amides increased.

The position of acetaldehyde as an intermediate in ethanol utilization (Scheme 2) was confirmed by studying its metabolic fate[15] in parallel feeding experiments (Table XII).   For example, acetaldehyde-C^{14} rapidly labeled the same products as ethanol-C^{14}, and only small amounts of $C^{14}O_2$ evolution occurred.   Chromatography of the organic acid fraction also revealed similarities; citrate and isocitrate were the first heavily labeled products, followed later by succinate and malate.   One difference between these two substrates was a greater incorporation of acetaldehyde into the neutral fraction.   The percentage of incorporated C^{14} in this fraction declined with time, which suggests that it contained an early product of acetaldehyde metabolism.   Paper chromatography revealed that all the C^{14} in the fraction migrated with authentic acetoin (3-hydroxybutan-2-one) in a variety of different solvent systems.   Acetoin is formed in yeast by condensation of acetaldehyde-thiamine pyrophosphate and another molecule of acetaldehyde.   So, it is likely that some of the labeled substrate was handled by pea cotyledons in this way.

Table XI. Sequence of Incorporation of C^{14} into Products of Ethanol-1-C^{14} and Ethanol-2-C^{14} Metabolism[a]

	% Incorporated C^{14}							
Time of Incubation (min)	5		15		30		60	
Fraction	Ethanol 1-C^{14}	Ethanol 2-C^{14}	Ethanol 1-C^{14}	Ethanol 2-C^{14}	Ethanol 1-C^{14}	Ethanol 2-C^{14}	Ethanol 1-C^{14}	Ethanol 2-C^{14}
Lipids	3.2	4.5	3.3	4.4	3.7	5.1	4.4	6.4
Neutral Compounds	6.4	5.9	5.8	5.5	3.6	3.9	3.6	3.9
Acidic Amino Acids								
Glutamate	22.8	17.5	36.6	30.4	39.7	34.8	33.2	28.0
Aspartate	7.4	5.7	8.3	3.7	9.3	4.6	8.5	5.9
Amides	1.5	1.6	2.2	3.6	4.4	7.6	5.6	8.7
Neutral and Basic Amino Acids	3.3	3.9	3.2	4.0	2.9	2.2	3.4	3.0
Organic Acids	52.8	54.3	38.2	45.1	32.2	37.9	32.1	36.7
Carbon Dioxide	0.5	0.2	0.9	0.1	1.8	0.1	5.2	0.3
Insoluble Residue	2.1	6.4	1.5	3.2	2.4	3.8	4.0	7.1
Total C^{14} Incorporated (counts/min)	90300	64400	294700	192600	688000	452300	992500	829200

[a]Used with permission.[15] Slices (0.5 g) of pea cotyledons were incubated at $30°C$ with 50 μmol of phosphate buffer, pH 5.5, and 1 μmol of ethanol (containing 5 μc of C^{14}) in a total volume of 0.6 ml.

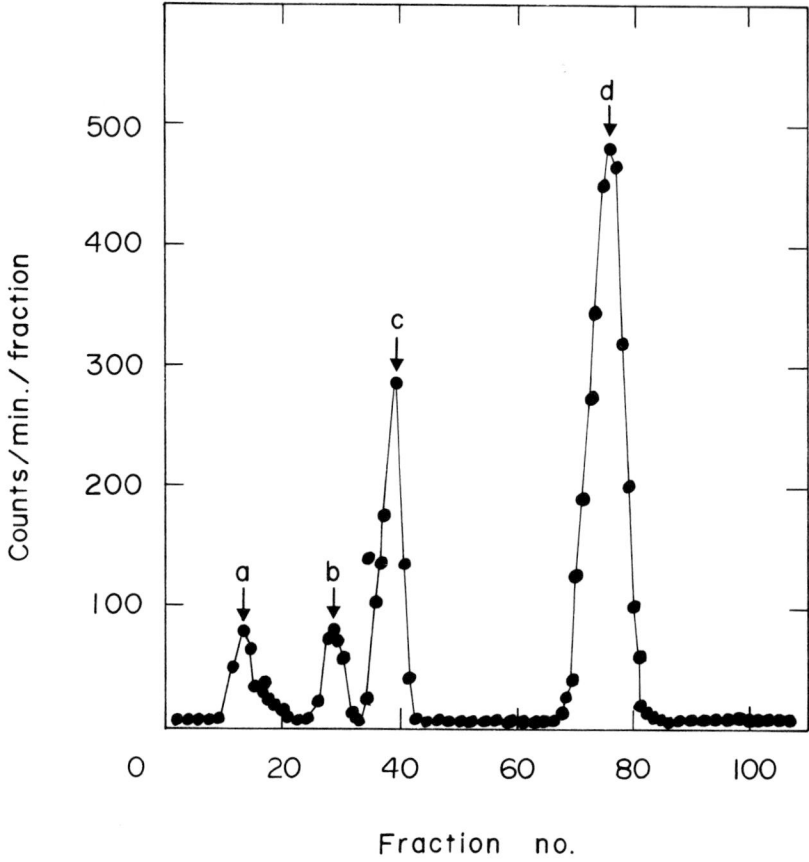

Figure 1. Organic acids labeled after incubation with ethanol-1-C^{14} for 5 min. a. Lactic acid; b. succinic acid; c. malic acid; d. citric acid and isocitric acid. Used with permission.[15]

Further evidence for the reactions of Scheme 2 was derived by Cameron and Cossins[15] when acetate-C^{14} was supplied to pea cotyledon slices (Table XIII, Figure 2). The data were similar to those obtained with ethanol-C^{14} and acetaldehyde-C^{14}, thus emphasizing the close metabolic relationships of these C_2 compounds in this tissue. For example, acetate carbon rapidly entered citrate, isocitrate and other intermediates of the tricarboxylic acid cycle (Table XIII). In experiments of longer duration (Figures 3 and 4), evidence was obtained for carbon flow from the organic acids into the acidic amino acids (mainly glutamate) and amides. Carbon dioxide was evolved by the tissues but contained very small amounts of C^{14} at all times.

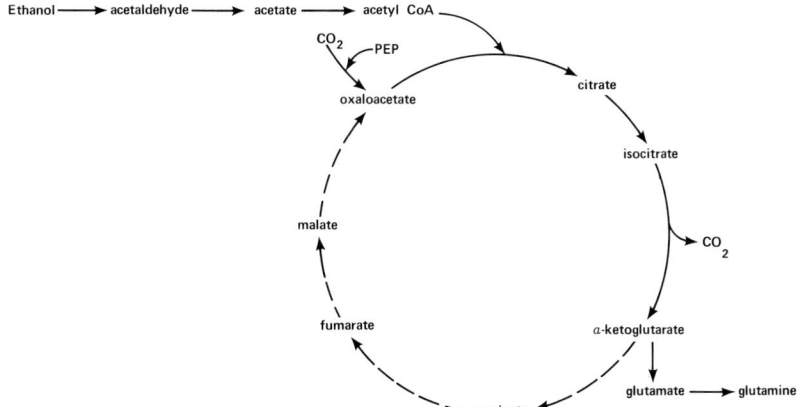

Scheme 2. The major pathway of ethanol metabolism in pea cotyledons. Note: Solid arrows denote the predominant flow of ethanol carbon. Some recycling occurs via succinate and malate as is evident from data in Figures 1 and 2.

Table XII. Sequence of Incorporation of C^{14} into Products of Acetaldehyde-1,2-C^{14} Metabolism[a]

Fraction	Time of Incubation (min)	% Incorporated C^{14}		
		5	15	30	60
Lipids		3.7	2.7	2.0	1.6
Neutral Compounds		24.4	16.7	12.1	9.4
Acidic Amino Acids					
Glutamate		15.3	19.6	32.4	23.9
Aspartate		2.5	2.5	4.5	5.8
Amides		1.2	4.7	6.3	8.9
Neutral and Basic Amino Acids		7.0	7.0	3.4	4.4
Organic Acids		39.7	39.3	31.8	35.2
Carbon Dioxide		0.8	1.9	2.4	3.6
Insoluble Residue		5.4	5.6	5.1	7.2
Total C^{14} Incorporated (counts/min)		24200	48400	114600	207000

[a]Used with permission.[15] Slices (0.5 g) of pea cotyledons were incubated at 30°C with 50 μmol of phosphate buffer, pH 5.5, and 2.5 μmol of acetaldehyde (containing 2.5 μc of C^{14}) in a total volume of 0.6 ml.

Table XIII. Labeled Products Isolated after Short Periods of Acetate-1-C^{14} Metabolism[a]

Fraction	Time of Incubation (min) % Incorporated C^{14}			
	1	2	3	5
Lipids	0.9	1.3	2.3	2.3
Neutral Compounds	1.7	1.7	1.6	2.4
Glutamate	1.3	1.7	3.0	7.3
Glutamine	0.4	0.5	0.5	0.7
Organic Acids				
Citrate + isocitrate	52.7	52.6	54.0	54.2
Succinate	21.0	21.9	22.2	15.6
Malate	6.5	5.6	3.8	5.8
Other organic acids	15.5	14.7	12.6	11.7
Total C^{14} Incorporated				
(counts/min)	110600	121900	134300	146100

[a]Used with permission.[15] Slices (0.5 g) of pea cotyledons were incubated at $30^{\circ}C$ with 50 μmol of phosphate buffer, pH 5.5, and 1 μmol of sodium acetate-1-C^{14} (containing 5 μc of C^{14}) in a total volume of 0.6 ml.

In other related experiments, these workers showed that pea cotyledons also utilized C^{14}-labeled lactate, isocitrate, a-ketoglutarate, and glutamate forming products which were consistent with metabolism centered on an active tricarboxylic acid cycle. A number of key enzymes necessary to mediate these interconversions were also detected in the cotyledons. However, the small yields of $C^{14}O_2$ derived from ethanol, acetaldehyde and acetate argue against extensive oxidation of these compounds. Rather, it appears that ethanol carbon is extensively converted to glutamate and glutamine and, in this sense, the tricarboxylic acid cycle in pea cotyledons must function to support such biosynthesis. Degradation of glutamate-C^{14} showed that withdrawal of carbon from the cycle was appreciable, and even after 1 hr of ethanol metabolism it was clear that little cycling of C^{14} had occurred. However, Cameron and Cossins[15] found that the total organic acid pool of pea cotyledons remained relatively stable during this period of ethanol metabolism. It follows, therefore, that conversion of ethanol → a-ketoglutarate → glutamate did not occur at the expense of the organic acids.

To account for this apparent anomaly, Cameron and Cossins[15] showed that oxaloacetate, needed to sustain generation of a-ketoglutarate, could be generated by transamination of aspartate and by dark CO_2 fixation. Thus, the pathway of ethanol metabolism in pea cotyledons is fairly complex (Scheme 2) and produces glutamate and glutamine via the partial reactions of the tricarboxylic acid cycle. Both these end products would have importance in the nitrogen metabolism of the developing seedling.

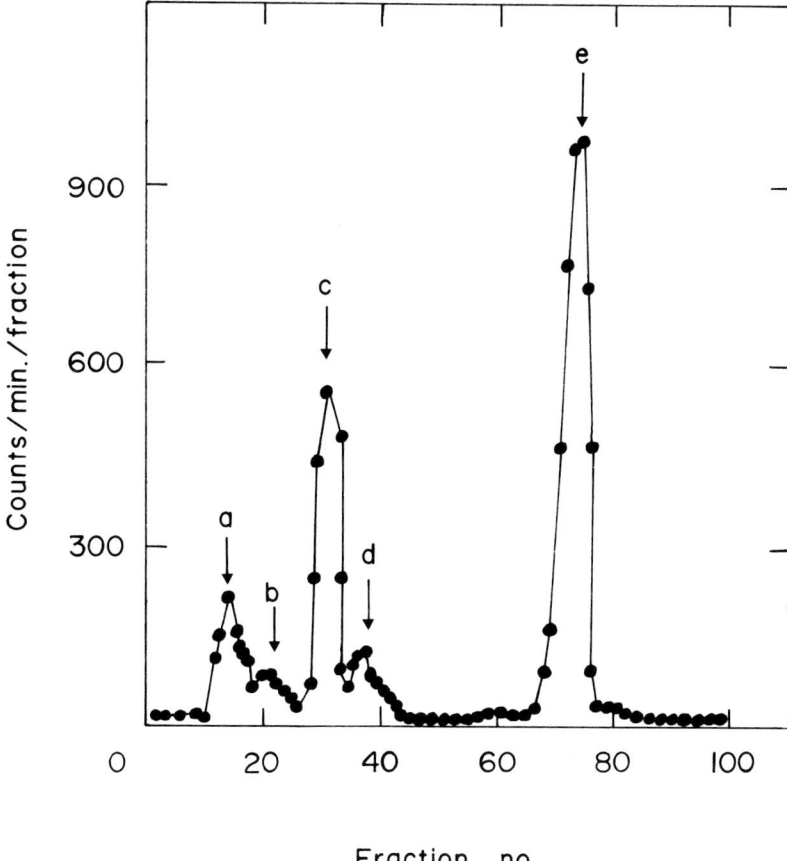

Figure 2. Organic acids labeled after incubation with acetate-1-C^{14} for 1 min. a. Lactic acid; b. glycollic acid; c. succinic acid; d. malic acid; e. citric acid and isocitric acid. Used with permission.[15]

The kinetics of ethanol metabolism in castor bean endosperm slices have been examined by Peterson and Cossins[27] and were found to be markedly different from pea cotyledons. During germination, fats stored in the endosperm are rapidly converted to acetyl CoA, which is metabolized via the glyoxylate cycle.[26] Oxaloacetate, arising from this sequence, is then used to support sucrose biosynthesis. It is clear from the work of Peterson and Cossins[27] that this is also the major route for ethanol utilization in this tissue. Time course experiments with ethanol-2-C^{14} supplied to 7-day-old endosperm showed heavy labeling of the organic

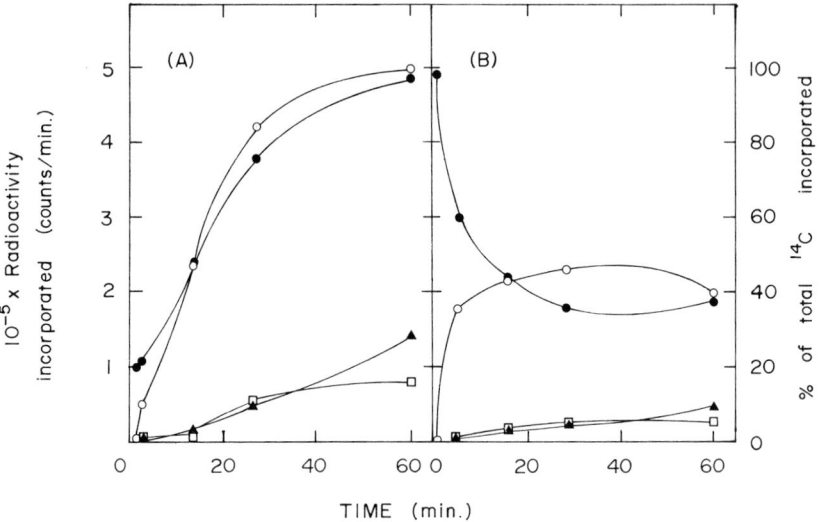

Figure 3.   Incorporation of acetate-1-C^{14} by pea cotyledons. (A) Incorporation of C^{14} into individual fractions; **(B)** Incorporation of C^{14} expressed as a percentage of total recovered in the isolated fractions: o, acidic amino acids; •, organic acids; ▲, amides; □, carbon dioxide.   Used with permission.[15]

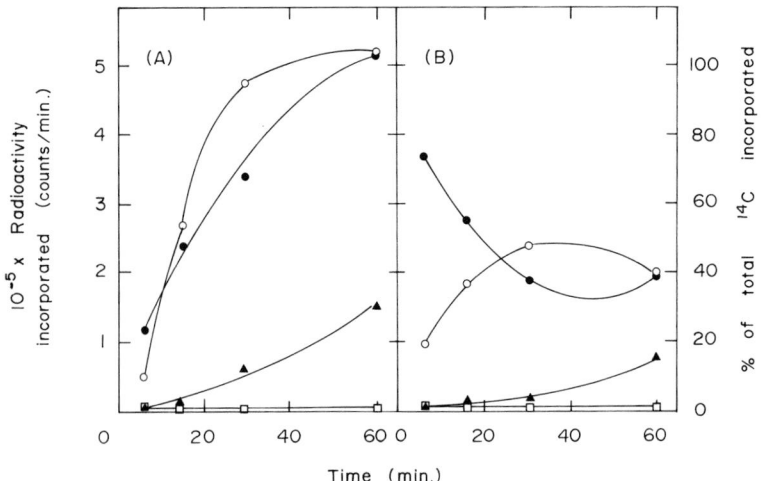

Figure 4.   Incorporation of acetate-2-C^{14} by pea cotyledons. (A) Incorporation of C^{14} into individual fractions; **(B)** Incorporation of C^{14} expressed as a percentage of total recovered in the isolated fractions: o, acidic amino acids; •, organic acids; ▲, amides; □, carbon dioxide.   Used with permission.[15]

acids within 60 sec, followed by a gradual accumulation of the label in the sugar fraction. At 5 min, the distribution of C^{14} in the organic acids was malate 67%, citrate 14%, succinate 6%, glycollate 2% and other acids 11%. When longer periods of ethanol-2-C^{14} feeding were examined, the percentage of C^{14} incorporated into the organic acids continued to decline (Figure 5b), while those for the amino acids and sugars rose. In this connection the total C^{14} in the organic acids reached a plateau after 30 min of ethanol-2-C^{14} metabolism, but C^{14} was still accumulating in the sugars after 1 hr at which point it was the chief respository of ethanol carbon (Figure 5a). Paper chromatography showed that sucrose accounted for the bulk of the labeling. These data show that ethanol is rapidly metabolized by a sequence that involved organic acids as primary products and leads to sugar synthesis as the period of metabolism is extended.

Unlike pea cotyledons, castor bean endosperm converted appreciable amounts of ethanol-1-C^{14} to $C^{14}O_2$. This production, however, only occurred after the C^{14} content of the organic acids leveled off. Furthermore, the contributions of the methyl and carbinol carbons of ethanol to respiratory CO_2 were markedly different, as shown in Figure 6. The individual carbons of ethanol were also incorporated into the sugar fraction in widely differing amounts (Table XIV). When unlabeled glyoxylate was added to the tissue prior to ethanol-2-C^{14}, the total incorporation of label was doubled with malate-C^{14}, $C^{14}O_2$ and sucrose-C^{14}, accounting for much of the increase (Table XV). It is clear that in many respects the kinetics of ethanol utilization parallel those of acetate metabolism in this tissue[52] and suggest a central role of the glyoxylate cycle (Scheme 3). However, incorporation of ethanol-1-C^{14} into CO_2 and into sugars was not equal (Table XIV) as would be expected if all ethanol carbon flowed through this cycle en route to sucrose synthesis. This additional CO_2 could be produced by some operation of the tricarboxylic acid cycle. Partially supporting this, Peterson and Cossins[27] found that glutamate-C^{14} was an important constituent of the acidic amino acid fraction even after relatively brief periods of ethanol metabolism. It seems possible, then, that some isocitrate carbon was converted to a-ketoglutarate in addition to that subject to isocitrate lyase activity.

PATHWAYS OF ETHANOL METABOLISM IN ANIMALS AND MICROORGANISMS

This discussion of alcohol metabolism would not be complete without some reference to the extensive work on this subject in other organisms. Considering the socioeconomic importance of ethanol and of alcoholism

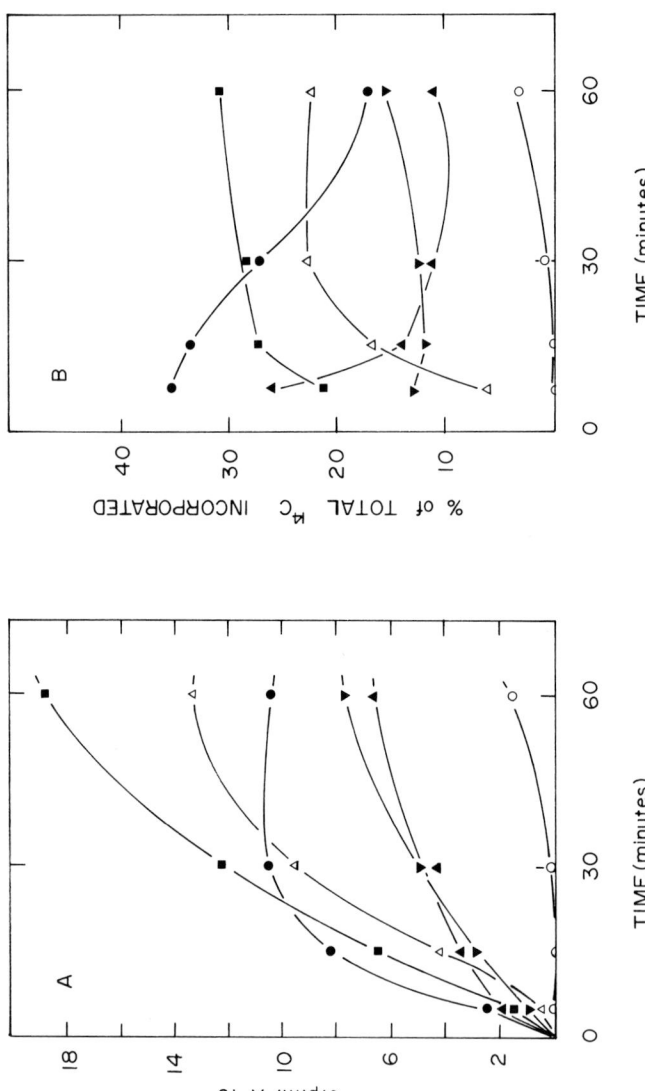

Figure 5. A. Incorporation of ethanol-2-C^{14} into various fractions of endosperm slices when incubated at 25°C: ■, sugars; △, neutral and basic amino acids; ●, organic acids; ▼, lipids; ▲, acidic amino acids; o, carbon dioxide; B. Percentage of incorporated C^{14} in various fractions of endosperm slices when incubated with ethanol-2-C^{14} at 25°C: ■, sugars; △, neutral and basic amino acids; ●, organic acids; ▼, lipids; ▲, acidic amino acids; o, carbon dioxide. Used with permission.[27]

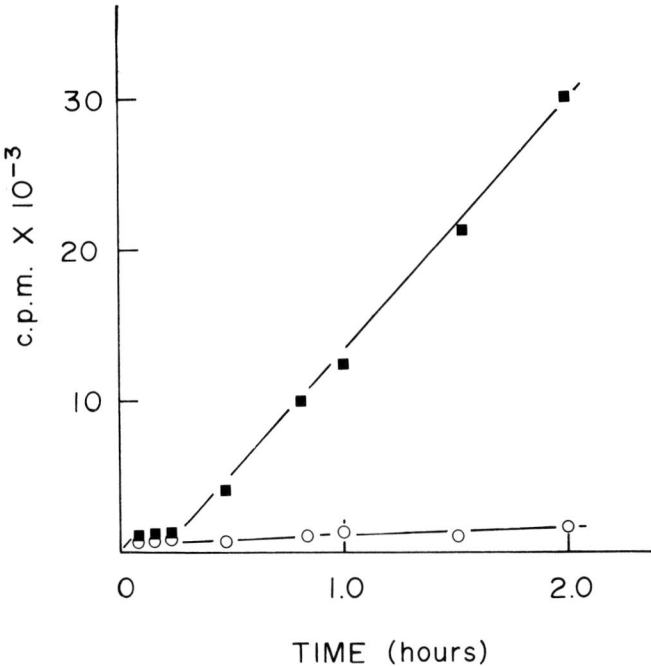

Figure 6. Incorporation of ethanol-1-C^{14} and ethanol-2-C^{14} into the respiratory carbon dioxide of endosperm slices. One gram of endosperm slice was incubated at $25^\circ C$ with equal specific radioactivities (2 $\mu C/0.32$ μmol) of ethanol-1-C^{14} (■) and ethanol-2-C^{14} (o). Used with permission.[27]

Table XIV. Utilization of Ethanol-1-C^{14} and Ethanol-2-C^{14} by
Castor Bean Endosperm Slices[a]

Fraction	% Incorporated C^{14}	
	Ethanol-1-C^{14}	Ethanol-2-C^{14}
Carbon Dioxide	49	4
Lipids	12	13
Sugars	4	35
Organic Acids	14	13
Neutral and Basic Amino Acids	11	26
Acidic Amino Acids	10	9

[a]Used with permission.[27] Slices of 7-day-old castor bean endosperm were incubated for 2 hr at $30^\circ C$ with ethanol-1-C^{14} and ethanol-2-C^{14} solutions of equal specific radioactivities (2 $\mu c/0.32$ μmol).

Table XV. Effects of Glyoxylate on the Utilization of Ethanol-2-C^{14} by Castor Bean Endosperm Slices[a]

| Fraction | ^{14}C in Products Isolated, cpm x 10^{-3} | |
	Ethanol-2-C^{14}	Ethanol-2-C^{14} plus Glyoxylate
Carbon Dioxide	11.2	33.0
Lipids	10.7	10.9
Sugars	14.0	42.6
Organic Acids		
Glycollate	2.5	3.0
Succinate	2.8	4.0
Malate	4.9	14.0
Citrate	0.4	4.3
Others	5.4	5.2
Amino Acids		
Neutral and basic	12.6	14.3
Acidic	4.3	5.8
Total	68.8	137.1

[a]Used with permission.[27] Slices of 7-day-old endosperm were preincubated with 50 μmol phosphate buffer (pH 5.5) and 10 μmol of glyoxylate as indicated, for 30 min at 30°C. Ethanol-2-C^{14} (1 μmol, 5 μc of C^{14}) was then added, and the incubation continued for a further 30 min. Total volume, 0.7 ml.

in man, it is not surprising that considerable research effort is being directed towards a better understanding of this topic in mammals. The reader interested in the biochemical or medical aspects of such work is referred to *Medical Consequences of Alcoholism*, edited by Seixas, Williams and Eggleston.[53] Excellent reviews of a less specialized nature also have appeared.[54,55]

In mammals, ingested ethanol is mainly oxidized in the liver by a cytostolic alcohol dehydrogenase to form acetaldehyde. Hepatic aldehyde dehydrogenases convert this intermediate to acetate, which may be incorporated into fats within this organ or released into the hepatic venous bloodstream to be oxidized finally via the tricarboxylic acid cycle in peripheral tissues. A microsomal ethanol oxidizing system which requires NADPH (Equation 1) has been studied in mammals, and there is also excellent evidence for a peroxidatic reaction (Equation 2), both of which form acetaldehyde from ethanol by reactions independent of alcohol dehydrogenase.[56]

$$CH_3CH_2OH + NADPH + H^+ + O_2 \xrightarrow{\text{microsomes}} CH_3CHO + NADP^+ + 2H_2O \qquad (1)$$

$$CH_3CH_2OH + H_2O_2 \xrightarrow{\text{catalase}} 2H_2O + CH_3CHO \qquad (2)$$

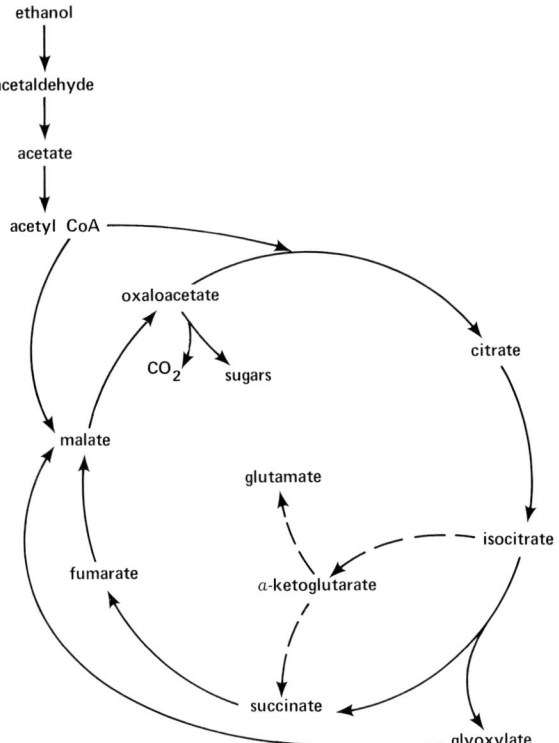

Scheme 3. Metabolism of ethanol via the glyoxylate cycle in castor bean endosperm. A minor flow of ethanol carbon denoted by broken arrows occurs via α-ketoglutarate as indicated by glutamate labeling.

Several of the effects of ethanol on hepatic intermediary metabolism can be traced to alteration of liver NADH/NAD ratios, which accompanies the conversion of ethanol to acetate.

Danforth and his colleagues have carried out extensive work on ethanol metabolism in *Euglena gracilis* var. *bacillaris*.[57] When ethanol is supplied as a sole source of energy and carbon, 0.97 mol of O_2 are taken up and 0.29 mol of CO_2 are produced per mol of ethanol consumed. This stoichiometry suggests that ethanol is not exclusively metabolized through the glyoxylate cycle, although key enzymes of this sequence are found in the cells. In this connection, Heinrich and Cook[58] found that ethanol triggered synthesis of malate synthetase in *Euglena* cultures and concluded that such growth conditions cause derepression of the glyoxylate cycle.

More recently, Morosoli and Bégin-Heick[59] have examined ethanol-C^{14}
metabolism in *Astasia longa* grown in 5% CO_2 and differing levels of O_2.
High oxygen concentrations resulted in rapid ethanol oxidation to acetate,
which accumulated in the medium. This acetate was rapidly utilized via the
glyoxylate cycle if the cells were then incubated in air containing 5% CO_2.
A comparison of acetate-C^{14} and ethanol-C^{14} feedings experiments showed
that 27% and 2% respectively of these substrates were metabolized via the
tricarboxylic acid cycle. Thus, Danforth's suggestion[57] that this cycle is
not operative during ethanol metabolism appeared valid for *Astasia.*

In *Rhizopus nigricans*, aerobic growth on ethanol-C^{14} is accompanied by
heavy labeling of fumarate.[60,61] Ethanol-1-C^{14} was incorporated in the
carboxyl carbons of this product. By contrast, ethanol-2-C^{14} labeled all
four positions of the fumarate molecule. Foster and Carson[61] suggested
that ethanol gave rise to C_2 units which, by the Thunberg reaction, gave
rise to succinate. Fumarate was formed by dehydrogenation of this intermediate. However, Kornberg,[62] in a review of this and related data for
acetate utilization, shows that these labeling patterns could arise if ethanol
carbon was metabolized via the tricarboxylic acid and glyoxylate cycles.
This latter cycle is, in fact, the major route for biosynthesis of cellular
constituents when *Escherichia coli* and *Pseudomonas* are cultured on media
containing ethanol as a sole source of carbon.[63]

In *Streptococcus faecalis*, 10Cl growth on ethanol occurs under anaerobic conditions, and ethanol-1-C^{14} is almost exclusively incorporated into
lipids.[64] After such culture, cell-free extracts were found to contain
alcohol dehydrogenase and a CoA-dependent aldehyde dehydrogenase.
Synthesis of this latter enzyme was repressed by acetate and lipoate, so
a regulation of ethanol consumption can be visualized when requirements
for lipid biosynthesis are met.

Dunstan *et al.*[65,66] have presented detailed evidence for operation of
a novel pathway during ethanol metabolism in *Pseudomonas* AM1. Although ethanol was converted to acetate, this intermediate was, to some
extent, converted to glycollate. Evidence for this hydroxylation reaction
was obtained by use of mutants whose growth was stimulated when ethanol and glycollate were both present in the growth medium. Attempts to
demonstrate this reaction in cell-free extracts of the wild type were not
successful, but the results of detailed C^{14} feeding experiments indicated
its operation *in vivo*. As the cells also converted glycollate to glyoxylate
and contained high levels of malate synthetase, ethanol carbon would tend
to label all four carbons of malate. Further clarification of this pathway
appears desirable, however, as an unidentified intermediate of ethanol and
acetate metabolism was reported. This compound was not an amino
or common carboxylic acid, but the kinetics of its labeling in ethanol-C^{14}

experiments clearly implicate it as an intermediate in the pathway of alcohol utilization.

FUTURE WORK ON ETHANOL UTILIZATION IN PLANTS

There are still several aspects of ethanol metabolism in plants that require elucidation. One of basic interest to plant physiologists would be a description of the factors that control the flow of ethanol carbon through the metabolic pathways described above. These factors are conceivably complex, but, as in animal tissues, would logically include factors that modulate the levels of alcohol dehydrogenase.

Some investigators have commenced study of this problem. The reported induction of enzyme activity by acetaldehyde[67,68] could clearly affect rates of ethanol accumulation. On the other hand, induction of this enzyme by ethanol[69,70] could effectively increase rates of ethanol oxidation. However, many plant alcohol dehydrogenases exist as separate isoenzymes (see Chapter 16), which could have differing substrate specificities. It is thus of some importance to identify possible changes in isoenzyme complement during induction by the oxidized and reduced substrates of this reaction.

Crawford and his colleagues have reported some interesting biochemical differences between species that are tolerant and those intolerant to flooding[71] (see Chapter 4). The alcohol dehydrogenases of species able to tolerate flooding were not induced by acetaldehyde and, therefore, their tendency to accumulate ethanol would remain relatively small. By contrast, intolerant species contained alcohol dehydrogenases which were induced by acetaldehyde and displayed a greater affinity for this oxidized substrate as a result of flooding. McManmon and Crawford,[71] therefore, distinguished between tolerant and intolerant species largely by their tendencies to accumulate ethanol, but possible differences in rates of ethanol oxidation have not been considered. It would also be interesting to determine the actual *in vivo* rates of ethanol accumulation in these ecologically distinct species and to characterize the isoenzymes of alcohol dehydrogenase which are induced in roots subjected to flooding.

Other reactions along the pathway for ethanol utilization may also be regulated. The allosteric nature of acetaldehyde dehydrogenase in higher plants[48] implies that effectors exist for controlling the rates of acetaldehyde oxidation and, hence, of ethanol breakdown. In animals, the levels of several common metabolites can stimulate ethanol oxidation.[72] In most instances, this can be traced to an associated increase in the capacity of the mitochondrial respiratory chain to oxidize NADH arising from ethanol. The pools of several metabolites fluctuate when seedlings are transferred from anoxia to air[28] and the NADH/NAD ratio rises as pea seedlings are utilizing ethanol.[73] Conceivably, rates of ethanol oxidation in plant tissues may be regulated by such factors, but detailed investigations will be necessary to elucidate this fairly complex aspect of alcohol utilization.

In animal tissues, ethanol oxidation, as noted earlier, is partially catalyzed by reactions independent of alcohol dehydrogenase. There is some evidence that the peroxidatic process would have physiological significance at high ethanol concentrations.[74] As catalase is ubiquitous to higher plants, it follows that some ethanol carbon might be oxidized by this route. If this did occur *in vivo*, it would have the advantage of allowing partial oxidation without accompanying NADH production and, hence, the need for oxygen uptake.

ACKNOWLEDGMENTS

The author acknowledges Mr. Benedict Blawacky's assistance in the translation of several German publications cited in this chapter. Mrs. Elizabeth Ford kindly helped in research of the literature and typed the final manuscript.

REFERENCES

1. LeChartier, G. and F. Bellamy. "Etude sur le gaz produit par les fruits," *C. R. Acad. Sci., Paris* 69:356-360 (1869).
2. LeChartier, G. and F. Bellamy. "De la fermentation des fruits," *C. R. Acad. Sci., Paris* 75:1203-1206 (1872).
3. LeChartier, G. and F. Bellamy. "De la fermentation des fruits," *C. R. Acad. Sci., Paris* 79:1006-1009 (1874).
4. Pfeffer, W. *The Physiology of Plants*, English Translation (Oxford: Oxford University Press, 1910).
5. Kostychev, S. P. *Kostychev's Plant Respiration*, C. J. Lyon, Ed. and translator (Philadelphia: Blackiston Son and Comp., 1927).
6. Thomas, M. "The Controlling Influence of Carbon Dioxide. V. A. Quantitative Study of the Production of Ethyl Alcohol and Acetaldehyde by Cells of Higher Plants." *Biochem. J.* 19:927-947 (1925).
7. Beevers, H. *Respiratory Metabolism in Plants* (New York: Row, Peterson and Co., 1961), pp. 13-44.
8. Fidler, J. C. "Studies in Zymasis. IV. The Accumulation of Zymasic Products in Apples During Senescence," *Biochem. J.* 27:1614-1621 (1933).
9. Neish, A. C. "Chemical Procedures for Separation and Determination of Alcohols" in *Methods in Enzymology*, S. P. Colowick and N. O. Kaplan, Eds., Vol. III (New York: Academic Press, 1957), pp. 255-263.
10. Kaplan, N. O. and M. M. Ciotti. "Enzymatic Determination of Ethanol," *Methods in Enzymology*, S. P. Colowick and N. O. Kaplan, Eds., Vol. III (New York: Academic Press, 1957), pp. 253-255.
11. van Straten, J. and E. L. Schmidt. "Volatile Compounds Produced During Acetylene Reduction by Detached Soybean Nodules," *Soil Biol. Biochem.* 6:347-351 (1974).

12. Cossins, E. A. and H. Beevers. "Ethanol Metabolism in Plant Tissues," *Plant Physiol.* 38:375-380 (1963).
13. Liu, T., A. Oppenheim and P. Castelfranco. "Ethyl Alcohol Metabolism in Leguminous Seedlings," *Plant Physiol.* 40:1261-1268 (1965).
14. Ragland, T. E. and D. P. Hackett. "Radioactive Tracer Studies of the Metabolic Fates of Intracellularly Generated NADH and NADPH in Higher Plant Tissues," *Plant Physiol.* 40:1191-1197 (1965).
15. Cameron, D. S. and E. A. Cossins. "Studies of Intermediary Metabolism in Germinating Pea Cotyledons. The Pathway of Ethanol Metabolism and the Role of the Tricarboxylic Acid Cycle." *Biochem. J.* 105:323-331 (1967).
16. Sawa, S. "Can Alcohols of the Methane Series be Utilized as Nutrients by Green Plants?" *Bull. Coll. Agric. Tokyo Imp. Univ.* 5:247-251 (1903).
17. Nagai, I. "Some Studies on the Germination of the Seed of *Oryza sativa*," *J. Coll. Agric. Tokyo Imp. Univ.* 5:109-159 (1916).
18. Taylor, D. L. "Influence of Oxygen Tension on Respiration, Fermentation and Growth in Wheat and Rice," *Am. J. Bot.* 29:721-738 (1942).
19. Phillips, J. W. "Studies on Fermentation in Rice and Barley," *Am. J. Bot.* 34:62-72 (1947).
20. Cossins, E. A. and E. R. Turner. "Losses of Alcohol and Alcohol Dehydrogenase Activity in Germinating Seeds," *Ann. Bot.* 26:591-597 (1962).
21. Bugajewsky, M. F. "Alkoholverbrauch bei der Atmung des Erbensamens," *Biochem. Zeitschr.* 238:60-66 (1931).
22. Cossins, E. A. and E. R. Turner. "Utilization of Alcohol in Germinating Pea Seedlings," *Nature* 183:1599-1600 (1959).
23. Cossins, E. A. and E. R. Turner. "The Metabolism of Ethanol in Germinating Pea Seedlings," *J. Exp. Bot.* 14:290-298 (1963).
24. Leblová, S., I. Zimáková, D. Sofrová and J. Barthová. "Occurrence of Ethanol in Pea Plants in the Course of Growth Under Normal and Anaerobic Conditions," *Biol. Plant.* 11:417-423 (1969).
25. Cossins, E. A. "Metabolism of Ethanol by Higher Plant Tissues," Ph.D. Thesis, University of London (1961).
26. Beevers, H. "Metabolic Production of Sucrose from Fat," *Nature* 191:433-436 (1961).
27. Peterson, C. A. and E. A. Cossins. "Participation of the Glyoxylate Cycle in the Metabolism of Ethanol by Castor Bean Endosperm Tissues," *Can J. Biochem.* 44:423-432 (1966).
28. Effer, W. R. and S. L. Ranson. "Respiratory Metabolism in Buckwheat Seedlings," *Plant Physiol.* 42:1042-1052 (1967).
29. Ruhland, W. and K. Ramshorn. "Aërobe Gärung in aktiven pflanzlichen Meristemen," *Planta* 28:471-514 (1938).
30. Ramshorn, K. "Zur partiellen 'aeroben' Gärung in der Wurzel von *Vicia faba*," *Flora* 145:1-36 (1957).
31. Betz, A. "Der Athanolumsatz in meristematischen Wurzelspitzen von *Pisum sativum*," *Flora* 146:532-545 (1958).
32. Betz, A. "Die aerobe Gärung in aktiven Meristemen hoherer Pflanzen," in *Encyclopedia of Plant Physiology*, W. Ruhland, Ed. Vol. XII (Berlin: Springer Verlag, 1960), pp. 88-113.

33. Ludwig, C. A., F. E. Allison, S. R. Hoover and F. W. Minor. "Biochemical Nitrogen Fixation Studies. III. Production and Oxidation of Ethyl Alcohol by Legume Nodules," *Bot. Gazette* 102:417-436 (1941).
34. Castelfranco, P., R. Bianchetti and E. Marré. "Difference in the Metabolic Fate of Acetate and Ethanol Fed to Higher Plant Tissues," *Nature* 198:1321-1322 (1963).
35. James, W. O. *Plant Respiration* (Oxford: Oxford University Press, 1953).
36. Bourne, D. T. and S. L. Ranson. "Respiratory Metabolism in Detached Rhododendron Leaves," *Plant Physiol.* 40:1178-1190 (1965).
37. Clarke, I. D. "An Investigation into the Possible Oxidative Consumption of Ethanol and Acetaldehyde by Higher Plants," Ph.D. Thesis, University Library, Newcastle upon Tyne (1957).
38. Fidler, J. C. "A Comparison of the Aerobic and Anaerobic Respiration of Apples," *J. Exp. Bot.* 2:41-64 (1951).
39. Spencer, M. "Ethylene in Nature," in *Progress in the Chemistry of Organic Natural Products,* I. Zechmeister, Ed., Vol. XXVII (Berlin: Springer Verlag, 1969), pp. 31-80.
40. Meheriuk, M. and M. Spencer. "Studies on Ethylene Production by a Subcellular Fraction from Ripening Tomatoes. I. Effects of Several Substrates, Cofactors and Cations," *Phytochem.* 6:535-543 (1967).
41. Gibson, M. "The Biogenesis of Ethylene," Ph.D. Thesis, Purdue University (1963).
42. Shimokawa, K. and Z. Kasai. "Biogenesis of Ethylene in Apple Tissue. I. Formation of Ethylene from Glucose, Acetate, Pyruvate, and Acetaldehyde in Apple Tissue," *Plant Cell Physiol.* 7:1-9 (1966).
43. Shimokawa, K. and Z. Kasai. "Ethylene Formation from Pyruvate by Subcellular Particles of Apple Tissue," *Plant Cell Physiol.* 8:227-230 (1967).
44. Wetzel, K. "Zur Physiologie der anaeroben Atmung höhere Pflanzen," *Ber. Dtsch. Bot. Ges.* 51:46-51 (1933).
45. Lowe, C. and W. O. James. "Carrot Tissue and Ethanol," *New Phytol.* 59:288-293 (1960).
46. Newcomb, E. H. and P. K. Stumpf. "Fat Metabolism in Higher Plants. I. Biogenesis of Higher Fatty Acids by Slices of Peanut Cotyledons *In Vitro*," *J. Biol. Chem.* 200:233-239 (1953).
47. Barker, J. and A. F. el-Saifi. "Studies in the Respiratory and Carbohydrate Metabolism of Plant Tissues. III. Experimental Studies of the Formation of CO_2 and Changes in Lactic Acid and Other Products in Potato Tubers in Air Following Anaerobic Conditions," *Proc. Roy. Soc. (London)* 140:Ser. B, 508-522 (1953).
48. Oppenheim, A. and P. A. Castelfranco. "An Acetaldehyde Dehydrogenase from Germinating Seeds," *Plant Physiol.* 42:125-132 (1967).
49. Cossins, E. A. "Utilization of Ethanol-2-C^{14} by Pea Slices," *Nature* 194:1095-1096 (1962).
50. Ulrich, R. "Post Harvest Physiology of Fruits," *Ann. Rev. Plant Physiol.* 9:385-416 (1958).
51. Cossins, E. A. "The Metabolism of [1,3-^{14}C] acetone by Higher Plant Tissues," *New Phytol.* 63:28-33 (1964).

52. Canvin, D. T. and H. Beevers. "Sucrose Synthesis from Acetate in the Germinating Castor Bean: Kinetics and Pathway," *J. Biol. Chem.* 236:988-995 (1961).
53. Seixas, F. A., K. Williams and S. Eggleston, Eds. "Medical Consequences of Alcoholism," *Ann. N.Y. Acad. Sci.* 252:1-399 (1975).
54. Krebs, H. A. "The Effects of Ethanol on the Metabolic Activities of the Liver," in *Advances in Enzyme Regulation*, G. Weber, Ed., Vol. 6 (Oxford: Pergamon Press, 1968), pp. 467-480.
55. Thurman, R. G., S. Hesse and R. Scholz. *Alcohol and Aldehyde Metabolizing Systems*, R. E. Thurman, B. Chance, T. R. Williamson and T. Yonetani, Eds. (New York: Academic Press, 1974), pp. 257-270.
56. Lieber, C. S. "Interference of Ethanol in Hepatic Cellular Metabolism," *Ann. N.Y. Acad. Sci.* 252:24-50 (1975).
57. Danforth, W. F. "Respiration" in *The Biology of Euglena*, D. E. Buetow, Ed., Vol. II (New York: Academic Press, 1968), pp. 55-71.
58. Heinrich, B. and J. R. Cook. "Studies on the Respiratory Physiology of *Euglena gracilis* Cultured on Acetate or Glucose," *J. Protozool.* 14:548-553 (1967).
59. Morosoli, R. and N. Bégin-Heick. "The Effect of Oxygen on Metabolism in *Astasia*," *Can. J. Biochem.* 51:1402-1411 (1973).
60. Foster, J. W., S. F. Carson, D. S. Anthony, J. B. Davis, W. E. Jefferson and M. V. Long. "Aerobic Formation of Fumaric Acid in the Mold *Rhizopus nigricans:* Synthesis by Direct C_2 Condensation," *Proc. Natl. Acad. Sci.* 35:663-672 (1949).
61. Foster, J. W. and S. F. Carson. "Metabolic Exchange of Carbon Dioxide with Carboxyls and Oxidative Synthesis of C_4 Dicarboxylic Acids," *Proc. Natl. Acad. Sci.* 36:219-229 (1950).
62. Kornberg, H. L. "Aspects of Terminal Respiration in Microorganisms," *Ann. Rev. Microbiol.* 13:49-78 (1959).
63. Kornberg, H. L. and H. A. Krebs. "Synthesis of Cell Constituents from C_2 Units by a Modified Tricarboxylic Cycle," *Nature* 179:988-991 (1957).
64. Kamihara, T. "Ethanol Utilization by *Streptococcus faecalis*," *Arch. Biochem. Biophys.* 133:137-143 (1969).
65. Dunstan, P. M., C. Anthony and W. T. Drabble. "Microbial Metabolism of C_1 and C_2 Compounds. The Involvement of Glycollate in the Metabolism of Ethanol and of Acetate by *Pseudomonas* AM1," *Biochem. J.* 128:99-106 (1972).
66. Dunstan, P. M., C. Anthony and W. T. Drabble. "Microbial Metabolism of C_1 and C_2 Compounds. The Role of Glyoxylate, Glycollate and Acetate in the Growth of *Pseudomonas* AM1 on Ethanol and on C_1 Compounds," *Biochem. J.* 128:107-115 (1972).
67. Hageman, R. H. and D. Flesher. "The Effect of an Anaerobic Environment on the Activity of Alcohol Dehydrogenase and Other Enzymes of Corn Seedlings," *Arch. Biochem. Biophys.* 87:203-209 (1960).
68. Crawford, R. M. M. and M. McManmon. "Inductive Responses of Alcohol and Malic Dehydrogenases in Relation to Flooding Tolerance in Roots," *J. Exp. Bot.* 19:435-441 (1968).

69. App, A. A. and A. N. Meiss. "Effect of Aeration on Rice Alcohol Dehydrogenase," *Arch. Biochem. Biophys.* 77:181-190 (1958).

70. Kollöffel, C. "Activity of Alcohol Dehydrogenase in the Cotyledons of Peas Germinated Under Different Environmental Conditions," *Acta. Bot. Neerl.* 17:70-77 (1968).

71. McMannon, M. and R. M. M. Crawford. "A Metabolic Theory of Flooding Tolerance: the Significance of Enzyme Distribution and Behavior," *New Phytol.* 70:299-306 (1971).

72. Scholz, R. and H. Nohl. "Mechanism of the Stimulatory Effect of Fructose on Ethanol Oxidation in Perfused Rat Liver," *Eur. J. Biochem.* 63:449-458 (1976).

73. Brown, A. P. and J. L. Wray. "Correlated Changes of Some Enzyme Activities and Cofactor and Substrate Contents of Pea Cotyledon Tissue During Germination." *Biochem. J.* 108:437-444 (1968).

74. Oshino, N., R. Oshino, and B. Chance. "The Characteristics of the 'Peroxidatic' Reaction of Catalase in Ethanol Oxidation," *Biochem. J.* 131:555-567 (1973).

8

METABOLISM OF ORGANIC ACIDS OF
PLANTS IN THE CONDITIONS OF HYPOXIA

A. A. Zemlianukhin and B. F. Ivanov
Voronezh State University
Department of Plant Physiology and Biochemistry
Voronezh, USSR

INTRODUCTION

This chapter presents a general picture of changes caused by the transfer of plant cells into anaerobic conditions and the effect on the main oxidizing mechanisms of the plant cell—the citric acid cycle. Conclusions are made on the basis of the results recently obtained by the authors and other investigators. The picture is far from complete. Some parts contradict each other. Still, its general features are becoming more evident, which transforms contradictions into the impetus for the development of further investigations.

There is no need to dwell on the significance to agriculture of investigations of plant metabolism. Still, it should be noted that unfavorable environmental conditions may become a factor, which makes it possible to reveal certain peculiarities of metabolism of a plant cell not observed under normal conditions.

First we shall briefly touch upon those features of the present-day picture that are the basis for the interpretation of the facts dealt with in this chapter.

For the majority of higher plants the long periods of anaerobic conditions are fatal. Under such conditions, tolerant plants change their metabolism to solve a number of problems connected with such changes of environment in a more or less coordinated way.

First, the plant must protect itself from the accumulation of suboxidized toxic products of anaerobic metabolism. For this purpose, roundabout metabolic pathways come into action and/or the system of detoxification of such products. Another important problem is a deficit of high-energy products and a change of redox balance in cells. It becomes quite evident that the transition to energetically ineffective glycolysis cannot be regarded as the only alternative for the normal aerobic plant. In hypoxia, of special significance is the system of succinate oxidation operating through flavine oxidases capable of retaining an oxidized state despite an oxygen deficit. The role of some other metabolites changes too; the latter play the part of acceptors of reducing equivalents in the sbsence of oxygen. Specific shoftened metabolic pathways of tricarboxylic acid cycle (TCAC) intermediates acquire greater specific weight.

It should be assumed that accumulation of the number of metabolites accompanying this metabolic reconstruction is not an ordinary accumulation of by-products of unbalanced metabolism. The plant cell under conditions of ineffective anaerobic oxidizing metabolism should also provide itself with the necessary amount of readily mobilized substrates for rapid rehabilitation of disturbed plastic processes in the posthypoxic period. Succinate and gamma aminobutyric acid (GABA) appear to be of special significance in solving the problem.

A separate important problem is that of regulating agents that control metabolism under hypoxia. The change of gas composition of external and internal medium seems to give a signal to such a transition, much earlier than do the deficits of high-energy compounds and accumulation of suboxidized intermediates. Two components of the medium, oxygen and carbon dioxide, are both active cell metabolites and physiologically active agents. Simultaneously, they are closing links of one of the loops of a feedback interrelationship between an organism and the environment. Physiological activity and the effect of high concentrations of carbon dioxide on oxidizing metabolism of plants under hypoxia vary considerably in the presence of high concentrations of inert gases.

Lastly, it should be noted that high plasticity of oxidizing plant metabolism and the above-mentioned ability to mobilize reserve mode of metabolism in hypoxia rest upon the system of compartments that differentiates the pools of identical metabolites having different predestination. Investigations in this field are rather fragmentary and are far from complete.

Such are the basic points that form the basis of our investigations as well as of the exposition of the material in the chapter. In search of satisfactory biochemical mechanisms explaining the observed changes of the plant cell metabolism in hypoxy, we have to use extensively the models described in the experiments with animal tissues. We consider such analogy to be justifiable.

HYPOXIC STATES IN PLANT LIFE

Plants, to a greater degree than animals, are subjected to the effects of damaging agents of the environment. Devoid of ability to move and other active forms of behavior, plants during evolution developed a specific system of adaptational mechanisms, the main part of which is played by a complex chain of interdependent biochemical processes. The complex of metabolic reactions during the primary adaptation of plants to unfavorable environmental conditions are of particular importance.[1]

Many agents sharply activate physiobiochemical processes, proceeding weakly under favorable conditions or not at all. While adapating to unfavorable effects, some enzymic systems are activated while others are synthesized.[2]

Analogous to the term "stress" used in animal physiology to designate nonspecific responses of an organism to the damaging effect,[3] plant physiology also often uses the term "stress" in a similar way.[4]

Respiration and its component TCAC, coupled with biochemical processes, are, in a way, the focal point of damaging effects, changing either the activity of a number of enzymes associated with the production and consumption of energy in the cell or acting by distortion and breakdown of the membranous structures connected with these processes.[5] The organization of the mitochondrial apparatus of the cell keenly responds to the presence of different damaging agents.[6-10]

On the other hand, the composition of the gas medium and the availability of atmospheric air to the cells are factors subject to change, as well as illumination, temperature, moisture and mineral elements in the medium.

In turn, both temperature and water regime and other factors of the environment influence the gas composition of the atmosphere beyond the plant and inside the plants—the very chemism of the respiratory process.[11-13] As for the deficit of molecular oxygen, oxygen stress may involve secondary ionic stress.[4] At redox potential, which is the result of oxygen deficiency, accumulation of Mn^{2+} and Fe^{2+} ions up to toxic level may be observed in plants.[14]

In adapting to hypoxia, the resistance of the plant is increased against many other damaging factors as is the restoration ability of the organism.[15]

The composition of gases in equilibrium with the plant tissue varies with its physiological state and with such physical parameters as the location of the group of cells inside the plant.

The latter is particularly noticeable in such massive formations as fruits and roots. Soldatenkov[16] found that carbon dioxide content in maturing tomatoes increased to 25% when the oxygen content was reduced to 1%.

The maximum CO_2 content coincided with the climacteric rise of fruit respiration. Artsikhovskaia and Rubin[17,18] determined the dependence of the depth of tissue deposition in plants and the variation of carbon dioxide and oxygen content in the following range: in apples: oxygen from 0.9-14%, CO_2 from 8.8-32.7%; in lemons: oxygen from 1.0-13.7%; CO_2 from 14.5-39.8%; in sugar beet leaves: oxygen from 7.1-17.4%, CO_2 from 0.9-5.1%.

Major changes in the internal gas medium are caused by changes of illumination of photosynthetic organs, submergence of plants growing in marshy places, and the stay of plants under the ice cover during frosts.[19]

In addition, a number of technological procedures and methods associated with the storage and manufacturing of agricultural produce are attained by the use of modified gas media.[20-23] Metabolic variations of the objects under these conditions have attracted much attention by many investigators. (We shall consider this problem further.) The normal gas composition of submarines, cosmic and other apparatus, is difficult to maintain. Therefore, the interest in the influence of hypoxia and anaerobic conditions of different duration upon the metabolism of quite different living organisms, including plants, has received widespread attention. A special case, of interest both from the practical and scientific points of view, is anaerobiosis of fruit stored in the modified gas media.

ANAEROBIC METABOLISM OF FRUITS

After cropping and cessation of uptake of organic substances and water, the main process connecting fruit and the environment is respiratory gas exchange. At this time, the process of disintegration predominates. Enzymic chlorophyll destruction takes place, coloration and fruit consistency are altered, the content of acid decreases, sugar content increases, and amino acid and protein composition of tissues change. In essence, a set of coordinated biochemical processes takes place which, in the long run, leads to aging of fruit, physiological illnesses and cell destruction. Among other metabolites, oxygen and carbon dioxide gases change quantitatively, too. In apple, banana and asimina fruits during climacteric rise of respiration, oxygen content increases for a short time, but then, during the postclimacteric period, it falls rapidly. During the five days after the climacteric maximum at 29°C, oxygen in banana fruits decreases to about 2%, whereas CO_2 content increases to 18%; that is, the composition of the internal fruit medium approaches anaerobic conditions.[24-27] Of special note are the changes in the composition of the gaseous hormone ethylene, whose accumulation coincides with the climacteric rise of respiration.[28] Fruit cells do not come into contact with the environment directly, but by the buffer of intracellular spaces. These spaces are fairly vast: *e.g.*, in apples they constitute up to 20% of the total volume of the fruit. With fruit aging, the volume of intracellular spaces increases.[17,18,29-32] Studying the composition of intertissue gases of fruits and vegetables, many workers[16,29,33-35] have observed high carbon dioxide content and low oxygen content in tissues. During storage of apples, the total amount of gases and CO_2 content increase, whereas the amount of oxygen decreases. In the surface layers of the fruit, oxygen content is 13-19%, and carbon dioxide content is 2.5-7.0%. In the innermost portions of the central part of the fruit, in the fine intracellular spaces, the amount of oxygen is 1-3%, and carbon dioxide is 25-30%. Thus, hypoxia and even anaerobic conditions are normal for the deep tissues of fruits during long storage, not a pathological disturbance of usual conditions. In the bulky fruit tissue, the role of intracellular spaces favoring gas diffusion is certainly very great.[36] The ways of diffusion may be different even with the fruits of the same species and kind. Gas diffusion is also influenced by the age of the fruit.

Air permeability of apple tissue has been shown to decrease at maturity and with storage. The same holds true for pears.[37] The internal atmosphere of fruits is determined both by the environmental composition and metabolic character of their cells. When bananas were in the atmosphere with varying partial oxygen pressure at 12°C, there was a linear dependence between internal and external oxygen content, while CO_2 content remained the same. At temperatures close to the upper limit, at which the normal respiration is sustained, oxygen diffuses into the fruit tissues at a rate insufficient to fulfill its consumption in the process of increased respiration.[36] During the last stages of maturing, gas permeability of the fruit may decrease as the result of wax deposition and filling of the air spaces with liquids.[38,39]

Replacement of atmospheric nitrogen by other inert gases does not influence fruit behavior in any way; neither does the replacement of nitrogen by hydrogen.[36] Reduction of oxygen content in the medium inhibits fruit respiration and ethylene production. However, the degree of this inhibition is not proportional to the reduction of oxygen content. It greatly increases after a critical minimum is achieved. Respiration and ethylene production in apple slices was reduced by 50% with 2.5% oxygen content, but with 8% oxygen in the medium no appreciable inhibition of these processes occurred.[36]

Understanding the mechanism of similar processes is of great importance for any work aimed at the improvement of the quality of fruits and prevention of great losses in storage. Investigations in this field are of still greater importance due to the fact that the influence of gas composition of the environment upon metabolism of maturing fruits can be regarded as one factor that regulates the aging of all living tissues. Conditions of storage gas essentially influence the whole complex of maturing phenomena. Higher oxygen content in the storing chamber increases fruit respiration and accelerates development and aging. However, the negative consequence of such stimulation is the rapid decay and damage of fruits. High carbon dioxide content, on the contrary, paralyzes respiration and creates negative effects of another type: the fruit loses its taste and juices and becomes inedible.[20] Determining the optimal ratio of the components of gas media for storage is an important economic problem.

Replacement of oxygen by inert gases changes the qualitative character rather than the rate of utilization of respiratory substrates of stored fruit. The amount of titrated acidity in apples stored at $12°$ was found to decrease at the same rate in pure nitrogen and in air.[40]

In regulated gas composition media with long storage periods, the losses of gas production decreased several times, the danger of physiological illnesses of fruit decreased, and the quality of the fruit improved.

It was shown[41] that lowering the oxygen content reduced the indices of respiratory intensity, which approach their minimal level with oxygen concentrations of less than 8%. However, the respiratory coefficient remained relatively stable until oxygen concentration decreased to 3.5%. Oxygen concentration of 1-3% in the medium revealed noncompensated carbon dioxide liberation and resulted in a greater increase of the respiratory quotient. Oxygen concentrations of 5% delayed the climacteric rise of fruit respiration. Reduction of respiratory intensity decreased dry matter expenditures, including sugars and organic acids. The storage of fruit in regulated gas media has been studied by many authors,[32,42-51] who showed that fruits stored in the gas medium at higher carbon dioxide content and lower oxygen content are characterized by greater acidity compared to the fruits in the usual atmosphere. Dynamics of organic acid content in fruits stored in a modified atmosphere are influenced by two factors: 1) inhibition of their utilization in the citric acid cycle, and 2) their resynthesis from intermediate products of fruit metabolism and heterotrophic CO_2 assimilation from the environment.[52] The change of gas composition of the environment essentially influences transformations of ketoacids in the stored fruits. At $0°C$, pyruvate content in apples decreased from 125 to 75 mcg/100 g during first five weeks of storage both in air and in the medium containing 3% oxygen. However, in the air medium, pyruvate content then rapidly increased up to 200 mcg/100 g. Similarly, alpha-ketoglutarate content decreased at the beginning and increased at the end of the storage period, the amount of this acid being less in the atmosphere with low oxygen content than under air storage conditions.[53] Green banana plants, maturing under a deficit of oxygen, showed a marked increase of ketoacid content and a decrease of oxyacid content.[54] These authors theorized that low oxygen concentration inhibits intermediate reactions of TCAC at pyruvate-acetate and alpha-ketoglutarate-succinate stages. A series of investigations concerning the influence of oxygen deficiency upon the metabolism of stored apples was made by Zemlianukhin et al., [55-62] at the Voronezh University. The disks from the pulp of the apple pericarp were treated exogenously with 1-C malate, 1, 2-C[14]-alpha-ketoglutarate and 5-C[14] glutamate. These substances were actively incorporated into the oxidizing metabolism of the fruit tissue and radioactive labels were distributed among citric acid cycle components and amino acids. Meanwhile, convincing proof was obtained from Fidler[40] and Kidd et al.,[63] that oxygen deprivation in itself does not inhibit organic acid utilization in fruits, but only changes the character of its utilization. Apple disks in nitrogen atmosphere injected with radioactive substrate showed that the radioactivity of citric acid cycle components (aconitate, alpha-ketoglutarate, succinate) were greatly reduced. Not only were radioactive labels reduced in separate compounds, but their content was reduced also. The oxygen deficit influenced this process, too. In the apple disks in air, malate content decreased simultaneously with that of aspartate and glutamate. Disks stored in 100% nitrogen atmosphere showed a decrease of the content of a wider range of amino acids— aspartate and glutamate. Alanine and GABA decrease also. In apples stored in hypoxic conditions (3% O_2 in helium atmosphere), an increase of ketoacid content—pyruvate, glyoxylate and alpha-ketoglutarate, and a decrease of oxaloacetate were observed. The function of organic acid metabolism inhibition is related primarily to the carbon dioxide content, which will be discussed further.

The difficulties in interpreting the results connected with the specificity of the stored fruits, as quite peculiar biological objects, can be overcome by sampling a wide range of plants. First we shall consider the problem of the effect of anaerobiosis on the utilization of the main substrates of respiratory metabolism.

HYPOXIA AND THE MECHANISM OF GLUCOSE UTILIZATION

Plant respiration under hypoxic conditions has been considered by many workers. Kostichev[64] investigated the Pasteur effect in plants, that is, an increase of glycolysis under conditions of suppression of normal respiration. He demonstrated equimolecular output of carbon dioxide and ethanol formation by a number of species of oxygen-free plants. During the whole period, investigations of the Pasteur effect in plant organisms were accompanied by speculations about the significance of this phenomenon, its quantitative characteristics and nature of its starting mechanisms. In an early study, Leach[65] measured the amount of carbon dioxide liberated by buckwheat seedlings at different air/nitrogen ratios in the medium and found that the critical value of this ratio, at which carbohydrate oxidation into CO_2 and water is wholly substituted by a fermentation process forming ethanol, CO_2 and lactate as its end products, equals 0.33. The rate of carbohydrate consumption under these conditions, according to Leach, was not reduced. These results assume that high oxygen content in the atmosphere does not lead to a more economic consumption of the main respiratory substrates. This assumption is correct only when the initial assumption that CO_2 and ethanol are produced in plants in equimolecular amounts under anoxia conditions, is also correct. Ranson's[66] study proved this was not the case. When the air/nitrogen ratios in the medium equal 0.33, the ratio between CO_2 and ethanol production by buckwheat seedlings ranged from 0.5-0.6. From the point of view of utilization of respiratory substrates it should be admitted that anaerobic conditions possess conserving effects. Thomas,[67] however, justifies the comparison of anaerobic CO_2 consumption with that at the extinction point of anaerobiosis, not with the consumption at the highest level of respiratory intensity. In this case, the difference between the rate of substrate utilization at normal respiration and anaerobic metabolism is again concealed.

If one follows Turner's[68] definition, according to which the "Pasteur effect operating in a plant cell if upon passage of respiring cell from the oxygen concentration, known as the extinction point, or from air, or from some higher concentration of oxygen to anaerobiosis the carbon loss is initially increased," one should assume that the Pasteur effect was only observed in a few plants. In particular, buckwheat seedlings, which were used for many investigations in this field, including Effer and Ranson's experiments,[69,70] found the absence of the Pasteur effect.[71] This again turns our attention to the fact that anaerobic reconstruction of metabolic systems of a plant is not equivalent to switching over to the glycolytic way of carbohydrate consumption. Effer and Ranson,[69,70] having introduced radioactive substrates of carbohydrate metabolism ($1\text{-}C^{14}$-and $6\text{-}C^{14}$-glucose) and the citric acid cycle (C^{14}-acetate), in combination with numerous quantitative determinations of different metabolites, showed that the process of anaerobic utilization of respiratory substrates is formed from their simultaneous consumption along several metabolic pathways (along glycolytic and pentosophosphate ways as well as suppressed citric acid cycle). Carbon dioxide output decreases to one third from an aerobic level simultaneously with an increase of ethanol and lactate content (Figure 1). Ethanol accumulation constituted only one half the amount that would stoichiometrically correspond to the amount of liberated CO_2. Respiratory quotient (RQ) decreased to 0.76, and accumulation of sub-oxidized products together with CO_2 only partially covered carbon consumption of carbohydrates. On return to anaerobic conditions, CO_2 output and ethanol and lactate content became normal again, and RQ was restored.

Contrary to the classic objects in biochemical investigations of animal tissue respiration, plant objects have somewhat more complex and less frequently studied systems of respiratory enzymes. A set of alternative pathways of oxidation of respiratory materials and transition of electrons to the acceptors of a different nature are present in plants.[72,73] Contrary to land animals, the majority of plants are able to resist absolute oxygen deprivation for 24-48 hr.[4] Only certain representatives of mollusks and reptiles inhabiting the tidal zone (most hypoxia-resistant are tortoises[74]) can be compared with plants in this respect. On the basis of similarity of accumulated metabolic products under hypoxia conditions, one can assume that the basis of hypoxia resistance of all living organisms is achieved by more or less similar biochemical reactions. According to the theories of the origin and evolution of life on the earth, living organisms first existed in an oxygen-free atmosphere, i.e., they had to possess the type of bioenergetics capable of operating in the conditions of permanent anoxia.[75-77] Apparently, the ability of some systematic groups of living organisms inhabiting modern earth to reproduce in this or that modification of enzymatic mechanisms produced in the period of anaerobic exposure provides the molecular basis of biochemical adaptation to oxygen deprivation. An absolutely necessary condition for respiration is the presence of a pool of acceptors for electron or hydrogen taken from oxidizing substrate. Under aerobic conditions, oxygen is such an acceptor. In the absence of oxygen, the use of internal chemically bound oxygen released as the result of chemical reactions, e.g., oxidation of nitrate oxygen, may increase peroxides. Thus, it has been found that inducing nitrate respiration occurs in the absence of oxygen.[78] In the absence of oxygen, the last point on the terminal section of the pathway of reducing equivalents may be cell metabolites themselves,[79] and involves accumulation of different reduced products. Based on the differences in toxic properties of these products, Crawford et al.[80-85] developed the theory of differences in plant resistance to submerging on the basis of differences in

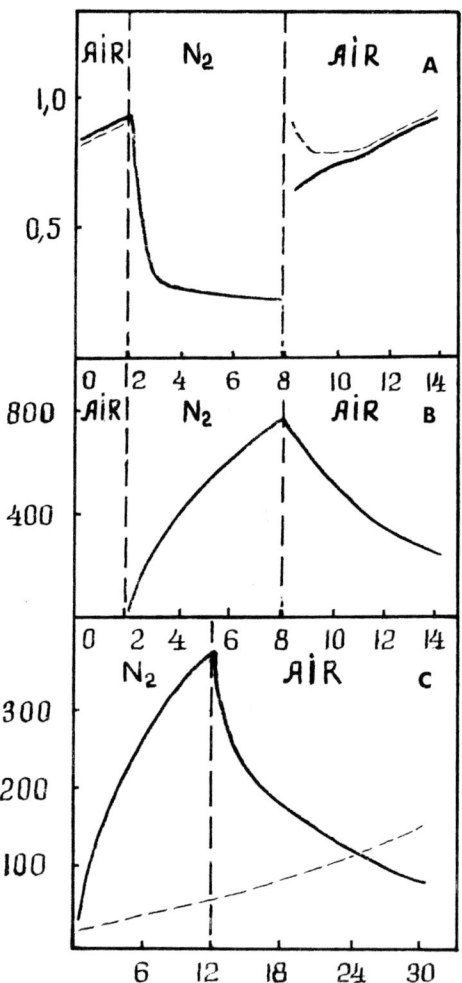

Figure 1. A. CO_2 liberation (–) and O_2 uptake (- -) by buckwheat seedlings (in %) in air, N_2 and in posthypoxic period; B. anoxic ethanol accumulation and its consumption on return into air in buckwheat seedlings in μmol per 100 g of fresh weight; (–) - experiment; (- -) - control (air); C. anoxic lactate accumulation and its consumption on return into air in buckwheat seedlings in μmol per 100 g of fresh weight; (–) - experiment; (- -) - control (air).[69]

their ability to reconstruct phosphoenol-pyruvate (the process is accompanied by its carboxylation) into toxic ethanol or into less harmful malate. Predominance of this way of reduction is determined by the presence of "malic" enzyme in the cell, which transforms malate into pyruvate.

Thus it was shown that intensity of malic enzyme activity in flood-resistant plants decreases under hypoxia conditions, resulting in lower accumulation of pyruvic acid than under normal aeration. In fact, pyruvic acid is used for malic acid synthesis (and perhaps succinic acid). Alcohol dehydrogenase in this group of plants possesses great similarity to ethanol, not to acetaldehyde, which accumulates under anaerobic conditions. In the resulting reaction, shifts take place not to formation but to oxidation of the end products of fermentation. Thus, an aerobic respiration is under metabolic control in this case.[78] A number of authors[86-88] observed changes in the ratio C-6/C-1 (ratio of specific radioactivities of carbon dioxide produced by 6-C^{14}- and 1-C^{14}- glucose utilization), indicative of the decrease of part of the glycolytic pathway of glucose catabolism, whereas the part of the pentosophosphate pathway increases with age of plant organs and plant organism as a whole during fruit maturation[89] as well as under the phytopathogenic effect,[90-92] i.e., under conditions of distorted respiration, hypoxia, and under the effect of 2,4-dichlorfenoxiacetate or fluoride-type substances.[93-95]

It is evident that the coefficient C-6/C-1 value cannot be used for quantitative characteristics of the ratio of glucose catabolism pathways, and widespread use of this index has been sharply criticized.[72,96] The possibility of exogenous glucose utilization via ways other than oxidation of endogenous pools should not be excluded. Taking into account all these circumstances, C-6/C-1 value variations can be regarded as qualitatively characteristic of variations in the ratio of respiratory pathways under experimental conditions.

Using specific inhibitors, some investigators[78,97,98] have studied the effect of hypoxia conditions on the ratio of glucose oxidation pathways among plants of different resistance to such conditions. They concluded that anaerobic conditions induce an increase in the pentosophosphate fraction of glucose oxidation in plants sensitive to prolonged hypoxia. The same was observed by Ivanov[99] in peas grown in helium and normal air atmosphere. This effect seems to some extent to protect plants from ethanol poisoning. In plants resistant to the lack of oxygen, the increase of glycolysis intensity, leading to an increase of its energetic effectiveness, is consistent with the lack of ethanol and lactic acid accumulation in their tissues. This is achieved either by the liberation of products into the medium surrounding the root and aerial parts of the plant or to a change in direction of enzymatic reactions, resulting mainly in organic acid formation instead of ethanol accumulation.

Convincing proof has been obtained showing that long-term hypoxia causes reconstruction of the respiratory chain of electron carriers and distortion of mitochondrial apparatus.[100-106] Increased alcohol and carbonic acid content arising under anaerobic conditions indicate the intensification of glycolytic processes, which, at some stages, may compensate for lack of energy from aerobic activity under oxygen deficit. Still, excessive alcohol accumulation in cells is fatal to plants. Any hypoxia condition during ontogeny affects ethanol content in tissues.

An essential feature of metabolic control under hypoxia is the protective transformation of toxic products into organic acid, which increases under conditions of anaerobic metabolism and posthypoxia regeneration. Even with this capability in plants, in our opinion, anaerobic respiration cannot fully provide sufficient energy and redox balance to maintain cell activity.

ENERGETIC AND REDOX BALANCE OF CELL IN HYPOXIA

Although the detoxification of anaerobiotic respiration products is important to resistance to hypoxia, in itself it cannot be the only source of energy-rich compounds in intact tissue.[107,108]

Investigations on metabolism of facultative anaerobic and aerobic organisms under hypoxia have shown that coupled carbohydrate and amino acid utilization is an effective adaptation mechanism in these objects.[108] Glycolysis products undergo transamination and form alanine as their end products. Free amino acids undergo changes, as shown in Figure 2, with the production of ATP and acid products, mainly glutamate-formed succinate. Another method of succinate accumulation under hypoxia is reduction of fumarate, which serves as a terminal electron acceptor in the restored chain of electron carriers. This process includes coupled phosphorylation. The remaining ATP is produced during substrate phosphorylation by phosphoglyceratekinase, pyruvatekinase, phosphoenolpyruvatekinase and thyokinase.

According to traditional concepts, consumption of one mole of hexose during glycolysis leads to the formation of two ATP moles. But, in facultative anaerobic organisms (invertebrates) coupled utilization of one glucose mole, two aspartate moles and two alpha-ketoglutarate moles leads to the synthesis of seven ATP moles with the maintenance of stable redox balance of the cell. By these metabolic schemes, the authors argue that accumulation of succinate, alanine and some other products is also confirmed for plants, according to results of investigations with different objects.[55,69,70,109-113] Kondrasheva[15,114-116] attaches much greater

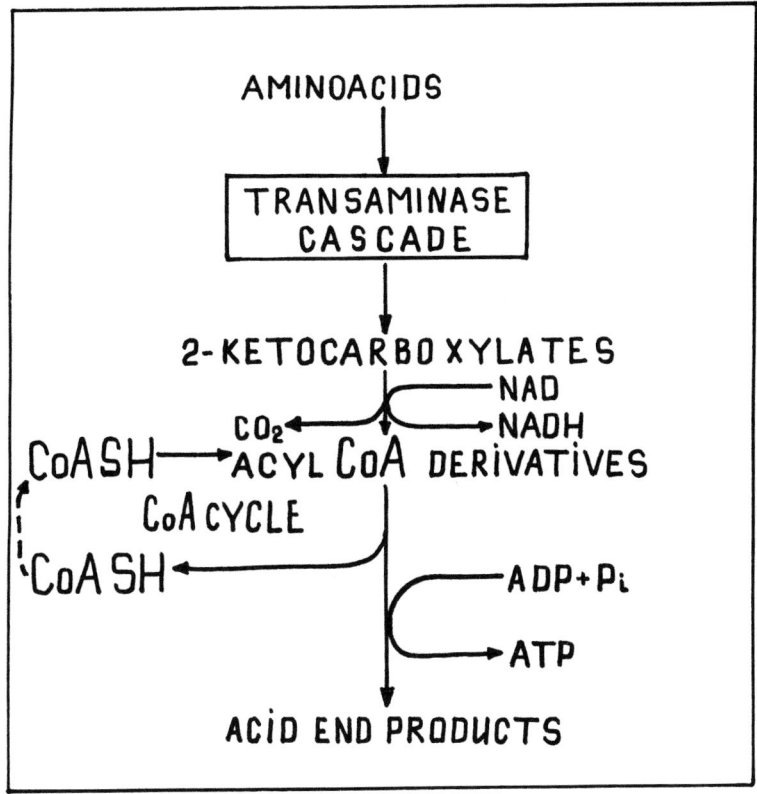

Figure 2. Anaerobic metabolism of amino acids.[108]

importance to succinate than its role as an end product of anaerobic metabolism warrants. These authors suggest that the role of succinate accumulation as respiratory substrate in hypoxia is vastly more important because it has much higher oxidation rate via flavoproteins as compared with NAD-linked substrates. Succinate is also a tetracarbonic compound, participating in a large number of biosynthetic processes, such as in the synthesis of porphyrins (*i.e.*, oxidation system components of the cell), fatty acids and corticosteroids (in animals), which promote the increase of the reserve of the nonspecific organism's resistance. In other words, succinate accumulation gives rise to a complex of phenomena, defined as "posthypoxic activation."[117-119] Blagoveschenski[119] pointed out an important role of succinic acid not only as a metabolite but also as a regulating agent.

The following pathways of succinate accumulation in tissue are possible under hypoxia:

1. resynthesis from pyruvate via reversed TCAC, accompanied by the utilization of NADH in hypoxia;
2. breakdown of fatty acids;
3. succinate formation from glutamic acid via GABA and SSA. The basis of the latter pathway in plants has been thoroughly documented.[120] All three pathways have common features, namely: (a) all escape the "weak point" of oxidation process in hypoxia NAD^+ deficit; (b) all are stimulated by metabolic shifts occurring in hypoxia (an increase of NADH, CO_2, fatty acid content and a reduction of intercellular pH); and (c) all are pathways of urgent mobilization of respiratory substrates, such as splitting, transfer and isomerization reactions, which are not associated with the bulky coupling apparatus occurring in the three pathways.

TCAC INTERMEDIATES EXCHANGE
IN ANAEROBIOSIS

As previously mentioned, toxic ethanol and malate are not the only products accumulating in plant tissues during anaerobiosis. We note from Effer and Ranson[69,70] that in air with a normal oxygen content, metabolic substrates may be liberated as CO_2. Under anaerobiosis, the loss of carbon by respiratory substrates is composed not only of CO_2 but of the remaining excess amounts of suboxidized products in plants—ethanol, lactate and succinate, which constituted (for buckwheat seedlings) 50%, 25% and 10% of carbon taken by CO_2 formation, respectively. Carbon dioxide liberation in anoxia is reduced by two thirds as compared with that in aerobiosis (extinction point of anaerobiosis). Thus, carbon loss by plants under anaerobiosis is believed to be smaller than in aerobic conditions. If amino acids (and their amides) accumulated under anoxia are considered in the category of consumed carbon, the picture changes greatly. After 12 hr of anoxia, the increase of amino acid and amide content reaches 3000 A of carbon per 100 g of fresh weight; hence, carbon consumption in air and anoxia are equal. The authors cited above consider the differences between the loss of carbon into carbon dioxide and extra metabolite pools temporary and, to a certain degree, conditional. Upon transition to aerobic type of respiration, these pools will oxidize into CO_2 and water by the citric acid cycle.

Makeev and Zemlianukhin,[121,122] almost simultaneously with Effer and Ranson, showed that in corn, exposure to short periods of 100% carbon dioxide atmospheres increased the formation of pyruvate, succinate and

GABA and reduced the content of some dicarbonic amino acids and malate. An increase of succinate and malate was also observed in pea and sunflower in atmospheres of 100% CO_2 and 100% N_2 (Table I). Malate and alpha-ketoglutarate content decrease under these conditions. Effer and Ranson also noted the fall of carbonic acid content with extended periods of anoxia. After an initial short-term increase, the content of pyruvate and alpha-ketoglutarate in buckwheat seedlings decreased in air (Figure 2).

Table I. The Content of Organic Acids in Pea and Sunflower Leaves (mg/g of fresh weight)[126]

Variant of the Experiment	Alpha-Keto-Glutarate	Pyruvate	Glyoxalate	Malic Acid	Succinic Acid
		Pea			
Air	0.170	0.126	0.535	0.428	0.110
Nitrogen	0.139	0.141	-	0.402	0.125
CO_2	0.076	0.515	0.350	0.370	0.493
		Sunflower			
Air	0.113	0.031	0.098	0.249	0.130
Nitrogen	0.105	0.033	0.102	0.238	0.185
CO_2	0.059	0.076	0.079	0.187	0.304

The authors explain such reduction of the content of a number of important tricarboxolic acid cycle intermediates by its complete inhibition. Figure 3 shows that CO_2 liberation increased on return from anoxia into air. Ethanol, lactate and succinate were utilized, and other components of the TCAC returned to their original levels. However, as seen in Table II, regeneration of the utilization rate of exogenic C^{14}-acetate occurred rather slowly. Judging by $C^{14}O_2$ liberation, it reached about one-third of the normal aerobic level after 3 hr of aerobic rehabilitation. Still, it would be erroneous to judge by this index as to the overall intensity of TCAC operation. It may be argued that the radioactive label from exogeneous C^{14}-acetate was greatly diluted, as active metabolic pools of TCAC intermediates in the posthypoxia period were filled by the utilization of excess endogenous pools of respiratory substrates (succinate and others) formed during the anaerobic period.

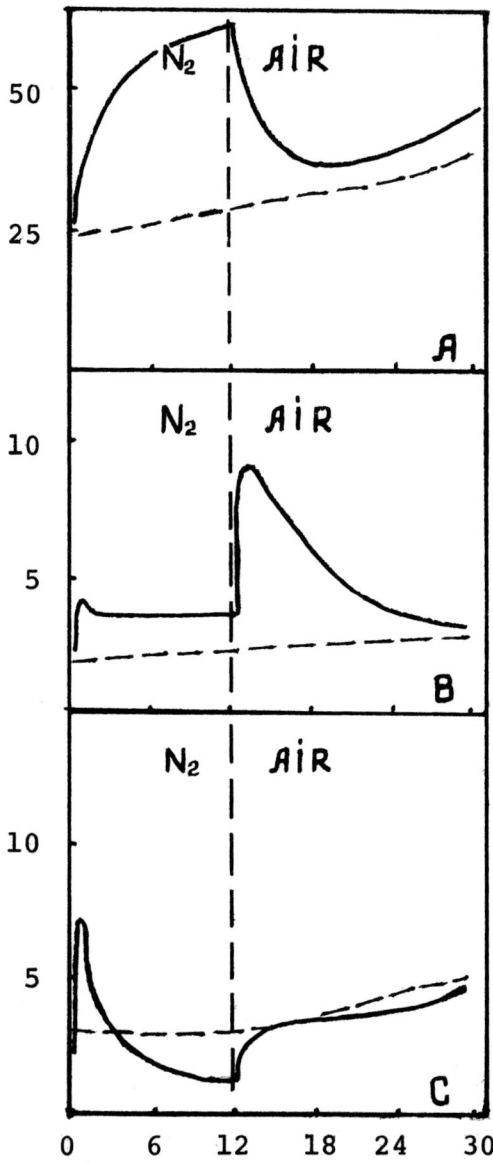

Figure 3. The effect and posteffect of anoxia on organic acids content in buckwheat seedlings: A - succinate; B - pyruvate; C - alpha-ketoglutarate. (−) - experiment; (- -) - control (air). Along the vertical line, content of acids in μmol per
100 g of fresh weight; along the horizontal, time in hours is given.[69]

Table II. Radioactivities of Fractions Isolated from 2-Day-Old Seedlings Supplied with Na Acetate-1-C^{14} [69]

	Air		Air after 3 hr in N_2 Anoxia		N_2	
	30 min	3 hr	30 min	3 hr	30 min	3 hr
	(counts/sec/g of fresh weight)					
CO_2	14,300	60,800	2,600	23,000	4,480	10,380
Alcohol-Soluble	2,430	3,100	500	1,930	625	1,925
Alcohol-Insoluble	6,500	12,300	1,650	8,500	1,550	3,030
Total Consumption	23,210	76,210	4,760	33,480	6,650	15,330
Alcohol-Soluble Fractions						
Basic	1,010	1,290	240	650	290	980
Acid	1,290	1,690	230	1,270	320	800
Neutral	30	40	20	35	5	45
Alcohol-Insoluble Fractions						
Basic	650	2,670	170	1,550	150	420
Acid	270	630	120	370	140	180
Neutral	10	40	10	0	15	0
Residue	1,360	2,450	340	1,420	320	560

Succinate accumulation in anaerobiosis is of particular interest because of the number of specific roles it plays in the anaerobiotic metabolism of cells. The rate of its formation along different metabolic pathways depends on the physiological state of the plant tissue.[123] To elucidate metabolic pathways of this compound, Zemlianukhin and Makeev[124] introduced 2,3-C^{14}-succinate into the slices from the leaves of the lower (10th leaf and upper 3-d leaf) layer of corn in its blossoming phase by the method of vacuum infiltration, i.e., under conditions imitating plant submerging.

As seen from Table III, both endogenous and exogenous succinic acid undergo transformation: which is indicated by the change of its specific radioactivity. Specific radioactivity of acids of the lower leaf layer decreased most (30% during a 3-hr period), which indicates some greater degree of stabilization of endogenous succinic acid in the older leaves than in the younger ones (compared to the change of succinate content). Such difference in metabolic activity of endogenous and exogenous succinic acid may be explained by their spatial separation. Also of interest

Table III. The Content of Organic Acids in Corn Leaves[109]

Variant	Layer	Succinic	Citric	Malic	Glyco-litic	Oxalic	Aconitic
		(mg/g of dry matter)					
Control	Upper	1.032	0.607	0.413	0.785	0.571	1.067
Control	Lower	1.065	0.626	0.638	0.421	0.510	1.154
Experiment	Upper	0.423	0.671	0.384	0.795	2.182	0.571
Experiment	Lower	0.667	0.690	0.485	0.438	1.345	0.765
Distribution of Radioactivity Among Organic Acids from C^{14}-Succinate							
		(counts/min/g of acid)					
Control	Upper	3921	none	none	none	635	none
Control	Lower	3584	none	none	none	635	none
Experiment	Upper	3435	195	596	148	148	none
Experiment	Lower	2665	271	527	417	220	none

are the data on label distribution in other acids of TCAC. There is no radioactive label in aconitic acid, though quantitatively it changed more than any other substrate. Pathways of exogenous succinate and endogenous aconitate seem to be independent. The possibility of reversible conversion of citrate into isocitrate was pointed out by Aronov and Hearon.[125] The absence of isocitrate among the acids proved to be due to its intensive utilization along the glyoxylic pathway, which is indicated by the great content of glycolic acid in the leaves and its relatively high radioactivity, particularly in the leaves of the lower layer. During a 3-hr period, malate content decreased at the same time high specific radioactivity was observed in it. The appearance of radioactive labels in malate is quite natural and is due to tricarboxylic and glyoxylic cycle operation. The same holds true with the label in citric acid. It should be noted that during the experimental period, oxalate accumulation occurred, whose content increased 3-4 times during a period of 3 hr; its total radioactivity remained practically unchanged, resulting in the decrease of the specific radioactivity of this compound. This indicates dissociation of the metabolic pathway of oxalate produced in the process of anaerobic incubation from exogenous succinate and oxalate, synthesized endogenously. In the described experiment, oxalate appears to be labeled as the result of glyoxylic pathway operation. During succinate metabolism, some part of it is converted into amino acids (Table IV). The intensity of label incorporation in amino acids was highest in the leaves of the upper layer.

Table IV. The Content of Amino Acids and their Radioactivity after the Introduction of C^{14}-Succinic Acid into Corn Leaves[109]

Leaf Layer	Glutamate	Aspartate	Alanine	GABA
	mg of acid/g dry weight			
Upper	0.343	0.726	0.833	0.206
Lower	0.377	0.709	0.697	0.326
	counts/min/mg of acid			
Upper	119	55	425	655
Lower	68	21	236	222

Production of labeled glutamate, aspartate and alpha-alanine is readily explained by the activity of corresponding transaminases, forming amino acids from their keto-precursors produced in TCAC during radioactive succinate utilization. The nature of GABA, produced in the anaerobic incubation period, will be studied more closely in the next section.

Thus, the study of anaerobic metabolism of succinate shows that for the utilization of this compound in the condition of submerged plants, shortened metabolic **pathways** are induced: glyoxylic cycle, amination and so on, which seem to be coupled with the decrease of this compound by respiratory metabolism. Succinate utilization is higher in the young upper leaves. According to Effer and Ranson[69,70] it is necessary to pay attention to the increase of the content of the aminated derivatives and amides of TCAC intermediates in anaerobiosis. During a 12-hr period of anoxia, the amount of many derivatives was doubled (Table V). The total amount of amine and amide nitrogen of amino acid fractions increased by no less than 50%.

Proceeding from generalized considerations, it is readily assumed that some part of this accumulation is due to the great retardation of protein synthesis (and hence the utilization of amino acids) in anoxia. On the other hand, catabolic consumption of amino acids via deamination and burning an TCAC, is sharply reduced. However, such an increase in the content of free amino acids in anoxia may be explained by the fact that equivalents reduced in the process of TCAC substrates in the deficit of an acceptor (oxygen) are spent to aminate its products. Thus, the content of amine and amide nitrogen in the common pool of transamination is increased. Operating the wide transaminase set provides its relatively uniform distribution among free amino acids. At the same time, operating TCAC at some minimal level explains how the utilization of acetate and

Table V. The Content of Amino Acids in 48-Hour-Old Buckwheat Seedlings in Air, in Anoxia and Upon Return to Air From Anoxia[69]

	In Anoxia after 48 hr Growth in Air (mM/100 g of fresh weight)			In Air after 12 hr of Anoxia (mM/100 g of fresh weight)		
	Hours			Hours		
	0	2	12	0	2	12
Aspartate	67.0	18.0	21.1	21.1	47.7	44.5
Treonine	58.3	48.5	83.2	83.2	57.2	49.0
Seringlutamine, Asparagine	90.0	68.5	88.0	88.0	70.0	64.0
Glutamate	310.0	239.0	174.0	174.0	167.0	224.0
Proline	7.5	3.4	7.0	7.0	3.7	6.3
Glycine	44.8	38.5	100.2	100.2	49.2	37.5
Alanine	126.0	265.0	510.0	510.0	211.0	105.0
Valine	38.0	37.8	75.8	75.8	60.0	29.3
Isoleucine	27.7	29.0	58.2	58.2	39.0	18.8
Leucine	39.0	35.5	85.0	85.0	57.2	25.5
Thirosine	24.0	24.7	46.0	46.0	29.5	20.0
Phenilalanine	20.7	22.5	52.8	52.8	21.6	18.3
GABA	23.7	89.0	262.0	262.0	168.0	68.0
Lisine	28.5	34.5	57.0	0.57	26.3	16.6
Histidine	40.0	25.1	51.4	51.4	33.7	40.0
Arginine	84.5	69.8	113.0	113.0	69.5	99.5

other TCAC substrates in anaerobiosis is automatically maintained (see above). On the other hand, this ensures constant regeneration of catalytic amounts of TCAC intermediates in the cell, necessary to restore the work of the most important metabolic process in the posthypoxic period. Free amino acids are mobile and capable of rapid utilization of substrates upon restoration of respiration. The content of free amino acids and amides caused a rapid fall to the normal level of respiration upon returning the plants to air (Table IV). A sharp increase in accumulation of alanine and GABA in anoxia attracts particular attention. This is the basis for theorizing that these compounds are the "gates" through which amine nitrogen enters an amino acid pool during anaerobiosis (in case of GABA this function may be performed by its precursors—glutamate and glutamine). As to rapid alanine accumulation, Effer and Ranson[69] argue that availability of pyruvate for transaminases increases under conditions of suppressed aerobic respiration. This appears to explain why these authors did not observe the increase of pyruvic acid content in anoxia despite TCAC inhibition by anaerobiosis and activation of the maintenance of the level of glycolysis. In experiments by Zemlianukhin and Makeev[121,124,126] with corn, pea and sunflower, a lowering was observed in the activity of pyruvate amination, the latter accumulating simultaneously with alanine. Of interest is the similarity of curves of pyruvate and alpha-ketoglutarate content in buckwheat seedlings in anaerobiosis, which may testify that the role of these compounds as "gates" for amine nitrogen entry into the amino acid pool is also similar. Makeev[127] explains alanine accumulation in hypoxia by activation of decarboxydase of aspartic acid. This concept is backed by the works of some authors who found the corresponding enzyme in bacteria[128] and later in higher plants[129]. The reduction of aspartate content under the effect of anoxia also speaks in favor of this chance (Table V).

ANAEROBIC GABA METABOLISM

GABA accumulation is a very interesting aspect of the redistribution of metabolic products of plants under conditions of anoxia. Such GABA accumulation in hypoxia is characteristic of a broad spectrum of biological objects. Some tumorous tissues of animals and plants differ from the normal ones by GABA content and glutamate decarboxylase activity.[130,131] Hypoxia and hyperoxy challenge GABA accumulation in the brain slices of mammals.[132] The response of GABA system to hypoxia appears to be of general biological significance.

As seen from Table V, GABA accumulation markedly increases accumulation of other amino acids and amides, which is readily explained by the cessation of protein synthesis and the increase of reductive amination. The content of this compound increased tenfold.

Attention should be directed to the fact that the observed GABA increase (238 mM/100 g fresh weight during 12 hr) was accompanied by the decrease of compensating the content of the two metabolically allied GABA compounds—glutamate and glutamine (136 and 110 mM/100 g of fresh weight, respectively). The presence of such balance in the content of three allied aminated compounds enables us to assume two possible pathways of hypoxic GABA production. It may be explained by glutamic acid decarboxylase (GAD) activation. Such activation could reduce glutamate content and readily convert it into glutamine in an amount equal to the increase of GABA content. However, the same equimolar consumption of glutamate content and its amide may be attributed to their increased transamination with SSA, leading to the production of GABA. The question of the main direction of metabolic pathways of GABA in plants is disputable to a certain degree. One argument speaks in favor of GABA production, that is during glutamate decarboxylation, which is later utilized in the TCAC via SSA and succinate.[120,133-135] Another argument is that GABA is an intermediate in glutamate production from succinic acid.[136-139] It is not clear whether any of these arguments are valid for plant metabolism as a whole or whether they just show the variability of results among teams working with the objects possessing different types of GABA metabolism. GABA production and utilization in plant cells is regulated not only due to the presence of the system of corresponding enzymes but by means of organizing their substrates and products into the complex system of the separate compartments. The study of compartmentalization is in its embryonic stage.

We believe that the pools of GABA precursors and utilization products are subdivided into metabolically isolated compartments. This concerns, above all, the organic acids and carbohydrates, which in general seem organized into pools.[140-146] In different objects, the presence of malate and isocitrate pools (one from TCAC intermediates, the other from carbon dioxide) having different origins, has been found. As tissues age, the specific weight of stored pools of these compounds increases significantly. It should be pointed out that heterogeneous subpools of the succinate, a substance close to GABA which seems to be associated with high availability of the compound to the internal mitochondria spaces, have not been found.[147] The whole supply of GABA in the cell is equally active metabolically.

Many studies on amino acids have been focused on elucidating peculiarities of the pool organization of protein synthesis.[148] We are concerned mainly with the processes of respiration and transamination.

Fletcher[149] demonstrated the presence of at least two types of mitochondria in the cells of rose tissue, which greatly differ from one another in relation to their metabolic functions and the link of organic acid and amino acid metabolism in them. The first type of mitochondria utilizes carbon skeletons of organic acids in the process of respiration, oxidizing them to carbon dioxide and water. The second type of mitochondria diverts carbon-containing material normally entering TCAC into the branches of this cycle and concentrates it in the carbon skeletons of amino acids by transamination, particularly the carbon skeleton of glutamate-GABA precursor. Amino acids synthesized by the second type of mitochondria are heavily incorporated into the *de novo* synthesized protein.

Pathways of tetracarbonic compound formed in the cells of legumes were studied by Mitchell.[150] Mitchell concluded that there are two different sequences of reaction with tetracarbonic compounds participating in them. One sequence is part of TCAC and is located in mitochondria. The other sequence is incorporated in the process of biosynthesis of aspartate, asparagine and homoserin and is located outside the mitochondria. Thus, if cytoplasmic carbonic acids serve as a substrate for aspartate synthesis, a substantial portion of glutamate is produced in mitochondria during alpha-ketoglutarate transamination.

The problem of pool organization of amino acid metabolism, in particular of glutamate and GABA, has been studied thoroughly by Ismailov[151-154] using legumes as experimental objects. While feeding isolated roots of *Vicia sativa L.* and their callus tissues with C^{14} sucrose, glutamine, proline and GABA showed different patterns of saturation of their pools by radioactive label, contrary to GABA's metabolic precursor-glutamate (Figure 4). The author concluded that glutamic acid accumulates in the vetch roots predominantly due to the sucrose utilization. However, along with this main glutamate pool, there is another pool of different origin. Via this pool, carbon of exogenous glutamic acid is rapidly incorporated into GABA, whose specific radioactivity remains much lower than that of glutamate during the whole period of radioactive sucrose introduction. The third glutamate pool, which seemed to have been produced from radioactive glucose earlier than the other pools and poorly mixed with them, served as a proline source. Its radioactivity markedly increased specific radioactivity of the total glutamate pool. In alfalfa roots, the radioactive label from sucrose was distributed among the products of glutamate transformations in a different way than aspartate label. Aspartate, in turn, differs from glutamine label. Of special interest is the fact that specific GABA radioactivity decreased or did not change with the introduction of labeled asparagine and glutamine when plants were placed into a greater GABA content (on the medium without C^{14}-asparagine). Consequently, the rate of GABA production under these conditions greatly exceeded the rate of increase of its pools via radioactive precursors. At the same time, label incorporation into proline from asparagine occurred at a relatively greater rate, despite its less intensive quantitative production. When glutamine was the source of radioactive label, specific proline radioactivity increased and even exceeded specific glutamate radioactivity by more than threefold during a 16-hr period, if placed in conditions favoring the fall of specific GABA radioactivity (in medium without C^{14}-glutamine). Such increase of specific radioactivity of the product over specific radioactivity of the precursor is usually interpreted as a sign of the division of the latter into specific radioactivity compartments.[155,156] Thus GABA and prolines pool saturations had quite different pathways in the experiments by Ismailov, who excludes the idea of the homogenous pool of glutamate-precursor of both compounds. Furthermore, the same studies showed that glutamate pool GABA precursor is readily available to dicarbonic amino acids, while the glutamate pool–glutamine and proline precursor–is readily available to carbohydrates (sucrose and glucose).[152,154] Low GABA availability to

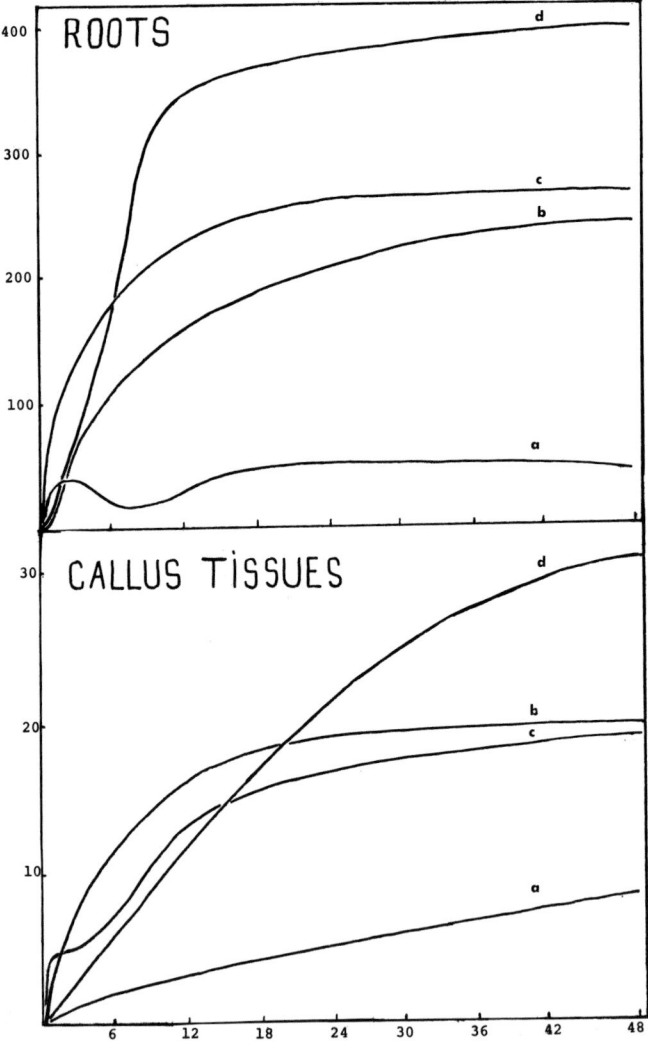

Figure 4. Dynamics of the specific radioactivity of free amino acids of vetch plants with the introduction of C^{14}-sucrose. Along the vertical line, specific radioactivity in nCi per microatom of carbon; along the horizontal line, time in hours. a - glutamate; b - GABA; c - glutamine; d - proline.[152]

glucose carbon atoms and relatively high availability for TCAC intermediates and their aminated derivatives were shown by Ivanov.[99] We assume that this ability of rapid transfer of some part of carbon and nitrogen of glutamic acid into GABA promotes the maintenance of glutamate concentration in cells at a constant level, thus stabilizing this important metabolic junction. Thus, there is convincing evidence that GABA production and further utilization takes place via the systems of metabolically divided pools of metabolites involved in this process. There is also an indication that there are separate pools of GABA itself in the roots of legumes.[157] The fact is that the exact site (cellular or tissue) of the localization of compartments of glutamate is not always clear.

It had been noted that GABA content in plant tissues as well as the concentration ratio of free glutamate and GABA show great fluctuations. Tobacco mosaic virus infection, low temperatures, water and oxygen stresses greatly alter GABA content in plant tissues; as a rule they cause it to increase.[158-162] There are no data that give a detailed picture of the effect of the main physiological agents on GABA pathway, including pH of the medium. There is some literature on the changes of GABA content in corn in the process of germination, growth and aging.[163-165] Having considered corresponding data for the leguminous plants, Khavkin[163] concluded that the glutamate/GABA ratio may be an index of the intensity of the plant growth and protein synthesis. GABA content regulation is associated in his opinion with the changes of the activity of the glutamate decarboxylase system. Inathoni and Slaughter,[166] investigating the activity of this enzyme in the process of barley germination simultaneously with GABA level, found that these two parameters do not correlate positively. They attributed changes of GABA content to a change in the rate of its utilization and/or changes of glutamate availability to the glutamate decarboxylase system. Differences in the conclusions of these authors and those of Khavkin, Inathoni and Slaughter may be attributed to the fact that they worked with different objects at different stages of development. The statement that the blockade of GABA utilization is the only reason for its accumulation in the unfavorable environment is misleading. In this respect, the findings of Lane and Stiller[167] are of interest. They found that intensive decarboxylation of endogenous glutamate in *chlorella* suspension is induced not only by anaerobic conditions but by metabolic poisons and by subsequent freezing and thawing of the cells. They stipulate that all these factors terminate in the sharp increase of cellular permeability, thus increasing glutamate availability to the glutamate decarboxylase system.

Streeter and Thompson[168,169] did a series of experiments in which aerobic and anaerobic GABA metabolism were investigated using a large set of radioactive indicators and enzymatic methods. Radish leaves were used as experimental material. Solutions of radioactive substrates were introduced into the leaves during a 15-min period with the subsequent incubation in air or nitrogen for up to 375 min. In the normal gas medium, introduced C^{14}-glutamate was readily transformed into GABA, which had greater radioactivity than glutamine or succinate. Also, radioactive label from exogenous 1-C^{14}-GABA rapidly entered succinate and malate. Malate showed increased label activity before glutamate. When 2,3-C^{14}-succinate and 2,3-H^3-succinate were introduced simultaneously into the leaves, C^{14}/H^3 ratio approached 4 in malate, and 2 in aspartate, as would be expected in case of succinate utilization via TCAC. In GABA, the ratio exceeded that for pure succinate by 16-fold, while in glutamate the C^{14}/H^3 ratio was only 1.5 times that for GABA. Streeter and Thompson estimated only 10% GABA production in the plants via succinic semialdehyde.

The same authors first discovered SSA dehydrogenase in higher plants. This enzyme was determined as NAD-dependent, with pH optimum equal to 9. Their findings explain the reverse reaction in an extremely minute way, which illustrates the trend of plant metabolism to utilize GABA and not to produce it via SSA. The main pathway of GABA production is, according to such results, decarboxylation of glutamate, not SSA amination, which is confirmed by the results previously obtained with bacterial objects,[170] plants,[135] brain tissue of mammals[171] and also by the negative results obtained while trying to observe the reverse of GAD action *in vitro*.[172]

Equilibrium constants of the reactions GABA + pyruvate SSA + alanine were found by Streeter and Thompson[168,169] to be equal to 0.15 on the basis of previously determined constants by the reactions GABA + alpha-ketoglutarate SSA + glutamate[173] and glutamate + pyruvate alpha-keto-glutarate + alanine.[174] Still, this cannot be taken as proof that succinate may convert into GABA in an appreciable quantity under normal conditions, as the real direction of the cell reactions depends on the metabolite concentration in the reaction site and on their availability to enzymes. GABA:pyruvate transaminase of the higher plants is about 15 times as active as GABA:alpha-ketoglutarate transaminase.[120,168,175] In higher plants, pyruvate appears to act as a physiologically important acceptor of amino groups, though in bacteria and mammals[171,173,176,177] GABA transamination occurs with alpha-ketoglutarate, as a rule. Anaerobiosis markedly influenced the metabolism of both endogenous and exogenous compounds in the experiments of Streeter and Thompson. Most rapid and prominent changes occurred in the content of five free amino acids (Figure 5a). They are in full agreement with the results of Effer and Ranson,

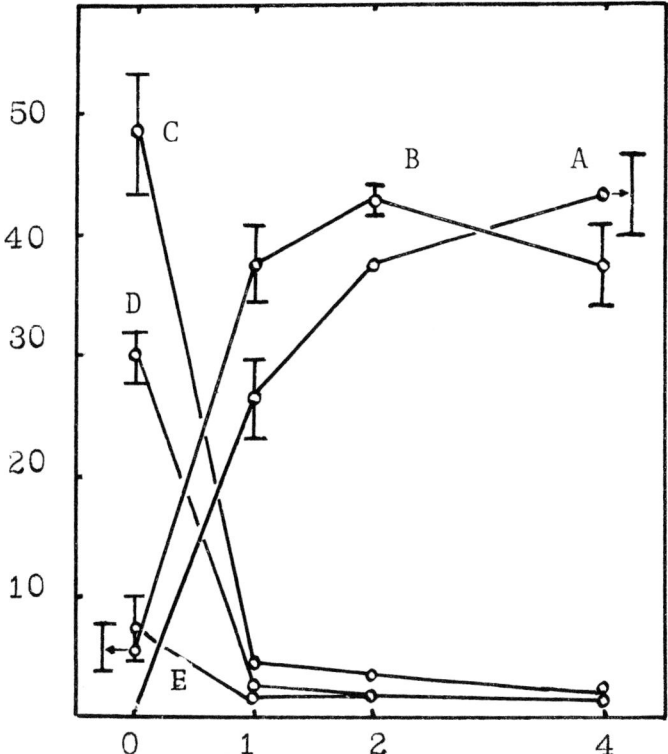

Figure 5. Effect of anaerobic incubation on the uncombined amino acid content. A - GABA; B - alanine; C - glutamate; D - aspartate; E - glutamine. Along the vertical line, μmol of amino acids; along the horizontal, time in hours.[168]

cited below (Table V). Glutamate and aspartate pools decreased significantly after 1 hr of anaerobic incubation; after 4 hr 90% of the glutamine was utilized. The content of alanine and GABA increased 7 and 100 times, respectively. Such results differ greatly from those obtained under lengthy anaerobiosis (1-7 days) on the plant roots.[178,179] In such experiments, there was an increase of the free glutamate content while aspartate level was stable. Under such lengthy anaerobic incubation, protolytic processes appear to contribute to the free amino acid content.

Streeter and Thompson studied amino acid content of radish leaves in an experimental system in which anaerobiosis was induced gradually. In this case, GABA accumulation and the fall of glutamate content was preceded by the decrease of aspartate content and alanine accumulation.

Changes in alpha-ketoglutarate and pyruvate content were much less than those in the amino acid content of plants (Figure 5b). Pyruvate accumulation appears to be associated with the fact that its oxidative utilization under anoxia ceases, while oxaloacetate formation via glycolysis and decarboxylation increases.

Both in aerobiosis and anaerobiosis, Streeter and Thompson failed to discover SSA in radish leaves in quantities greater than 0.03 M (the limit of sensitivity of the method used), which demonstrates that the role of this compound in GABA production is no more in anaerobiosis than aerobiosis. In anoxia, GABA metabolism is likely to be blocked at the transamination stage. This is indicative of the results of the experiments in which 1-C^{14}-GABA was introduced into the radish leaves in aerobiosis and anaerobiosis.[169] In aerobiosis, 95% of labeled GABA was utilized during the period in which GABA pool did not change to any appreciable degree, but after 2 hr in anaerobiosis, 95% of radioactive label remained in the GABA. After 6 hr of anaerobic incubation, 90% of introduced radioactivity remained in GABA. In anoxia, GABA is likely to be isolated from its utilization site. There are no grounds to believe that in anaerobiosis GABA intake in the cell ceases. Similar results were obtained by Zemlianukhin and Ivanov[99,180] with pea seedlings. Anaerobiosis almost entirely blocked 1-C^{14} and 4-C^{14}-GABA utilization.

In the experiments of Streeter and Thompson, the rate of decarboxylation of exogenous C^{14}-glutamate was great enough to observe an increase of its rate in anoxia. However, labeled glutamate, endogenously produced from C^{14}-acetate, transformed into GABA much faster than under the normal conditions (Figure 6). Exogenous and endogenous glutamate pools did not prove to be identical in this respect.

Stimulation of exogenous C^{14}-alpha-ketoglutarate utilization seemed to occur because of an increase in the reduced amination and transamination rate. This is also confirmed in that the addition of unlabeled alpha-ketoglutarate accelerated utilization of exogenous C^{14}-aspartate.

Table VI shows that anaerobiosis aids transfer of succinate carbon atom from TCAC into alanine, and oxygen content in the medium slightly altered malate radioactivity. However, when aspartate radioactivity increased in aerobiosis, alanine showed quite the opposite trend. Streeter and Thompson assumed that aerobiosis induced oxaloacetate transamination with aspartate production, whereas in anoxia, oxaloacetate is decarboxylated with pyruvate production. In anaerobiosis, oxaloacetate decarboxylation predominates over its condensation with acetyl CoA because of a deficit in the latter.

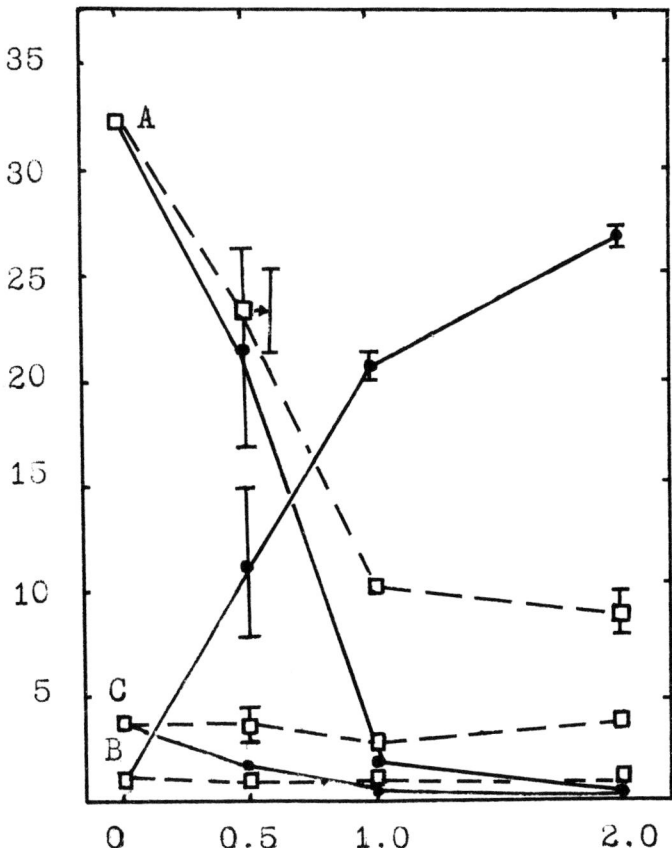

Figure 6. Effect of anaerobiosis on the radioactivity of metabolites labeled endogenously from acetate 1,2-C[14]. A - glutamate; B - GABA; C - glutamine. (−) - anaerobiosis; (- -) - air. Along the vertical line, radioactivity in % of total; along the horizontal, time in hours is given.[168]

Streeter's and Thompson's investigations showed a rather rapid fall of the enzymatic activity of GABA metabolism: GABA pyruvate: transaminase and GAD in anaerobiosis (Figure 7). Increased GABA production (and glutamate utilization) seems to be the result of its greater availability to the substrate rather than GAD activation by anaerobiosis (plant tissues contain potentially active GAD with a large excess over its really active portion). Alanine and aspartate transaminase activity decreased much less than the enzyme activity of GABA metabolism.

Table VI. The Effect of Anaerobiosis on the Metabolism of
Succinate-2,3-C^{14} in Intact, Excised Radish Leaves[168]

Time of Incubation (hr)	Atmosphere	Succinate Consumption		Radioactivity of Separate Compounds in % from Succinate Consumed		
		counts/min x 10^{-3}	in % of the total radioactivity	Malate	Aspartate	Alanine
0	Air	2230 ∓ 240	79 ∓ 7	15 ∓ 3	40	13 ∓ 2
	Nitrogen	1590 ∓ 50	61 ∓ 1	21 ∓ 3	2.6	22 ∓ 1
0.5	Air	2050 ∓ 240	83 ∓ 4	31 ∓ 3	20	7.6 ∓ 1
	Nitrogen	1600 ∓ 250	60 ∓ 5	32 ∓ 2	2.5 ∓ 1	20

Thus, Streeter and Thompson found that in anaerobiosis, reactions of oxaloacetate and glutamate decarboxylation as well as transamination between aspartate and alpha-ketoglutarate and glutamate and pyruvate were intensified. GABA transamination with pyruvate was completely blocked. The authors argue that the primary effect of anaerobiosis on the plant cell metabolism is the destruction of the compartmental organization of their metabolism, resulting in metabolic inactivity, in that under normal conditions glutamate and aspartate pools become available to react with the corresponding decarboxylases and aspartate-alpha-ketoglutarate transaminase. Such argument is indirectly supported by the investigation data, in which there was a loss of ions, organic acids and amino acids from the roots in anaerobiosis[181,182] and disorganization of their cell cytoplasm.[183] This is based on the assumption that cytoplasmic pH in anaerobiosis is reduced because of suboxidized products and approaches pH optimum for GABA: pyruvate transaminase activity equals 8.9. Nevertheless, the shift of cytoplasmic pH may result from the distortion of compartmental organization of cells. In view of such a possibility, there is no ground to oppose these two explanations.

Data on 1,4-C^{14}-succinate metabolism in corn leaves in different gas media were studied by Zemlianukhin and Makeev.[110] In these experiments, with the leaves removed the roots and endosperm were placed by their base into 0.05 M solution of 1,4-C^{14}-succinate sodium. The plants were in dark chambers connected to alkaline traps for CO_2. Either (1.5 liter/min) atmospheric air, or 100% CO_2, or 100% nitrogen were blown through the chambers. After 5 hr, the plant material was fixed to study radioactive label distribution among soluble metabolites. The analysis of

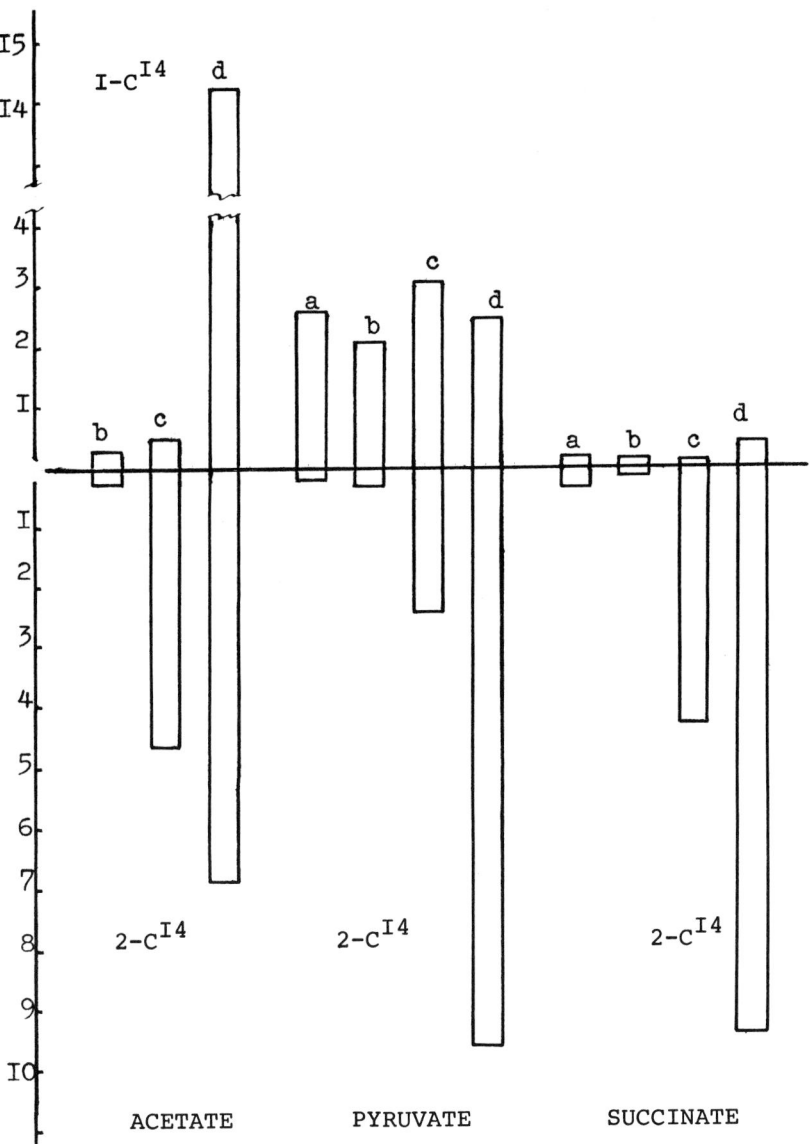

Figure 7. GABA/glutamate radioactivities ratio in pea seedlings with introduction 1-C[14] and 2-C[14] acetate and pyruvate, as well as 1,4-C[14] and 2,3-C[14]-succinate.
a - after 30 min introduction of labeled substrate in the air,   b - exposed 3 hr in air after substrate introduction,   c - exposed 3 hr in helium after substrate introduction,   d - exposed 3 hr in CO_2 after substrate introduction.

the label distribution among different groups of organic compounds (Table VII) showed that labeled succinate consumption was retarded, contrary to its utilization in air (0.05% CO_2) and nitrogen. In the latter two variants, there was a certain intake of radioactive label by the carbohydrate fraction, which may be the result of glycolysis conversion. This possibility was pointed out by Sikes and Boswell.[184] Short-duration anaerobiosis, induced by a replacement of air with nitrogen, did not affect the rate of succinate utilization. These workers found that C^{14}-succinate introduction into the potato discs appears to be incorporated in sugars. Dinitrophenol addition into the incubation medium prevented label transfer into carbohydrates.

Table VII. The Effect of Different Conditions of Gas Medium on the Distribution of Radioactivity According to Fractions from 2,3-C^{14}-Succinate[110]

Variant	Entire Alcohol Extract	Organic Acids	Amino Acids	Keto-Acids	Sugars
	(counts/min/g of dry weight)				
Air	17802	10058	1975	298	185
Nitrogen	16463	9785	2149	286	180
CO_2	18369	12958	820	-	-

Table VIII shows the data on the content of some organic acids and their radioactivity in corn leaves. Attention is first directed to the CO_2 atmosphere, in which introduced succinate was utilized twice as slowly as in other variants. Blocking of succinic acid converstion led to the reduction of malic acid, citrate and the sum of alpha-ketoglutaric and oxalacetic acids. This was espeically obvious in the change of specific malate radioactivity: in the CO_2 variant it was almost three times lower than in other variants.

Pyruvate accumulation in corn leaves indicates a Kreb's cycle disturbance and an increase of glycolysis processes in anaerobiosis. The latter is confirmed by the previous data of Makeev,[121] which show that sucrose consumption in corn leaves greatly increases in CO_2-rich atmosphere. In corn leaves, no distinct changes of quantitative content of certain organic acids occurred in an atmosphere of nitrogen. Succinate content may be found in greater quantities than in the control variant, which indicates partial blockade of TCAC.

Table VIII. The Content of Organic Acids[a] and Their Radioactivity[b] with 2,3-C[14]-Succinate Introduction[110]

Variant	Succinic		Malic		Citric		Group of Starting Acids		α-ketoglutarate, Oxalacetate		Pyruvic		Glyoxylic	
	a	b	a	b	a	b	a	b	a	b	a	b	a	b
Air	0.190	2201	0.763	$\frac{2067}{2708}$	0.388		0.204	$\frac{327}{1603}$	0.294	$\frac{267}{908}$	0.262		0.513	
Nitrogen	0.267		0.758	$\frac{2138}{2823}$	0.374		0.218	$\frac{313}{1436}$	0.298	$\frac{249}{835}$	0.274		0.510	
CO$_2$	0.418	6390	0.265	$\frac{171}{645}$	0.169		0.223	$\frac{215}{964}$	0.132		0.683		0.549	

[a]Content - mg/g of dry weight.

[b]Radioactivity - total (in numerator) - counts/min/g of dry weight - specific (in denominator) - counts/min/mg of acid.

In corn leaves in air and nitrogen, introduced succinic acid is consumed in large quantities for malate production. This is substantiated by relatively high specific radioactivity in the variant with air and nitrogen in the overall fraction of alpha-ketoglutaric and oxalacetic acids and in the "starting" acids group, whose main component is oxalic acid in the air and nitrogen variants.[185]

The effect of different gas media on the incorporation of succinate radioactive label into free amino acids is given in Table IX. In all variants, the label was mainly distributed between alpha-alanine, glutamate and aspartate. It was observed in GABA to a much smaller degree, despite the large quantitative accumulation of the latter. In all variants, both overall and specific glutamate radioactivity greatly exceeded the corresponding values for GABA. This indicates that glutamate is the substance precursor among these two compounds. As expected from the transition into TCAC, the label from the lateral succinate carbon atoms is found in the lateral carbon atoms of alpha-ketoglutarate and glutamate formed from it. This results in a large loss of radioactive label during the latter decarboxylation with the production of GABA and carbon dioxide. There are no grounds to believe that GABA production during its anaerobic decarboxylation occurs via SSA.

In all experimental variants, two-thirds of the radioactive label of amino acids was concentrated in alpha-alanine, and one-third in other amino

Table IX. The Effect of Gas Medium on the Content of Free Amino Acids[a] and Their Radioactivity[b] with $2,3\text{-}C^{14}$-Succinate Introduction[110]

Variant	Alpha-alanine		Glutamate		Aspartate		γ-aminobutyrate	
	a	b	a	b	a	b	a	b
Air	2.483	$\dfrac{374}{150}$	1.561	$\dfrac{443}{289}$	2.951	$\dfrac{594}{201}$	0.415	$\dfrac{18}{43}$
Nitrogen	2.509	$\dfrac{310}{123}$	1.506	$\dfrac{583}{387}$	3.023	$\dfrac{374}{123}$	0.428	$\dfrac{32}{74}$
CO_2	2.765	$\dfrac{392}{140}$	0.750	$\dfrac{56}{74}$	2.351	$\dfrac{22}{9}$	1.880	$\dfrac{14}{7}$

[a]Content - mg/g of dry weight.

[b]Radioactivity - total (in numerator) - counts/min/g of dry weight specific (in denominator) - counts/min/mg of acid.

acids. However the data described do not allow conclusions to be made on the route of production of this compound.

More thorough studies of the pecularities of GABA and alanine production and accumulation in different gas media were made by Zemlianukhin et al.[186] with pea seedlings.

Table X shows the data on the incorporation of radioactive label from $1,4\text{-}C^{14}$ and $2,3\text{-}C^{14}$ succinate into amino acids and into the fraction of neutral compounds in pea seedlings, incubated for 3 hr in air, helium or carbon dioxide, 35 min after introducing the labeled succinate. As seen from this table, GABA radioactivity with $1,4\text{-}C^{14}$ succinate introduction in all variants of the experiment was negligible. Contrary to this, $2,3\text{-}C^{14}$ succinate introduction into pea seedlings induced marked label incorporation into GABA. This shows that GABA production from succinate is coupled with TCAC operation and corresponding loss of the extreme carbon atoms in decarboxylation processes. The same data show that the metabolic pathway—succinate SSA—GABA, does not contribute to the overall GABA pool. Poor label intake by GABA from $1,4\text{-}C^{14}$ succinate was observed in the studies with corn leaves cited above.[109] Further analysis of Table X shows that anaerobiosis caused by helium and carbon dioxide did not affect GABA accumulation in plants equally. Anaerobiosis, caused by air replacement with carbon dioxide resulted in a 30-fold accumulation of labeled GABA as compared to the normal and more than 2-fold excess of the analogous effect caused by helium, in case of $2,3\text{-}C^{14}$-succinate introduction. In the variant with $1,4\text{-}C^{14}$-succinate, carbon dioxide causes slightly more than 2-fold label accumulation in GABA as compared to normal, whereas helium did not cause any appreciable changes in the label content in GABA. There is no decrease of radioactive label from alanine in helium in case of $2,3\text{-}C^{14}$-succinate introduction and its 2-fold reduction in the variant with $1,4\text{-}C^{14}$-succinate, whereas in CO_2, alpha-alanine activity exhibited 5-fold fall with the introduction of both $1,4\text{-}C^{14}$-succinate and $2,3\text{-}C^{14}$-succinate. In anaerobiosis, endogenous aspartate and glutamate produced from $2,3\text{-}C^{14}$-succinate have much lower activity than normal, which may indicate blocking of their synthesis by inhibition of TCAC or by activation of the corresponding decarboxylases. The change in gas conditions does not alter to any appreciable degree the label transition from succinate into GABA via SSA. It may be noted that changes in the overall amino acid content, given in Table X, are less pronounced than those in radioacitivity, which demonstrates their pool division into compartments of different metabolic activity. The fraction of substances not absorbed on ion-exchange resins, which contains mainly sugars and other water soluble compounds of the

Table X. Incorporation of Radioactive Label from 1,4-C[14]-Succinate and 2,3-C[14]-Succinate into Some Amino Acids and into Sugar Fraction in Pea Seedlings in Different Gas Conditions[186a]

Compound	After Pumping 1,4-C[14]	After Pumping 2,3-C[14]	In Air 1,4-C[14]	In Air 2,3-C[14]	After 3-hr Exposure In Helium 1,4-C[14]	After 3-hr Exposure In Helium 2,3-C[14]	In CO$_2$ 1,4-C[14]	In CO$_2$ 2,3-C[14]
GABA	42.5 ±2.5 / 1.81 ±0.18	739.5 ±74.5 / 1.89 ±0.05	35.5 ±3.5 / 0.94 ±0.2	238 ±14 / 0.87 ±0.02	22.5 ±2.5 / 3.58 ±0.33	2572 ±78.5	97 ±12 / 8.70 ±58	6994 ±113.5 / 8.27 ±0.60
Glutamate	696 ±47 / 29.64 ±0.86	2865 ±139 / 30.97 ±0.78	292 ±18 / 32.30 ±0.10	2358 ±270 / 31.54 ±0.28	237 ±27 / 24.5 ±0.14	606 ±13 / 24.70 ±0.56	206 ±13 / 29.50 ±0.56	741 ±14 / 25.46 ±0.56
Alanine	767 ±25 / 4.07 ±0.26	2177 ±29 / 4.49 ±0.07	48 ±8 / 2.99 ±0.13	80 ±2 / 2.54 ±0.20	327 ±27 / 4.94 ±0.13	2027 ±235 / 5.07 ±0.05	82.5 ±7.5 / 2.54 ±0.20	475 ±21 / 2.28 ±0.20
Aspartate	768 ±104 / 29.35 ±0.29	1310 ±45 / 30.49 ±2.90	261 ±13 / 22.49 ±3.00	1056 ±4 / 28.79 ±0.29	307 ±44 / 22.80 ±0.90	502.5 ±25.5 / 25.65 ±0.10	114 ±3 / 25.37 ±0.98	213.5 ±22.5 / 24.51 ±0.15
Sugar Fractions	85 ±3	245 ±59	58 ±1	904 ±112	77 ±19	295 ±76	53 ±1	259 ±5

[a]Numerator - activity in counts/min /g of fresh weight, denominator - content of the substance in μmol/g of fresh weight, sugar fraction - counts/min /g fresh weight.

neutral nature, was subjected to radio chromatographic analysis. Its radioactivity belonged primarily to two nonidentified components. Anaerobiosis caused both by carbon dioxide and helium induces large decreases of
radioactivity of the neutral compound fraction, indicating the dependence
of the synthesis of the cited compounds on respiratory metabolism intensity. It is characteristic that the label from central carbon atoms of
succinate molecule is incorporated into the neutral fraction to a much
greater degree. The lack of radioactivity in the products of glycolysis
conversion (glucose, fructose and sucrose) indicates that the products
of some other metabolic pathway are involved. The percentage of radioactive label transition into the neutral fraction is particularly great for
GABA, that is 36% of the overall extract radioactivity for GABA and
no more than 10% for all other introduced C^{14} compounds. Radioactive carbon incorporation into the compounds in question from the introduced metabolites and TCAC intermediates may occur via GABA along
a pathway other than TCAC and reversed glycolysis. Along with anaerobiosis, radioactivity intake of these compounds is retarded by transaminase and decarboxylase-inhibitor hydroxylamine (5×10^{-3} M) with 5-C^{14}
-glutamate introduction. Ershova[187] extracted these compounds from
the 2-week-old pea seedlings of the variety "Ramonsky-77."

Previous analysis of the extracted substances using infrared and ultraviolet spectroscopy and other physico-chemical methods show that these
compounds are related to each other. Both substances have close
maxima of ultraviolet absorbance: compounds in the butanol: acetic:
water fraction (4:1:5) with an Rf = 0.36-2120 Å, and with an Rf =
0.74-2080 Å. Both compounds react with reagents sensitive to sugars
(most pronounced reaction occurs with diphenylamine interaction. However, coloration produced is not identical to the colored reactions of the
familiar plant sugars. The relatively better interaction with reagents to
ketoses (orcin, resorcine) indicates ketogroup may be present in them.
Decoded infrared spectra sorption in the area 1684-1530 sm^{-1} indicates
the presence of COH link in these compounds. The presence of such link
is confirmed by a weak positive reaction with Lowry reagent, which both
substances exhibit. During hydrolysis of one of these substances, succinate and glucose were formed. These results confirm the identity of
the two compounds with those previously found by Vitak in pea plants.[188]
They were later identified by Lui and Castelfranco[189] as isosuccinimide-
glucoside compound, which plays an important part in ethanol metabolism
as the donor of glucosil during the synthesis of ethyl-glucoside.[190] The
role of ethanol in anaerobic metabolism has been discussed previously.
Thus, another aspect of GABA relation to anaerobic plant metabolism
is revealed.

The fact that succinate does not undergo direct intensive conversion into GABA in hypoxia, as does glutamate, shows, in our opinion, that the reversed metabolic pathway serves for the drain of gamma-aminobutyrate into TCAC. Hence, the TCAC is always capable of using mobile respiratory substrate, which is accumulated during the period of unfavorable effect in the form of GABA.

Posthypoxic GABA utilization via SSA is important in that it may serve for oxaloacetate regeneration, which is necessary to "trigger" TCAC. Ivanov[99] investigated radioactive label intake from 1-C^{14}-acetate into GABA and some other amino acids in the roots of the seedlings of the pea, corn and barley under conditions imitating submergence. In the course of experiments, roots removed from the green part of the plant and endosperm tissue were submerged into the solution of 1-C^{14}-acetate in *tris*-buffer at pH 7.4 and a specific radioactivity 1.7 mC/μM. After specified internals, the plant parts were washed in the same buffer without radioactive substrate and extracted with ethanol and subjected to radio-chromatographic analysis. The data concerning 2-hr exposure of the roots in 1-C^{14}-acetate solution are given in Table XI.

It is seen that in corn and barley, a considerable portion of radioactive label from acetate entered GABA and that radioactivity ratio in GABA and glutamate for these crops is much more than unity, whereas the case is reversed for pea. For corn and barley, radioactive label distribution is close to that of plants in hypoxia. It is evident that, submerged into the buffer solution, roots of these crops experience hypoxia conditions much more rapidly than pea roots. Such a shift in radioactive label distribution in the chain of reactions appears to be associated with the stimulation of metabolic pathway of acetate TCAC→alpha-ketoglutarate→glutamate→GABA during its last stages.

Table XI. Distribution of Radioactive Label from 1-C^{14}-Acetate Among Free Amino Acids of the Roots of Three Crop Plants[99]

Variant	GABA	Glutamate	Alanine	GABA/Glutamate
		(counts min/ g of fresh weight)		
Pea	91 ± 2	550 ± 10	330 ± 10	0.17 ± 0.03
Corn	1380 ± 130	100 ± 10	125 ± 5	14.07 ± 2.71
Barley	1100 ± 100	370 ± 50	85 ± 5	3.31 ± 0.44

Quantitative determination of GABA and glutamate content in barley and corn roots with the help of ninhydrin showed that their molar ratio approached unity, which indicates that GABA specific activity greatly exceeds that of glutamate— its metabolic precursor. Such formal distortion of the well known radiochemical rule of radioactive label distribution in the products of subsequent chain reaction[155,156] argues the presence of a small, metabolically active glutamate pool, specifically oriented to GABA production and hardly mixed with the overall glutamate pool in the plant.

In the other experiments these authors introduced $1\text{-}C^{14}$ and $2\text{-}C^{14}$ -acetate solutions into pea seedlings by the method of continuous introduction in the dark, in the atmospheres of air, helium and carbon dioxide.

Amino acid fraction analyses showed (Table XII) that in hypoxia there was a stimulation of GABA accumulation and inhibition of anaerobic alanine accumulation by CO_2. Alanine content constituted 1.15 $\mu M/g$ of fresh weight of the plant tissue in air, 4.00 $\mu M/g$ in helium, and only 2.36 $\mu M/g$ in CO_2. GABA intake of radioactive label is abnormal and differs from its intake by other amino acids. Because this radioactive label of methyl carbon acetate produced more radioactivity in GABA in hypoxia and less radioactivity in CO_2 atmospheres, we assume that the chain reaction analogous to TCAC reactions feeds the pool GABA precursor. The analyses of specific radioactivity of GABA and glutamate shows that GABA specific radioactivity greatly exceeds that of glutamate— its metabolic precursor. Thus, the assumption concerning the existence of a separate pool of glutamate-GABA precursor with high specific radioactivity in plants is again confirmed. The same holds true for the experiments with pyruvate, given below. This assumption can be verified only by the experiments investigating the behavior of the radioactive labels pointed out in specific dynamics. Analysis of GABA and glutamate radioactivity ratio (Figure VII) shows that it is the second carbon atom of acetate that is mobilized for GABA synthesis in hypoxia, and the carboxyl undergoes such mobilization only under CO_2. $1\text{-}C^{14}$ and $2\text{-}C^{14}$-pyruvate were introduced into the plants during a period of 35 min in darkness, after which the plants were placed in the atmosphere of air, helium, and CO_2 for 3 hr. The pathways of the first and second carbon atoms of pyruvate differed. These compounds labeled GABA in a different way (Table XIII).

For the first carbon atom of pyruvate, the pattern observed resembles that for the experiments with radioactive glucose and acetate (4500 counts/min/g of the fresh weight after introduction and 6600,3300 and 2100 after exposure in air, helium and CO_2). For the second atom, there was an increase of the label content in GABA in hypoxia on the order of

Table XII. Incorporation of Radioactive Label into some Amino Acids of Pea Seedlings from 1-C^{14} and 2-C^{14}-Acetate in Different Gas Environments[99],a

Variant	Air		Helium		CO_2	
	1-C^{14}	2-C^{14}	1-C^{14}	2-C^{14}	1-C^{14}	2-C^{14}
GABA	$\dfrac{3000\pm70}{0.80\pm0.05}$	$\dfrac{3900\pm210}{0.59\pm0.05}$	$\dfrac{3000\pm300}{1.28\pm0.10}$	$\dfrac{5840\pm780}{1.96\pm0.10}$	$\dfrac{1420\pm30}{5.20\pm0.80}$	$\dfrac{2310\pm210}{3.31\pm0.57}$
Glutamate	$\dfrac{9700\pm100}{6.40\pm0.20}$	$\dfrac{12800\pm260}{6.78\pm0.18}$	$\dfrac{7750\pm110}{6.80\pm0.20}$	$\dfrac{1300\pm100}{5.63\pm0.18}$	$\dfrac{110\pm30}{6.40\pm0.10}$	$\dfrac{340\pm10}{5.85\pm0.17}$
Alanine	$\dfrac{930\pm60}{1.15\pm0.35}$	$\dfrac{2900\pm130}{0.97\pm0.13}$	$\dfrac{1240\pm160}{4.00\pm0.60}$	$\dfrac{290\pm10}{2.79\pm0.01}$	$\dfrac{110\pm10}{2.75\pm0.20}$	$\dfrac{290\pm10}{2.36\pm0.10}$
Asparagine	$\dfrac{1300\pm200}{2.78\pm0.02}$	$\dfrac{4960\pm220}{2.70\pm0.60}$	$\dfrac{110\pm20}{2.67\pm0.16}$	$\dfrac{125\pm15}{2.59\pm0.07}$	2.67 ± 0.27	$\dfrac{122\pm10}{2.28\pm0.24}$

aNumerator: counts/min/g of fresh weight, denominator content of amino acids in μM/g of fresh water.

Table XIII. Incorporation of Radioactive Label into some Amino Acids of Pea Seedlings from 1-C^{14} and 2-C^{14}-Pyruvate in Different Gas Conditions[99],[a]

Variant	After Pumping		In Air		In Helium		In CO_2	
	1-C^{14}	2-C^{14}	1-C^{14}	2-C^{14}	1-C^{14}	2-C^{14}	1-C^{14}	2-C^{14}
GABA	3450±740 / 1.30±0.01	2800±700 / 1.45±0.16	4500±1100 / 1.52±0.10	950±50 / 1.52±0.03	1100±300 / 3.01±0.25	4600±1200 / 3.11±0.57	310±90 / 6.21±0.02	5300±700 / 5.86±0.64
Glutamate	354±120 / 5.10±0.04	6070±280 / 4.84±0.06	670±110 / 4.94±0.06	1200±200 / 5.10±0.06	300±100 / 4.94±0.01	1600±200 / 4.73±0.09	190±30 / 4.50±0.10	700±200 / 4.63±0.19
Alanine	1400±180 / 2.78±0.01	17000±2200 / 2.80±0.08	2300±1100 / 1.99±0.31	3000±700 / 1.99±0.01	700±100 / 3.90±0.42	10800±800 / 3.42±0.13	900±300 / 2.30±0.011	9200±2800 / 2.45±0.15
Proline	7400±800 / 0.48±0.02	17600±7900 / 0.50±0.07	23000±7000 / 0.35±0.08	21300±4300 / 0.31±0.01	5800±500 / 0.41±0.04	19000±6000 / 0.46±0.06	6600±1300 / 0.37±0.03	10800±3800 / 0.37±0.05
Aspartate	1400±600 / 4.02±0.30	1510±360 / 3.96±0.01	7600±2600 / 4.30±0.01	600±180 / 4.42±0.59	3700±100 / 3.33±0.07	1500±100 / 3.10±0.27	1800±500 / 3.67±0.02	800±300 / 3.83±0.13

[a]Numerator—specific radioactivity in counts/min/μM and denominator—the content of amino acids in μM/g of the fresh weight.

1,440 counts/min/g of the fresh weight in air, 13,600 in helium, and 29,500 in CO_2. Hence, CO_2 intensified the hypoxic effect. Label content in the glutamate-GABA precursor was small, its specific radioactivity was 2-fold lower than GABA. The label from the first carbon atom of pyruvate readily entered the GABA synthesis system.

Spatial differentiation, but not allosterical effects, were the regulating links for the process of GABA synthesis from glutamate. Based on the literature of the intercellular localization of glutamatedecarboxylase and transamination enzymes,[191,192] we concluded that the glutamate pool serving for emergency GABA synthesis was concentrated in the cytoplasm of the plant cell, and the main pathways for the final utilization of glucose and acetate were concentrated in mitochondria.

Comparisons of GABA/glutamate radioactivity ratios, obtained by introduction of differently labeled substrates into the pea seedlings, shows (Figure VII) that the compounds were located in the metabolic "knot" of glycolysis and TCAC; acetate and pyruvate were partially incorporated into GABA, escaping decarboxylation processes of the first carbon atoms in TCAC, but the typical Krebs cycle substrate-succinate did not enter this metabolic shunt and labeled GABA only by its central carbon atoms.

Anaerobic metabolism of the pyruvate precursor (a very important component of anaerobic metabolism), alanine, has been studied by Zemlianukhin et. al.[186]

Racemic $1\text{-}C^{14}$ and $3\text{-}C^{14}$-alanine were introduced into pea seedlings in the atmosphere of air and CO_2 for 7 hr while seedlings were actively, transpiring. As seen from Table XIV, CO_2 effect produced a sharp increase in radioactive label incorporation into GABA and inhibited its transfer from alpha-alanine to other compounds, namely the glutamate and aspartate fractions of organic acids and neutral compounds. In aerobiosis, $3\text{-}C^{14}$-alanine introduction into the pea seedlings produced greater intake of radioactive label by the compounds given above, as compared to $1\text{-}C^{14}$-alanine. Thus, the main route of incorporation of labeled carbon from the molecules of alpha-alanine into organic acids and sugars, aspartate and gamma-aminobutyrate under normal conditions was associated with the following reaction chain: alpha-alanine→pyruvate→acetyl KoA→TCAC, which is in full agreement with the data of Mokronosov et al.[193]

The unexpectedly high glutamate activity in the variant with $1\text{-}C^{14}$-alpha-alanine introduction in aerobiosis was not consistent with the reaction sequence mentioned as the only metabolic pathway of exogenous alpha-alanine.

Of great interest was the fact that under complete anaerobiosis caused by CO_2, label from exogenous C^{14}-alpha-alanine increases sharply in GABA. This indicates certain discrepancy and autonomy of some reactions of the Krebs cycle, hindering full blocking of this vital process for the organism. Di- and tricarboxylic cycle refers to amphibolic systems,[194] for which the turn of reaction flows in various pathways depending on the environment. Dixon and Fowdon[120] argue that GABA plays an important part as a circumventing pathway, which operates when TCAC is somehow blocked. Due to the presence of a number of enzymic systems influencing the cytoplasm, the course of some TCAC reactions outside mitochondria[191] should be considered carefully when judging the extent of GABA pathway association with purely respiratory mitochondrial metabolism. Carbon dioxide activation of the label entering GABA from exogenous C^{14}-alanine as compared to CO_2 activation of label's consumption from endogenous alanine pools (Table X), as well as the stimulation of GABA production from pyruvate (Table XIII), in our opinion indicates the activation of the common reaction link in both cases, by combining alpha-alanine and GABA pools. In studying the effect of gas media on alpha-ketoglutarate and glutamate metabolism and metabolism of GABA, labeled alpha-ketoglutarate was introduced by vacuum infiltration, after which plants were placed into different gas conditions.

To elucidate the role of exogenous alpha-ketoglutarate on GABA production, hydroxylamine (5×10^{-3} M), a powerful inhibitor of transaminases and other enzymes associated with pyridoxalphosphate, particularly decarboxylases,[176] was introduced into pea seedlings along with labeled alpha-ketoglutarate. After vacuum infiltration[176] of the original solutions, the plants were placed into darkness for 5 hr under different gas conditions. Radioactivity of water alcohol extract, represented mainly by labeled substrate in the variants without inhibitor, was 22,100±300 and 23,900±600 counts/min/g of the dry material in air and in CO_2, respectively. In the variants with inhibitor, this radioactivity was 42,000±2,100 and 38,800± 400 counts/min/g of dry material in air and in CO_2, respectively. The latter was readily explained and was the consequence of inhibition of the reactions of alpha-ketoglutarate transamination by the inhibitor hydroxylamine. Table XIV gives the data on radioactive label content in glutamate and gamma-aminobutyrate, depending on the variant of the experiment. Maximum glutamate radioactivity was observed immediately after infiltration. Although localization of enzymes of alpha-ketoglutarate transamination has been shown in mitochondria,[120,191] the cytoplasm also has corresponding enzymatic systems.[191] It is the first intercellular compartment along the exogenous alpha-ketoglutarate route. Taking into account certain difficulties of alpha-ketoglutarate transport into mitochondria, we consider the cytoplasm to be the main site of exogenous alpha-ketoglutarate transamination and accumulation of glutamate. The 5-hr exposure of plants, both in air and CO_2, caused great reduction of glutamate radioactivity. It is characteristic that in CO_2, hydroxylamine almost completely eliminates label accumulation of GABA via hydroxylamine-blocking transaminase reactions.

Taking into account that hydroxylamine is an inhibitor for decarboxylation reactions, hydroxylamine, together with 5-C^{14}-glutamate, was introduced into the seedlings. The data from Table XIV show that immediately after vacuum infiltration GABA exhibited highest radioactivity, influenced primarily by temporary anaerobiosis, caused by tissue saturation with infiltration solution. During the 5-hr exposure of plants in air, reduction of GABA radioactivity occurred, hydroxylamine having inhibited label entering GABA twofold.

Table XIV. The Effect of Hydroxylamine in the Incorporation of Radioactive Label into the Fraction of Sugars, Glutamate and GABA from Pea Leaves Introduced with 5-C^{14}-Glutamate[186]

Variant Compound	After Vacuum Infiltration	After 5-hr Exposure Without and With an Inhibitor			
		In Air		In CO_2	
		Without	With	Without	With
		(counts/min/g of fresh weight)			
Sugar Fraction	traces	255∓25	112∓22	38∓6	10∓1
Glutamate	2131∓54	815∓35	1064∓64	1085∓25	808∓80
GABA	202∓14	27∓1	14∓1	556∓34	390∓70
Glutamate/GABA	10.5	30.2	76.0	1.9	2.07

Glutamate radioactivity was higher in plants with an inhibitor than in those without an inhibitor. Prolonged exposure of plants to CO_2 caused sharp accumulations of labeled GABA both with and without the inhibitor. At the same time, in the variant with an inhibitor, GABA radioactivity was somewhat lower, which indicates partial inhibition of the process of glutamate decarboxalation. It should be pointed out that although 20-fold increases of GABA activity occurred in CO_2, no appreciable decrease of glutamate radioactivity was observed. On the contrary, there was a trend to use it more slowly. These data indicate that GABA accumulation was not associated with glutamate-decarboxylase activation but appeared to be the consequence of slow GABA utilization. These results differ from the data of glutamate radioactivity (Table X). The difference may be explained by an alternate route of exogenous glutamate utilization instead of glutamate endogenously formed from succinate. Glutamate pool compartmentalization has been clearly shown by many authors.[142,152,154] The total radioactivity of the fraction of neutral compounds in this experiment (Table XV) belonged to two different compounds. Both carbon dioxide and hydroxylamine greatly inhibited label incorporation into the neutral fraction.

1-C^{14}-GABA obtained by the biosynthesis method[99] was injected into pea seedlings in the darkness for 10 hr in different gas media. 1-C^{14}-GABA uptake by seedlings in air was accompanied by its conversion primarily into organic acids (mainly malate). Carbon dioxide inhibited this conversion. At the same time, GABA specific radioactivity in anaerobiosis markedly decreased. This may be explained by dilution of exogenous

Table XV. The Effect of Carbon Dioxide on the Incorporation of Radioactive Label into Glutamate and GABA from 1,2-C^{14}-Alpha-Ketoglutaratic Acid Injected into Pea Leaves[186]

Amino Acid	3 hr in Air	3 hr in 30% CO_2	3 hr in 100% CO_2	3 hr in Air, then 3 hr in 100% CO_2
(counts/min/g of fresh weight)				
Glutamate	569∓5	348∓7	90∓4	102∓6
GABA	80∓3	145∓11	730∓9	834∓5
Glutamate/GABA	7.1	2.4	0.12	0.22

radioactive label by endogenous GABA. Certain radioactivity discovered in glutamate (during CO_2 exposure) may be attributed to the weak reversal reaction of GAD.

In general, all three main groups of experiments described above[69,168,169,186] are in agreement. Based on these experiments, GABA appears to function as a convenient, readily mobilized and nontoxic substrate, which serves for rapid posthypoxic respiratory and plastic processes. It also functions in rehabilitation of plant cells by succinate accumulation and utilization (see page 214).

However, the results of the work of Zemlianukhin's group makes it necessary to explain differences in the effect of nitrogen and carbon dioxide anaerobiosis on the organic acid metabolism and to estimate the effect of high CO_2 concentrations as an independent physiologically active agent. This is the topic of the final section of our chapter.

CARBON DIOXIDE AS A REGULATING AGENT IN PLANT REACTION TO ANAEROBIOSIS

Dealing with plant metabolism in hypoxia, one must consider the role of such a gas medium component as CO_2. As seen from the facts previously discussed, oxygen and carbon dioxide content fluctuate widely depending on physiological state and depth of tissue. Carbon dioxide is a very active cell metabolite,[195] but because of its pure regulating effect it is difficult to differentiate among the metabolic effects of its participation in the process of carboxylation, photosynthetic fixation, etc. The effect of carbon dioxide on the uptake of water and mineral substances by plant roots has been known for some time.[196] Agaverdiev et al.[197] demonstrated that higher carbon dioxide content in the atmosphere causes outbursts of super light luminescence in plants, caused by the oxidating reactions of a radical nature. In their experiments, carbon dioxide induced acceleration of organic matter consumption in wheat seedlings. The effect of carbon dioxide on plant metabolism cannot be limited to the creation of hypoxia. It has specific interaction with cell structures which is of phase character. By anaerobiosis, we mean deprivation of investigation objects of oxygen not only as an acceptor of reductive equivalents, but as an effector influencing mitochondria state and the course of plastic processes in the cell.[102,198] Oxygen and carbon dioxide seem to be important elements for regulating cellular respiration systems, ensuring optimal operation of this process.

Carbon dioxide content in plant tissues regulates respiratory process at different levels. Gas exchange process in leaves is controlled by the extent of stomata opening which, in its turn, depends on the content of carbon dioxide in intracellular spaces of leaves.[199] Carbon dioxide appears to have a significant effect on the succinate oxidizing system of plant mitochondria.

The mitochondria of castor-bean revealed specific competitive inhibition by CO_2 of succinate dehydrogenase.[200,201] Similar results were obtained in the experiments with apples.[202] Thus, the gas medium component, whose accumulation in tissues is a direct consequence of hypoxia, is also regarded as an effector, inducing reconstruction of plant enzymic systems associated with succinate accumulation.

According to Wager,[203] at least 10 plant enzymes are inhibited by CO_2; however, only succinate dehydrogenase changes, observed in organic acid content in plants subjected to higher CO_2 concentrations, are in agreement with results obtained on enzymatic mixtures. On the whole, the role of carbon dioxide in the metabolism of TCAC intermediates has not been as thoroughly studied as the role of CO_2 in photosynthesis. With comparatively low increases of CO_2 fraction in the atmosphere and prolonged exposure in cucumber and bean leaves[204] and in tomato leaves,[205] increases of ascorbic acid content have been observed.

A cycle of experiments aimed at elucidating the effect of higher CO_2 concentrations on different phases of plant metabolism were made by Madsen.[204-209] He found that the content of malate, citrate, aconitate, isoaconitate, succinate and fumarate in tomato leaves grown at different CO_2 concentrations in the atmosphere increased together with the rise of CO_2 content in the medium until it reached 0.10% (0.22% in the morning). The main fraction of the gain was due to aconitic acid, whereas isocitrate, fumarate and succinate were present in small quantities. Malate and citrate content comprised slightly more than one-third of the total acid content. These results represent a complex picture, influenced by the sum of the effects of carbon dioxide as a substrate of nonphotosynthetic and photosynthetic carboxylation on one hand, and as a physiological effector on the other.

Investigations dealing with carbon dioxide effect as a modified gas media component used in the storage of fruit and other agricultural products are of great practical significance. Even under these conditions, it is difficult to differentiate effects caused by high CO_2 and low O_2 content.[210,211] Fidler's data[212] speak in favor of the different effect of these two agents. In the absence of carbon dioxide, the fall of oxygen concentration by 2% did not influence respiratory quotient value; hence, it did not initiate an increase in anaerobic processes in the fruit. In the presence of higher carbon dioxide concentrations, even with comparatively high oxygen content (13-16%), there was a fall of RQ which, however, increased with further reduction of oxygen concentration. Data for apples[213] and lemons[214] show an increase in respiratory intensity in the medium with 10% CO_2.

The effect of carbon dioxide and oxygen depends on the physiological state of plant objects. Carbon dioxide in concentrations of 5-10% caused strong depression of respiration in mature apples, but did not affect respiration of newly harvested fruits.[215]

However, in an atmosphere enriched by CO_2, plants not only liberate CO_2 in respiration, but incorporate carbon dioxide of the environment into the composition of its organic material. By introducing $C^{14}O_2$ into the modified gas medium of the stored apples, some authors[216-218] discovered labels in citric, malic and oxalic acids. The main route of labeled CO_2 entry appears to be by nonphotosynthetic pyruvate carboxylation. Malate accumulation under high CO_2 concentrations, according to some authors,[44,52,219-222] is associated with inhibition of the effect of decarboxylating malate dehydrogenase by excess CO_2 in conformity with the law of effective masses.

Popov[60] convincingly demonstrated that carbon dioxide greatly alters malate metabolism and, at the same time, that it was actively incorporated into organic acid metabolism. In apples, the greater portion of $1\text{-}C^{14}$-malate was subjected to decarboxylation with labeled carbon dioxide liberation. Within two days after labeled malate introduction, the total radioactivity of extracted components of apple tissue was reduced more than 20 fold. In fruits placed into a medium containing 40% CO_2, organic acid content, mainly malate, greatly increased with the reduction of total and specific radioactivity; this indicates the presence of two partially uncoupled malate pools. In fruits kept in the CO_2-enriched atmosphere, the content of dicarbonic amino aicds—aspartate and glutamate—fell, and alanine accumulated. Apples incorporate $C^{14}O_2$ into organic acids, amino acids and carbohydrates. The maximum radioactivity was found in malic acid and in dicarbonic amino acids. With extended storage of fruit, radioactivity of extracted fruit tissue components falls the same as the radioacitivity fraction of organic acids. Radioactivity of amino acids and sugars increased under the same conditions. At the beginning of the storage of apples in atmosphere modified with higher carbon dioxide content (8-10%), CO_2 blocked the entry of radioactive label from malic acid into other compounds. Later, however, radioactivity of organic acids and amino acids in this variant was higher than in control fruits and those stored in the atmosphere of helium, containing 3% oxygen.

Not all plant objects respond similarly to the effect of high carbon dioxide concentrations. The respiration of apples, avocado and some other fruits[223-225] was inhibited by CO_2, whereas in lemons, stimulation occurred. Usually, the carbon dioxide effect was coupled with succinate accumulation but citrate was not always accumulated.[226-232] With very high carbon dioxide concentrations, ethanol production was initiated in plants.[233,234] A series of works is devoted to the studies of the effect of high CO_2 concentrations on the metabolism of organic acids of pea.[113,203,232,235,236]

Wager[232] succeeded in differentiating carbon dioxide effect from hypoxia effect since the oxygen content in the gas mixture he used did not differ from that in the normal atmosphere. But in our opinion, specific effect of carbon dioxide on the stomata apparatus has not been taken into account. Respiration intensity was increasingly inhibited with higher CO_2 content in tissues. When a medium contained 10-15% carbon dioxide, plants initiated ethanol production, which continued to increase until carbon dioxide concentration approached 37%. Under CO_2, the content of malate, pyruvate and alpha-ketoglutarate in plants was reduced and succinate content increased, while citrate content remained unchanged (Figure 8A). On return to a medium with normal CO_2 content, malate content in plants rapidly increased, and succinate concentration slowly returned to its normal level (Figure 8B). Phosphoenol pyruvate content sharply decreased, then increased again, whereas a reverse pattern was observed for alpha-ketoglutarate. The author cited considers that the main route for malate resynthesis is phosphoenolpyruvate carboxylation with oxaloacetate production followed by its reduction. This sequence of reactions seems to be stimulated to a greater degree by the changes of pH contents of plant cells subjected to CO_2 effect than by the direct effect of the latter. Differences in plant responses to carbon dioxide effect thus greatly depend on the characteristics of their buffer systems. Wager[113,232] felt that the changes of TCAC intermediates content, observed in his experiments, could not be entirely explained by the concept of the system of unique pools within the TCAC circuit. It seems more logical that organic acids were redistributed between different pools, to a different extent associated with respiratory mechanisms on one hand and systems of enzymatic transamination on the other.

Lastly, returning to the analysis of the investigation data by Zemlianukhin et al.,[109,112,185,186] given above (Tables I, VII-XV), we should examine the conclusions of this team of workers concerning the CO_2 effect on the metabolism of organic acids. In these experiments, CO_2 effect was not limited to the production of anaerobiosis. As to GABA accumulation, carbon dioxide has a more pronounced effect than anaerobiosis alone. Carbon dioxide effect on different respiratory substrates, particularly succinate via TCAC, was also more pronounced than usual oxygen deprivation. Contrary to the normal anaerobic conditions, high carbon dioxide concentrations did not induce appreciable alanine accumulation. In CO_2, the transformation of carbon skeletons of alanine and pyruvate into GABA was intensified (Tables XV and XVI). Evidently, this was due to the operation of the isolation from the main respiratory mechanism sequence of small pools of intermediate products, as there was no increase of organic acid radioactivity in the route of alanine and pyruvate to GABA via TCAC.

The general picture of anaerobiosis and CO_2 effect on organic acid metabolism, observed in these experiments, is plotted in Figure 9.

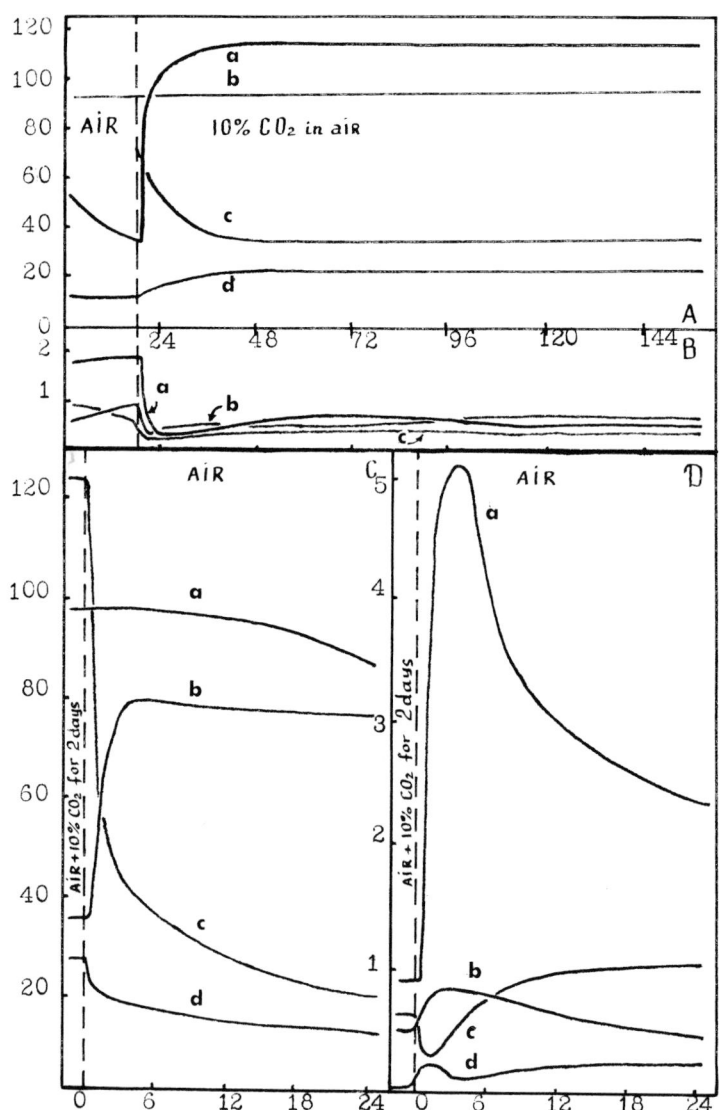

Figure 8. The effect and post effect of CO_2 on the content of organic acids in plants.[203]
A - T_{CO_2} (a); citrate (b); malate (c); succinate (d).   B - alpha-oxoglutarate (a); pyruvate
(b) and phosphoenol-pyruvate (c).   C - citrate (a); malate (b); T_{CO_2} (c); succinate (d).
D - alpha-oxoglutarate (a); pyruvate (b); phosphoenolpyruvate (c); oxaloacetate (d).
Along the horizontal line, time in hours; along the vertical line, μmol of acids per 10 g
fresh weight.

Table XVI. Activity of Radioactive Label in some Amino Acids, Liberated CO_2, Organic Acids, and Sugars incorporated with $1\text{-}C^{14}$ and $3\text{-}C^{14}$ Alanine from Pea Seedlings in Air and CO_2 [186]

Compound	$1\text{-}C^{14}$-Alanine		$3\text{-}C^{14}$-Alanine	
	Air	CO_2	Air	CO_2
	(counts/min/g of fresh weight)			
GABA	traces	102 ∓ 24	traces	794 ∓ 145
Glutamate	3880 ∓ 40	traces	4520 ∓ 400	traces
Alanine	4820 ∓ 1420	5500 ∓ 500	1098 ∓ 242	6225 ∓ 995
Aspartate	85 ∓ 42	traces	245 ∓ 123	traces
Organic acids	804 ∓ 182	286 ∓ 80	3315 ∓ 425	889 ∓ 54
Sugars	80 ∓ 16	48 ∓ 1	943 ∓ 143	238 ∓ 30
CO_2	57100	----	5700	----

CONCLUSIONS

In this review, we considered and compared the results of different authors concerning the effect of hypoxia and high CO_2 concentrations on the intermediates of the citric acid cycle and their close derivatives in plant tissues.

The basic conclusions made from the exposed material are as follows:

1. Anaerobic effect on plant metabolism is not equal to the inhibition of their respiration. In anoxia, the effect of specific metabolic pathways, leading to transfer of respiratory substrates into the pools of relatively nontoxic and readily utilizable compounds—succinate, GABA and alanine—is initiated. GABA appears as a protected succinate form, which has little influence on the pH of the cell contents and is readily mobilized under conditions where reactivation of TCAC occurs.

2. Reductive amination consumes an excess of reducing equivalents, which are produced in cells under anoxia, and temporarily transfers it into a less dangerous form for cell operation with an excess of free amino acids.

3. Protective cell reaction manifests itself in the decrease of production of toxic substances of glycolysis at the expense of increasing part of pentosophosphate pathway and/or altering of glycolysis to malate synthesis (which provides one more reserve TCAC substrate).

4. Carbon dioxide, acting both as metabolite and physiological effector, blocks (in high concentrations) utilization of organic acid pools, aids malate resynthesis from phosphenolpyruvate and

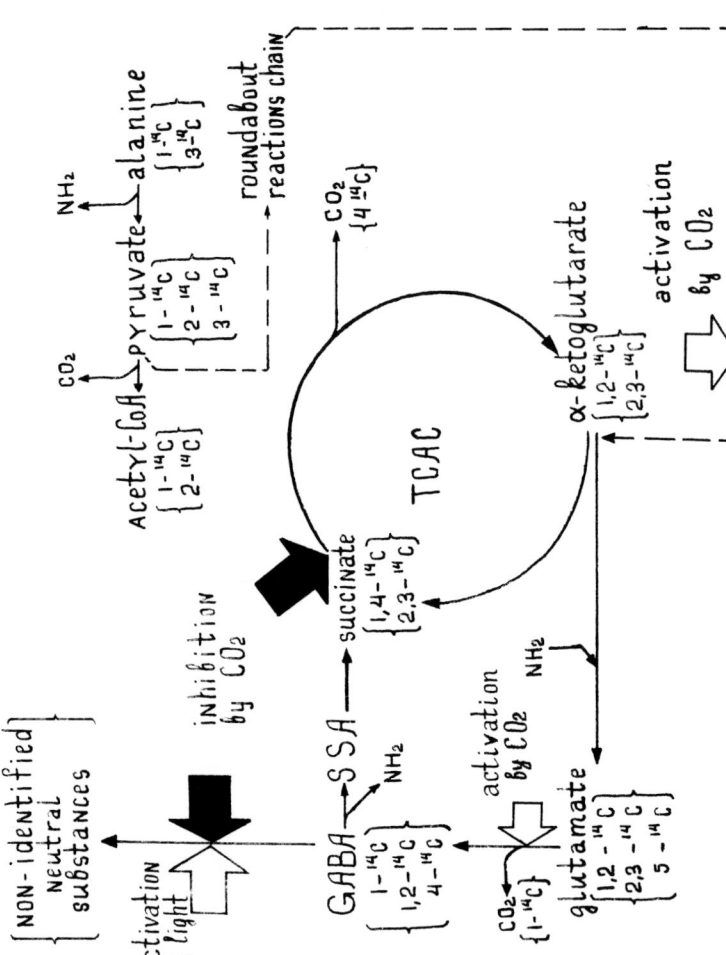

Figure 9. Scheme of GABA metabolism and some of its labeled precursors in plants.

puts into operation specific sequences of reactions, leading to alanine and pyruvate transformation into GABA. On the whole, the role of this compound may be regarded as regulative in the process of anoxia reconstruction of the respiratory mechanisms of plants.

REFERENCES

1. Ziegler, M. "Biochemische Anpassungen der Pflanzen an extreme Standortbedingungen," *Biol. Rund.* 12(2):81 (1975).
2. Filner, P., J. L. Wray and J. E. Warner. "Enzyme Induction in Higher Plants," *Science* 165(3891):358 (1969).
3. Selije, H. *On the Level of the Whole Organism* (Moscow: Nauka, 1972) (In Russian).
4. Levitt, J. *Responses of Plants to Environmental Stresses* (New York: Academic Press, 1972).
5. Tairbekov, M. G. "Structural and Functional Aspects of the Plant Cell Resistance," *Uspekhi Sovr. Biol.* 75(3):406 (1973) (In Russian).
6. Meerson, F. Z. "Molecular Mechanisms of Adaptation to the Hypoxia," *Priroda* 6(74) (1973) (In Russian).
7. Andreeva, I. N. and B. B. Vartapetian. "Ultrastructure of Rice Coleoptile Mitochondria during Biogenesis," 1975, in *Abstracts of the Twelfth International Botanical Congress* (Leningrad: Nauka, 1975), p. 347.
8. Tsuji, H., T. Katoh and K. Ueda. "Growth and Metabolism in Plants Under Anaerobic Conditions," in *Abstracts of the Twelfth International Botanical Congress* (Leningrad: Nauka, 1975), p. 373.
9. Vartapetian, B. B. and A. I. Maslov. "Respiratory Activity of Mitochondria in Rice Coleoptiles Grown Anaerobically," in *Abstracts of the Twelfth International Botanical Congress* (Leningrad: Nauka, 1975), p. 375.
10. Andreeva, I. N., G. I. Koslova and B. B. Vartapetian. "Specificity of Mitochondria Formation in Rice Coleoptiles During the Process of Germination in Aerobic and in Anaerobic Conditions," *Fiziol. Rast.* 23(1):111 (1976) (In Russian).
11. Rubin, B. A. and E. V. Artsikhovskaia. *Biochemistry and Physiology of Plant Immunitet* (Moscow: Vischaia schcola, 1968) (In Russian).
12. Rubin, B. A. and M. E. Ladigina. *Physiology and Biochemistry of Plant Respiration* (Moscow: Moscow University Publishing House, 1974) (In Russian).
13. Semikhatova, O. A. *Energetics of Plant Respiration Under High Temperatures.* (Leningrad: Nauka, 1974) (In Russian).
14. Conway, V. M. "Aeration and Plant Growth in Wet Soils," *Bot. Rev.* 6(4):149 (1940).
15. Kondrasheva, M. N., E. I. Maevski, G. V. Babaian and I. R. Saakian. "Adaptation to the Hypoxia by Turning Metabolism on Succinate Conversions," in *Mitochondria: Biochemistry and Ultrastructure* (Moscow: Nauka, 1973), p. 112 (In Russian).

16. Soldatenkov, S. V. *Role of Oxygen in Ripening of Fruits* (Leningrad: Leningrad University Publishing House, 1941) (In Russian).
17. Artsikhovskaia, E. V. and B. A. Rubin. "Respiration of Plants as Adaptative Function." *Uspekhi Sovr. Biol.* 37(2):136 (1954).
18. Artsikhovskaia, E. V. and B. A. Rubin. "Oxidative Processes and Their Role in Biology of Different Plant Organs. V. The Value of the Oxidative System for Adaptation of Generative Storaging Organs to the Environmental Conditions," *Biochemistry of Fruits and Vegetables* 3, 5 (1955) (In Russian).
19. Rakitina, Z. G. "Influence of the Ice Crust on the Gaseous Content of the Internal Atmosphere of Winter Wheat," *Fiziol. Rast.* 17(5): 907 (1970) (In Russian).
20. Tserevitinov, F. B. *Chemistry and Science of Commodites of Fresh Fruits and Vegetables* (Moscow: Gostorgisdat, 1949). (In Russian).
21. Metlitski, L. V. *Biochemistry of Fruits and Vegetables* (Moscow: Economika, 1970) (In Russian).
22. Metlitski, L. V. *Principles of the Biochemistry of Fruits and Vegetables* (Moscow: Economika, 1976) (In Russian).
23. Ulianov, A. M. and N. V. Popov. "Gathering, Trade Treating and Storage of Fruits and Berries," in *Gardening in Central Zone of Russia* (Moscow: Rosselkhosizdat, 1973), p. 246 (In Russian).
24. Hulme, A. C., J. D. Johnes and L. S. C. Wooltorton. "The Respiration Climacteric of Apple Fruits. Biochemical Changes Occurring During the Development of the Climacteric in Fruits on the Tree *New Phytol.* 64, 158 (1965).
25. Wardlaw, C. W. and E. R. Leonard. "Studies in Tropical Fruits. III. Preliminary Observations on Pneumatic Pressures in Fruits," *Ann. Bot.* 2(6):301 (1949).
26. Leonard, E. K. and C. W. Wardlaw. "Studies in Tropical Fruits. XII. The Respiration of Bananas During Storage at 53°F and Ripening at Controlled Temperatures," *Ann. Bot.* 5(19):379 (1941).
27. Wardlaw, C. W. and E. R. Leonard. "Studies in Tropical Fruits. I. Preliminary Observations on Some Aspects of Development, Ripening and Senescence with Special Reference to Respiration," *Ann. Bot.* 50(199): 621 (1936).
28. Burg, S. P. and E. A. Burg. "Role of Ethylene in Fruit Ripening," *Plant Physiol.* 37(1):179 (1962).
29. Banaitis, Y. I. "Investigations of Respiratory Gas Exchange in Some Varieties of Apples of the Latvian SSR," in *Biochemistry of Fruits and Vegetables,* 4 (Moscow: Academy of Sciences of the USSR, 1958), p. 59 (In Russian).
30. Ivanova, T. M. "On the Factors Defining the Interrelation Between the Aerobic and Anaerobic Respiration in the Ontogenesis of Fruits. Communication 4," in *Biochemistry of Fruits and Vegetables,* 4 (Moscow: Publishing House of AN SSSR, 1958). p. 24 (In Russian).
31. Kolesnik, A. A. *Factors of Long-Term Storage of Fruits and Vegetables* (Moscow: Torgovaia Literatura, 1959) (In Russian).
32. Metlitski, L. V., E. G. Salkova, I L. Volkind, V. I. Bondarev and V. J. Janiuk. *Storage of Fruits in Controlled Gaseous Medium* (Moscow: Ekonomika, 1972) (In Russian).

33. Rakitin, U. V. "Intensity of Ethanol and Acetaldehyde Accumulation in Ripening Fruits," *Biochem.* 10(5-6):373 (1945) (In Russian).
34. Artsikhovskaia, E. W., B. A. Rubin and T. M. Ivanova. "Oxidative Processes and Their Role in Biology of Different Plant Organs," in *Biochemistry of Fruits and Vegetables,* 2 (Moscow: Academy of Sciences of the USSR, 1951), p. 136 (In Russian).
35. Troian, A. V., P. C. Golian and C. C. Kedish. "The Rate of Gas Exchange in Apples and Peaches," *Priklad. Biochim. Microbiol.* 3(3):255 (1967) (In Russian).
36. Spenser, M. "Ripening of Fruits," in *Plant Biochemistry*, J. Bonner, Ed. (Moscow: Mir, 1968). p. 484.
37. Marcellin, P. *Rev. Gen. Bot.* 64:322 (1967).
38. Burg, S. P. and E. A. Burg. "Role of Ethylene in Fruit Ripening," *Plant Physiol.* 37(1):179 (1962).
39. Ben-Yehoshua, S., R. N. Robertson and J. B. Biale. "Respiration and Internal Atmosphere of Avocado Fruit," *Plant Physiol.* 38(2): 194 (1963).
40. Fidler, J. C. "A Comparison of the Aerobic and Anaerobic Respiration of Apples," *J. Exp. Bot.* 19(1):41 (1968).
41. Jurin, V. and M. Karel. "Studies on Control of Respiration of McIntosh Apples by Packaging Methods," *Food Tech.* 17(6):104 (1963).
42. Tsiprush, R. J. "Internal Tissue Gas Content of Fruits and Its Changing During the Storage," *Proc. Kishinev SHI* 11(9):183 (1960) (In Russian).
43. Tsiprush, R. J. "Storage of Apples in Synthetic Packing," in *Storage of Fruits in Polymeric Packing Materials* (Moscow: CINTI Pischeprom, 1966), p. 30 (In Russian).
44. Tsiprush, R. J. and G. G. Tomash. "Dynamics of Organic Acids in Apples var. Renet Simirenko in Relation to Conditions of Growth and Regime of Storage," *Proc. Kishinev SHI* 105:81 (1973) (In Russian).
45. Saburov, N. V. and A. G. Guseinov. "Film—for Storaging of Apples," *Sadovodstvo* 8:20 (1964) (In Russian).
46. Saburov, N. V. and A. G. Guseinov. "Long-Term Storage of Apples in the Packing from Polyethylene Film," in *Storage of Fruits in Polymeric Packing Materials* (Moscow: CINTI Pischeprom., 1966). p. 22 (In Russian).
47. Makaschvili, G. A. "Storage of Fruits in Polyethylenic Packing," in *Storage of Fruits in Polymeric Packing Materials* (Moscow: CINTI Pischeprom, 1966), p. 27 (In Russian).
48. Makaschvili, G. A. and G. J. Chinkotadze. "Storage of Fresh Fruits in Polymeric Packing," *Konservn. i. Ovoschesushiln. Prom.* 1: 20 , (1968) (In Russian).
49. Ignateiv, B. D. and Z. A. Sedova. *Account on the Mission in Ditton's Laboratory (England)* (Moscow: Ministry of Agriculture USSR, 1969) (In Russian).
50. Kolesnik, A. A., M. A. Fedorova and E. H. Osenova. *Storage of Fruits in Controlled Atmosphere* (Moscow: Kolos, 1967) (In Russian).
51. Metlitski, L. V., A. P. Ivanovskaia, V. J. Janiuk, V. I. Bondarev and I. I. Nikiforov. *Experience of France in Storage of Fruits in Controlled Gaseous Medium* (Moscow: Centrosoiuz, 1969) (In Russian).

52. Metlitski, L. V. and E. G. Salkova. "Biochemical Aspects of Storage of Fruits in Controlled Gaseous Medium," *Priklad. Biochim. Microbiol.* 5(4):387 (1969) (In Russian).

53. Hulme, A. C., W. H. Smith and L. S. C. Wooltorton. "Biochemical Changes Associated with the Development of Low-Temperature Breakdown in Apples," *J. Sci. Food Agric.* 15, 303 (1964).

54. McGlasson, W. B. and W. B. Wills. "Effect of Oxygen and Carbon Dioxide on Respiration, Storage Life and Organic Acids in Green Bananas," *Aust. Biol. Sci.* 25(1):35 (1972).

55. Zemlianukhin, A. A., A. M. Makeev and N. V. Popov. "Influence of the Gaseous Content of Medium on C^{14}-Organic Acids Metabolism in the Fruits of Apple Trees," *Prilk. Biochim Microbiol.* 9(3): 443 (1973) (In Russian).

56. Popov, N. V. "Influence of Gaseous Content of Medium on Organic Acids Conversions in Apples," in *Sbornik Stud. Nauchn. Rabot.,* (estestv. nauki) (Voronezsh-VGU Publishing House, 1972), p. 28.

57. Popov, N. V. "Influence of Gaseous Content of Medium on the Conversions of Organic Acids, Sugars and Amino Acids in Apples During the Storage After Cropping," in *Materiali VIII nauchnoi stud. conf.,* (Novosibirsk: Novosibirsk University Publishing House, 1970), p. 55 (In Russian).

58. Popov, N. V. "Influence of CO_2 of Environment on the Metabolism of Different Varieties of Apples During their Storage After Cropping," in *Sborn. stud. nauchn.rabot (estestv. nauki)* (Voronezsh: VGU Publishing House, 1971), p. 30 (In Russian).

59. Popov., N. V. "Fruits Storage in Controlled Gaseous Medium," in *Gardening in Central Zone of Russia* (Moscow: Rosselchosisdat,, 1973), p. 67 (In Russian).

60. Popov, N. V. "Influence of Gaseous Content of Medium on Metabolism of Respiratory Substrates of Apples During Storage," Ph.D. thesis, Voronezsh, 1975 (In Russian).

61. Popov, N. V. and T. G. Popova. "Conversions of Sugars and Organic Acids During the Ripening and Storage of Apples," *Fiziol. Physico-Chim. Mechanism: Regul. Obm. Proc. Org.* 2, 67 (1972).

62. Popov, N. V. and N. P. Suprunova. "Dynamics of Organic Acids, Sugars and Amino Acids in Apples During the Storage in Different Packages," in *Fiziol. Phisico-Chim. Mechanismi Regul. Obm. Proc. Org.* 3, 41 (1974) (In Russian).

63. Kidd, F., C. West, D. G. Griffith and N. A. Potter. "Metabolism of Malic Acid in Apples," *J. Hort Sci.* 26:169 (1951).

64. Kostichev, S. P. *Plant Physiology. I. Chemical Physiology* (Moscow: Gos. isd. kolch. sovch. lit., 1933) (In Russian).

65. Leach, W. "Researches on Plant Respiration. IV. The Relation Between the Respiration in Air and in Nitrogen of Certain Seeds During Germination. (b) Seeds in which Carbohydrates Constitute the Chief Food Reserve," *Proc. Royal Soc. (London)* 119(815): Ser. B, 507 (1936).

66. Ranson, S. L. "Some Aspects of the Effect of Oxygen on Respiratory Metabolism in Higher Plants," Ph.D. Thesis, Newcastle upon Tyne, 1947.

67. Thomas, M. *Plant Physiology* (London: J. and A. Churchill, 1947).
68. Turner, J. S. "Respiration. The Pasteur Effect in Plants," *Ann. Rev. Plant Physiol.* 2:145 (1951).
69. Effer, W. R. and S. L. Ranson. "Some Effects of Oxygen Concentration on Levels of Respiratory Intermediates in Buckwheat Seedlings," *Plant Physiol.* 42(8):1053 (1967).
70. Effer, W. R. and S. L. Ranson. "Respiratory Metabolism in Buckwheat Seedlings," *Plant Physiol.* 42(8):1042 (1967).
71. Turner, J. C. "Fermentation in Higher Plants; its Relation to Respiration; the Pasteur Effect," in *Handbuch der Pflanzenphysiologie* Bd. 12, 1 (Berlin: Springer Verlag, 1960), p. 42.
72. Semikhatova, O. A. *Change of Respiratory Systems* (Leningrad: "Nauka," 1969) (In Russian).
73. Rubin, B. A. and L. N. Loginova. *Alternative Pathways of Biological Oxidation* (Lubertsi: VINITI, 1973) (In Russian).
74. Belkin, D. A. "Anoxia: Tolerance in Reptiles," *Science* 139(3554): 492 (1963).
75. Kenyon, D. H. and G. Steinman. *Biochemical Predestination* (New York: McGraw-Hill Book Company, 1969).
76. Leopold, A. C. *Plant Growth and Development* (New York: McGraw-Hill Book Company, 1964).
77. Oparin, A. I. *Life, its Nature, Origin and Development* (Moscow: "Nauka," 1968) (In Russian).
78. Chirkova, T. V. "Specific Features of Plant Metabolism in Connection with their Adaptation to Hypoxia," in *Problems of Evolutionary Physiology of Plants* (Leningrad: "Nauka," 1974), p. 131 (In Russian).
79. Krebs, H. A. and I. M. Lowenstein. "The Tricarboxylic Acid Cycle," in *Metabolic Pathways, VI.,* D. M. Greenberg, Ed. (New York: Academic Press, 1960), p. 305.
80. Crawford, R. M. M. "The Control of Anaerobic Respiration as a Determining Factor in the Distribution of the Genus Senecio," *J. Ecol.* 54(2):403 (1966).
81. Crawford, R. M. M. "Alcohol Dehydrogenase Activity in Relation to Flooding Tolerance in Roots," *J. Exp. Bot.* 18(56):458 (1967).
82. Crawford, R. M. M. "The Physiological Basis of Flooding Tolerance," *Ber. Dtsch. Bot. Ges.* 82:111 (1969).
83. Crawford, R. M. M. and P. D. Tyler. "Organic Acid Metabolism in Relation to Flooding Tolerance in Roots," *J. Eid.* 57(1):235 (1969).
84. Crawford, R. M. M. "Metabolic Adaptations to Anoxia in Plants and Animals," in *Abstracts of the Twelfth International Botanical Congress* (Leningrad: "Nauka," 1975).
85. McManmon, A. and R. M. M. Crawford. "A Metabolic Theory of Flooding Tolerance: The Significance of Enzyme Distribution and Behaviour," *New Phytol.* 70(2):299 (1971).
86. Gibbs, M. and H. Beevers. "Glucose Dissimilation in the Higher Plant. Effect of Age of Tissue," *Plant Physiol.* 30(4):343 (1955).
87. Lustinec, J., V. Pokorna and J. Ruzicka. "Activation of Glycolysis and Inhibition of Glucose Transport into Leaves by Fluoride," *Biol. Plant.* 4(2):126 (1962).

88. Lustinec, J. and V. Pokorna. "Alternation of Respiratory Pathways During the Development of Wheat Leaf," *Biol. Plant. Acad. Scient. Bohemosl.* 4(2):101 (1962).
89. Ramsay, J. C. and C. H. Wang. "Catabolic Changes in Ripening Tomato Fruit," *Nature* 193(4817):120 (1952).
90. Shaw, M. and D. Samborski. "The Physiology of Host-Parasite Relations. III. The Pattern of Respiration in Rusted and Mildewed Cereal Leaves," *Can. J. Bot.* 35(3):389 (1957).
91. Shaw, M., D. Samborski and A. Oaks. "Some Effects of Indoleacetic Acid and Maleic Hydrazide on Respiration and Flowering of Wheat," *Can. J. Bot.* 36(2):233 (1958).
92. Daly, J. M., L. R. Krupka and A. A. Bell. "Influence of Hormones on Respiratory Metabolism of Healthy and Rust-Affected Tissues," *Plant Physiol.* 37(2):130 (1962).
93. Humphreys, T. E. and W. M. Dugger. "The Effect of 2,4-Dichlorphenoxyacetic Acid on Pathways of Glucose Catabolism in Higher Plants," *Plant Physiol.* 32(2):136 (1957).
94. Lüstinec, J., V. Pokorná and J. Růžička. "Activation of Glycolysis and Inhibition of Glucose Transport into Leaves by Fluoride," *Biol. Plant Acad. Scient. Bohemosl.*, 4(2):126 (1962).
95. Ross, C. W., H. Wiebe and G. W. Miller. "Effect of Fluoride on Glucose Catabolism in Plant Leaves," *Plant Physiol.* 37(3):305 (1962).
96. Landau, B. R., G. E. Bartsch, J. Katz and H. G. Wood. "Estimation of Pathway Contributions to Glucose Metabolism and the Rate of Isomerization of Hexose-6-Phosphate," *J. Biol. Chem.* 239(3):686 (1964).
97. Chirkova, T. V., I. V. Khasova and T. P. Astafurova. "On the Question of Metabolic Regulation of Adaptation of Plants to the Conditions of Transitive Anaerobiosis," *Fiziol. Rast.* 21(1):102 (1974).
98. Chirkova, T. V. and I. V. Khasova. "Relation Between the Pathways of Glucose Oxidation under Different Supply of the Plant by Oxygen," *Vestn. LGU* 21(4):111 (1974) (In Russian).
99. Ivanov, B. F. "Metabolism of Gamma-aminobutyric Acid in Plants in the Conditions of Different Gaseous Content of the Medium," Ph.D. Thesis, Voronezch (1975) (In Russian).
100. Grineva, G. M. "Accumulation and Releasing of Alcohol by Plant Roots Under Oxygen Deficit," *Dokl. Akad. Nauk. SSSR* 156(5):1225 (1964). (In Russian).
101. Grineva, G. M. "Metabolism of Ethanol in Plants and Influence of Hypoxia on it," *Usp. Sovr. Biol.* 76(1):68 (1973) (In Russian).
102. Grineva, G. M. "Mechanism of Adaptation and Resistance to the Oxygen Deficit in Plants," *Usp. Sovr. Biol.* 80(2)(5):238 (1975) (In Russian).
103. Grineva, G. M. and Z. S. Burkina. "Influence of Atmospheric Conditions on Consumption and Distribution of Water with High-Weight Oxygen (H_2O^{18}) in Corn and Sunflower," *Fiziol. Rast.* 13(4):682 (1966) (In Russian).
104. Grineva, G. M., I. N. Andreeva and V. A. Lipasova. "Respiratory Metabolism and Structure of the Corn Root Cells in the Conditions of Anaerobiosis," *Biochim.* 33(4):766 (1968) (In Russian).

105. Grineva, G. M., and L. A. Frolova. "Influence of Oxygen Deficiency on the State of Mitochondria of Corn Roots During the Inhibition of Electron Transport," in *Mitochondria. Molecular Mechanisms of Enzymatic Reactions*, S. E. Severin, Ed. (Moscow: "Nauka," 1972), p. 193 (In Russian).

106. Grineva, G. M. and L. A. Frolova. "Influence of Hypoxia on the State of Mitochondria of Corn Roots and on the Inhibition of Electron Transport," *Fiziol. Rast.* 19(2):348 (1972) (In Russian).

107. Hochachka, P. and G. M. Somero. *Strategies of Biochemical Adaptation* (Philadelphia: W. B. Saunders Company, 1973).

108. Hochachka, R. W., J. Fields and T. Mustafa. "Animal Life Without Oxygen: Basic Biochemical Mechanisms," *Am. Zool.* 13(2):543 (1973).

109. Zemlianukhin, A. A. and A. M. Makeev. "Metabolism of the Organic Acids in Plants in the Conditions of Different Gaseous Content of Medium," *Theses of the II Vsesousn. Biochem. Congr.* (Taschkent, 1969), p. 97 (In Russian).

110. Zemlianukhin, A. A. and A. M. Makeev. "Influence of the Gaseous Content of the Medium on the Metabolism of $1,4\text{-}^{14}$C-succinic Acid in Maize Leaves," *Dokl. Akad. Nauk SSSR* 186(4):964 (1969) (In Russian).

111. Zemlianukhin, A. A. and A. M. Makeev. "Metabolism of $1,4\text{-}C^{14}$-succinate in the Conditions of Short-Time Anaerobiosis," in *Some Problems of Biology and Soil Science* (Voronezsh: VGU Publishing House, 1969), p. 73 (In Russian).

112. Zemlianukhin, A. A. and A. M. Makeev. "Metabolism of $2,3\text{-}C^{14}$-succinate, Introduced in Maize Leaves," *Fiziol. Rast.* 16(2):344 (1969) (In Russian).

113. Wager, H. G. "The Effect of Anaerobiosis on Acids of Tricarboxylic Acid Cycle in Peas," *J. Exp. Bot.* 12(34):34 (1961).

114. Kondrasheva, M. N. "Metabolic States of Mitochondria and Basic Metabolic States of Living Tissue," in *Properties and Functions of Macromolecules and Macromolecular Systems* (Moscow: "Nauka," 1969), p. 151 (In Russian).

115. Kondrasheva, M. N. "Accumulation and Utilization of Succinic Acid in Mitochondria," in *Mitochondria. Molecular Mechanisms of Enzymatic Reactions* (Moscow: "Nauka," 1972), p. 151 (In Russian).

116. Kondrasheva, M. N. and A. A. Ananenko. "Examination of the State of the Isolated Mitochondria," in *Handbook for Investigation of Biochemical Oxidation by Polarographic Technique* (Moscow: "Nauka," 1973), p. 106 (In Russian).

117. Lukianova, L. D. in *Oscillating Processes,* 2 (Puschino: "Nauka," 1971), p. 153 (In Russian).

118. Karnaukhov, V. N. and V. P. Zinchenko. "Luminescence Spectral Investigations of the Pathways of the Terminal Oxidation of the Single Muscle Cell During the Change of its Functional Activity," *Zitol.* 13(10):1243 (1971) (In Russian).

119. Blagoveschenski, A. V. *Theoretical Bases of Succinic Acid Action on Plants* (Moscow: "Nauka," 1968) (In Russian).

120. Dixon, R. and L. Fowden. "γ-Aminobutyric Acid Metabolism in Higher Plants," *Ann. Bot.* 25(100):513 (1961).

121. Makeev, A. M. "Influence of Gaseous Content of the Medium on the Organic Acids Metabolism in Corn Leaves," in *Sborn. Stud. Rabot.* 2 (Voronezsh: VGU Publishing House, 1968), p. 80 (In Russian).

122. Makeev, A. M. and A. A. Zemlianukhin. "Influence of Gaseous Content of Medium on the Content of Organic Substances in the Leaves of Pea and Sunflower," in *Problems of Botany* (Voronezsh: VGU Publishing House, 197), p. 57 (In Russian).

123. Zemlianukhin, A. A. and O. V. Alekseeva. "Daily Dynamic of Organic Acids Content in the Sunflower Leaves," *Biol. Nauki.* 1:176 (1966) (In Russian).

124. Zemlianukhin, A. A. and A. M. Makeev. "Metabolism 2,3-C^{14}-succinate, Introduced in Corn Leaves," *Fiziol. Rast.* 16(2):344 (1969) (In Russian).

125. Aronov, S. and J. Z. Hearon. *Arch. Biochem. Biophys.* 88:302 (1960).

126. Zemlianukhin, A. A. and A. M. Makeev. "Influence of Gas Composition of the Medium on the Content of Organic Substances in Leaves of the Pea and Sunflower," *Trudi VGU* 78:57 (1971) (In Russian).

127. Makeev, A. M. "Influence of the Gaseous Content of Medium on the Organic Acid Metabolism in Plants," Ph.D. Thesis, Voronezsh, 1970 (In Russian).

128. Meister, A., H. A. Sober and S. V. Tice. "Enzymatic Decarboxylation of Aspartic Acid to Alpha-alanine," *J. Biol. Chim.* 189(2):577 (1951).

129. Secenska, M. and S. Vaklinova. "Synthesis of Glutamic Acid and Alanine as a Result of Transaminase Reactions in Isolated Pea Chloroplasts," *Dokl. Bolg. Akad. Nauk.* 17(10):153 (1964) (In Russian).

130. Gonsales, P. and M. Cascales. "Sistema del 4-aminobutirato en Tejidos Normales y Tumorales (Crown-gall.) de *Helicantus tuberosis,*" *An. Real. Acad. Formac.* 32(2):215 (1966).

131. Beresov, T. T. *Amino Acid Metabolism of Normal Tissues and Malignant Tumors* (Moscow: "Medicina" Publishers, 1969) (In Russian).

132. Wood. J. D. and W. J. Watson. "The Effect of Hyperoxia and Hypoxia on Free and Bound γ-aminobutyric Acid in Mammalian Brain," *Can. J. Biochem.* 47(994) (1959).

133. Pietruszko, R. and L. Fowden. "γ-aminobutyric Acid Metabolism in Plants. Part 1. Metabolism in Yeasts," *Ann. Bot.* 25(100):491 (1961).

134. Naylor, A. and N. Tolbert. "Glutamic Acid Metabolism in Green and Etiolated Barley Leaves," *Plant Physiol.* 9(2):220 (1956).

135. Bidwell, R. G. S. "Pathways Leading to the Formation of Amino Acids and Amides in Leaves," *Can. J. Bot.* 41(12):1623 (1963).

136. Steward, F., H. Bidwell and E. Yemm. "Protein Metabolism, Respiration and Growth. A Synthesis of Results from Use of ^{14}C-Labelled Substrates and Tissue Cultures," *Nature* 178(4536):734 (1956).

137. Steward, F. C. and D. J. Durzan. "Metabolism of Nitrogenous Compounds," in *Plant Physiology,* F. C. Steward, Ed. (New York: Academic Press, 1965), p. 379.

138. Kretovich, W. L., T. I. Kariakina, N. V. Lubimova and A. N. Neronova. "Succinic Semialdehyde–Precursor of Glutamine in Plant," *Dokl. Akad. Nauk. SSSR* 170(5):1212 (1966) (In Russian).

139. Kretovich, W. L., T. I. Kariakina and N. V. Lubimova. "On the Metabolism of the Succinic Semialdehyde of the γ-aminobutyric Acid in Plants," *Fiziol. Rast.* 14(5):919 (1967) (In Russian).

140. MacLennan, D. M., H. Beevers and J. L. Harley. "Compartmentation of Acids in Plant Tissues," *Biochem. J.* 89(2):316 (1963).

141. Lips, S. H. and H. Beevers. "Compartmentation of Organic Acids in Corn Roots. I. Differential Labelling of 2-malate Pools," *Plant Physiol.* 41(4):709 (1966).

142. Lips, S. H. and H. Beevers. "Compartmentation of Organic Acids in Corn Roots. II. The Cytoplasmic Pool of Malic Acid," *Plant Physiol.* 41(4):713 (1966).

143. Kholodova, V. P. "Sucrose Localization in the Storaging Root Tissues of Sugar Cane," *Fiziol. Rast.* 14(3):444 (1967) (In Russian).

144. Yudina, O. S. "Investigation of Glucose Catabolism in Higher Plant Respiration," Ph.D. Thesis, Leningrad, 1974 (In Russian).

145. Outlow, W. H., Jr., and D. B. Fisher. "Compartmentation in *Vicia faba* Leaves. I. Kinetics of ^{14}C in the Tissues Following Pulse Labelling," *Plant Physiol.* 55(4):699 (1975).

146. Outlow, W. H., Jr., D. B. Fisher and A. L. Christy. "Compartmentation in *Vicia faba* Leaves. II. Kinetics of ^{14}C-sucrose Redistribution Among Individual Tissues Following Pulse Labelling," *Plant Physiol.* 55(4):704 (1975).

147. Lehninger, A. L. *The Mitochondrion* (New York: W. A. Benjamin Inc., 1964).

148. Oaks, A., and R. G. S. Bidwell. "Compartmentation of Intermediary Metabolites," *Ann.Rev. Plant Physiol.* 21:43 (1970).

149. Fletcher, J. S. "Heterogenous Population of Mitochondria in Higher Plant Cell," *Nature* 238(5365):466 (1972).

150. Mitchell, D. "The Metabolism of Amino Acids in Pea Seedlings," Ph.D. Thesis, Case Western Reserve University, 1968.

151. Ismailov, S. F., A. M. Smirnov and L. A. Arman. "Heterogeneity of Sucrose Pools and Amino Acid Synthesis in Isolated Roots," *Isv. ANSSSR,* ser. biol.(6):909 (1973) (In Russian).

152. Ismailov, S. F. "Amino Acid Metabolism in Isolated Roots and Root Callus Tissues of *Vicia sativa L.,*" *Fiziol. Rast.* 21(6):1217 (1974) (In Russian).

153. Ismailov, S. F., L. A. Arman and A. M. Smirnov. "Metabolism of Asparagine and Glutamine in Isolated Roots of *Medicago sativa L.,*" *Fiziol. Rast.* 21(3):518 (1974) (In Russian).

154. Ismailov, S. F. and A. M. Smirnov. "Regulation of Glutamic Acid Compartmentation in Isolated Leaves and Root Callus Tissues," in *Abstracts of the Twelfth International Botanical Congress* (Leningrad: "Nauka," 1975), p. 384 (In Russian).

155. Aronoff, S. *Techniques of Radiobiochemistry* (Ames, Iowa: Iowa State College Press, 1956).
156. Khodos, V. N. "Different Approaches to the Investigation of Compartmentalization of the Metabolism," *Fiziol. Biochim. Kult. Rast.* 6(1):23 (1974) (In Russian).
157. Obrucheva, N. V. "Specificity of Metabolism of the Root," in *Plant Physiology, Vol. I. Physiology of the Root*, N. V. Obrucheva, Ed. (Moscow: VINITI, 1973), p. 107 (In Russian).
158. Steward, F., H. Bidwell and E. Yemm. "Protein Metabolism, Respiration and Growth. A Synthesis of Result from the Use of ^{14}C-Labelled Substrates and Tissue Cultures," *Nature* 178(4536):734 (1956).
159. Gubanski, M. "Dekarboksylacija Kwasu Glutaminovego w Lisciach Tytonin Zakazonago Wirusem Mosaiki Tytonowei," *Acta Soc. Bot. Polon.* 31(1):77 (1962).
160. Thompson, J. F., C. R. Stewart and C. J. Morris. "Changes in Amino Acid Content of Excised Leaves During Incubation. I. The Effect of Water Content of Leaves and Atmospheric Oxygen Level," *Plant Physiol.* 41(12):1578 (1966).
161. Tomlison, H. and S. Rich. "Metabolic Changes in Free Amino Acids of Bean Leaves Exposed to Ozone," *Phytopathol.* 57(9):972 (1967).
162. Segeta, V. and E. Vedralová. "Content of Free Amino Acids and their Exosmose from Maize Kernels in Relation to Cold Resistance," *Biol. Plant.* 12(5):315 (1970).
163. Khavkin, E. E. "Changes in Free Amino Acids and Accumulation of Aminobutyric Acid in the Leaves of Legumes During the Aging," *Fiziol. Rast.* 11(5):862 (1974) (In Russian).
164. Weinberger, P. and C. Godin. "Glutamic Decarboxylase in Relation to Leaf Ontogeny and Vernalisation," *Can. J. Bot.* 42(3):329 (1964).
165. Ländesmaki, P. "The Amount of γ-aminobutyric Acid and Activity of Glutamic Decarboxylase in Aging Leaves," *Physiol. Plant.* 21(6):1322 (1968).
166. Inathoni, K. and J. C. Slaughter. "The Role of Glutamate Decarboxylase and γ-aminobutyric Acid in Germinating Barley," *J. Exp. Bot.* 22(72):561 (1971).
167. Lane, R. T. and M. Stiller. "Glutamic Acid Decarboxylation in Chlorella," *Plant Physiol.* 45(5):558 (1970).
168. Streeter, J. G. and J. F. Thompson. "Anaerobic Accumulation of γ-Aminobutyric Acid and Alanine in Radish Leaves (*Raphanus sativus* L.)," *Plant Physiol.* 49(4):572 (1972).
169. Streeter, J. G. and J. F. Thompson. "*In Vivo* and *In Vitro* Studies on γ-aminobutyric Acid Metabolism with the Radish Plant (*Raphanus sativus* L.)," *Plant Physiol.* 49(4):579 (1972).
170. Tsunoda, T. and I. Shiio. "Glutamic Acid Formation from γ-aminobutyric Acid by *Bacillus pumilus*. II. On the Pathway of Glutamic Acid Formation," *J. Biochem.* (Tokyo) 46:1227 (1959).
171. Roberts, E. and M. M. Bregoff. "Transamination of γ-aminobutyric Acid and β-alanine in Brain and Liver," *J. Biol. Chem.* 201:393 (1953).

172. Clendenning, K. A., E. R. Waygood and P. Weinberger. "The Carboxylases of Leaves and Their Role in Photosynthesis," *Can. J. Bot.* 30:395 (1952).

173. Scott, E. M. and W. B. Jacobi. "Soluble γ-aminobutyric-glutamic Transaminase from *Pseudomonas fluorescens,*" *J. Biol. Chem.* 334: 922 (1959).

174. Krebs, H. A. "Equilibria in Transamination Systems," *Biochem. J.* 54:83 (1953).

175. Kretovich, V. L., E. A. Morgunova, I. I. Kariakina and N. V. Lubimova. "Transamination of Keto Acids with Gamma-aminobutyric Acid and its Interaction with Glyoxilic Acid," *Dokl. Akad. Nauk SSSR* 161(2):479 (1965) (In Russian).

176. Sitinsky, I. A. "Some Drugs for Brain Treatments, Influencing the Gamma-aminobutyrate Metabolism," *J. Vses. Chim. Obsch. Im. D. I. Mendeleeva* 18(2):182 (1973) (In Russian).

177. Sitinsky, I. A. *Gamma-aminobutyric Acid in the Action of Nervous System* (Biochemistry, Pharmacology, Physiology and Clinic) (Moscow: "Nauka," 1972) (In Russian).

178. Dubinina, I. M. "Metabolism of Roots Under Various Levels of Aeration," *Fiziol. Rast.* 8(2):359 (1961) (In Russian).

179. Guinn, G. and L. A. Brinkerhoff. "Effect of Root Aeration on Amino Acid Levels in Cotton Plants," *Crop Sci.* 10(1):178 (1970).

180. Zemlianukhin, A. A. and B. F. Ivanov. "Accumulation of Gamma-aminobutyrate and Compartmentation of Glutamate in Pea Seedlings," *Fiziol. Rast.* 24(2):298 (1977) (In Russian).

181. Grineva, G. M. "Excretion by Plant Roots During Brief Periods of Anaerobiosis," *Fiziol. Rast.* 8(3):686 (1961) (In Russian).

182. Hiatt, A. J. and R. H. Love. "Loss of Organic Acids, Amino Acids, K and Cl from Barley Roots Treated Anaerobically and with Metabolic Inhibitors," *Plant Physiol.* 42(12):1731 (1967).

183. Marschner, H., R. Handley and R. Overstreet. "Potassium Loss and Changes in Fine Structure of Corn Root Tips Induced by H-ion," *Plant Physiol.* 41(12):1725 (1966).

184. Sykes, L. and I. C. Boswell. "Succinate Metabolism in Potato Tuber Tissue," *Ann. Bot.* 27(107) (1963).

185. Zemlianukhin, A. A. "Phiological Role of Ascorbic Acid and Acids of Tricarbonic Cycle in Plants," Ph.D. Thesis, Voronezsh, 1964 (In Russian).

186. Zemlianukhin, A. A., A. M. Makeev, B. F. Ivanov and A. N. Ershova. "Metabolism of Gamma-aminobutyric Acid Studied in Pea Seedlings under Conditions of Different Gaseous Composition of the Atmosphere," *Fiziol. Rast.* 21(5):1025 (1974) (In Russian).

187. Ershova, A. N. "Some Specific Features of Organic Acids and Amino Acid Metabolism in Pea Seedlings in the Dark," *Physiol. Physico-Chem. Mech. Regul. Obm. Proc. Organ.* 3:15 (1974) (In Russian).

188. Vitek, V. "Sugar-Like Diphenilamine-Positive Metabolites in Pea Seedlings," *Biochim. Biophys. Acta* 93(2):429 (1964).

189. Lui, T-Y. and P. Castelfranco. "The Biosynthesis of Ethyl-glucoside in Extracts of Pea Seedlings," *Plant Physiol.* 45(4):425 (1970).

190. Lui, T-Y, A. Oppenheim and P. Castelfranco. "Ethyl Alcohol Metabolism in Leguminous Seedlings," *Plant Physiol.* 40:1261 (1965).

191. Roodin, D. B. "The Mitochondria," in *Enzyme Cytology,* D. B. Roodin, Ed. (New York: Academic Press, 1968), p. 84 (from Russian translation, Moscow: "Mir," 1971).

192. Kretovich, W. L. *Metabolism of Nitrogen in Plants* (Moscow: "Nauka," 1972) (In Russian).

193. Mokronosov, A. T., S. K. Naparov and G. I. Rakhimova. "Metabolism of Exogeneous Alanine in the Plant Leaves," *Fiziol. Rast.* 20(4): 759 (1973) (In Russian).

194. Kafiani, K. A. "Regulation of the Enzymatic Apparatus of the Cell by Metabolites," in *Enzymes,* A. E. Braunstain, Ed. (Moscow: "Nauka," 1964), p. 269 (In Russian).

195. Guli, M. F. and D. A. Melnitchuk. "Role of the Carbon Dioxide in the Animal Metabolism," *Ukr. Biochim. J.* 45(4):489 (1973). (In Russian).

196. Chang, H. T. and W. E. Loomis. "The Effect of Carbon Dioxide on the Absorption of Water and Minerals by Roots," *Plant Physiol.* 20(1):221 (1945).

197. Agaverdiev, A. S., V. V. Mesentsev and O. R. Bart. "Extra Releasing of Energy by Plants Under Influence of High Concentration of CO_2," *Trudi MOIP* 39:179 (1972) (In Russian).

198. Grineva, G. M. *Regulation of Plant Metabolism Under Deficit of Oxygen* (Moscow: "Nauka," 1975) (In Russian).

199. Raschke, K. "The Stomatal Feedback System: Responses to Carbon Dioxide and Abscisic Acid," *Fiziol. Biochim. Kult. Rast.* 8(3):242 (1976) (In Russian).

200. Ranson, S. L., D. A. Walker and J. D. Clarke. "Effects of Carbon Dioxide on Mitochondrial Enzymes from *Ricinus,*" *Biochem. J.* 76(2):216 (1960).

201. Bendall, D. S., S. L. Ranson and D. A. Walker. "Effects of Carbon Dioxide on the Oxidation of Succinate and Reduced Diphosphopyridine Nucleotide by *Ricinus* Mitochondria," *Biochem. J.* 76(2):221 (1960).

202. Shipway, M. R. and W. J. Bramlage. "Effects of Carbon Dioxide on Activity of Apple Mitochondria," *Plant Physiol.* 51(6):1095 (1973).

203. Wager, G. G. "The Effect of Subjecting Peas to Air Enriched with Carbon Dioxide. I. The Path of Gaseous Diffusion, the Content of CO_2 and the Buffering of the Tissue," *J. Exp. Bot.* 25(85):330 (1974).

204. Madsen, E. "Effect of CO_2 Concentration on the Accumulation of Starch and Sugar in Tomato Leaves," *Physiol. Plant.* 21(1):168 (1968).

205. Madsen, E. "The Effect of Carbon Dioxide Concentration on the Photosynthetic Rate in Tomato Leaves," in *Royal Vet. Agric. Univ. Yearb.* (1971), p. 195.

206. Madsen, E. "Cytological Changes Due to the Effect of Carbon Dioxide Concentration on the Accumulation of Starch in Chloroplasts of Tomato Leaves," in *Royal Vet. Agric. Univ. Yearb.* (1971), p. 191.

207. Madsen, E. "The Influence of CO_2 Concentration on the Content of Ascorbic Acid in Tomato Leaves," *Ugeskr. Agron.* 116(28):592 (1971).
208. Madsen, E. "The Effect of CO_2 Concentration on Development and Dry Matter Production in Young Tomato Leaves," *Acta Agric. Scand.* 23(4):235 (1973).
209. Madsen, E. "The Effect of CO_2 Concentration on the Occurrence of a Number of Acids from the Citric Acid Cycle in Tomato Leaves," *Physiol. Plant.* 32(1):10 (1974).
210. Ito, T. "Plant Growth and Physiology of Vegetable Plants as Influenced by Carbon Dioxide Environment," *Trans. Fac. Hortic. Chiba Univ., Japan* 7:1 (1973).
211. Krasinskij, N. P. "The Cultivation of Plants in Artificial Light under Glass, and the Increase in the Effectiveness of Illumination by CO_2 Nutrition," *Tr. Inst. Fiziol. Rast., Timiryazeva* 10:64 (1955) (In Russian).
212. Metlitski, L. V. "Biochemistry of Fruits and Vegetables," in *Technological Biochemistry* (Moscow: Vischaia schcola, 1973), p. 116 (In Russian).
213. Kidd, F. and C. West. "Respiratory Activity and Duration of Life of Apples Gathered at Different Stages," *Plant. Physiol.* 20(4) (1945).
214. Biale, J. B. "Respiration of Fruits," in *Handbuch der Pflanzenphysiologie, Vol. XII, part II* (Berlin: Springer-Verlag, 1960), p. 536.
215. Kidd, F. and C. West. "Recent Advances in the Work of Refrigerated Gas-Storage of Fruit," *J. Pom. Hort. Sci.* 14:299 (1936).
216. Allentoff, N., W. R. Phillips and F. B. Johnston. "A ^{14}C-Study of CO_2 Fixation in the Apple," *J. Sci. Food. Agric.* (5):231 (1954).
217. Rakitin, U. V., A. V. Krilov and A. A. Kolesnik. "On the Dark Fixation of Carbon Dioxide by Plants," *Fiziol. Rast.* 3(3):225 (1956) (In Russian).
218. Kollas, D. "Preliminary Investigation of Influence of the Controlled Atmosphere Storage on the Organic Acids in Apples," *Nature* 204: 758 (1964).
219. Ulrich, R. and I. Landry. "Formation d'acides Organiques dans les Fruits Conservées en Atmosphere Riche en Gaz Carbonique," *C. R. Acad. Sci., Paris* 23:242 (1956).
220. Dilley, D. R. "Malic Enzyme Activity in Apple Fruits," *Nature* 196:387 (1962).
221. Salkova, E. G. and T. A. Nikifirova. "Content and Dynamics of Nonvolatile Organic Acids in Ripening Apples," *Dokl. Akad. Nauk SSSR* 179(1):218 (1968) (In Russian).
222. Salkova, E. G. and T. A. Nikiforova. "Conversions of Organic Acids and Their Role in the Development of Physiological Disease in Fruits," in *Biochemistry of Immunity and Quietness of Plants. 3* (Moscow: Publishing House of Akad. Nauk. SSSR, 1969) (In Russian).
223. Kidd, F. and C. West. "Effect of Oxygen and Carbon Dioxide on the Chemical Changes in Stored Apples," *New Phytol.* 38:105 (1939).

224. Young, R. E., J. R. Rogers and J. B. Biale. "Carbon Dioxide Effects on Fruit Respiration. II. Response of Avocados, Bananas and Lemons," *Plant Physiol.* 37(3):416 (1962).
225. Lebermen, K. W., A. I. Nelson and M. P. Steinberg. *Td. Technol., Champaing* 22 (1962).
226. Ranson, S. L. "Zymasis and Acid Metabolism in Higher Plants," *Nature* 172:252 (1953).
227. Hulme, A. C. "Carbon Dioxide Injury and the Presence of Succinic Acid in Apples," *Nature* 48:218 (1956).
228. Williams, M. V. and M. E. Patterson. "Nonvolatile Organic Acids and Core Breakdown of Bartlett Pears," *J. Agric. Food Chem.* 12(1):80 (1964).
229. Le Roux, P. *Mushr. Sci.* 7:31 (1968).
230. Wankier, B. N., D. K. Salunkhe and W. F. Campbell. *J. Am. Soc. Hort. Sci.* 95:604 (1970).
231. MacGlasson, W. B. and W. B. Wills. "Effect of Oxygen and Carbon Dioxide on Respiration, Storage Life and Organic Acids of Green Bananas," *Aust. J. Biol. Sci.* 25(1):35 (1972).
232. Wager, H. G. "The Effect of Subjecting Peas to Air Enriched with Carbon Dioxide. II. Respiration and Metabolism of the Major Acids," *J. Exp. Bot.* 25(85):338 (1974).
233. Thomas, M. "The Controlling Influence of Carbon Dioxide," *Biochem. J.* 19:427 (1925).
234. Smith, W. H. "Reduction of Low-Temperature Injury to Stored Apples by Modulation of Environmental Conditions," *Nature* 181:275 (1958).
235. Walker, P. A. and J. M. A. Brown. "Physiological Studies on Acid Metabolism. 5. Effects of Carbon Dioxide Concentration on Phosphoenolpyruvic Carboxylase Activity," *Biochem. J.* 67:78 (1957).
236. Wager, H. G. "The Incorporation of $^{14}CO_2$ into Green Peas in the Dark," *J. Exp. Bot.* 18(57):672 (1967).

9

ROOT AERATION IN THE WETLAND CONDITION

W. Armstrong

Department of Plant Biology
The University of Hull
Hull HU6 - 7RX, United Kingdom

INTRODUCTION

Although temperature differentials, barometric changes and unequal gas-phase exchange may cause some mass flow of gases between the soil and the aerial environment, soil aeration is considered to be largely a diffusion-controlled process.[1-3] Diffusion is also thought to be the overriding mechanism in root aeration.

In the gas phase of unsaturated soils, the balance between the rate of oxygen supply and the rate of its consumption favors high oxygen concentrations because of the high gas-phase oxygen diffusivity.* In flooded soils, because of the very low oxygen diffusivity in aqueous media,** the situation is reversed. Water preferentially displaces air from the soil pore space, and although the flooded soil retains a proportion of gas-filled space this is discontinuous.[4,5] Consequently, the rate of oxygen supply in the flooded soil falls to insignificant levels, and the equilibrium swings rapidly towards anaerobiosis. As the detailed physico-chemical characteristics of wet soils are discussed in Chapters XIII and XIV, it will suffice here to note that not only are wetland soils essentially devoid of free oxygen,[6] but they may be considered also as negatively aerated because of the accumulated quantities of readily oxidizable compounds

---

*The diffusion coefficient of oxygen in air, D_{O_2}/air, is 2.05×10^{-2} $cm^2 s^{-1}$ @ $23°C$.
**$D_{O_2}/H_2O = 2.267 \times 10^{-5}$ $cm^2 s^{-1}$ @ $23°C$.
   $D_{O_2}/wet \; soil$ can be 1×10^{-5} $cm^2 s^{-1}$ or less.

they contain—compounds that have arisen from the activities of bacterial anaerobes. These compounds, together with the facultative anaerobes, may constitute an appreciable latent oxygen sink within the wetland soil,[7] while among the reductants there are chemicals of proven phytotoxicity.

Despite these potentially harmful properties of the waterlogged soil, many plants are endemic to wetland sites. Many others can withstand some degree of soil anoxia yet are unable to compete successfully with wetland species. Anaerobic metabolism may account for some tolerance of anoxia in wetland conditions[8-10] (see Chapters IV, V, VII, VIII, XIV and XV). Submerged and leafless rhizomes may survive some winter waterlogging in this way, and it is possible also that nondormant species may benefit from anaerobic metabolism during temporary flood periods. However, anaerobic forms of metabolism have not been found that will support the active growth of higher plants for which a sustained supply of molecular oxygen seems to be essential.[11,12] (See also p. 277).

Therefore, two major problems confront the submerged root: the permanency of anaerobiosis in the wetland soil and the presence of soluble phytotoxins. Deficiencies of essential nutrients, such as nitrate, and the inactivity of aerobic bacteria and fungi are associated difficulties.

The reliance on aerobic respiration for growth and the permanency of the anoxic soil environment clearly demonstrate the need for some form of ventilation in submerged plant organs; the presence of phytotoxins, many of which can be rendered harmless by direct oxidation or biological destruction, also indicates the desirability for ventilation.[13]

A great deal of research into the subject of root aeration has taken place in recent years, and with better understanding of the mechanism of aeration and its complexities, there has been a movement towards mathematical and other forms of modeling.[13-16] Internal ventilation has emerged as a vital property of the wetland root system, as has the phenomenon of rhizosphere amelioration by radial oxygen loss through the root wall. With this interest in root aeration has also come the realization that a ventilating pathway is also not an insignificant property of nonwetland plants.[14,15,17,18]

This chapter is concerned primarily with the ventilating mechanism in higher plants and the adequacy of internal aeration in both the wetland and nonwetland root types. The concluding remarks include some suggestions for future research.

THE VENTILATING MECHANISM

It is not certain whether the liquid-phase movement of gases across cells and into mitochondria is primarily diffusion-controlled or is enhanced by cytoplasmic activity. However, there seems to be little doubt that diffusion plays a major role in the aeration of submerged roots and that the gross aeration of the plant can be treated in diffusion terms. Molecular oxygen freely enters the aerial parts of the plant through stomata or lenticels, and in herbaceous species, will move rapidly by gas-phase diffusion towards the extremities of submerged organs via the intercellular gas space continuum of the ground tissues; likewise, carbon dioxide will diffuse away from the respiratory sites to the atmosphere. Within trees a more circuitous route may be followed along secondary rays, through empty elements of the vascular system and air spaces in the root pericycle.[19,20]

The simplest analogue of diffusion in a tubular structure such as a root is the diffusion equation

$$\text{Diffusion rate} = -\frac{DA\,(C_2-C_1)}{L}\ \text{g s}^{-1} \tag{1}$$

The gas source (C_2) is located at one end of the tube, the sink represented by the concentration C_1 (g cm^{-3}) at the other end; A is the tubular cross-sectional area (cm^2) and L, the length of the tube (the diffusion path length: cm); D, the diffusion coefficient, quantifies the diffusivity of the gas in air $(\text{cm}^2\,\text{s}^{-1})$. Although living roots only rarely, if ever, approximate this analogue, the equation does illustrate two fundamental aspects of diffusion. Solution of the equation gives the *net* diffusive flow from source to sink, and this depends, on the one hand, upon the concentration difference between source and sink, and, on the other, upon the value of the expression L/DA. L/DA represents the diffusional impedance of the system (s cm^{-3}). Usually, the living submerged root deviates from this simple system in several ways. In the first place, the gas space system is often distributed nonuniformly along the root.[14,21] Also, the intercellular channels of low-porosity roots can be sufficiently tortuous to lengthen considerably the diffusion path and hence the nonmetabolic resistance of the system.[22] To accommodate this feature in a root segment of uniform porosity, the expression L/DA can be rewritten as follows:

$$R_p = \frac{L}{\epsilon\tau DA} \tag{2}$$

where R_p is the nonmetabolic resistance (s cm^{-3}),
    L is the length of the segment (cm),
    τ is a tortuosity factor,[22]
    ϵ is the fractional porosity of the segment.

Sink activity differs most markedly from that in the simple analogue equation and it is this which makes the mathematical description of root aeration so complex.[14] In the submerged root, respiratory activity and radial leakage to the soil both contribute to the total oxygen sink; the former is largely independent of concentration,[23] the latter perhaps more concentration-dependent,[24] and both tend to be nonuniformly distributed along the root.[14,18,21,25] Lateral roots also contribute to the general sink activity.[18] This distribution of activities along the whole length of the diffusion path has a considerable influence on the oxygen balance of such remote and physiologically active parts as the root apices. The oxygen sinks behave synergistically with the nonmetabolic diffusional resistance: they may be looked upon as effective resistances and quantified in units of diffusional resistance.[16,18,26] For any given level of sink activity, the higher the nonmetabolic resistance the higher will be the effective resistance of these sinks.

The oxygen status of the submerged root is therefore determined by the interaction of a number of factors, which may be categorized as real or as effective resistances in the diffusion system. The magnitude of these resistances and their synergistic effects are examined in some detail in the following sections.

THE WETLAND PLANT

The Adverse Effects of Anoxia

Vartapetian[27] has demonstrated that in the roots of at least one wetland species there is a high sensitivity to anoxia: electron microscope studies revealed destructive changes in the cellular organelles of rice roots after only 4-5 hr of exposure to a nitrogen atmosphere, and after 7 hr of anaerobiosis the cell ultrastructure appeared grossly impaired. After 24 hr of anaerobiosis, the separated roots were placed in continuously aerated water. Not only was there no sign of recovery, but there was also evidence of even further organelle damage and breakdown. It seems reasonable to assume that the roots of other successful wetland species may be equally sensitive to anoxia.

Gas Space Development

Extensive gas space development in roots, together with a reciprocal decrease in oxygen demand, is the typical response to soil anaerobiosis of most wetland species. Some degree of oxygen stress in the meristematic regions of the root may always be necessary to trigger gas space development,[28,29] but there is nevertheless ample evidence to show that by means of their well developed ventilating structure, wetland roots not only avoid anoxia but are relatively well aerated.

Most roots, whether of wetland or nonwetland plants, develop small but continuous intercellular gas spaces in cortical parenchyma. These spaces are first obvious just behind the meristem and when fully developed may occupy up to 12% of the total root cross-sectional area. The much larger gas spaces developed in the wetland root can arise in several ways and at varying distances from the root meristem. In the grasses and sedges, the normal cortical intercellular gas space system becomes greatly enlarged about 2-3 cm behind the apex by the separation and collapse of cells occupying substantial sectors of root. Sometimes collapse will be so extensive as to leave intact only the cortical cells immediately adjacent to the endodermis and hypodermis. In other herbaceous species, rhizogeny without lysigeny is common; the number of cells is not reduced and a honeycomb structure is formed. Detailed accounts of the ventilating structure of wetland plants may be found elsewhere.[30-34]

The Diffusional Resistances

Because of this pattern of relatively low apical porosity and very high subapical porosity, the wetland root as it grows does not accumulate a high nonmetabolic resistance. The example in Figure 1 will serve to illustrate this point; excluding the nonporous tissues of the meristematic region we find that the pore space resistance of the apical 4 cm of root totals 0.210×10^5 s cm^{-3}, whereas each succeeding centimeter adds little more than 0.013×10^5 s cm^{-3} to the accumulating resistance.*

Respiratory activity in plant tissues usually declines with age; consequently, roots naturally exhibit their greatest oxygen demand at the apex followed by a basipetal decline. In nonwetland species, the decline may be very gradual,[14] but in wetland plants, it can be most pronounced because of the cell collapse and aerenchyma formation. Luxmoore et al.[14] recorded an 80% decline over the apical 4 cm in rice. Reduced oxygen demand in the subapical region naturally reduces the effective resistance to oxygen flow to the root apex; the juxtaposition of low respiratory demand and low nonmetabolic resistance means that the effective resistance imposed by respiratory activity in the subapical regions is extremely low. Similarly, because of their subapical position on the root, the oxygen uptake by lateral roots will add little to the total longitudinal resistance.

---

*A tortuosity factor of 0.75 was used for the apical centimeter.

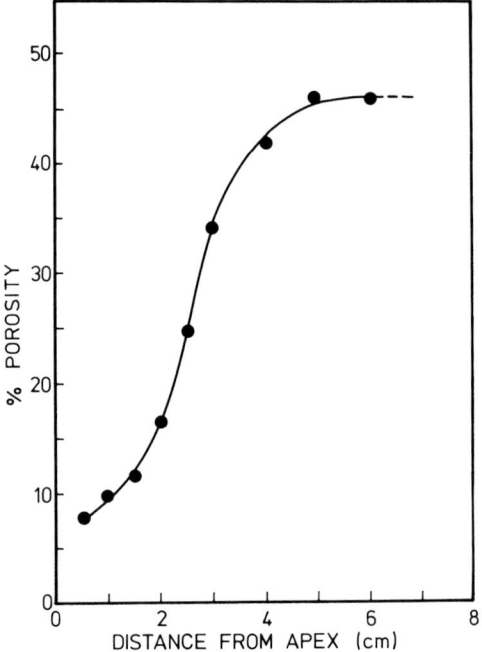

Figure 1. Changes in total gas-space volume (porosity, %) which occur along rice roots grown in saturated soil.[21]

The Leakage Characteristics of the Root Wall

A further characteristic of the wetland root, which at first sight might seem very beneficial to the ventilation of remote parts, is the rapid reduction in the oxygen permeability of the root wall in subapical regions.[14,21,35] Clearly, it will conserve some oxygen, but again, because of its subapical position, it will not greatly affect the total effective longitudinal resistance in the root. It is possible that its chief functions are to form a barrier to the influx of phytotoxic products from the soil, and to curtail the period over which oxygen is released to any fixed point in the soil. It is of interest that the boundary of observable rhizosphere oxidation around static artificial Si-roots placed in wet soil can diminish rapidly after an initial peak and the oxidation boundary eventually approaches the root wall (unpublished results). If the living wetland root did not quickly lose its permeability to oxygen and phytotoxins in subapical regions, such a contraction in the rhizosphere boundary might lead to a critical inflow of phytotoxins. Hence, it may be that rhizosphere oxygenation at any point in the soil may·be most advantageous for a relatively short period only and, thereafter, may become disadvantageous. The dynamics of rhizosphere oxidation and oxygenation require much more thorough investigation.

Respiratory Activity and Internal Oxygen Pressure

The relatively high oxygen status attained by wetland root systems has been demonstrated frequently. Our own analyses, which are nondestructive and based on measurements of radial oxygen loss from the root to an encircling tubular Pt-electrode sink have shown that even the apical nonlacunate gas spaces can be well furnished with oxygen: the mean apical concentration recorded in a series of experiments on rice was about 12% where the root length lay between 5 and 10 cm.[21]

However, concentrations of this magnitude are not, in themselves, sufficient justification for assuming adequate internal aeration; it is also necessary to establish the relationship that exists between respiration and oxygen concentration in the intact plant. Similarly, one must know the optimal levels of oxygen required to achieve root growth and sustain the oxidizing activity of the rhizosphere (see pp. 277 and 285). Unfortunately, little is known of the respiratory responses of intact plants; it has proved much more convenient to measure the oxygen uptake of tissue blocks in manometric or membrane electrode assemblies. Under these circumstances, oxygen uptake increases hyperbolically as the oxygen pressure is raised, until a point is reached—the critical oxygen pressure (COP)[25]—at which the respiration becomes constant. A search of the literature reveals few instances in which the COP obtained by *in vitro* methods lies below 0.10 atm (10% oxygen), and Luxmoore *et al.*[14] have recorded values in excess of 0.2075 atm for rice root respiration.

However, one must seriously question the significance of COP data obtained by *in vitro* analysis. In most instances, these methods cause a flooding of the intercellular gas space system of the sample. The infilling of gas spaces will substantially increase diffusional impedance and, in these circumstances, an abnormally high oxygen pressure at the boundary of the sample will be necessary to sustain its respiratory activity; consequently, an abnormally high COP will be recorded.

In the intact plant, the unflooded gas space system greatly enhances oxygen diffusivity, and consequently, the COP of the *in vivo* condition should be substantially lower than that normally detected *in vitro*. Some of our recent experiments[23] have shown that the COP for root respiration in the intact wetland plant may be an order of magnitude lower than found by *in vitro* analyses: rice (cv. Norin 36) 0.024 ± 0.001 atm (cv. Norin 37) 0.026 ± 0.002 atm, and *Eriophorum angustifolium* 0.02 ± 0.004 atm. (NB: *in vitro*, a value of 0.14 atm was recorded for *E. angustifolium*.)

The pattern of our results has also indicated that the COPs recorded for the intact roots are chiefly a function of high diffusional impedance in the meristem and stelar regions, and that the COP of porous tissue such as the cortex may lie below 0.001 atm and thus approach the K_m value for the activity of the cytochromes.[36,37] We have suggested that respiratory activity in the intact wetland root does not show the normally accepted hyperbolic relationship with oxygen partial pressure (Figure 2), but adheres instead to the type of curve shown in Figure 3. Below the COP, the fall in respiratory activity proceeds in two clearly defined stages: the first is thought to correspond with the declining activities of the stele and meristematic regions and be indicative of the progressive enlargement of anaerobic centers within these tissues; the second stage is believed to represent the decline in cortical cell respiration.

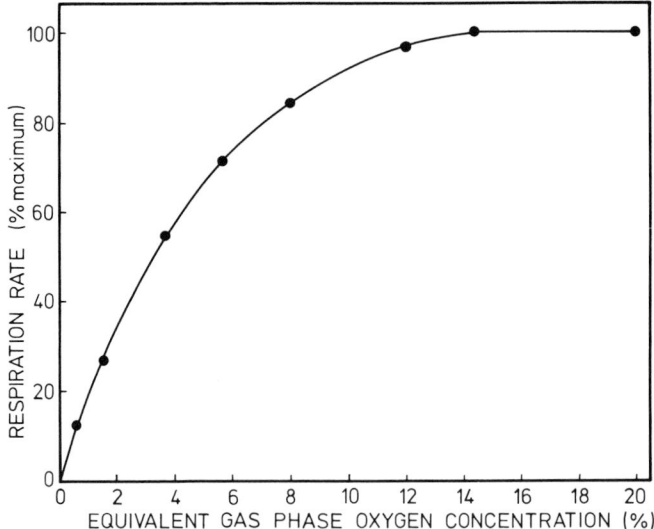

Figure 2. *Eriophorum angustifolium.* Root respiration as a function of oxygen concentration. Measurements were made *in vitro* in a Rank membrane-electrode assembly using slices taken from 2-cm apical segments.[23]

Undoubtedly, the attainment of full respiratory activity at such low oxygen pressures would be a beneficial feature in roots subjected to the wetland condition; it would be particularly so when combined with the high oxygen diffusivity and low respiratory demand of the wetland plant root system.

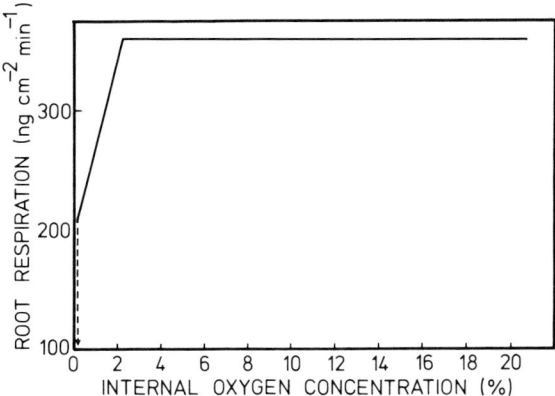

Figure 3. Postulated relationship between respiration rate and cortical oxygen concentration in the apical centimeters of roots.[23]

Internal Oxygen Pressure and Root Growth

It is not possible to more than comment briefly on the relationship between internal oxygen pressure and root growth, for nothing has yet appeared on this subject in the literature, and our own studies are at a very early stage. Our preliminary studies indicate, however, that growth is unaffected by a lowering of the internal oxygen pressure until the COP is reached.[60] Whenever the oxygen concentration in the cortical gas spaces of the root apex has been allowed to fall below the COP, we have found that growth immediately declines and eventually is halted. Surprisingly perhaps, this seems to occur at all concentrations below the COP, but this would be in accordance with the suggested development of anaerobic centers in the root at values immediately below the COP (see above). It will be interesting to learn for what period the root may be held below the COP and still retain its capacity for growth; because of low oxygen diffusivity in the meristematic and stelar regions, these tissues may rapidly suffer the organelle damage noted by Vartapetian[27] when the oxygen concentration in the cortical spaces falls below the COP.

Photosynthesis and Internal Aeration

It has been reported that rice roots are much more readily damaged by sulfides in the wetland soil during periods of sunless weather or if the aerial parts of the plant are deliberately shaded.[38] It has been suggested that the low light intensity interferes in some way with the oxidizing activities of the roots.

The oxygen produced in photosynthesis can undoubtedly improve the oxygen status of the submerged root.[60] The extent to which this occurs depends upon two factors: the degree of immersion of the leafy parts, and the availability of CO_2 at the leaf surface. The bicarbonate ion, it seems, is ineffective where cuticular resistance necessitates entry of the carbon source by the stomata. If the photosynthetic parts of the plant are wholly above the free-standing water, the likelihood of increased oxygen pressure in the roots, even at high light intensity, is remote, except in the presence of astomatal but photosynthetic leaf bases. Stomatal surfaces of leaves offer little restraint to the escape of gases, and hence the oxygen produced in photosynthesis is lost rapidly to the atmosphere.

If, on the other hand, a portion of the leaf system is immersed in free water and illuminated, the high diffusional resistance of the water will seriously impede the escape of oxygen from the surfaces of the immersed parts and may lead to a significant increase in oxygen pressure. This effect will be transmitted to the root system.

Using *E. angustifolium,* we have found that at light intensities up to $100~\mu Em^{-2}s^{-1}$ photosynthetic oxygen production has an insignificant effect on root aeration in the nonsubmerged plant and in plants immersed to the top of the outermost leaf sheath (approx. one quarter submerged). In the dark, oxygen levels in the root were depressed by all degrees of immersion above one quarter because of the blocking of stomata, and hence increased internal diffusion path length. With no free CO_2 in the submerging medium, a light flux of $100~\mu Em^{-2}s^{-1}$ gave internal oxygen pressures at the root apex to about 110% of the nonsubmerged condition at a half submergence; 30-40% for three quarters to full submergence. With added CO_2, oxygen concentrations in the root increased sigmoidally with light flux. In fully submerged plants at a light flux of $100\mu Em^{-2}s^{-1}$, $150^{+}\%$ of the nonsubmerged value was reached at 0.8 mM CO_2/l; 140% at three quarters submergence; 110% at half submergence. At $300~\mu E$, the latter value rose to 120%.

However, the leaf bases in *E. angustifolium* are virtually nonphotosynthetic; with its green leaf sheaths one would anticipate a much greater response in rice. These results do suggest that for any fixed level of submergence, sunny weather may have advantages over sunless weather for the oxygen status and oxidizing activities of wetland roots. It is not clear, however, whether a reduced level of submergence might, in the long term, be just as advantageous. During periods of darkness, submergence of leafy parts will add both real and effective resistance to the system, and this clearly will be detrimental for root aeration.

Stomata, Lenticels and Internal Aeration

Results indicate that internal aeration is not normally restricted by stomata or a paucity of lenticels. Relatively few stomata are required for the normal aeration of the root system of *E. angustifolium*.[60] Darkness, antitranspirants and increased CO_2 levels all failed to reduce the oxygen status of the root until the effective stomata had been reduced to small numbers by lanolining or by partial submergence of the leaf system. Similar effects can be demonstrated in woody species, the aeration of the root system being little affected by the blocking of lenticels lying more than a few millimeters away from the waterline (personal observations). However, if the lenticels immediately above the water level are blocked, the effect is considerable.[39,40]

The Gas Space System as an Oxygen Reservoir

It is generally believed that the air spaces of wetland plants form an invaluable oxygen reservoir. It remains to be shown if and under what circumstances this is so. The substantial gas space provision in *E. angustifolium* did not sustain the normal respiratory activity of the submerged plant for longer than 40 min at 23°C, neither does there appear to be any feedback mechanism, which might lower the respiratory rate in response to a lowering of the internal oxygen concentration: respiratory rate remained constant until the COP was reached (Figure 4).

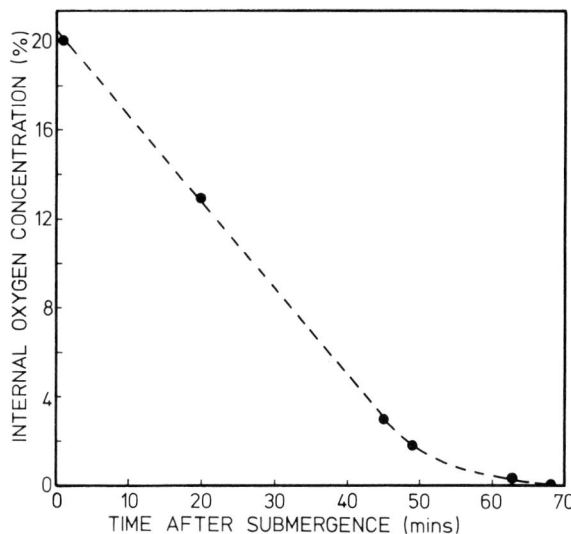

Figure 4. *Eriophorum angustifolium.* Internal oxygen concentration in the leaves following the submergence of the intact plant in anaerobic medium.[23]

THE NONWETLAND ROOT

Anoxia

It appears that the root cells of some nonwetland plants may be somewhat less sensitive to anoxia than those of rice. Vartapetian[27] found little evidence of changes in the mitochondrial apparatus of pumpkin, tomato and bean roots after 24 hr of nitrogen bubbling; in rice roots, damage was evident after 4 hr. However at 50 hr there were signs of irreversible damage to the mitochondrial ultrastructure and of extensive destructive changes occurring throughout the cells of all species. The advantages or disadvantages of a more delayed response to anoxia have yet to be elucidated.

Growth

The growth of roots in wet soil is probably a good indication of the sufficiency of internal aeration in the nonwetland plant; recorded observations also demonstrate some degree of adaptation to the wetland environment among nonwetland species, although the plasticity of response found in wetland species is never realized.

Yu et al.[41] observed the root growth of several nonwetland crop species under a whole range of soil treatments, which included full flooded, half flooded and drained. With the exception of barley, the roots produced in the drained treatments were invariably of lower porosity than those that grew in the fully flooded soil (Table I).

Table I. Root Porosities Developed Under Drained and Flooded Conditions[41]

Species	Root Porosity (%)	
	Drained	Fully Flooded
Corn (1)	6.5	15.5
Corn (2)	11$^+$	17
Sunflower (1)	5$^+$	11$^+$
Barley	3.5	2$^+$
Sunflower (2)	7	11
'Inia' Wheat	3$^+$	7.5
'Pato' Wheat	5.5	14.5
Tomato	4$^+$	7.5

Also of note was the poor penetration of the water table by the low porosity roots of the half flooded treatment and the death of low porosity roots (when fully flooded). Most plants responded to full flooding by the production of new adventitious roots of higher porosity which penetrated up to 17 cm depending upon the species; penetration in drained soil occurred to 47 cm. Corn showed the greatest degree of adaptation to waterlogged conditions, but notably poor penetration occurred in tomato (5 cm), wheat (4-10 cm) and barley (12 cm). Penetration by the barley was nevertheless more than normally commensurate with its low root porosity of 2.4%. This could be accounted for by a particularly low respiratory rate. Yu et al.[41] were thus able to explain their observations and the variations discovered in terms of the factors likely to control the internal ventilation of the roots. Similarly, others have related the depth of wet soil penetration by nonwetland species to the oxidizing activities of their roots. (See p. 273).

Diffusional Resistance

Wetland plants usually adapt to soil wetness by the development of roots in which there is an exceptional reduction in diffusional resistance. This is brought about by the development of cortical zones of high porosity and a correspondingly low oxygen demand in subapical regions (see p. 273). Plants of normally well aerated habitats do not achieve the same degree of response, remaining structurally and physiologically ill-adapted and hence less competitive in the wetland condition. The basipetal changes in porosity and respiratory demand that characterize wetland species are not nearly so pronounced; neither is there the same tendency for reduced wall permeability in the subapical regions.

In the previous section, barley was described as showing the least structural adaptation to the wetland environment with a root porosity of 2.4%; the tortuosity of the gas channels at such low porosities can mean an effective porosity of less than half this value. Low porosity is synonymous with high diffusional resistance, and we may illustrate the effects of porosity *per se* on ventilation as follows. The nonmetabolic diffusional resistance of a rice root of length 10 cm, radius 0.05 cm, and from a wetland soil, is unlikely to exceed 0.29×10^5 s cm^{-3}; the comparable value for a barley root of similar length and radius, but with a porosity of 2.4% (effective value 1.04%) would be about 5.98×10^5 s cm^{-3}. If these roots neither respired nor lost oxygen through their walls, the oxygen pressure in the apices of both would be that of a moist aerial environment (0.2075 atm, 20.75%). The positioning of simple tubular electrode oxygen sinks around the apices would, however, be sufficient to lower the oxygen concentration in the apex of the barley root to 6.2%, while in the rice root, the concentration would only fall to 17.6%. An additional centimeter of root would lower the concentration in barley to 5.7%, but a fall of only 0.1% would be experienced in rice.

If the oxygen status of the barley root can be reduced by 75% after only 10 cm of growth, without the agency of root respiration or a more extensive external sink, then it is not difficult to imagine the possible consequences of a uniform respiratory demand along the root and a relatively undiminished subapical leakage of oxygen to a wet soil. The synergistic effects of these activities could be enormous and the total effective diffusional resistance probably approaches infinity at the limits of root penetration.

Pisum sativum L.

It is evident from the very nature of the structural and respiratory characteristics of the nonwetland root type that considerable variation in ventilating efficiency should be expected even during the growth of individual plants. This is demonstrated very convincingly in a recently completed study of oxygen transport in pea roots.[18] In this, the first investigation of its type, diffusional resistances (real and effective) were assessed from measurements of oxygen flux from living roots recorded by means of the cylindrical Pt electrode technique.[26] Changes in nonmetabolic resistance and the synergistic effects of respiratory activity and leakage during root elongation were quantified, as were the effects of secondary root production and the partial submergence of the developing plumule. Also, changes in oxygen status, brought about by aging effects in the roots, were documented.

The magnitudes of the nonmetabolic resistances recorded for pea roots were consistent with the known characteristics of nonwetland plants, but the changing pattern of resistance during the earlier stages of root elongation was not entirely anticipated. From 3.5-8.5 cm, there was no net gain in the resistance of the roots (Figure 5). This observation, attributable to the changing shape of the developing root, illustrates how ventilation may be effectively improved in nonaerenchymatous roots. A root developing as an inverted cone (carrot-shaped) will, if the respiratory demand is reduced in the basal regions, have an effectively greater porosity than one in which the narrow apical diameter persists into the basal regions.

The peas were grown, for the most part, in sterile 1% agar medium in 250-ml glass measuring cylinders, and the plants were removed for assay by extracting the agar core intact. To minimize oxygen leakage from subapical regions, only the agar enclosing the primary apex was trimmed away before assay; plant and agar core were immersed in anaerobic liquid medium to the cotyledon junction before moving the Pt electrode into position around the primary apex. The effects of subapical oxygen leakage were studied by trimming away the whole of the agar jacket, and the

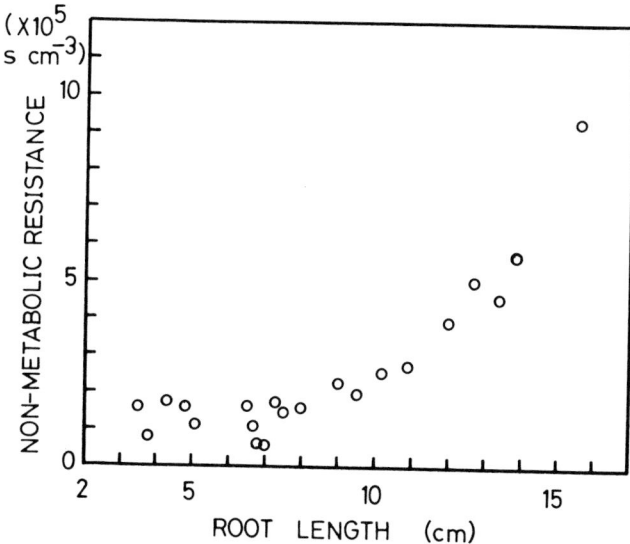

Figure 5. Changes in nonmetabolic diffusional resistance that take place during the growth of the primary pea root (unpublished results).

influence of lateral roots on the oxygen status of the primary was determined by assaying the oxygen flux from the primary apex before and after excision of the laterals.

In roots up to 10 cm long, jacketed and with lateral roots removed, oxygen flux from the primary apex showed a smooth curvilinear decrease with increasing length. This corresponded with an increase in effective diffusional resistance from 7.5×10^5 s cm^{-3} at 3 cm to 25×10^5 s cm^{-3} at 10 cm, attributable to the synergism of primary root respiration and nonmetabolic resistance. Retention of secondary roots was associated with a bimodal flux pattern. Initially, the decrease in flux was as in the previous example, but then a steep decline in the oxygen status within the primary root apex accompanied the emergence of the laterals: effective diffusional resistance increased from about 13×10^5 s cm^{-3} at 5.5 cm immediately prior to emergence, to a value of 270×10^5 s cm^{-3} at 10 cm. This is the first record of the possible influence of secondary roots on internal ventilation of a major root, and the effect can be seen to be enormous. One would anticipate a minor role for the secondaries of wetland roots.

The role of lateral roots is more pronounced when leakage is permitted; in pea, the effective diffusional resistance approached infinity, and the apical oxygen status became immeasurably small at a root length of about

8 cm. These observations agree well with the finding that the root growth in periodically degassed liquid medium ceases abruptly at about 8.5-9 cm. Within a 1% agar core, root growth continues considerably beyond 9 cm.

We have suggested that the initial 8-9 cm of root growth in pea is sustained by internal ventilation, provided there is some impedance to radial oxygen loss,[17] and provided also that the level of the culture solution does not come above the root shoot junction. Occlusion of stomata on the plumule can considerably increase the effective diffusional resistance of the ventilating path.[18] Static oxygen-free culture solution will impede radial oxygen loss, but if growth is to proceed beyond 9 cm a more effective 'jacketing' appears to be essential (viz. 1% agar). For prolonged growth, it will eventually become necessary for the rooting medium to receive forced aeration.

In a true wetland situation, it seems unlikely that pea will initially attain a root length of 8-9 cm, because of the intensity of soil sink activity. It is interesting that in continuously degassed medium, growth ceases at 4-5 cm.[18] However, the final root length attained by the non-wetland plant in wetland conditions might also depend upon aging effects in the root system. Healy[18] noted that a respiratory decline in the basal regions that accompanied aging brought about the elevation of oxygen pressure and a resumption of growth in the primary root of the pea.

VENTILATION AND THE IMMOBILIZATION OF PHYTOTOXINS

Internal ventilation not only serves the normal respiratory requirements of roots, but is probably essential, directly or indirectly, for the protective oxidation reactions in and around roots.[13] These reactions remove from circulation the potentially harmful phytotoxins of wetland soils, inorganic substances such as the reduced form of manganese, ferrous iron, S^{2-} and possibly organic products of various types.[6] The immobilization of substances takes place within roots, but depositions on the root surface and in the rhizosphere are the more obvious. Immobilization may occur simply because of direct oxidation by the molecular oxygen within, or leaking from, the root system.[24,42] It may be catalyzed by plant enzymes,[43] or it may require the activities of the rhizosphere and rhizoplane microbe populations.[44]

Yoshida et al.[45] have suggested that the ventilation of roots might encourage the activities of N_2-fixing bacteria in the rhizosphere: Read and Armstrong[46] have shown that fungal symbionts can be supported by oxygen leaking from the roots of conifer seedlings.

A close relationship between the oxidizing power of roots, as measured by the staining of α-naphthylamine, and their extension into reduced paddy soil was observed by Fukui.[47] Among crops producing little or weak staining were alfalfa, white, black-bur and crimson clovers, and lupin. Included among medium stainers were barley, wheat, oats, rye, peas, red clover, alsike and ladine. Tall oatgrass and Italian ryegrass produced strong staining, as did rice. Bartlett,[48] also working with forage crops and rice, found that there was always a tendency for roots to effect an improvement in the oxidative situation in their rhizospheres. Oxidation of iron was demonstrated directly in solutions, and indirectly in soil, by a method involving the extraction of ferrous iron. Of particular significance were the observations that oxidizing power appeared to have some relation to the recognized ability to tolerate less than optimum aeration in the field. Poor oxidizers generally took up the most iron into their tops, while efficient oxidizers tended to be iron excluders with greater quantities of oxidized iron remaining in the rhizospheres. Among the species studied, reed canary grass, timothy and trefoil were consistently amongst the most successful oxidizers; alfalfa was consistently the lowest in oxidizing power.

Rice and other wetland species can effectively immobilize both iron, S^{2-} and reduced forms of Mn. Enzymes very active in Fe^{2+} oxidation have been extracted from rice roots.[43] Ethylene degradation occurs in the rice rhizosphere[49] and may require oxygen enrichment by the root system.[50] Species and varietal differences in tolerance to root rots have been correlated both with oxidizing activities[47,51,52] and with differences in ventilating efficiency.[53-55] Martin[56] found that Fe^{2+} exclusion by woodland species was related to the porosities of the root system; Hook and Brown[57] found that rhizosphere oxidation occurred only in the better aerated of tree species.

As oxidation processes are foremost in the toxin immobilization process, it is not surprising to find that wetland species with their superior ventilating powers also appear to be the more efficient oxidizers in wetland soil. Calculations show that oxygen leakage from the roots of wetland plants is sufficient to create zones of rhizosphere oxygenation comparable with the zones of oxidized iron which have been observed in wetland soil.[24] The better ventilated the root system, the broader should be the zone of oxygenation (see also p. 275). This should advantageously prolong the period in which slowly oxidizable compounds may be immobilized during their passage to the root. From what has been noted of their ventilating characteristics, it would seem that only during the early stages of root elongation might rhizosphere oxygenation by nonwetland plants compare with wetland species.

MODELING

Introduction

Solutions to scientific problems are frequently aided by building models, and although we have learned much from the experimental studies of root aeration, we can also learn a great deal from the use of physical, mathematical and electrical models. Information can be gained from these models which otherwise would be difficult or impossible to obtain experimentally, and the range of factor interaction that may be examined is enormous.

A more simple mathematical treatment, used by Greenwood,[58] gives a broad indication of the potential limits of oxygen diffusion within the plant body. Assuming that respiration and pore space resistance are homogeneously distributed and constant along the full length of the diffusion path, that there is no radial oxygen loss, and that the rate of oxygen uptake in the plant is unaffected by declining oxygen concentration until extremely low values are reached, then the synergism between respiration and pore space resistance is expressed in the equation

$$C_0 = \frac{ML^2}{2D_0 \epsilon \tau} \qquad (3)$$

where C_0 =   the oxygen concentration where oxygen enters the plant (g cm^{-3})

M =   the rate of oxygen uptake by the plant (g cm^{-3} s^{-1})

D_0 =   the diffusion coefficient of oxygen in air (cm^2 s^{-1})

  =   the effective porosity ($\epsilon\tau$; see Equation 2)

L =   aerated path length, *i.e.,* distance (from the point of entry) at which the oxygen concentration becomes zero (cm).

The data presented in Figure 6 have been computed using this equation and demonstrate clearly how the aerated diffusion path in a plant may vary enormously within the recognizable limits of respiration and pore space resistance.

Luxmoore et al.[14,15] have demonstrated the usefulness of the mathematical approach for studying internal root aeration. Their predictions concerning wetland roots in conditions of soil oxygen demand proved very informative, but their respiratory input data are considerably at variance with our more recent findings (see p. 275). Although their model was not applied to the oxygen relations of nonwetland plants in the wetland condition, they showed, nevertheless, that internal aeration might contribute significantly to nonwetland root respiration, even in aerated soils.

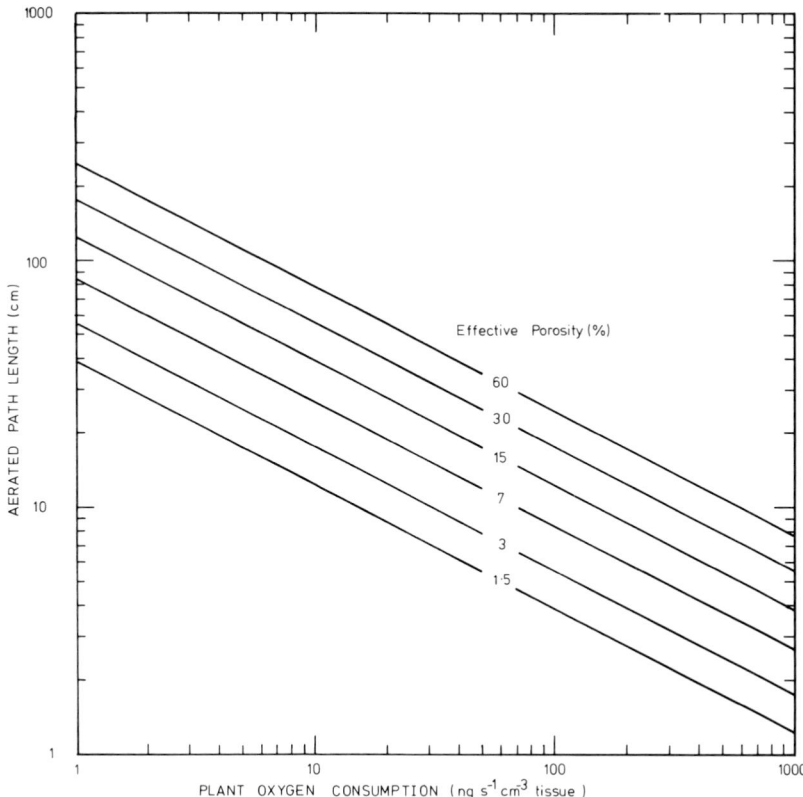

Figure 6. Data computed from the Greenwood equation (Equation 3) to show the probable relationship between tissue porosity, tissue respiration and aerated path length in plants. Uniform tissue porosity and uniform tissue respiration are assumed.

The fundamental similarity between electrical and diffusion laws forms the basis for the electrical modeling of diffusion systems, and we have recently built an electrical analogue to simulate the internal oxygen relations of roots in the wetland condition.[16,61] This model, which is described briefly below, has proved especially useful for the interpretation of certain experimental data.[23] It is used here to demonstrate more fully the limits of internal ventilation in nonwetland plants and to contrast further the internal aeration of the wetland and nonwetland root type.

The Electrical Analogue

The basic features of the analogue are presented in Figure 7. Resistors simulate root porosity, root wall permeability and diffusion path length. Voltmeters and milliammeters indicate, respectively, internal oxygen concentration, and radial oxygen loss and respiratory activity. The latter functions are controlled by variable resistance leakage to earth. Circuit units are linked in series, each unit representing a centimeter length of root-soil system.

Although essentially similar in its aims to the model of Luxmoore *et al.*, the electrical analogue differs in one major respect. In the mathematical model, soil sink activity is simulated by a hypothetical root-bounding shell of liquid of fixed radius with an outer surface oxygen concentration of zero. The electrical model, however, is programmed for specific levels of soil oxygen demand, and the radius of the hypothetical bounding liquid shell varies as a nonlinear function of the oxygen concentration within the root.

Programming Details:

Nonwetland Root Type. To simulate the nonwetland root type, maximum respiration was programmed at all positions along the root, provided that the internal oxygen concentration exceeded 1% (COP, 0.001 atm). Root wall permeability was considered to be at a maximum (100%) in the submeristematic apical centimeter, falling to a minimum (60%) at 6 cm from the apex. Root radius was constant ($r = 0.05$ cm); the simulated soil oxygen demand was 4×10^{-5} cm^3 cm^{-3} s^{-1}. It was assumed that the rate of soil oxygen consumption can be sustained down to very low oxygen concentrations.[59]

Wetland Root Type. Programming was carried out in accordance with known characteristics of rice roots. Root wall permeability was assumed to decrease from a maximum (100%) at the apex to a minimum (0) 5 cm from the apex.[21] Similarly, respiratory activity was assumed to decline from a maximum of 120 ng cm^{-3} s^{-1} at the apex to a minimum of 55%, 5 cm from the apex. Root radius and soil oxygen demand remained as before.

Results and Discussion

Nonwetland Root Type. Four levels of effective porosity ($\tau\epsilon$) together with two levels of respiratory activity (120 ng cm^{-3} s^{-1} and 30 ng cm^{-3} s^{-1}) have been examined. The results are arranged in Figure 8. If one accepts

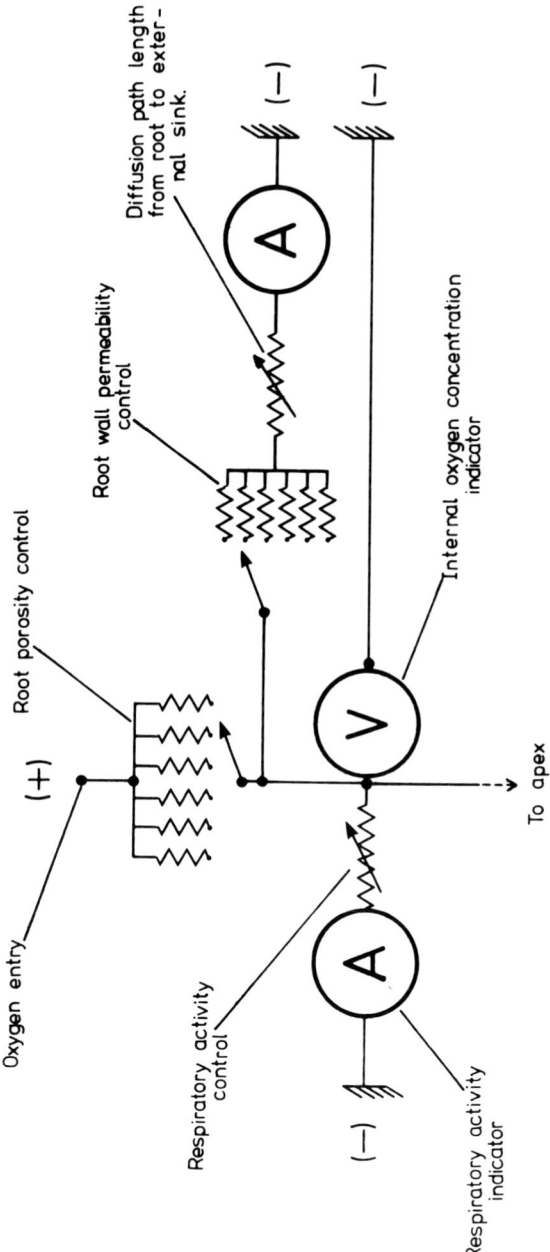

Figure 7. One of a series of analogue circuits to simulate root aeration in wet soils.[16]

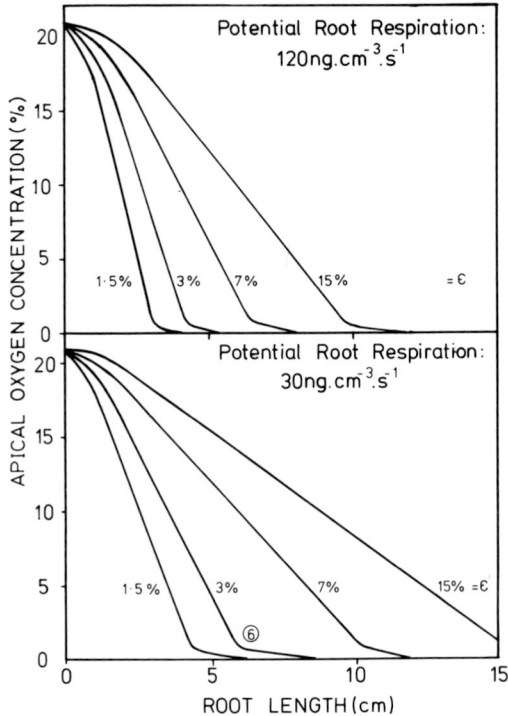

Figure 8. Electrical analogue predictions. Apical oxygen concentration with increasing root length for hypothetical nonwetland roots in saturated soil.
Soil oxygen consumption $= 4 \times 10^{-5}$ cm^3 (O_2) cm^{-3} (soil) s^{-1}; root
radius $= 0.05$ cm; $\tau\epsilon$ = effective root porosity (%).

a tortuosity factor of approximately 0.43,[22] an effective porosity of 7%
would equate approximately with a porosity of 15%. On the other hand,
tortuosity probably decreases significantly as the porosities approach 10%;
where lacunae form it is probably near to unity. Consequently, an effective porosity of 7% might be more likely to equate with a porosity of
about 12%. In Figure 8, the 3% effective porosity level is probably near
the upper limit for nonwetland roots growing in aerated soil; the value of
15% is included to represent an absolute upper limit for the nonwetland
root type.

It is abundantly clear from the upper part of Figure 8 that many roots
developed in aerated soil may be regarded as totally unsuited to a wetland
environment. The data accord with the findings and interpretation of
Yu *et al.*[41] (see also p. 280) regarding their half-flood treatment. Where
$\tau\epsilon = 1.5\%$, internal aeration would only fully support potential respiratory

activity of roots shorter than 3 cm. If root growth ceases below the COP (see p. 275) then the limit of root growth would be 3 cm at a COP of 1%; if the COP was 2.5% (see p. 275) growth might cease at about 2.75 cm. If effective porosity is doubled ($\tau\epsilon = 3\%$), only those roots not exceeding 4 cm in length could be fully maintained at preflood levels of respiration; irreparable damage might occur in longer roots. Some improvement accompanies the substantial reduction in respiratory activity to 30 ng $cm^{-3} s^{-1}$ (Figure 8 lower) but some degree of anaerobiosis still may be expected if root length exceeds 6 cm at an effective porosity of 3%. With a COP of 2.5%, this value decreases to 5 cm.

Reduction in the base respiration rate or further increases in porosity lead to improvements in oxygen balance. However, with no reduction in respiratory rate, the higher porosity limits characteristic of the "wet-adapted" adventitious roots would seem to place a limit on root extension beyond 5.5-6 cm ($\tau\epsilon = 7\%$) or 8.5-9.5 cm ($\tau\epsilon = 15\%$) at the soil oxygen demand considered. However, it must be noted that a particularly high rate of soil respiration has been used for these examples.

If adventitious root formation under waterlogged conditions is accompanied by substantial reductions in base respiratory rate, or if aging of the roots or a reduction in soil temperature lowers the respiratory demand, then more extensive root penetration can be predicted (Figure 8 lower). Depending upon the COP (1 or 2.5%), roots of between 9.5 and 16 cm could be expected, with a respiratory level of 30 ng $cm^{-3} s^{-1}$ and effective porosities between 7 and 15%. These rooting depths compare closely with the depths of penetration recorded by Yu et al.[41] in their full-flood treatment.

Conversely, penetration might be improved at both respiratory levels by a considerable reduction in soil oxygen demand. The complete cessation of soil oxygen consumption would, for example, be accompanied by a rise in the apical oxygen concentration in root (6) to 7.75% at a length of 7 cm.

Wetland and Nonwetland. In Figure 9 the internal oxygen balance and radial oxygen loss (ROL) of a rice root is contrasted with an "adapted" nonwetland root under conditions of soil oxygen demand.

The superiority of the oxygen balance in the rice root is clear. Moreover, rhizosphere oxygenation in the absorbing region of the rice root is shown to be nearly 13 times greater than in the nonwetland root type; with a slight increase in root length, it would approach infinity. In common with the predictions of Luxmoore et al.[14] the analogue forecasts internal apical oxygen concentrations in rice in excess of 5% at a root length of 20 cm.

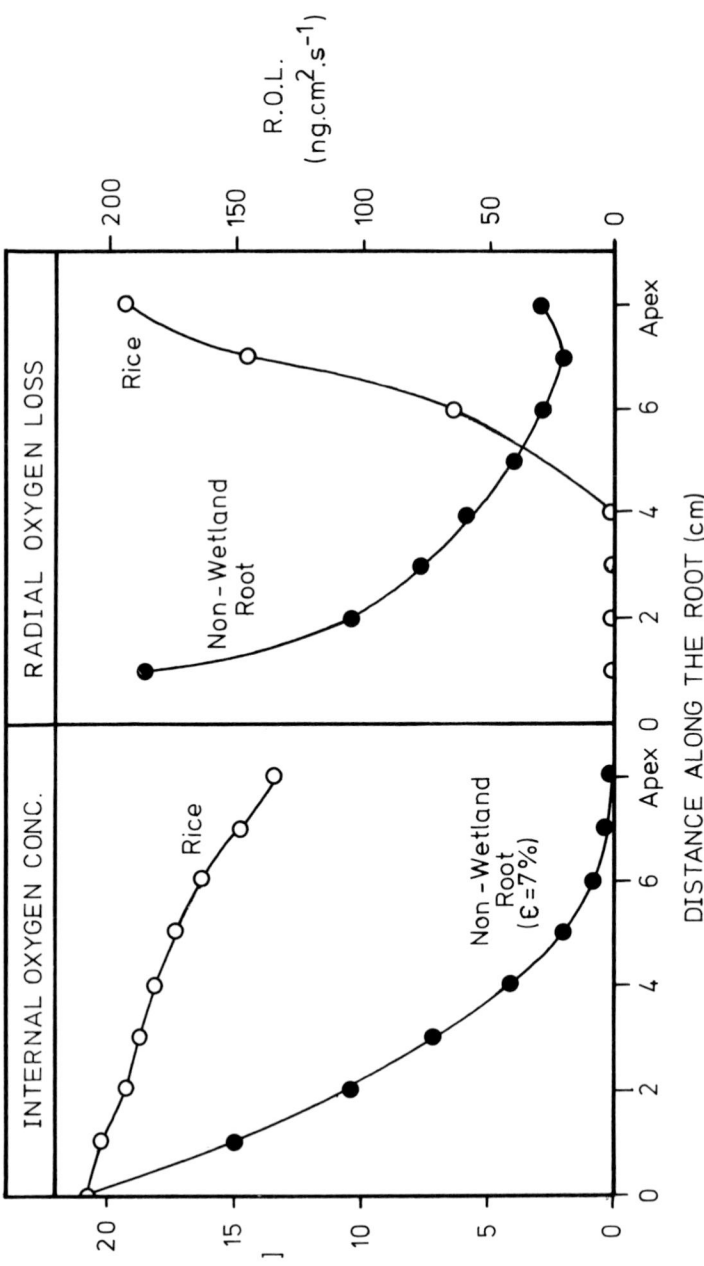

Figure 9. Electrical analogue predictions. Internal oxygen balance and radial oxygen loss along a nonwetland 'adapted' root ($\tau\epsilon = 7\%$), and a rice root. Root length = 8 cm; root radius = 0.05 cm; soil oxygen consumption = 4×10^{-5} cm^3 (O$_2$) cm^{-3} (soil) s^{-1}; $\tau\epsilon$ = effective root porosity (%).

In compiling this data, I have deliberately resorted to some simplifications. The nonwetland root type was considered to be uniformly porous with length, and no account was taken of any basipetal lowering of respiratory demand. A constant radius was adhered to in all cases, and lateral root formation was ignored. These conditions are, however, easily accommodated and are to be considered in some detail elsewhere.[61]

FINAL COMMENTS

Although the study of internal root aeration has advanced considerably in the last 20 years or so, there still remains much to discover. We have been made aware of internal ventilation in the nonwetland plant and have progressed considerably in analyzing the adequacies and inadequacies of ventilation in both wetland and nonwetland species.

As a means of examining aeration patterns in roots, models are proving beneficial in providing data not easily obtained experimentally, and we can begin to predict the relative importance of the various factors that control the oxygen balance within the root. We have found our electrical analogue invaluable for interpreting experimental data.

At this stage, we are not limited by the technical difficulties of modeling, but rather by our incomplete knowledge of the characteristics of the living plant. More information is urgently required concerning root wall permeability, gas space linings, pore space distribution (*i.e.,* tortuosity) and the aeration of the vascular system. We know little of the dynamics of rhizosphere oxygenation and have much to learn of the chemical and biological reactions in the rhizosphere. Our recent findings regarding the COPs in roots need considerable amplification, together with studies of longevity and damage in roots at suboptimal levels of aeration. Finally, we should all wish to learn more concerning the mechanism of gas space production in response to wetland conditions and of possible feedback mechanisms which might lower respiratory rates in response to a wetland environment.

ACKNOWLEDGMENTS

My thanks are due to my wife for reading and correcting the manuscript and to Miss E. M. Sharpe, Miss S. Lythe and Mr. T. J. Gaynard for assisting in its preparation. Our studies have been aided by the U. K. Science Research Council.

REFERENCES

1. Currie, J. A. "Diffusion within Soil Microstructure: A Structural Parameter for Soils," *J. Soil Sci.* 16:279 (1965).
2. Greenwood, D. J. "The Distribution of Carbon Dioxide in the Aqueous Phase of Aerobic Soils," *J. Soil Sci.* 21:314 (1970).
3. Grable, A. R. "Soil Aeration and Plant Growth," *Adv. Agron.* 18:57 (1966).
4. Currie, J. A. "The Importance of Aeration in Providing the Right Condition for Plant Growth," *J. Sci. Food Agric.* 13:380 (1962).
5. Grable, A. R. and E. G. Siemer. "Effects of Bulk Density, Aggregate Size, and Soil Water Suction on Oxygen Diffusion, Redox Potentials, and Elongation of Corn Roots," *Soil Sci. Soc. Amer. Proc.* 32:180 (1968).
6. Ponnamperuma, F. N. "The Chemistry of Submerged Soils," *Adv. Agron.* 24:29 (1972).
7. Teal, J. M., and J. Kanwisher. "Gas Exchange in a Georgia Salt Marsh," *Limnol. Oceanogr.* 6:388 (1961).
8. Rowe, R. N. and D. V. Beardsell. "Waterlogging of Fruit Trees," *C.A.B. Hort. Abstr.* 43:533 (1973).
9. John, C. D. and H. Greenway. "Alcoholic Fermentation and Activity of Some Enzymes in Rice Roots Under Anaerobiosis," *Aust. J. Plant Physiol.* 3:325 (1976).
10. Crawford, R. M. M. "Some Metabolic Aspects of Ecology," *Trans. Bot. Soc. Edinburgh* 41:309 (1972).
11. Amoore, J. E. "Arrest of Mitosis in Roots by Oxygen-Lack or Cyanide," *Proc. Roy. Soc. B.* 154:95 (1961).
12. Huck, M. G. "Variation in Taproot Elongation Rate as Influenced by Composition of the Soil Air," *Agron. J.* 62:815 (1970).
13. Armstrong, W. "Waterlogged Soils," in *Environment and Plant Ecology* by J. R. Etherington (New York: John Wiley and Sons, 1975), p. 181.
14. Luxmoore, R. J., L. H. Stolzy and J. Letey. "Oxygen Diffusion in the Soil Plant System I-IV," *Agron J.* 62:317 (1970).
15. Luxmoore, R. J. and L. H. Stolzy. "Oxygen Diffusion in the Soil Plant System V and VI," *Agron J.* 64:720 (1972).
16. Armstrong, W. and E. J. Wright. "An Electrical Analogue to Simulate the Oxygen Relations of Roots in Anaerobic Media," *Physiol. Plant.* 36:383 (1976).
17. Healy, M. T. and W. Armstrong. "The Effectiveness of Internal Oxygen Transport in a Mesophyte (*Pisum sativum* L.)," *Planta (Berlin)* 103:302 (1972).
18. Healy, M. T. "Oxygen Transport in *Pisum sativum* L.," Ph.D. Thesis, University of Hull, Great Britain (1975).
19. Hook, D. D., C. L. Brown and R. H. Wetmore. "Aeration in Trees," *Bot. Gaz.* 133:443 (1972).
20. Coutts, M. P. and W. Armstrong. "Role of Oxygen Transport in the Tolerance of Trees to Waterlogging," in *Tree Physiology and Yield Improvement,* M. G. R. Cannel and F. T. Last, Eds. (New York: Academic Press), p. 361.

21. Armstrong, W. "Radial Oxygen Losses from Intact Rice Roots as Affected by Distance from the Apex, Respiration and Waterlogging," *Physiol. Plant.* 25:192 (1971).

22. Jensen, C. R., L. H. Stolzy and J. Letey. "Tracer Studies of Oxygen Diffusion through Roots of Barley, Corn and Rice," *Soil Sci.* 103:23 (1967).

23. Armstrong, W., and T. J. Gaynard. "The Critical Oxygen Pressures for Respiration in Intact Plants," *Physiol. Plant.* 37:200 (1976).

24. Armstrong, W. "Rhizosphere Oxidation in Rice and Other Species: a Mathematical Model Based on the Oxygen Flux Component," *Physiol. Plant.* 23:623 (1970).

25. Berry, L. J. and W. E. Norris Jr. "Studies of Onion Root Respiration. I. Velocity of Oxygen Consumption in Different Segments of Root at Different Temperatures as a Function of Partial Pressures of Oxygen," *Biochem. Biophys. Acta* 3:593 (1949).

26. Armstrong, W. and E. J. Wright. "Radial Oxygen Loss from Roots: the Theoretical Basis for the Manipulation of Flux Data Obtained by the Cylindrical Platinum Electrode Technique," *Physiol. Plant.* 35:21 (1975).

27. Vartapetian, B. B. "Aeration of Roots in Relation to Molecular Oxygen Transport in Plants," *Proc. Uppsala Symp.*, 1970 (Ecology and Conservation 5) (New York: Unesco, 1973).

28. Boeke, J. E. "On the Origin of the Intercellulary Channels and Cavities in the Rice Root," *Ann. Jardin Bot. Buitenzorg* 50:199 (1940).

29. van der Heide, H., M. H. van Raalte and B. M. de Boer-Bolt. "The Effect of a Low Oxygen Content of the Medium on the Roots of Barley Seedlings," *Acta. Bot. Neerl.* 12:231 (1963).

30. Arber, A. *Water Plants.* (Cambridge: Cambridge University Press, 1920).

31. Sifton, H. B. "Air-Space Tissue in Plants," *Bot. Rev.* 11:108 (1945).

32. Sifton, H. B. "Air-Space Tissue in Plants II," *Bot. Rev.* 23:303 (1957).

33. Sculthorpe, C. D. *The Biology of Aquatic Vascular Plants* (London: Arnold, 1967).

34. Arikado, H., and Y. Aduchi. "Anatomical and Ecological Responses of Barley and some Forage Crops to the Flooding Treatment," *Bull. Fac. Agr. Mie Univ.* 11:1 (1955).

35. Armstrong, W. "Oxygen Diffusion from the Roots of some British Bog Plants," *Nature, London* 204:801 (1964).

36. Griffin, D. M. "A Theoretical Study Relating the Concentration and Diffusion of Oxygen to the Biology of Organisms in Soil," *New Phytol.* 67:561 (1968).

37. Yocum, C. S. and D. P. Hackett. "Participation of Cytochromes in the Respiration of Aroid Spadix," *Plant Physiol.* 32:186 (1957).

38. Vámos, R. and E. Kóves. "Role of the Light in Prevention of the Poisoning Action of Hydrogen Sulphide in the Rice Plant," *J. Appl. Ecol.* 9:519 (1972).

39. Armstrong, W. "Oxygen Diffusion from the Roots of Woody Species," *Physiol. Plant.* 21:539 (1968).

40. Hook, D. D., C. L. Brown and P. P. Kormanik. "Inductive Flood Tolerance in Swamp Tupelo (*Nyssa sylvatica* var biflora (Walt.) Sarg)" *J. Exp. Bot.* 22:78 (1971).

41. Yu, P. T., L. H. Stolzy and J. Letey. "Survival of Plants under Prolonged Flooded Conditions," *Agron. J.* 61:844 (1969).

42. Engler, R. M. and W. H. Patrick Jr. "Stability of Sulphides of Manganese, Iron, Zinc, Copper and Mercury in Flooded and Non-Flooded Soil," *Soil Sci.* 119:217 (1975).

43. Yamada, N. and Y. Ota. "Study on the Respiration of Crop Plants. Part 7 Enzymatic Oxidation of Ferrous Iron by Root of Rice Plant," *Proc. Crop Sci. Soc. Japan* 26:205 (1958).

44. Pitts, G., A. I. Allan and J. P. Hollis. "Beggiatoa: Occurrence in the Rice Rhizosphere," *Science* 178:990 (1972).

45. Yoshida, T., T. Takai and D. C. del Rosario. "Molecular Nitrogen Content in a Submerged Rice Field," *Plant Soil* 42:653 (1975).

46. Read, D. J. and W. Armstrong. "A Relationship Between Oxygen Transport and the Formation of the Ectotrophic Mycorrhizal Sheath in Conifer Seedlings," *New Phytol.* 71:49 (1972).

47. Fukui, J. "Studies on the Adaptability of Green Manure and Forage Crops to Paddy Field Conditions," *Proc. Crop Sci. Soc. Japan* 22:110 (1953).

48. Bartlett, R. J. "Iron Oxidation Proximate to Plant Roots," *Soil Sci.* 92:372 (1961).

49. Yoshida, T. and M. Suzuki. "Formation and Degradation of Ethylene in Submerged Rice Soils," *Soil Sci. Plant Nutr.* 21:129 (1975).

50. Cornforth, I. S. "The Persistence of Ethylene in Aerobic Soils," *Plant Soil* 42:85 (1975).

51. Goto, Y. and K. Tai. "On Differences of Oxidizing Power of Paddy Rice Seedling Roots Among some Varieties," *Soil Plant Food* 2:198 (1957).

52. Armstrong, W. "The Oxidizing Activity of Roots in Waterlogged Soils," *Physiol. Plant.* 20:920 (1967).

53. Armstrong, W. and D. J. Boatman. "Some Field Observations Relating the Growth of Bog Plants to Conditions of Soil Aeration," *J. Ecol.* 55:101 (1967).

54. Armstrong, W. "The Use of Polarography in the Assay of Oxygen Diffusing from Roots in Anaerobic Media," *Physiol. Plant.* 20:540 (1967).

55. Armstrong, W. "Rhizosphere Oxidation in Rice: an Analysis of Intervarietal Differences in Oxygen Flux from the Roots," *Physiol. Plant.* 22:296 (1969).

56. Martin, M. H. "Conditions Affecting the Distribution of *Mercurialis perennis* L. in Certain Cambridgeshire Woodlands," *J. Ecol.* 56:777 (1968).

57. Hook, D. D. and C. L. Brown. "Root Adaptations and Relative Flood Tolerance of Five Hardwood Species," *Forest Sci.* 19:225 (1973).

58. Greenwood, D. J. "Studies on Oxygen Transport through Mustard Seedlings (*Sinapsis alba* L.)" *New Phytol.* 66:597 (1967).

59. Greenwood, D. J. "The Effect of Oxygen Concentration on the Decomposition of Organic Materials in Soils," *Plant Soil* 14:360 (1961).
60. Gaynard, T. J. and W. Armstrong. (University of Hull, Department of Plant Biology) (Unpublished data).
61. Armstrong, W., T. J. Gaynard and E. J. Wright. Soil Sink Activity and the Internal Aeration of Roots (in preparation).

10

ADAPTATIONS AND FLOOD TOLERANCE
OF TREE SPECIES

Donal D. Hook and John R. Scholtens

Belle W. Baruch Forest Science Institute
Clemson University
Georgetown, South Carolina 29440

INTRODUCTION

Although many arborescent species of the geologic past thrived in the low, shallow seas of the Carboniferous era, modern tree species are unlikely candidates to live in flooded soils where their roots must function for long periods without soil oxygen and where the soil is highly reduced and toxic compounds are present (Figure 1). Trees do not exhibit outwardly any of the better-known adaptations of aquatic species. They do not have well defined gas diffusion pathways such as exist in rice (*Oryza sativa* L.), *Menyanthes sp.*, and other nonwoody hydrophytes. Trees are long lived and their structure is complex and rigid. Also, trees must contend not only with the annual variations in the root environment but also with the long-term changes of a century or more. The presence and maintenance of large transpirational and photosynthetic surfaces of trees further complicates their ability to adapt to changing root environments.

In spite of these apparently limiting morphological and physiological characteristics, massive trees such as bald cypress (*Taxodium distichum* L.), water tupelo (*Nyssa aquatica* L.) and swamp tupelo [*N. sylvatica* var *biflora* (Walt) Sarg.] thrive in swamps in the southern United States. Black mangroves (*Avicennia nitida* Jacq.) thrive in the intertidal zone along the coast in semitropical and tropical regions of the world. It is apparent that certain tree species have evolved with adaptations which enable them to cope efficiently with flooded root environments. Adaptations in these large,

Figure 1. Buttress and knee development on bald cypress trees in a swamp environment.

specialized plants are apt to be more subtle and refined than in smaller, less specialized, nonwoody species.

AERATION SYSTEM

The internal tissues of higher plants are assumed to be aerated by gas exchange through the stomata of the leaves and stem lenticels[1] and by direct gaseous diffusion between the root and soil atmosphere in well aerated soils. For this ventilation system to function adequately in mature trees, the thick bark would have to be readily permeable to gas exchange or oxygen would have to diffuse great distances through limited intercellular spaces of the bole from leaves to active tissues of the stem. Results of recent studies involving both woody and nonwoody plants have raised some doubts as to the adequacy of the internal aeration system of vascular plants. End-products of anaerobic respiration have been found in stem and root tissue from well aerated environments.[2,3] These findings suggest that internal structure may restrict complete aeration of localized tissue or may be inadequate for overall aeration needs.

When the lower trunk and roots of tree species are inundated by water, normal gas exchange through the bark of the flooded portions is impeded.

Under these conditions the cambium, actively growing root tips, and other vital tissue must receive their aeration by transport of oxygen from the plant parts above the water level and/or adjust their metabolic processes accordingly.

Evolution

Morphological and anatomical features which appear to facilitate gas exchange with the atmosphere can be traced to vascular plants of the Middle and Upper Devonian ages. The first aeration canals in arboreal plants were found in the Lepidodendroids and Sigillarians of the Carboniferous era. Strands of arenchyma connecting the outer and middle cortex of the stem to the leaf mesophyll were defined as "parichnos."[4] Parichnos were connected to the rootlets at the base of the stem through the lacunar tissue of the mid-cortex. They apparently functioned in the Lipidodendraceae by taking air in through the leaf and conducting it through the stem to the Stigmarian rhizome and associated rootlets.[5]

In ancient plant forms, parichnos were found in the absence of cauline stomata. Similar structures were initially present in some members of the genus *Pinus*, certain *Abietineae*, and in *Ginkgo biloba*. They were transitory in nature, being quickly destroyed by radial growth; hence, other means of internal aeration were necessary. Primitive lenticels were first found in association with leaf scars and the intercellular system of the stomata-leaf mesophyll-stem cortical regions which formed the parichnos. These structures and later-formed lenticels on other parts of the stem provided open systems for gas exchange between internal tissue and the atmosphere.

There is no evidence that parichnos ever existed in angiosperms. Therefore, the anatomical and morphological conditions from which lenticels originated in angiosperms were quite different than in the gymnosperms. Cauline stomata were widespread in this phyletic group. Thus the stomata in the epidermis appear to provide adequate aeration of the cortical cells and ground parenchyma through the anastamosing network of intercellular spaces until the phellogen and periderm are differentiated. Deep-seated, vertically and horizontally oriented lenticels which originate in the phellogen provide continuity between the atmosphere and internal tissue exterior to the vascular cambium.[6-9]

Although the evolutionary history of aeration systems and anatomical descriptions of the aeration structures are fairly well documented, quantitative data are lacking on how efficiently these structures perform their aeration role.

Bark and Phloem

The tissues of the phloem are protected from exposure to the atmosphere by the differentiation of successive peridermal layers from the phellogen. Lenticels are produced in the periderm by the phellogen and provide avenues for gaseous continuity between the atmosphere and the larger intercellular spaces of the phloem through an otherwise impervious layer of corky bark.[9]

The outermost layer of parenchyma cells in the lenticel develops into a closing layer which protects the succulent inner parenchyma cells from desiccation. However, there are sufficient breaks in the closing layer to permit free gas exchange (Figure 2).

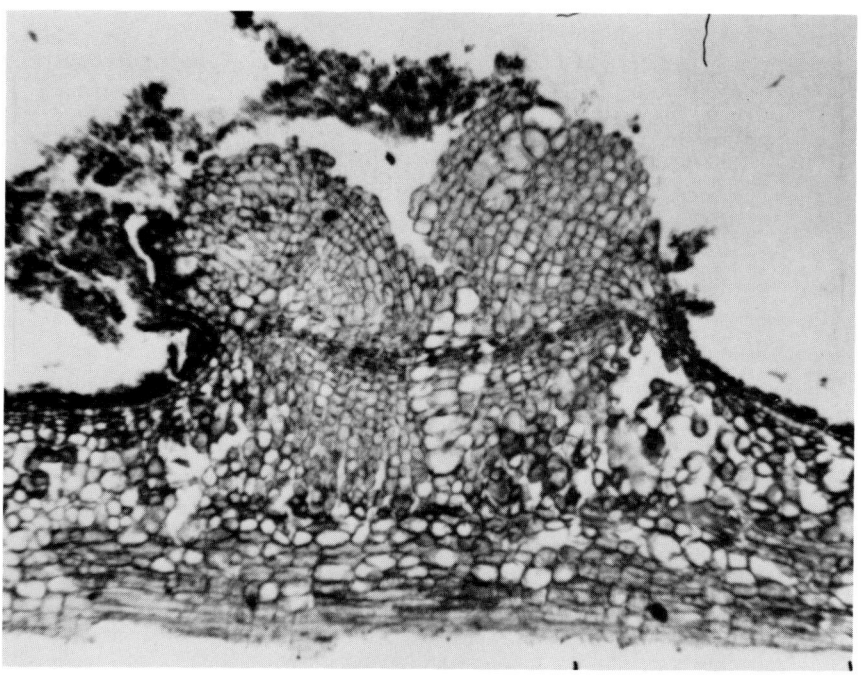

Figure 2. Cross section of lenticel with broken closing layer (185X).

Periodic or continuous flooding of the stem of certain tree seedlings results in the proliferation of lenticels on the flooded stem and roots (Figure 3). Inundation of the lenticels causes hypertrophy and they may protrude outward 2 or 3 mm from the stem (Figure 4). Flooded lenticels do not produce closing layers, and the increased size of the lenticel and

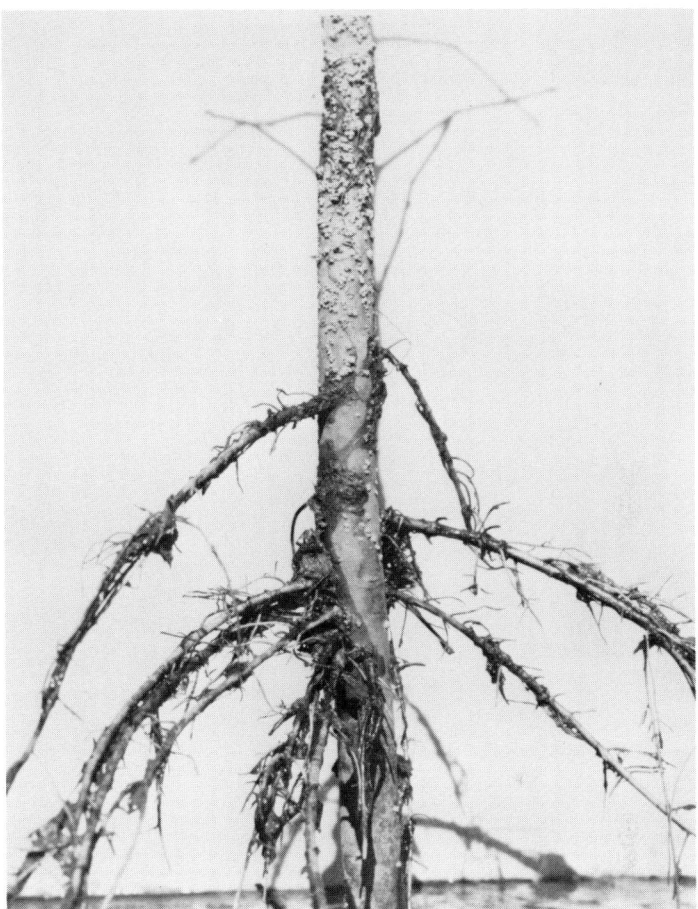

Figure 3. Swamp tupelo seedling with proliferation of lenticels and water roots on flooded section of stem.[10]

spherical cell structure would appear to facilitate gas exchange (Figures 5 and 6). Theoretically, the hypertrophic lenticels may enhance gas exchange by providing a greater surface area and a more pervious structure for the dissolved gases in flood water to permeate than nonflooded lenticels. Although hypertrophic lenticels collapse upon exposure to air and form a closing layer, they appear to be more pervious to gas exchange than non-flooded lenticels because of larger intercellular spaces and numerous breaks in the closing layers.[10,11]

The organized structure of lenticels is lost to the unaided eye as soon as the bark loses its smooth surface, but according to Esau[9] "...lenticels are continued through the whole thickness of the tissue, a feature well

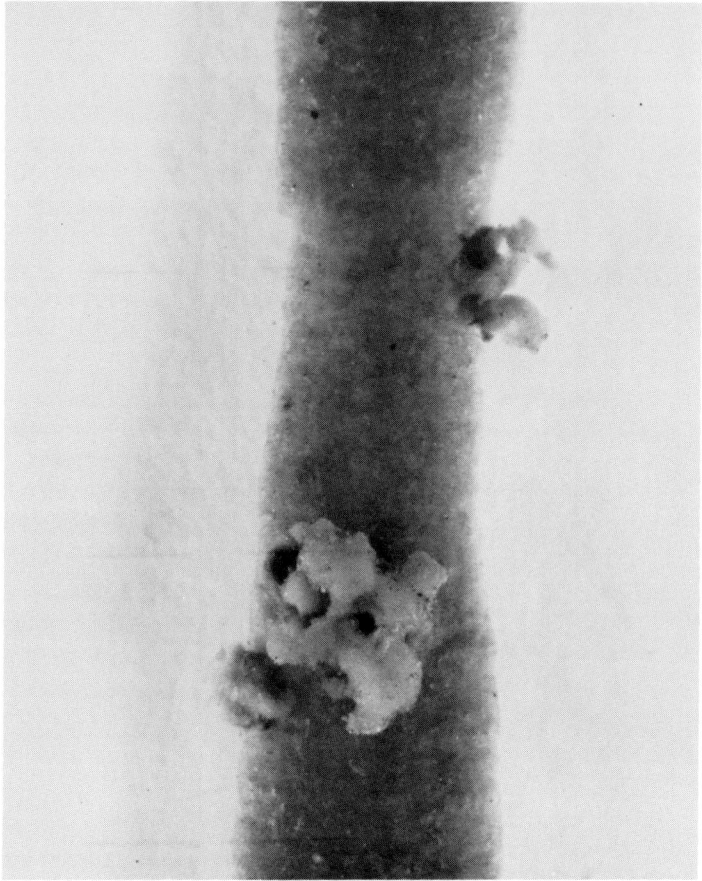

Figure 4. Tangential view of bark with protruding lenticels (5X).[10]

illustrated by the commercial cork oak (*Quercus suber*)...." Once bark
development has proceeded this far, the question arises of whether it is
pervious to gas exchange. Some experiments on gas exchange of cypress
knees and the bole of cypress and tupelo trees suggest, but do not prove,
that the bark is pervious to limited gas exchange.[12] Observations on the
surface coating of deeply flooded roots on mature trees (Roots, p. 311)
suggest that oxygen may diffuse into the mature bark just above the
floodline and through the intercellular spaces of the phloem and cortex
to the root. Such circumstantial evidence does not define how efficient
the bark of mature trees is in aerating the underlying tissue.

Air may be freely drawn through the lenticels of submerged small woody
cuttings of certain species.[13,14] When the stem lenticels of tupelo seedlings

Figure 5. Radial view of flooded lenticel without a closing layer (180X).[10]

are artificially covered, the seedlings do not oxidize their rhizosphere as do seedlings with lenticels exposed to air.[15]

Lenticels may play a dual role in aeration in some species, but not others. Chirkova and Gutman[3] found that on rooted cuttings of willow (*Salix alba* L.), oxygen from the atmosphere diffused into the plant through the lenticels and internal ethanol, acetaldehyde and ethylene were excreted through the same lenticels. By contrast, oxygen entered the lenticels on poplar (*Populus petrowskiana* Sch.), but no internal metabolites were excreted.

Cambium

Early experiments on the permeability of the cambium and gas content in tree stems led researchers to hypothesize that the cambium of trees was highly resistant to gas exchange, if not impermeable.[16-18] However, MacDougal[19] demonstrated that air could be drawn across the cambium of several tree species by only 5-30 mm Hg negative pressure. He concluded that the continuity of intercellular space across the cambium could not be documented by microscopic studies because the path was too small and tortuous to follow.

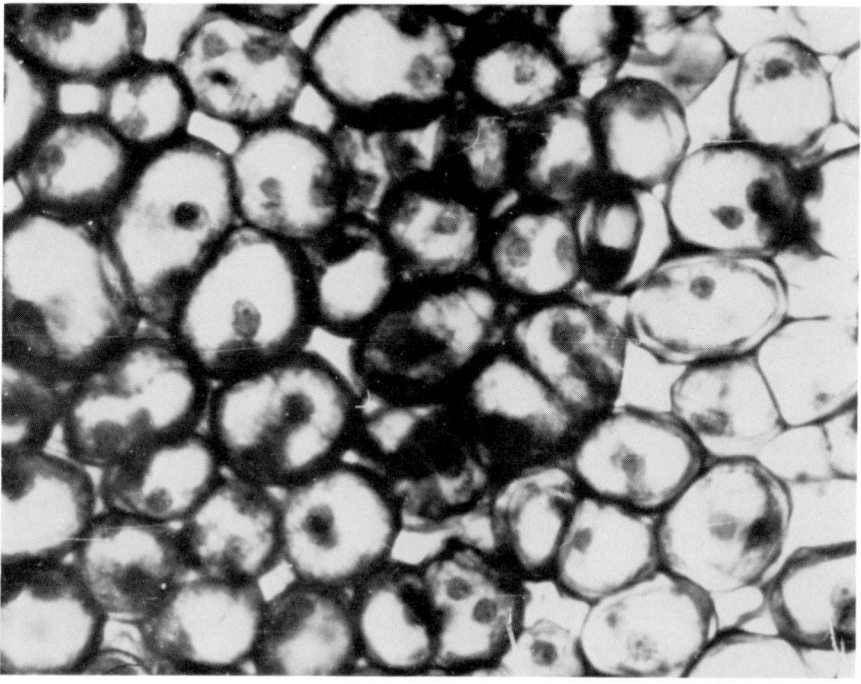

Figure 6. Tangential view of phellogen in a lenticel. Shows large intercellular space in phellogen (1750X).[10]

More recent experiments by Hook and Brown[14] have led us to hypothesize that the permeability of the cambium of tree species is a function of site-specific adaptive evolutionary mechanisms. Species from mesic and xeric sites where soil water may be limiting but air is not have cambia that are relatively impervious to gas exchange. On the other hand, species from hydric sites where soil air is limiting and water is plentiful have cambia that are relatively pervious to air.

Evidence to support this hypothesis is based on three types of experiments conducted by Hook and Brown:[14] (1) resistance of air movement across the cambium, (2) microscopic studies of the continuity of intercellular spaces (ICS) across the cambium, and (3) diffusion of gases across the cambium of live tree seedlings. Experiments with woody cuttings, as illustrated in Figure 7, have shown variability among species in resistance to air movement (Table I). Water tupelo and green ash (*Fraxinus Pennsylvania*) species which live in continuously and periodically flooded habitats, respectively, exhibit little resistance to air movement across the cambium. Conversely, species such as sweetgum (*Liquidamber styraciflua* L.), yellow poplar (*Liriodendron tulipifera* L.) and cottonwood (*Populus deltoides*(Marsh),

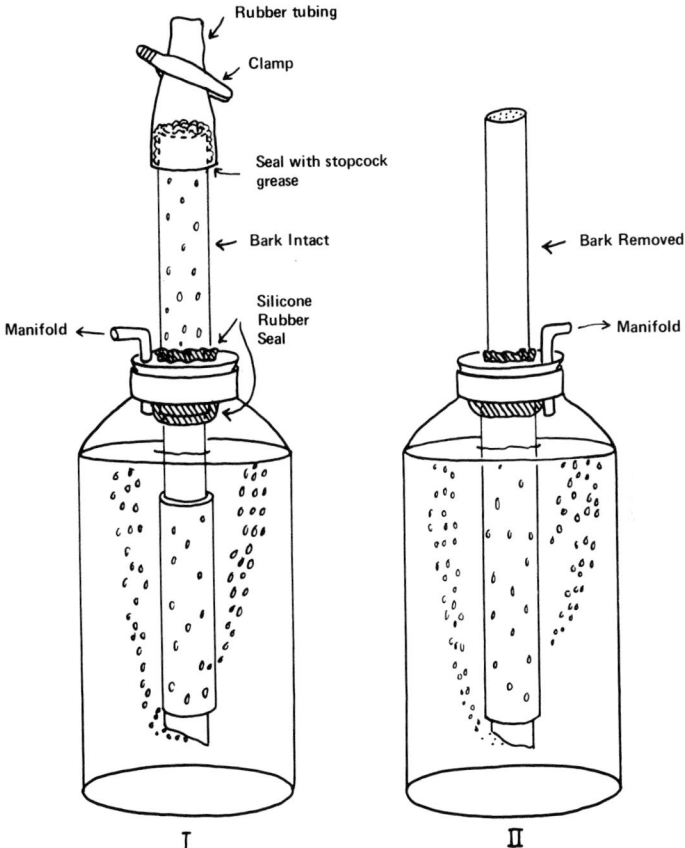

Figure 7. Apparatus for measuring resistance to air movement in woody stems. The outlet was attached to a manifold with a measured vacuum and air movement was detected by air bubbles escaping from lenticels or xylem. I. Stem ringed from silicone rubber seal to below water level. II. Portion of stem exposed to the atmosphere was ringed.[14]

which live in mesic habitats, exhibit high resistance to air movement across the cambium. Microscopic examinations of the ICS across the cambium of these species substantiated the results of the physical experiments (Figures 8-10).

The most conclusive tests to demonstrate free gas exchange in stems of woody plants were done with live seedlings. Under experimental conditions described by Hook and Brown,[14] it was demonstrated that sufficient oxygen could diffuse into the stem of tupelo seedlings via lenticels and across the cambium to oxidize the rhizosphere under normal atmospheric conditions.

Table I. Reduction in Atmospheric Pressure Necessary to Cause Air Bubbles to Escape from Lenticels and Xylem at the Base of Stem Segments under Treatments I and II[20]

Species	Observations (N)	Pressure Reduction before Air Bubbles Escape (mm Hg)	
		Lenticels	Xylem at Base of Stem
Treatment I[a]			
Water tupelo	2	10-10	55-100
Green ash	2	10-20	30-50
Yellow poplar	2	40[b]	15-45
Cottonwood	1	c	15
Sweetgum	1	180	125
Sycamore	2	30-40	30-45
Treatment II[a]			
Water tupelo	2	5-5	55-60
Green ash	2	10-15	30-50
Yellow poplar	1	15	80
Cottonwood	1	20	70
Sweetgum	2	20-80	150-200
Sycamore	2	50[b]	45-50

[a]See Figure 7 for identification of Treatments I and II.
[b]Only one observation on escape of bubbles from lenticels.
[c]No air bubbles escaped from lenticels at a tension of 110 mm Hg.

Xylem

The precise pathways by which the xylem tissues are aerated and the degree of aeration provided have yet to be clearly defined in tree species.

Measurements of the concentrations of gases in tree stems have yielded variable results.[21-24] Practically all investigators have confirmed that oxygen content in the stem is lower and carbon dioxide is higher than in the atmosphere. Variations in gas concentrations reported in woody stems are quite large. Some variations in internal gas concentrations can be attributed to experimental techniques, ambiguous sampling sites and analytical procedures. Generally, investigators have sought to determine the gross gas content of the stem. From a physiological standpoint, gas concentrations within the xylem mother cell zone become important because this is the site where most of the world's fiber source originates. Gross measurements of gas content in the stem may not adequately reflect aeration conditions proximal to the cambium zone. With the development and perfection of oxygen microelectrodes, such as used by Tjepkema and Yocum[25] to study oxygen partial pressures in soybean nodules, it seems

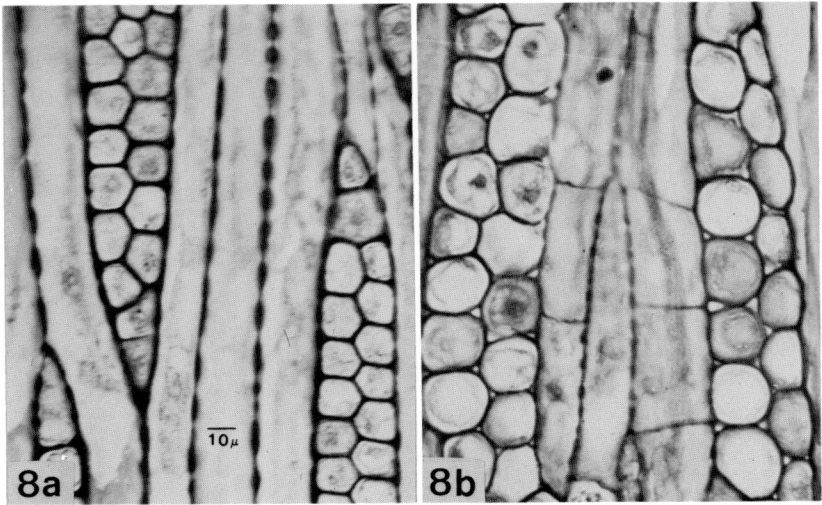

Figure 8. Tangential sections of sweetgum showing (a) absence of intercellular spaces among the cambial ray initials and (b) the presence of intercellular spaces among the ray cells of the differentiating phloem.[14]

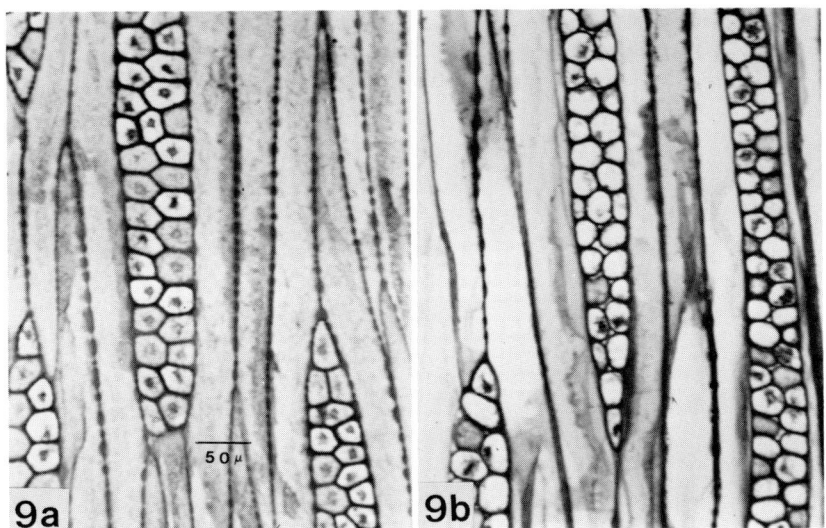

Figure 9. Tangential sections of yellow poplar showing (a) absence of intercellular spaces among the cambial ray initials and (b) the presence of intercellular spaces among the ray cells of the conducting phloem.[14]

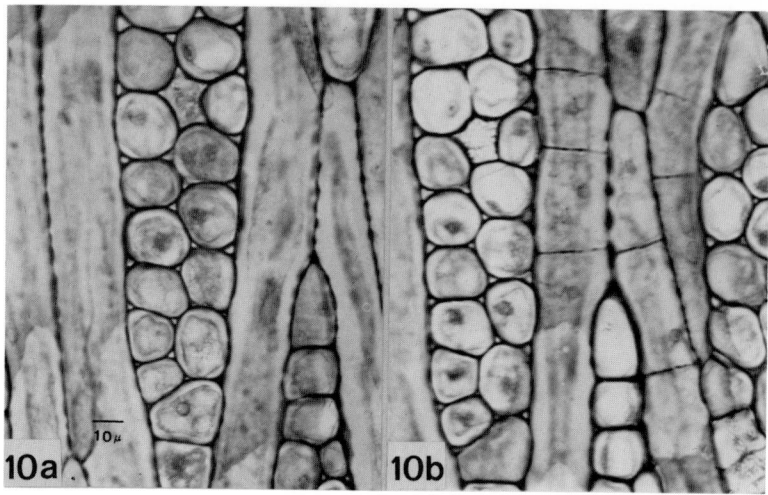

Figure 10. Tangential section of water tupelo stem showing (a) well defined inter-cellular spaces among the cambial ray initials and (b) among ray cells of the differentiating phloem.[14]

highly probable that meaningful *in situ* measurements of oxygen can be made at specific sites in tree stems today.

The transpirational stream probably plays a more vital role in aerating the xylem tissues of nonhydrophytes than in hydrophytes since the water taken up by roots of the latter is poorly aerated.[26] Other experimental results support the hypothesis that the transpirational stream functions in internal aeration. Toxic compounds produced in the roots of willow seedlings ascend to the leaves via the transpirational stream.[3] The ascending sap in birch trees (*Betula pubescens*) growing on wet sites has considerably higher concentrations of malate than does the sap of birch trees growing on drier sites.[2] Also, the hydrophyte *Lobelia dortmanna* takes up free carbon dioxide from the sediment of lakebeds which diffuses or rises to the leaves and is used in photosynthesis.[27] If such varieties of compounds are transported in the transpirational stream, there is no reason why dissolved oxygen would not be carried also. It may be argued that because of the low solubility of oxygen in water the amount transported via this process would be much less than that diffused through unsaturated tissue. Even so, it may be adequate to maintain aerobiosis in the xylem tissue. Also, the large volume of water moving daily through the tree stems may partially overcome this problem.

Roots

A critical factor in determining whether a tree species will tolerate root anaerobiosis is its ability to regenerate new "adapted roots" under the stimuli of inundation. In responding to hypoxia, usually the entire secondary root system dies and the plant dies also, unless new roots develop quickly. Seedlings of hydrophytes, such as swamp tupelo and water tupelo, regenerate new roots from the primary root within a few days after inundation. Differences between new adapted roots and roots grown in well aerated soils are listed in Table II. Adapted roots are succulent and less fibrous; the epidermis is weakly or nonsuberized along the distal 2 cm of the root, and the endodermis is poorly differentiated. Casparian strips are not commonly found within 2 cm of the root apex.[15]

Table II. Characteristics of Swamp Tupelo and Water Tupelo Roots that Develop under Well Aerated and Flooded Conditions

Root Characteristics	Root Environment	
	Well Aerated	Flooded
Morphology	Small diameter and fibrous except at the apex	Succulent with very little branching
Epidermis	Highly suberized	Little or no suberization
Endodermis	Highly organized with Casparian strips	Poorly organized, Casparin strips not evident
Adventitious Water Root	None	Prolific just below water line[a]
Intercellular Space in Cortex	Abundant	Abundant
Oxidizes Rhizosphere in Anaerobic Conditions	No	Yes

[a]May be absent on water tupelo under many types of flooding.

Many woody species develop adventitious or water roots under the influence of flooding at or just below the surface water level (Figure 3). Such roots are thought to aid the species in flood tolerance, but experimental evidence to support this supposition is absent in woody plants.

Development of water roots is stimulated by fluctuating or moving water but they seldom occur on trees in still or stagnant water.[11,28] The statement by Gill[28] that *Taxodium distichum* does not develop water roots is misleading. Hook has observed water roots on *Taxodium distichum* under controlled flooding and in its natural environment. However, both bald cypress and water tupelo, which are very tolerant to flooding, produce water roots only under very specific flooding conditions.

Although lenticels on the seedlings of several woody plant species are known to provide entrance for atmospheric oxygen and to excrete several volatile organic compounds and gases,[3,26,29,30] we still lack quantitative data on the permeability of bark on mature trees.

The gas diffusion pathway is potentially much longer in tree species than in nonwoody plants. In many cases when the tree trunk is flooded, oxygen must diffuse through a meter or more of stem and root tissue with poorly defined intercellular spaces to reach actively growing root tips. Based on data given in Chapter 9, it seems unlikely that a sufficient quantity of oxygen would be diffused to oxidize the rhizosphere. As a consequence, it is not clear whether sufficient oxygen diffuses to the roots to oxidize the rhizosphere or whether some other mechanism is involved.

Field observations on mature bald cypress, swamp tupelo and green ash trees under prolonged deep flooding have shown that they respond similarly to seedlings in that: (1) the secondary root systems die; (2) new adapted roots develop similar to those described above; and (3) the presence of substances on the surface of the new adapted roots (presumed to be iron oxides) indicate they oxidize their rhizosphere.[31] Therefore, it is assumed but not proved that oxygen diffuses through the bark of these species above the flood water in sufficient quantities to oxidize their roots—a distance of over 100 cm.

STEM AND ROOT MORPHOLOGY

Stem buttresses, knees, swollen bases and other odd developments on swamp tree species have intrigued botanists and researchers from the time of Linneaus to the present (Figure 1). Linneaus gave the tupelos the generic name *Nyssa* which has been interpreted to mean "water nymph."

Bald cypress trees develop large-ridged buttresses and numerous knees in most swamp environments (Figure 1). Under continuous deep flooding or very shallow flooding, these traits may be absent or greatly diminished. Swamp and water tupelos develop swollen bases and looped or kinked roots. Height of buttresses, knees and swollen bases are determined by the height of prevailing water levels. Knees and the hypertrophy of buttresses and swollen bases may extend 3-4 meters above the swamp floor.[32,33]

The wood of buttresses, knees, swollen bases and flooded roots is soft and spongy and has lower density than the bole wood on nonflooded trees. Such structures do not have arenchyma nor more intercellular space than similar woody tissue above the flood level. However, the diameter of parenchyma cells, tracheids and vessel elements of these structures are larger than in nonflooded woody tissue. The ratio of xylem parenchyma cells to tracheids and/or vessels in knees and swollen bases is also greater than in normal wood.[33]

Even though these structures add mystique to the trees in the swamp environment, their physiological significance is questionable. Kramer, Riley and Bannister[34] studied the gas exchange of intact and detached cypress knees and concluded that they did not provide aeration to the remainder of the tree. In recent studies, carbon dioxide analyzers have shown that limited gas exchange occurs through the bark of cypress boles and cypress knees. Since the rate of carbon dioxide exchange increased during the day, it may be assumed that internal respiration may have been measured.[12]

Kramer[35] hypothesizes that lack of oxygen near the water line may interfere with downward translocation of carbohydrates and auxins. Hence their accumulation near the flood line would stimulate hypertrophy of the tissue and water root development below the water line.

METABOLISM

The metabolism of roots of tree species under anaerobic conditions is quite similar to that of nonwoody plants. Roots of several woody species produce ethanol, lactic acid and other organic compounds as end-products of anaerobic respiration.[3,15] The concentration of the end-products accumulated is generally higher in flood-tolerant species than in nonflood-tolerant species. But the ratio of increase in the concentration of the compounds produced under anaerobic conditions versus the concentration under aerated conditions is much higher in most nonflood-tolerant tree species than in flood-tolerant species (Table III).[20] Green ash appears to be an exception to this generalization. Crawford[36] hypothesized that distribution of species in wetland habitats was determined by the species' ability to control anaerobic respiration under flooding. This theory was advanced further when McMannon and Crawford[37] found flood-tolerant species had low ADH activities and intolerant species had high ADH activities. They hypothesized that the relative flood tolerance of a species was inversely related to alcohol dehydrogenase activity in roots. By contrast, Hook et al.[15] showed that the roots of swamp tupelo produced about equal amounts of lactic acid whether grown in aerated or flooded soils. However, ethanol production was five times greater in roots from

Table III. Summary of Ethanol Accumulation in the Roots of Five Species Grown in Flooded and Drained Soil from All Experiments after Incubation in Air or Prepurified N_2 for Four Hours[20]

| | Ethanol Accumulation[a] Mol ETOH $\times 10^7$/mg ODW[b] roots/4 hr | | | | |
| | Air | | N_2 | | |
Species	$\bar{x}$	Range	$\bar{x}$	Range	N_2/Air
Drained					
Water Tupelo	.33	(.29-.40)	1.11	(.82-1.78)	3.4
Green Ash	.03	(.00-.07)	.62	(.26-1.51)	20.7
Sycamore	.16	(.09-.23)	.63	(.12-1.97)	3.9
Sweetgum	.11	(.05-.18)	1.54	(.56-2.14)	14.0
Yellow Poplar	.03	(.00-.05)	.78	(.58-1.18)	26.0
Flooded					
Water Tupelo	.59	(.29-.93)	2.36	(1.17-3.75)	4.0
Green Ash	.03	–	1.55	(.31-3.23)	51.7
Sycamore	.26	(.00-.69)	1.11	(.73-1.87)	4.3
Sweetgum	.14	(.11-.17)	1.94	(1.26-3.22)	13.8
Yellow Poplar	–	–	–	–	–

[a]Averages are based on 3-7 observations except for the green ash ethanol accumulation in air atmosphere. Here there were two samples in the drained treatment and only one in the flooded.
[b]Oven dry weight.

flooded than from aerated soils after four hours of incubation in N_2 (Figure 11). Other flood-tolerant species also accumulate high concentrations of ethanol.[20]

Davies et al.[38] fould large quantities of malic enzymes in the roots of flood-tolerant and nonflood-tolerant species. Therefore, the earlier theory of metabolic adaptation has been modified. It appears that some form of metabolic control and coupling of metabolic pathways is functional in flood-tolerant species but is lacking or is less refined in nonflood-tolerant species (Chapters 4 and 5).[39]

The presence of higher concentrations of ethanol and organic acids in the roots of flood-tolerant than in nonflood-tolerant species suggests that the internal tissues are better buffered and more tolerant to compounds that are known to be toxic to plant metabolism. Specific biochemical traits which provide such tolerances are elaborated on in Chapters 6, 7 and 8. Adapted roots of tree species which produce the highest concentration of ethanol (Table III; Figure 11) also oxidize their rhizosphere (Roots, p. 311).

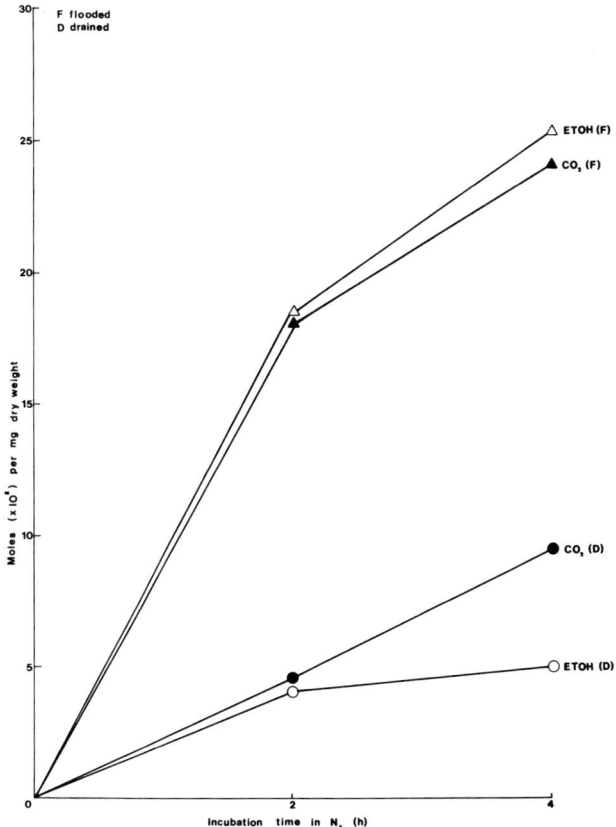

Figure 11.   Ethanol concentration and CO_2 evolution from swamp tupelo root tips from drained and flooded pots after incubation in nitrogen. Each point is the average of six seedlings.[15]

Fiscus and Kramer[40] have shown that the tissue inside the stele of corn (*Zea mays*) and jack bean (*Canaualis ensitormis*) roots operates under an oxygen deficit even when the external root is in the presence of 21% oxygen.   Unless the internal tissues are subjected to an oxygen deficit under flooding, there would be no stimuli to induce the mechanism for accelerated anaerobic respiration, which has been observed in tupelo roots.   Also, compaction of tissue in the meristematic region may reduce gas diffusion sufficiently to cause hypoxia (Chapter 9).

A dual adaptive mechanism in the roots has an obvious advantage to tree seedlings.   In the natural environment, flood water level usually fluctuates several centimeters within a few days.   It seems logical to assume that when the stem of a flood-tolerant seedling is exposed to air, oxygen diffuses to the roots as explained above and that nutrient and water uptake occurs essentially in an oxidized environment. When the flood water rises and covers most, if not all, of the stem lenticels, anaerobic respiration would be stimulated by the absence of oxygen in the root. This explanation leaves unanswered the question of whether interior aeration of the root is adequate for aerobic respiration when the root is oxidizing its rhizosphere.

Leaf chlorosis, epinasty and abscission become evident shortly after flooding.[41-43] These symptoms apparently are associated with the dieback of the secondary roots under flooding. Experiments with sunflower plants have shown that within 96 hours of flooding, roots are killed and lower leaves become chlorotic. In this case, chlorosis of detached leaves could be reversed by application of kinetin.[44] These authors concluded that flooding kills the roots, thereby preventing the synthesis of cytokinin and amino acids and their translocation to the leaves. This explains leaf chlorosis but not epinasty and reduced growth, which are also associated with flooding in most plants. However, Phillip[45,46] found that shoots of flooded plants had three times as much auxin as nonflooded plants. He hypothesized that flooding reduces the export of gibberellin-like substances, thus accounting for reduced stem elongation. Excess auxins in the flooded stem may be involved in stem and root anomalies reported above. Ethylene and abscissic acid are probably involved, but there are not data to substantiate their role.

The photosynthetic and transpirational rate of flooded cottonwood seedlings was only half that of nonflooded seedlings.[47] According to these investigators, rate of shoot elongation was not affected by flooding but rate of leaf maturation (as measured by size, color and gas exchange) was reduced. Although they attributed reduced photosynthetic rate to increased resistance (not measured) of roots to water uptake, their results could also be explained by reduced export of cytokinins from roots resulting in lower chlorophyll content in leaves. High carbon dioxide in the soil could have caused increased resistance to water uptake and accounted for a lower transpirational rate.

Based on the information reviewed, it is possible to develop descriptive models of how the aeration system of a tree functions. In Figures 12a and b, Model I illustrates how oxygen, carbon dioxide and other volatiles may move through a tree seedling that is partially inundated. Model I can be expanded into a conceptual model (Figure 13, Model IIa,b,c), which depicts the various avenues and barriers to gaseous diffusion in a tree species. Model IIa illustrates the theoretical pathways of gaseous flow in tree species and Model IIb shows the theoretical flow of oxygen and carbon dioxide in a tree growing in well aerated soil. By contrast, Model IIc shows

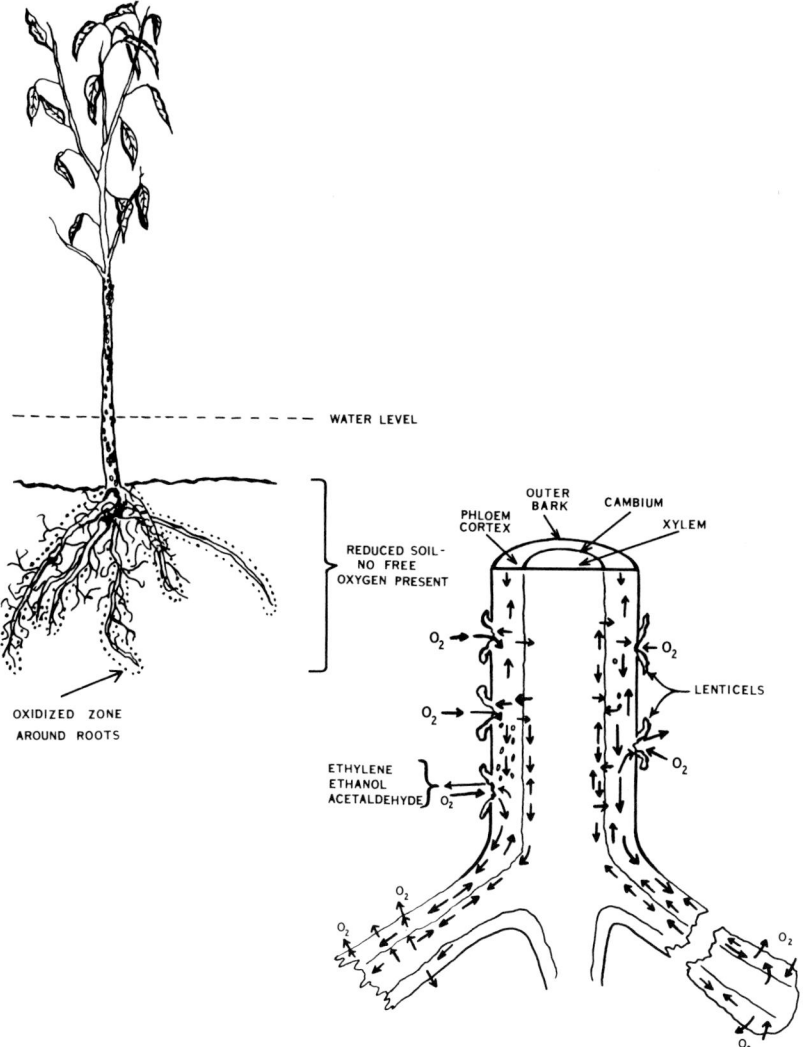

WATER LEVEL

REDUCED SOIL-
NO FREE
OXYGEN PRESENT

OXIDIZED ZONE
AROUND ROOTS

OUTER
BARK
PHLOEM
CORTEX
CAMBIUM
XYLEM

O_2

O_2

O_2

LENTICELS

O_2

ETHYLENE
ETHANOL
ACETALDEHYDE O_2

O_2

O_2

O_2

O_2

O_2

Figure 12. Aeration system of tupelo seedlings (Model I): (a) illustrates typical rhizosphere oxidation on an inundated tupelo seedling; and (b) longitudinal section illustrating the dual diffusion pathway of oxygen from the atmosphere to the rhizosphere and of volatile compounds from the root to the atmosphere.[30]

the critical nature of permeability of the cambium to a tree growing in poorly aerated water. The latter conditions is quite typical of the conditions that exist in many swamps in the southern United States during the active growing season. Under such conditions the xylem mother cells

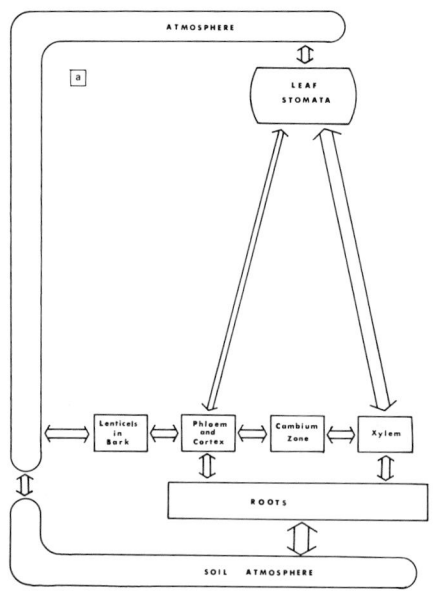

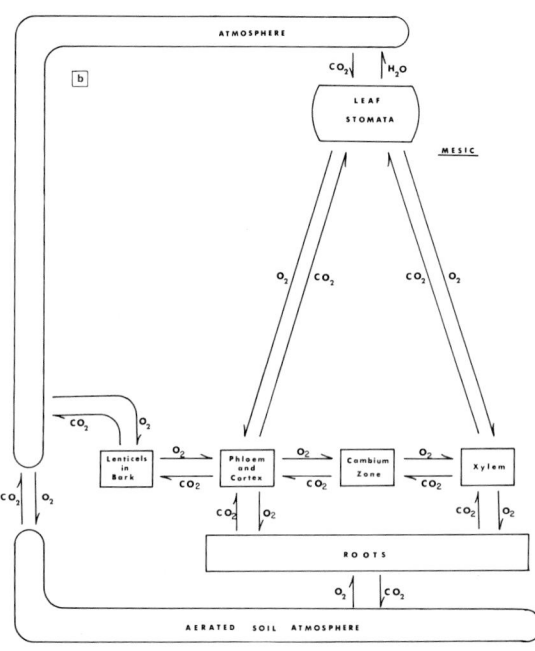

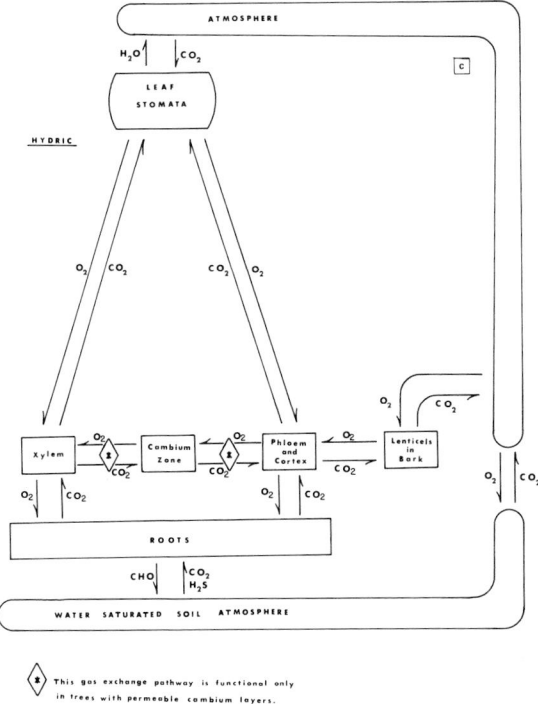

Figure 13. Generalized model illustrating gas exchange pathways in trees (Model II): (a) potential aeration pathways for trees living in well aerated soil; (b) gases and diffusion pathways between sinks and sources; and (c) gases and diffusion pathways where soil is poorly aerated. * This symbol indicates gas exchange pathway is functional only in trees with permeable cambial zones.

receive only aeration that diffuses against a transpirational gradient or which diffuses across the cambium.

SPECIES–SITE FACTORS

Although our understanding of the basic mechanisms of flood tolerance is still quite limited, a large number of field and laboratory studies have given us considerable insight into the scope of the problem and the variance in flood tolerances among tree species. Gil[28] has done an excellent job summarizing such research.

Flood tolerance is a function of both plant and site factors,[28] the more important of which are:

> Species
>   Intraspecific
>   Interspecific
>   Age and size of stand
> Edaphic
>   Soil type
>   Depth of flooding
>   Time of flooding
>   Duration of flooding
>   State of flood water

Species

The relative flood tolerance of a number of tree species can be found in Gil,[28] Hook and Brown[20] and Loucks and Keen.[48] Although many species exhibit some degree of flood tolerance, only a few live exclusively in swamp habitats. In the southern United States, bald cypress, swamp tupelo and water tupelo regenerate on sites that are very wet. They will grow in moist, well drained soils but cannot compete successfully with mesophytes on such sites. Their best growth usually occurs with moving flood water and fluctuating water levels.[49-51]

At the opposite extreme of swamp species are those species that cannot tolerate flooding even for short durations. Yellow poplar typifies this group in the southern United States. Although mature yellow poplars can withstand a few days of flooding during the growing season, the seedlings of this species may be killed by two to four days of flooding, particularly if the air temperature is high.[52,53] Occasionally, a yellow poplar tree may be found in the edge of a shallow swamp but generally they are restricted to well drained soils.

With the exception of *Taxodium distichum,* conifers are generally not considered to be very flood-tolerant. However, the data on conifer flood tolerance is not definitive. In comparing loblolly pine (*Pinus taeda*), shortleaf pine (*Pinus echinata)* and pond pine *(Pinus rigida* var. *serotina)*, Hunt[54] found no significant differences in their flood tolerance. In fact, pond pine, which grows on wet sites, appeared to be the least tolerant of the three species in height-growth and survival. Armstrong and Read[55] showed that of five conifers tested, three pine species diffused oxygen from their roots but two species of *Picea* diffused very small amounts, if any. Amounts of oxygen diffused from roots were inversely related to apparent respiration rates. When local seed sources of lodgepole pine (*Pinus contorta*) were compared with ponderosa pine (*Pinus ponderosa*), there were no differences in flood tolerance under controlled conditions, although in south central Oregon lodgepole pine grows in pure stands on essentially inundated sites, whereas ponderosa pine does not.[56] Huikari[57] found that anaerobic root media (sterilized and unsterilized) retard root growth of pine (*Pinus silvestris*) and spruce (*Picea abies*) but not the root growth of common birch *(Betula verrucosa)* and white birch *(Betula pubescens)*. An interesting trait is exhibited by loblolly pine. As its name infers it is frequently associated with wet sites. A species site test in South Carolina* indicated loblolly grew well and had high survival rates on sites where the growing season water table level averaged from 15-60 cm below the soil surface. However, a few individuals of this species survived (less than 5%) and grew as well on sites where the growing season water table level stayed above the soil surface most of the time. Individuals of this species are frequently found on hummocks in shallow swamps, and achieve exceptionally good growth on such sites. Obviously, there are interspecific flood-tolerant traits in conifer species as well as among hardwood species.

Edaphic

To the layman, swamps probably seem like a homogeneous lot, but as sites for growing tree species, they vary considerably because of soil and water conditions.

Swamps in the southern United States may be of nonalluvial (upland) or alluvial (river bottoms) origin; flood depth may exceed three meters; and the water may flow through the trees at rates of up to 5 kilometers per hour or be stagnant. Each of these variables has a significant effect on the soil in which the roots grow. In the Ocklawaha River Swamp in Florida, the redox potential of soils varied from +500 millivolts (mV) with natural shallow flooding to -100 to -240 mV with 1-2 meters of standing water flooding.[58] Hydrogen sulfide and reduced iron were considerably higher in the surface 15 cm in deep flooded soils than in the shallow flooded soils. All indications were that the deeply flooded soils were highly reduced and that reduced iron and sulfide concentrations near the soil surface were high enough to be toxic to the root growth of most plant species.

In semicontrolled flooding experiments, soil oxygen in stagnant water may reach concentrations near zero, and carbon dioxide may vary from 0.06-0.30 atmospheres (atm). Moving water mitigates reduced conditions, particularly in the upper 15 cm of the soil.[50,51]

Soil oxygen and carbon dioxide concentrations in two soil types and under different water regimes are given in Table IV.[51] CO_2 values given by Harms[51] are comparable to those found in nature. Soil characteristics other than aeration may override the effects of water regimes. Water tupelo seedlings grew better in soil from an alluvial swamp than in soil from a nonalluvial swamp, regardless

---

*Office Report S-46, Southeastern Forest Exp. Sta., USDA, Charleston, S.C.

Table IV. Average Partial Pressure of Dissolved CO_2 and O_2 in Water 8 cm Below
the Soil Surface in Three Water Regimes[a][51]

Water Regime	CO_2 (%)	O_2 (%)
Deep Flooding, Stagnant	3.3	6.1
Deep Flooding, Moving	0.9[a]	13.9[a]
Surface Flooding, Moving	0.9[a]	14.4[a]

[a]Within each column, values not followed by the same subscript are significantly different at the 5% level.

of water regimes and associated aeration. The alluvial swamp soil was heavy textured and had more available nutrients and different microorganism populations than the nonalluvial. It is interesting that there were interspecific differences in growth response of the two swamp species. Water tupelo seedlings appeared to be more sensitive to soil nutrition and microorganism populations and less sensitive to soil aeration than swamp tupelo.[51]

Flood Tolerance

To this point we have considered primarily the qualitative aspects of adaptations of tree species to flooding. Now we will discuss whether adaptations appear to be necessary and/or sufficient to account for a species flood tolerance and thereby attempt to assign some quantitative significance to various adaptations. In trying to develop a rationale to evaluate flood tolerance, Hook and Brown[20] have used the logic given below:

> Local patterns of species distribution are assumed to be the results of genetic processes operating under specific selective regimes over long periods of time. If this is true, and if a gradient exists in the specific selection regime, then there should be a recognizable sequence of adaptive mechanisms among the spatially separated species corresponding to their distribution along the gradient.

In bottomland forests of the eastern United States, such a gradient exists in soil moisture conditions from mesic to hydric sites. Although other selective pressures are operating within the bottomland ecosystem, moisture appears to be a dominant factor. Hosner and Boyce[60] and Dickson et al.[61] have shown interspecific differences in relative flood tolerance between species which normally grow within this moisture gradient. Therefore, it follows that species growing within the gradient in a local area should share some common adaptations or discrete adaptive mechanisms. If the adaptations are common among the species, then spatial separation should be a matter of degree of specialization. Presence of discrete adaptations among species would indicate specific differences in genetic make-up.

To test their theory, Hook and Brown[20] used five hardwood species—yellow poplar, sweetgum, sycamore *(Platanus occidentalis)*, green ash and water tupelo— which commonly occur along moisture gradients in bottomland forests of the southeastern United States.

Assumptions

Root adaptations were assumed to be the key factor in determining relative flood tolerance. Adaptations of other plant parts are known to be important but were assumed to be secondary to root adaptations in the process of natural selection.

Hypothesis

If root adaptations are critical in determining relative flood tolerance, then the presence of discrete adaptations and degree of refinement would appear to be correlated with flood tolerance.

Tests

To test their hypothesis, Hook and Brown selected three morphological and three physiological root adaptations and compared the ability of the five species to develop these traits with their survival and growth under controlled flooded conditions.

Morphological adaptations observed were ability of (1) secondary roots to survive prolonged flooding; (2) seedlings to regenerate new secondary roots from the primary root; and (3) seedlings to develop adventitious water roots on the submerged stem. Physiological adaptations observed were the ability of flood tolerant roots to (1) accelerate anaerobic respiration rate (ethanol accumulation) in the absence of oxygen; (2) oxidize their rhizosphere; and (3) tolerate high concentrations of CO_2.

Results

If tree species are compared with other vascular plants, they can be separated on the bases of their specialization (Figure 14). Furthermore, the same specializations can act as barriers or adaptive mechanisms to flood tolerance in tree species. Under the stress of prolonged inundation, secondary roots persisted on only one species (Figure 14). Of the four species on which the secondary roots died, new roots developed on three species within a few days. Yellow poplar developed no new roots and most of the seedlings died within the test, thereby confirming the known intolerance of this species to flooding. In comparing the four living species, only two developed adventitious water

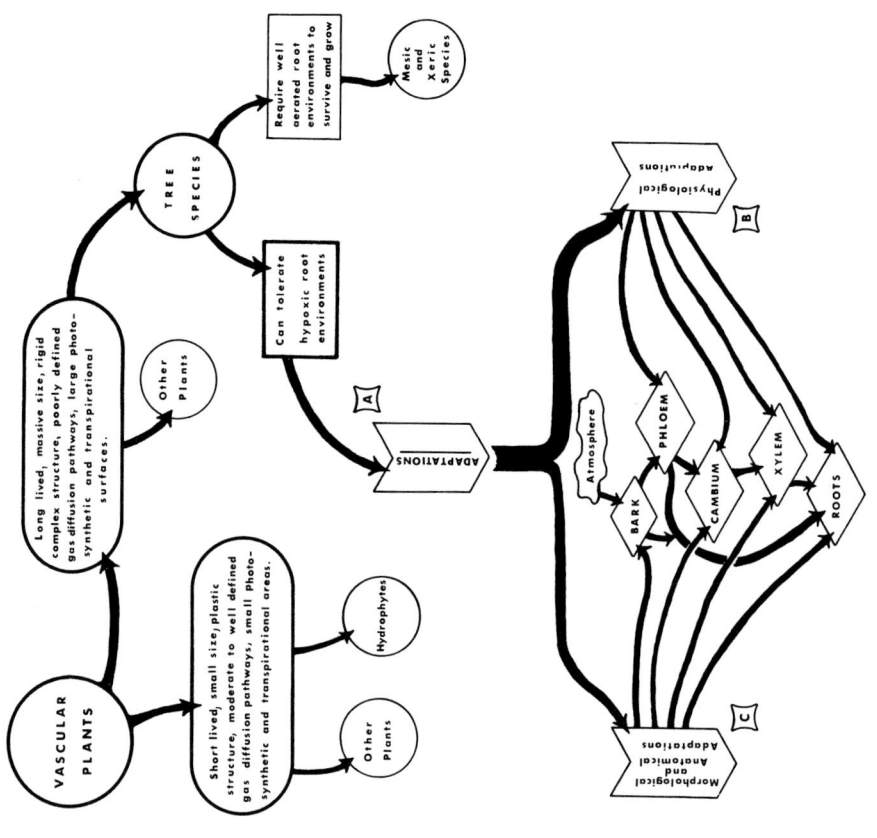

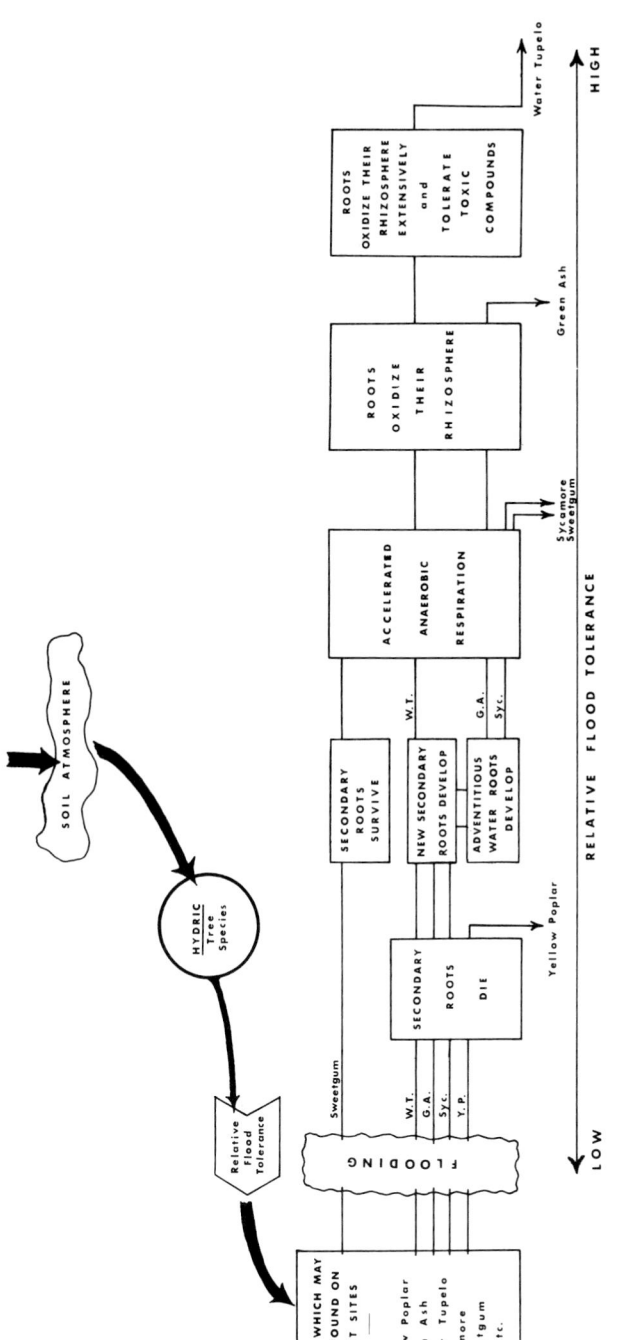

Figure 14. A flow diagram of the processes of adaptation to flooding in tree species and the relative flood tolerance of five tree species in relation to morphological and physiological traits. A. GENERAL HYDROPHYTIC CONSIDERATIONS: (a) aeration of roots and submerged parts via direct gas exchange not possible;[26] (b) degree of permeability of bark for gas exchange has not been

Figure 14. **(continued)** quantified;[9] (c) since water at roots is poorly aerated, oxygen transport must come from above water level. Mechanisms for this transport are poorly understood;[26] (d) field observations show significant quantities of oxygen are released from submerged roots of some species—a process of rhizosphere oxidation;[3,15,29] and (e) research also shows that certain compounds are carried upward by the transpiration system (TS) supporting the hypothesis that oxygen may be carried upward in the TS from roots on mesophytic species to aerate xylem mother cells.[2,3,27] B. PHYSIOLOGICAL ADAPTATIONS. *Bark and Phloem:* (a) proliferation of lenticels—hypertrophic without closing layers—have large spherical cells. Increased surface area and more previous structure may enhance gas exchange,[11] and (b) possible exchanges of dissolved oxygen and/or excretion of volatiles via lenticels. Research shows possible interspecfic variations of exchange potentials for submerged parts.[3,14,17-19,21] *Cambium:* (a) gas diffusion through cambial tissues—a site-specific adaptive mechanism—a result of evolutionary processes,[14] and (b) mesic-site species pervious.[14] *Xylem:* (a) highly variable data from measurements of stem oxygen concentrations. Variability among plants and variability between measurement techniques.[22-24] *Roots:* (a) rapid root adaptations—development of new roots with special properties;[15,20] (b) new roots more succulent, less fibrous, with little or no suberization at tips;[15] (c) new roots have poorly organized endodermis;[15] (d) adventitious water roots develop near water line after flooding, possibly to increase trees' ability to cope with flooding conditions. These water roots usually develop in moving water and at other times under special sets of conditions;[10] (e) rapid dieback of secondary roots occurs soon after flooding. Associated effects include epinasty, abscission and leaf chlorosis;[15,41-43] (f) adjustments in flooded tree roots similar to nonwoody species in relation to anaerobic metabolism;[15,20] and (g) paradox situation exists which is recognized but not fully understood (higher concentrations of toxic compounds are usually found in flood-tolerant species, an apparent result of superior buffering and toleration by internal tissues, in response to anaerobic respiration processes; however, in some species, flood-tolerant adapted roots which produce higher concentrations of ethanol also oxidize their rhizosphere). C. MORPHOLOGICAL AND ANATOMICAL ADAPTATIONS. *Stems and Roots:* (a) growth forms of roots and stems typically odd, swollen;[33,34] (b) bald cypress and water tupelo exhibit buttresses and "knees." These structures are without arenchyma tissue. Wood is of low density with numerous, large cells;[33,34] (c) odd growth forms develop primarily under fluctuating flooding conditions;[10] and (d) there is conflicting evidence to support that "knees" provide any aeration to the tree.[12,34]

roots—green ash and sycamore. Although these morphological traits have segregated the five species into four groups (Figure 14), only one of the traits has given any measure of flood tolerance. Obviously, if a plant's roots die under a specific stress and no new roots develop, as yellow poplar did under inundation, the plant species is very sensitive to the stress.

With the four surviving species under the experimental conditions, we know that secondary roots will persist on one, three will develop new secondary roots, and two will develop adventitious water roots. However, no quantitative measure of flood tolerance has been revealed.

If these four species are compared on the basis of physiological adaptations (Figure 14), they can be segregated into three groups that will be correlated with their relative flood tolerance. The roots of all four species accumulate more ethanol under N_2 incubation than in air incubation, but the new roots of sweetgum and sycamore do not oxidize their rhizosphere. These two species could be further segregated if adventitious water roots were considered to be a significant adaptation. However, the two species cannot be segregated on the basis of growth data from experiments of Hook and Brown or on the basis of their natural habitats. These two species occasionally share the same site in nature but their optimum growth is made on distinctly different sites.

New roots of green ash and water tupelo oxidize their rhizosphere but can be segregated at this point by degree of refinement. Water tupelo roots oxidize their rhizosphere several centimeters proximal to the root tip. In comparison, the roots of green ash showed some oxidation just below the root collar and very sporadic and indistinct oxidation near the root apex. Assuming that pronounced oxidation of the rhizosphere provides protection to roots in reduced soils from toxic gases and reduced compounds, the segregation of green ash and water tupelo along the relative flood tolerance scale is warranted.

The dichotomy of sweetgum from the other four species on the basis of secondary root survival under prolonged inundation indicates a distinctly different genetic make-up. As a consequence, under stress of flooding, sweetgum follows a different route in adapting to stress although it shares the common trait of accelerated anaerobic respiration under N_2 incubation.

SUMMARY

Tree species are very rigid in structure; therefore they appear to offer considerable obstacles to adapting to drastic changes in their environment. Despite such limitations, several tree species live and thrive in well aerated or poorly aerated soils, and some withstand rapid changes from one root environment to the other.

The aeration system of some tree species can be traced to the Carboniferous era. Parichnos were the earliest mechanism for gas exchange, but were limited to gymnosperms and were inadequate for aeration as they were destroyed by rapid radial growth. On present-day tree species, lenticels on the stem and roots appear to be the major avenue of gas exchange between the internal tissues of the bole and root and the atmosphere. Although lenticels are frequently undistinguishable on the bark of mature trees, they appear to remain functional. However, there are no quantitative data that evaluate the adequacy of the lenticels' role in gas exchange in mature trees. By contrast, seedlings of flood-tolerant species readily oxidize the rhizosphere via the diffusion of oxygen through stem lenticels to the root. Likewise, there are limited data on the internal gas concentrations of trees, particularly at the cellular level.

Under anaerobic conditions, metabolism of tree roots appears to function in a manner similar to metabolism of the roots of nonwoody species. Ethanol, lactic acid and other end products of anaerobic respiration can be found in the boles and roots of some tree species. Species that are tolerant to flooding appear to produce high quantities of anaerobic metabolites. But the production of these metabolites appears to be under metabolic control in flood-tolerant species and not in nonflood-tolerant species.

Tree species show a wide range in flood tolerance, varying from those species which live with their roots and lower stems flooded continuously to those tree species which cannot tolerate more than 3 or 4 days of flooding.

Relative flood tolerance appears to be correlated to specific physiological adaptations—namely, accelerated anaerobic root respiration and the ability of species to oxidize their rhizosphere.

ACKNOWLEDGMENT

The authors wish to express appreciation to Drs. Claude L. Brown and William R. Harms for their review and suggested improvements to this manuscript and to Mrs. Nadine Cunningham for typing and editorial assistance.

REFERENCES

1. Haberlandt, G. *Physiological Plant Anatomy.* (S-H Service Agency, Inc. New York. Reprint Edition as translated by M. Drummond 1965).
2. Crawford, R. M. M. "Physiologische Ökologie: ein Vergleich der Anpassung von Pflanzen und Tieren an sauerstoffarme Umgebung," *Flora* 161:209-223 (1972).
3. Chirkova, T. V. and T. S. Gutman. "Physiological Role of Branch Lenticels in Willow and Poplar under Conditions of Root Anaerobiosis," *Soviet Plant Physiol.* 19(2):289-295 (1972).
4. Bertrand, C. E. "Remarques sur le *Lepidrodendron Harcourtii*," *deWithan. Mon. Facultes de Lille* 2:1-59 (1891).

5. Weiss, F. E. "The Parichnos in the *Lepidodendraceae*," *Mem. and Proc. Manchester Lit. and Phil. Soc.* 51(8):1-22 (1907).

6. Wetmore, R. H. "Organization and Significance of Lenticels in Dicotyledons," *Bot. Gaz.* 82:113-131 (1926).

7. Sifton, H. B. "Air-space Tissue in Plants," *Bot. Rev.* 11(2):108-143 (1945).

8. Sifton, H. B. "Air-space Tissue in Plants II," *Bot. Rev.* 13:303-312 (1957).

9. Esau, K. *Plant Anatomy*, 2nd ed., (New York: John Wiley & Sons, Inc. 1965).

10. Hook, D. D. "Growth and Development of Swamp Tupelo [*Nyssa sylvatica* var. *biflora* (Walt.) Sarg.] under Different Root Environments," Ph.D. Dissertation, University of Georgia, Microfilm No. 69-9493 (1968).

11. Hook, D. D, C. L. Brown and P. P. Kormanik. "Lenticels and Water Root Development of Swamp Tupelo under Various Flooding Conditions," *Bot. Gaz.* 131(3):217-224 (1970).

12. Cowles, S. W. III. *Metabolism Measurements in a Cypress Dome*, M.S. Thesis, University of Florida, Gainesville (1975).

13. Thomas, M. T., S. L. Ranson and J. A. Richardson. *Plant Physiology*, 4th ed. (London: Churchill, 1956).

14. Hook, D. D. and C. L. Brown. "Permeability of the Cambium to Air in Trees Adapted to Wet Habitats," *Bot. Gaz.* 133:304-310 (1972).

15. Hook, D. D., C. L. Brown and P. P. Kormanik. "Inductive Flood Tolerance in Swamp Tupelo [*Nyssa sylvatica* var. *biflora* (Walt.) Sarg.]," *J. Exp. Bot.* 22(70):78-89 (1971).

16. von Hohnel, F. X. R. "Einige anatomische Bermerkungen über das räumliche Verhältniss der Interceullularräume zu den Gefässen," *Bot. Zeit* 37:541-545, (1879).

17. MacDougal, D. T, J. B. Overton and G. M. Smith. "The Hydrostatic-Pneumatic System of Certain Trees," *Carnegie Inst. Washington Pub.* No. 397 (1929).

18. MacDougal, D. T. "The Communication of the Pneumatic System of Trees with the Atmosphere," *Am. Phil. Soc. Proc.* 76(6):823-845 (1936).

19. MacDougal, D. T. and E. B. Working. "The Pneumatic System of Plants Especially Trees," *Carnegie Inst. Washington Pub.* No. 441 (1933).

20. Hook, D. D. and C. L. Brown. "Root Adaptations and Relative Flood Tolerance of Five Hardwood Species," *Forest Sci.* 19(3):225-229 (1973).

21. MacDougal, D. T. "The Pneumatic System of Trees," *Proc. Am. Phil. Soc.* 71:229-307 (1932).

22. Chase, W. W. "The Composition, Quality, and Physiological Significance of Gases in Tree Stems," *Minnesota Agric. Exp. Sta. Tech. Bull.* 99 (1934).

23. Jensen, K. F. "Measuring Oxygen and Carbon Dioxide in Red Oak Trees," *USDA, Northeast For. Exp. Sta. Res. Note NE-74* (1967).

24. Carrodus, B. B. and A. C. K. Triffett. "Analysis of Composition of Respiratory Gases in Woody Stems by Mass Spectrometry," *New Phytol.* 74:243-246 (1975).

25. Tjepkema, J. D. and C. S. Yocum. "Measurement of Oxygen Partial Pressure within Soybean Nodules by Oxygen Microelectrodes," *Plant (Berlin)* 119:351-360 (1974).

26. Hook, D. D., C. L. Brown and R. H. Wetmore. "Aeration in Trees," *Bot. Gaz.* 133(4):443-454 (1972).

27. Wium-Anderson, S. "Photosynthetic Uptake of Free CO_2 by Roots of *Lobelia dortmann*," *Physiol. Planta* 25:245-248 (1971).

28. Gil, C. J. "The Flooding Tolerance of Woody Species–A Review," *For. Abst. Leading Article Series No. 44*, 31(4):671-688 (1970).

29. Armstrong, W. "Oxygen Diffusion from the Roots of Woody Species," *Physiol. Planta* 21:539 (1968).

30. Hook, D. D. "Root (Botany)," *McGraw Hill Yearbook of Science and Technology* (1974), pp. 364, 365.

31. Harms, W. R., H. T. Schreuder, D. D. Hook, C. L. Brown and F. W. Shropshire. "The Effects of Flooding in Lake Ocklawaha," (In prep.).

32. Mattoon, W. R. "The Southern Cypress," *USDA Agric. Bull.* 272 (1915).

33. Penfound, W. T. "Comparative Structure of the Wood and the Knees; Swollen Bases, and Normal Trunks of the Tupelo Gum (*Nyssa aquatica* L.)," *Am. J. Bot.* 21:623-631 (1934).

34. Kramer, P. J., W. S. Riley and T. T. Bannister. "Gas Exchange of Cypress Knees," *J. Ecol.* 33(1):117-121 (1952).

35. Kramer, P. J. "Causes of Injury to Plants Resulting from Flooding Soils," *Plant Physiol.* 26:722-736 (1951).

36. Crawford, R. M. M. "The Control of Anaerobic Respiration as a Determining Factor in the Distribution of the Genus *Senecio*," *J. Ecol.* 54:403-413 (1966).

37. McMannon, M. and R. M. M. Crawford. "A Metabolic Theory of Flooding Tolerance: the Significance of Enzyme Distribution and Behaviour," *New Phytol.* 70:209-306 (1971).

38. Davies, D. D., K. H. Nascimiento and K. D. Patil. "The Distribution and Properties of NADP Malic Enzyme in Flowering Plants," *Phytochemistry* 13:2417-2425 (1974).

39. Crawford, R. M. M. "Metabolic Adaptations to Anoxia in Plants and Animals," *Proc. 12th Internat. Bot. Congress* Vol. 2 (1975), p. 353.

40. Fiscus, E. L. and P. J. Kramer. "Radial Movement of Oxygen in Plant Roots," *Plant Physiol.* 45:667-669 (1970).

41. Kolster, H. W. "High Water and Poplars," *Populier* 3(2):31-32 (1966).

42. Jackson, W. T. "The Role of Adventitious Roots in Recovery of Shoots Following Flooding of the Original Root Systems," *Am. J. Bot.* 43:816 (1955).

43. Yelonosky, G. "The Tolerance of Trees to Poor Soil Aerations," Abstract of Thesis in Dissertation, *Abstr.* 25:734-735 (1964).

44. Barrows, W. J. and D. J. Carr. "Effect of Flooding the Root System of Sunflower Plants on the Cytokinin Content in the Xylem Sap," *Physiol. Planta* 22:1105-1112 (1969).

45. Phillip, I. D. J. "Root-Shoot Hormone Relations. I. The Importance of an Aerated Root System in the Regulation of Growth Hormone Levels in the Shoot of *Helianthus annuus*," *Ann. Bot.* 38:17-35 (1964).

46. Phillip, I. D. J. "Root-Shoot Hormone Relations. II. Changes in Endogenous Auxin Concentration Produced by Flooding of the Root System in *Helianthus annuus*," *Ann. Bot.* 28:37-45 (1964).
47. Regehr, D. L., F. A. Bazzaz and W. R. Boggess. "Photosynthesis, Transpiration, and Leaf Conductance of *Populus deltoides* in Relation to Flooding and Drought," *Photosynthetica* 9(1):52-61 (1975).
48. Loucks, W. L. and R. A. Keen. "Submersion Tolerance of Selected Seedling Trees," *J. For.* 71(8):496-497 (1973).
49. Applequist, M. B. "Soil-Site Studies of Southern Hardwoods," in *Eighth Ann. Forestry Symp. LSU* (1960), pp. 49-63.
50. Hook, D. D., O. G. Langdon, J. Stubbs and C. L. Brown. "Effect of Water Regimes on the Survival, Growth and Morphology of Tupelo Seedlings," *For. Sci.* 16:304-311 (1970).
51. Harms, W. R. "Some Effects of Soil Type and Water Regime on Growth of Tupelo Seedlings," *J. Ecol.* 54:188-193 (1973).
52. McAlpine, R. G. "Flooding Kills Yellow Poplar," *For. Farmer* 19(3): 9, 13,14 (1959).
53. McAlpine, R. G. "Yellow Poplar Seedlings Intolerant to Flooding," *J. For.* 59:566-568 (1961).
54. Hunt, F. M. "Effects of Flooded Soil on Growth of Pine Seedlings," *Plant Physiol.* 26:363-368 (1951).
55. Armstrong, W. and D. J. Read. "Some Observations on Oxygen Transport in Conifer Seedlings," *New Phytol.* 71:55-62 (1972).
56. Cochran, P. H. "Tolerance of Lodgepole and Ponderosa Pine Seeds and Seedlings to High Water Tables," *Northwest Sci.* 46:322-331 (1972).
57. Huikari, O. "On the Effect of Anaerobic Media upon the Roots of Birch, Pine, and Spruce Seedlings," *Metsantutkimuslaitoksen Julkaisuja* Communications Instituti Forestalis Fenniae 50.9 (1959).
58. Forest Service. "Final Environmental Statement Proposal for Oklawaha River, Ocala National Forest, Florida," *USDA For. Serv. Vol. I, Appendices 1-14* (1972).
59. Kessler, G. D., J. T. May and D. D. Hook. "Soil-Tree Growth Relationships in an Upper Coastal Plain Swamp," in *Tree Growth and Forest Soils*, C. T. Youngberg and C. B. Davey, Eds. (Eugene, Oregon: Oregon State Univ. Press, 1971).
60. Hosner, J. F. and S. G. Boyce. "Relative Tolerance to Water Saturated Soil of Various Bottomland Hardwoods," *For. Sci.* 8:180-186 (1962).
61. Dickson, R. E., J. F. Hosner and N. W. Hosley. "The Effects of Four Water Regimes upon the Growth of Four Bottomland Tree Species," *For. Sci.* 11:299-305 (1965).

11

MORPHOLOGY AND FUNCTION OF ROOTS AND SHOOT GROWTH OF CROP PLANTS UNDER OXYGEN DEFICIENCY

M. C. J. de Wit

Biological Center for Plant Physiology
University of Groningen
Haren, The Netherlands

INTRODUCTION

The oxygen supply of higher plants is of the utmost importance for the aerobic respiration and thus for the supply of metabolic energy. No plant can live and thrive under complete anoxia for a prolonged period.[1,2] If oxygen deficiency occurs in any part of the plant this deficiency will interfere with all physiological processes in a direct or indirect way.

It is impossible to understand the influence of anaerobic conditions on the physiology and growth of the roots without also studying the performance of the roots under aerobic conditions. Within the extensive area of interactions between plant roots and oxygen, several phenomena of the behavior of plants confronted with a deficient oxygen supply in the root medium will be discussed. These concern plants that are not thriving optimally, when the root medium is oxygen-deficient.

Most of our crop plants belong to this group. In this category, I want to pay specific attention to plants that meet conditions of low oxygen content in the root medium for a prolonged period from early growth. Secondly, I want to discuss aspects of growth and physiology in plants exposed to a temporary oxygen deficiency after a growth period of sufficient oxygen supply.

STRUCTURAL MODIFICATIONS OF THE ROOTS UNDER
CONDITIONS OF PROLONGED OXYGEN DEFICIENCY

Anatomical Modifications

The presence of air-filled cavities and channels in the cortex of the roots of hydrophytes has been known for many years.[2] In addition, there is a group of plants that normally grows in well aerated soil, but which, under conditions of deficient oxygen in the root medium, is able to form voids or lacunae in the cortex of the roots.[3]

Dunn[4] started short-term experiments with representatives of the latter group of plants to analyze the conditions under which air space formation in the roots take place. She pointed out that maize (*Zea mays* L.) and wheat (*Triticum aestivum*) had no air spaces when cultivated in well aerated soil, whereas lacunae were formed in culture solutions, independent of the type of culture used. Similiar experiments by Bryant[5] with barley (*Hordeum vulgare*) over a two-month period showed that in nonaerated culture solution, in addition to the formation of air cavities in the root cortex, the roots were straighter, shorter, thicker and more numerous than in the aerated control solution. Shoot growth, however, was about equal under both conditions, and also the dry weight of the root system was practically identical.

Suppression of growth of root hairs has also been reported in non-aerated culture solutions.[6] Pitman[7] reported that the larger root diameter of barley roots from the nonaerated solution was due to the greater cell width and not to an increase in cell number.

McPherson[8] described the development of the lacunae formation in maize roots and presented a concept of the causal factors underlying this phenomenon. He stated that when the cortical cells are fully vacuolized, they lose their turgidity, collapse and deteriorate. He suggested that lack of oxygen starts a series of anaerobic processes which leads to a rapid breakdown of the cytoplasm and of the protein in the cell wall. Andreeva et al.[2] reported ultrastructural changes in the mitochondria of rice coleoptiles under anaerobiosis. The formation of lacunae in maize roots could be prevented by bubbling equal mixtures of oxygen and nitrogen through the culture solution. In addition, there were no air spaces produced when the pectic acid in the cell wall had been transformed to calcium pectate by a sufficient calcium supply to the culture solution.

Schramm[9] gives another explanation for the air space formation and concludes that the outer cells of the cortex, by having a better oxygen supply, tear apart the inner cells of the cortex.

Another aspect of the formation of air spaces was given by van der Heide et al.[10] They found that barley roots grown in anaerobic culture solution had a higher sugar content, an equal amount of protein and a lower polysaccharide content than the aerobic control plants. They suggested that the synthesis of polysaccharides, being the most important cell wall constituent, was inhibited by the low oxygen concentration and that inhibition of cell wall synthesis is an important factor in the formation of air spaces in roots. Following this line, de Wit[11] concluded that the inhibition of synthesis of cell wall polysaccharides in the roots present in the nonaerated culture solution is due to a diminished quantity of cell wall-producing enzymes, but that the inhibition is caused by the influence of oxygen deficiency on the activity of cell wall-producing enzymes itself. Luxmoore and Stolzy[12] suggested that the formation of air spaces is not dependent on oxygen content to a large extent, since they found no effect of the oxygen concentration of the solution on root porosity of maize and rice. Grable[13] suggested that lacunae formation is a process controlled by indoleacetic acid (IAA). He mentioned van Overbeek,[14] who showed that a treatment with indoleacetic acid softens the cell wall when the calcium content has been lowered, inducing methylation of pectic acid. In agreement with this suggestion, Pitman[7] reported that barley roots grown in a nonaerated $CaSO_4$ solution formed air cavities, but that such roots had a very low calcium content compared with roots from the aerated control solution. The content of K^+ and Na^+ in the roots was the same in both treatments.

Another aspect Pitman[7] reported on was the larger cell width of the root cells from roots growing in the nonaerated solution. This fits very well with the suggestion of cell wall softening by indoleacetic acid. Datko and MacLachlan[15] reported that pea segments treated with indoleacetic acid showed a marked increase in glucanase 1-4 (cellulase) activity, and the segments showed swollen cells and lacunae after treatment. As the inactivation of indoleacetic acid transported from the shoot to the roots by perioxidase is an oxygen-requiring process, it seems likely that oxygen deficiency could lead to a higher IAA content of the roots. Support for this hypothesis was given by Kefford,[16] who showed that in rice coleoptiles under oxygen–deficient conditions, the auxin content increased, possibly because of inhibition of the activity of IAA oxidase. Ethylene might be another factor important to air cavity formation in roots. Kawase[17] found a correlation between flooding damage symptoms and ethylene concentration in sunflower plants. One of these symptoms was the increased diameter of the hypocotyl. Anatomical investigation revealed that this increase was due to radially enlarged cells and increased intercellular spaces.

In conclusion, the formation of cavities in the cortex could be explained as follows. Lack of oxygen will locally start a series of biochemical processes that leads to death and destruction of clusters of cells. These processes and their interrelations are not sufficiently understood, but indoleacetic acid, ethylene and calcium appear to play important roles in these processes and affect lacunae formation in roots.

Morphological Modifications

Two types of modification of root growth under anaerobic conditions or conditions of low oxygen content are reported; that is, increased branching of the roots and the formation of adventitious roots.

Kleinendorst and Brouwer[18] reported that lack of aeration in culture solution promotes the emergence of branched roots. In general, a finer branched root system occurs under such conditions[19] with an increase of the number of roots and, as a consequence, the **area** for active ion absorption located close to the root tips will increase.[20] Geisler[21] also reported that a reduction of the oxygen supply leads to a higher number of lateral roots per unit of root length and an enhancement of the density of the root system.

The formation of adventitious roots is thought to be important for survival of the plant after flooding.[22] Shoot growth of tomato and sunflower plants is resumed as soon as adventitious roots are formed. Development of adventitious roots leads to an "escape" of the root system from places where oxygen deficiency is most severe.[23] Thus, it compensates for the loss of roots injured by oxygen deficiency. In maize and rice, the newly formed roots are more porous than the primary root systems,[12] and these roots are able to grow into water-saturated soil or into nonaerated solutions.

In most of our crop plants, however, this compensation is only partial.[5,12,24] In culture solution, the limited extent of the root system as a result of oxygen scarcity is less important, since the nutrient availability is not diminished. Vose[19] found that when light intensity was low and obviously growth-limiting, shoot growth of perennial ryegrasses, cultivated in a nonaerated culture solution, was better than shoot growth of the aerated control plants. Alberda[23] observed in rice that a mat of fine dense roots was formed at the surface of the water at the end of the tillering period. These roots supplied the more deeply located roots with oxygen. Finally, it should be noted that adventitious roots formed under aerobic conditions are less effective in transporting oxygen to the oxygen-deficient areas of the root system than adventitious roots developed under flooding.[25]

Root Growth

A distinction should be made between plants that grow from the beginning under conditions of little or no oxygen in the root medium and plants that start under aerobic conditions. Roots of maize start growing at an equal rate in nonaerated culture solution as well as in aerated solution, and maintain equal growth rates for a considerable time, depending on the environmental conditions (Brouwer, personal communication). On the contrary, barley roots are unable to maintain the original growth rate under nonaerated conditions.[9] The lacunae, formed in roots growing in the nonaerated solution, may occupy about 30 and 45% of the area of maize roots (cross section), while in barley roots this value is less than 10%. These air cavities facilitate oxygen transport to a large extent, but root growth of maize in diluted Hoagland solution at 23°C stops at about 20 cm in length and, in barley roots, at about 11 cm in length.[4] This cessation of root growth depends on temperature and the availability of nitrogen. In a nonaerated nutrient solution without nitrogen, root growth of maize will continue to about 40-50 cm instead of 20 cm (Brouwer, personal communication).

Greenwood and Goodman[26] showed that root elongation stopped in intact mustard seedlings *(Sinapis alba)* when the partial pressure of oxygen in the solution was about 0.005 atm and, under these conditions, oxygen transport from the leaves was prevented. It seems likely that in the above experiments with maize and barley the limiting factor for root growth was the length of the diffusive path for oxygen from the shoot to the root tip, to provide the root tip with adequate oxygen for growth. At lower temperatures, root growth can proceed for a longer period, resulting in longer roots, since respiration is more temperature-sensitive than the diffusion of oxygen.

Increased nitrogen uptake may stimulate protein synthesis in the roots and thus increase the respiration rate per cm root. Consequently, oxygen consumption increases along the root, resulting in a steeper decline of the oxygen concentration over the length of the root and reduced root growth.

According to Kordan,[27] the formation of root primordia in rice seedlings proceeds under anaerobic conditions, but there is no visible root growth. This observation is in accordance with van Overbeek,[14] who stated that nuclear division can proceed without oxygen, whereas cell elongation is highly oxygen–dependent.

EFFECT OF REDUCTION OF OXYGEN SUPPLY AFTER A PERIOD OF ADEQUATE OXYGEN SUPPLY

Physiological Responses

It has been frequently reported that a sudden stop of aeration or the event of a sudden flooding exerts a very harmful effect on a plant whose root system has been developed during a period of ample oxygen supply. Kramer[28] listed a series of symptoms resulting from a sudden change in oxygen availability. Symptoms such as yellowing and wilting of the leaves, epinasty of the leaves and complete inhibition of the shoot growth, finally result in the death of the plant unless development of adventitious roots takes place.

Brouwer[29] reported that elongation growth of maize roots after a period of adequate oxygen supply is inhibited immediately upon cessation of aeration of the root solution. Also, the absorption of potassium, nitrate and phosphate is inhibited at once, but water absorption is only influenced after a prolonged period of oxygen scarcity. In bean plants, however, water absorption by the roots is also immediately reduced upon anaerobiosis. The inhibition of root growth and of mineral absorption by the roots has been attributed to the inhibition of ATP synthesis by respiration.[20]

According to Brouwer (personal communication), transport of sugar to the root system does not stop immediately upon anaerobiosis, but the distribution of sugar over the length of the roots does change at once. Sugar transport to the apices of the roots, which is characteristic for aerobic conditions, was inhibited by anaerobiosis, and the carbohydrates were distributed more equally over the entire root system. This change in the distribution of sugars might be the result of cessation of root growth under anaerobiosis rather than the cause. Greenwood and Goodman[26] found that sucrose did not affect root growth at low oxygen tension, whereas there was an effect at high oxygen tension.

Wilting of the leaves has been attributed to a decrease of water permeability of the root system. It was suggested by Kramer and Jackson[30] that this decrease is a result of an increased viscosity of the protoplasm of the root cells. The diminished permeability can be restored if the period of oxygen scarcity is kept shorter than 24 hr, and little or no permanent injury occurs under such conditions. If, however, this period is prolonged, the root system gradually becomes damaged and the root tissue dies, resulting in increased water permeability. Finally, the roots decay and permeability drops again, probably because of plugging of the xylem vessels by decay of plants and/or bacteria.[31]

The susceptibility of plant roots is highest when the roots have an active metabolism—that, is when the need for oxygen is highest. Dormant roots of grasses are less susceptible to flooding than actively growing roots.[32] Roots can withstand a longer period of flooding when the root temperature is low.

Effect of the Carbon Dioxide Level

The significance of an increased CO_2 concentration on the physiology of the roots is still uncertain. It has been assumed that the carbon dioxide concentration seldom reaches a toxic level in most cropped soils.[33-35] Jackson[31] showed that removal of carbon dioxide from the flooded soil decreased wilting of the leaves of sunflowers only slightly. On the other hand, accumulation of CO_2 may affect the water permeability of the roots more directly than deficiency of oxygen.[30,36]

Mohammed and Williamson[37] suggested that in saline soils a short period of inundation may be less injurious to the plant roots than in nonsaline soils. This suggestion was based upon the observation that at high salinity levels CO_2 treatments stimulated mitotic divisions in the root tips of broad beans (Vicia faba L.).

Phytohormones

One of the effects of flooding of the tomato plant is epinasty of the leaves.[22] This effect is very similar to the effect of ethylene in the root environment.[38,39] Phillips[40] found that epinasty of the leaves of flooded sunflower plants was removed by decapitation. Addition of indoleacetic acid to the cut surface restored epinasty of the leaves. This experiment supports the idea that flooding of the root system induces an increased auxin content of the shoot either by preventing transport to the roots or by inhibition of the IAA-oxidase activity of the shoot. Increasing lignification of roots of barley and wheat as a result of flooding, as reported by Yamasaki,[41] is supposed to be controlled by indoleacetic acid.[13] Phillips[42] suggested that the development of adventitious roots after flooding is associated with both a raised level of indoleacetic acid of the shoot and an inhibition of synthesis of gibberellic acid in the roots.

The root system, and particularly the root tips, appear to be the site of synthesis of gibberellins and cytokinins.[43,44] In sunflower plants, flooding caused death of many root apices, and the content of cytokinins transported in the xylem was reduced to a minimum level.[45] Flooding of the root system of tomato plants resulted in a decreased stem growth and in a reduced level of gibberellic acid in the roots, shoot and bleeding sap. After 3-4 days of flooding and the appearance of adventitious roots, gibberellic acid accumulated in the shoot. This is suggested to originate from the newly formed roots. In the first days after flooding, application of gibberellic acid to the apical bud stimulated stem growth, but at a later stage gibberellic acid was less effective, indicating that other factors were limiting.[46] Different forms of stress, *e.g.,* water stress (drought), heat stress and, as mentioned above, flooding, lead to a reduced production of cytokinins in the roots.

The fact that formation of adventitious roots after flooding prevents senescence of the leaves[47] is associated with renewed synthesis of cytokinins and gibberellins in the newly formed roots.[48]

In anaerobic soils, ethylene may be present at a concentration that will affect root growth.[49] In addition, injured and diseased roots may produce ethylene.[39] A treatment of barley roots with a gas mixture of ethylene in air caused only a slight reduction of root and shoot growth.[50] Thus, a lowered oxygen concentration of the root environment may enhance the effect of ethylene. Extension of seminal roots was greatly inhibited by ethylene; however, lateral root growth was stimulated. This effect resembles the stimulation of the extension of lateral roots upon removal of the root apex. This phenomenon is also visible in roots upon a sudden stop of aeration.[18]

Epinasty of the leaves could also be caused by ethylene. Jackson and Campbell[51] showed that waterlogging increased the concentration of ethylene in the shoots of tomato plants. They indicated that the ethylene concentration in the soil can increase during waterlogging to a level that allows sufficient transport to the shoot to induce epinasty of the leaves. On the other hand, in plants grown in nutrient solution, a low oxygen concentration in the nutrient solution also promotes epinasty and increased ethylene concentration of the leaves. The mechanism of raising the level of endogenous ethylene of the shoot by oxygen deficiency in the root medium is largely unknown. The presence of a root system is required. However, ethylene biosynthesis requires oxygen,[52] and thus it is doubtful whether the root system is the major source of synthesis of ethylene.

SHOOT GROWTH AS AFFECTED BY A CONTINUOUS LOW OXYGEN CONTENT OF THE ROOT MEDIUM

Effects Upon the Vegetative Development

During vegetative development, the surfaces of both shoot and roots increase rapidly. Shoot and root growth are mutually dependent and generally proceed concurrently in a linear fashion though the shoot is growing at a faster rate.[53]. Under constant conditions during vegetative growth, the shoot:root ratio will remain constant.[54] The rate of leaf growth depends on the activity of the roots and on the amounts of mineral nutrients and water transported to the shoot. On the other hand, root growth depends on the quantity of carbohydrates and hormones transported to the roots. If one of the factors for shoot and root growth becomes limiting, a shift in the shoot:root ratio may be expected. For such a shift, the distance from source to sink becomes very important. In the case of water shortage, the growth of the root system is less affected than that of the shoot, because the source of water is closest to the roots and the shoot:root ratio will tend to decrease. At low light intensities, however, carbohydrates are more available for shoot growth, and the shoot:root ratio will increase.

As we have seen in the foregoing, lack of oxygen in the root medium reduces the extension growth of the root considerably and, as a result, the absorption rate is decreased. The absorbing surface per gram fresh weight of roots will increase by intensified branching of the roots and thus absorption may be improved. As a result, the shoot growth may also increase. In a study concerning the influence of aeration of culture solutions, Brouwer[55] found the rate of both shoot and root growth of bean seedlings to be reduced when the culture solution was not aerated. The shoot:root ratio was not affected. Ranson and Parija,[56] however, found opposite effects of the O_2 concentrations in the root medium upon the shoot:root ratio in wheat, rice and bean plants. In wheat, the shoot:root ratio decreased with decreasing O_2 concentrations, while in rice, the shoot:root ratio increased under such conditions. Brouwer (personal communication) obtained a reduction in root growth in maize in nonaerated solutions, while shoot growth was similar to that in the aerated controls. In perennial ryegrass, Troughton[57] did not observe an effect of aeration on the shoot:root ratio. In agreement with this observation, Luxmoore et al.[58] did not find an effect of oxygen supply on the shoot:root ratio in Mexican semidwarf wheat cultivars. Sojka et al.,[59] also working with Mexican semidwarf wheat, did find an increase in the shoot:root ratio when the oxygen level in the soil was reduced.

Using a series of light intensities, Brouwer[29] observed in bean plants in culture solution that the effect of oxygen deficiency of the root medium was more severe the higher the light intensity to which the plants were exposed (Figure 1). Since the K^+-content of the aerated and nonaerated seedlings was the same, Brouwer suggested that water uptake by the bean plants was affected. This was based on his observation that the rate of leaf elongation showed a diurnal pattern—that of being reduced

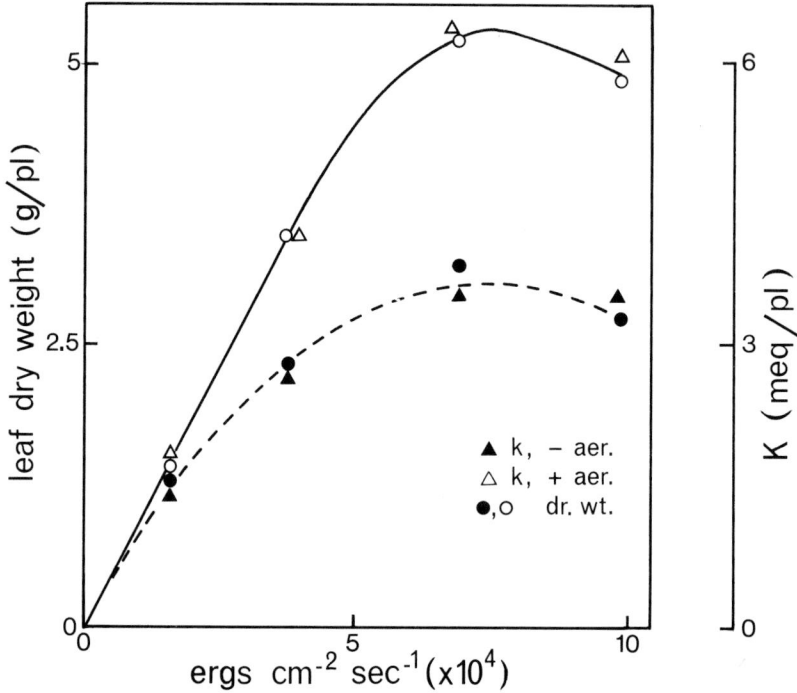

Figure 1. Effect of light intensity and aeration on dry weight and potassium accumulation of *Phaseolus* bean plants.[29]

by oxygen deficiency in the light period and showing only minor or no reduction at all during the dark period. The effect of oxygen deficiency on the leaf growth rate was diminished under conditions of a high relative humidity of the air (95%). Another argument for the involvement of the water balance was that growth of the bean plants depended upon the plant density (Figure 2). At a low light intensity and/or dense planting of the bean plants, the demand for water per root system decreased and, instead of aeration, light absorption became the limiting factor for shoot growth. At higher light intensities, the demand for water became stronger, and the oxygen supply of the roots may have limited shoot growth. As the K^+-content of both aerated and nonaerated plants was equal, it is unlikely that K^+-uptake was deficient. At higher light intensities, the leaf growth of the aerated control plants was no longer proportional to the plant density, perhaps because of mutual shading. Plants of the same age from the nonaerated group of plants were smaller, and the growth remained proportional to the density of the plants. As a consequence, the difference in the leaf growth rate between the aerated and nonaerated

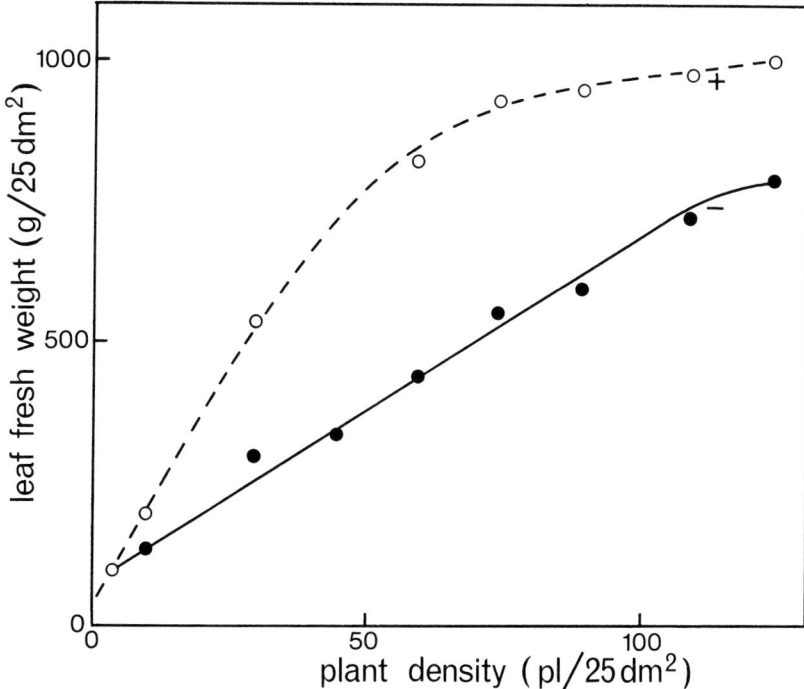

Figure 2. Effect of planting density and aeration on leaf fresh weight of *Phaseolus* bean plants. Age of plants at harvest time was 14 days.[29]

plants tended to become smaller, but did not disappear completely. The fact that this influence of aeration did not completely disappear in a closed plant canopy indicates the existence of an additional factor determining shoot growth. Since photosynthesis per unit of leaf area was the same, differences in photosynthesis could not explain this observation, and Brouwer suggested that positioning of the leaves as a factor in light absorption was responsible for the observed difference between aerated and nonaerated plants with dense plant canopies.

Several workers have shown, however, that a low oxygen concentration in the culture solution did not exclusively affect the leaf area but that it also affected the net assimilation rate of the leaves. Luxmoore and Stolzy[12] found, for maize, that the net assimilation rate decreases in culture solution at a low oxygen concentration as compared with the aerated control, and Troughton[57] made the same observation in perennial ryegrass. Under agricultural conditions, growth of a crop is strongly dependent on the total leaf area, and a change in the net assimilation rate is supposed to occur only under extreme environmental conditions.[60]   By contrast to Brouwer, Sojka et al.[59] observed no consistent effect of aeration on the leaf water potential in a Mexican semidwarf wheat variety cultivated in soil. The soil atmosphere was varied by passing humidified gases with varying oxygen content through the soil. They found an interaction between a low soil temperature and a low level of oxygen in the soil. Both features tended to retard shoot development, indicating a dependence of shoot growth on root metabolism.

The Porosity of the Roots

The formation of air spaces in the cortex of roots growing in an oxygen-deficient soil, or in a nonaerated culture solution, has been mentioned earlier. The significance of these air spaces for the transport of oxygen from the shoot to the roots will be discussed in Chapter 9. Oxygen transport through the plant, often called internal aeration, has been demonstrated several times, and the question arises about the efficiency of internal aeration for satisfying the need for oxygen of the roots when there is no oxygen supply from the soil or from the culture solution. Agronomists are often quite skeptical about the significance of air cavities in the root cortex of crop plants for internal oxygen supply to the roots because the cortex of older roots is often sloughed off[61] and thereby makes diffusion of oxygen throughout the plant to the roots impossible. The significance of the air spaces in the roots might be restricted to the younger parts of the roots and also to oxygen transport from well aerated soil layers to poorly aerated areas.

It is quite clear that root growth has a direct influence on the growth of the shoot. Shoot growth may be related to the root porosity under such conditions. Troughton,[57] comparing several clones of perennial ryegrass in culture solution, found that nonaeration of the culture solution increased air cavity formation. The formation of air spaces in nonaerated culture solutions depended on the genetic constitution of the clone used. Clones with the highest number of air spaces had the highest rate of shoot growth.

Using a series of crop plants of barley, corn, sunflower (*Helianthus annuus*), tomato (*Lycopersicum esculentum*) and two varieties of wheat in pot culture, Yu et al.[6] observed that a full-flooding treatment for two and four weeks reduced the growth rate and dry weight of all plants used, as compared with the nonflooding treatment. Corn, sunflower and Pato wheat were less severely damaged by flooding, and the roots showed a higher porosity than roots of barley, tomato and Inia wheat. However, on a dry weight basis, no relation between porosity of the roots and yield was evident. Arikado and Adachi[24] theorized that under conditions of prolonged flooding, growth of the shoot and the crop depend on the ability of the plants to improve their "ventilating system" and to form new adventitious roots. It has been shown by Luxmoore and Stolzy[12] that adventitious roots, which originated after flooding, are more porous than the primary roots in rice and maize and also that adventitious roots form the larger part of the root system. Improvement of the ventilating system might be the result of the formation of new, more porous adventitious roots. Thus, improvement of the ventilating system of the older parts of the root system might only be of minor importance. Hook et al.[62] reported that in tree species tolerant to flooding, an entirely new root system is formed after flooding. Nevertheless, Arikado and Adachi[24] observed that the yield of grains correlated with the ability of the plant to improve the ventilating system of the entire plant, especially of the root system.

SHOOT GROWTH AS RELATED TO A FLUCTUATING OXYGEN LEVEL IN THE ROOT MEDIUM

We have already mentioned that plants growing in well aerated soil are very susceptible to a sudden oxygen deficiency. When root systems of plants are flooded, the lower-positioned leaves begin to yellow, and this yellowing proceeds up the stem in tobacco and sunflower plants. In tobacco, epinasty of the leaves occurs after 24 hr. At last, wilting symptoms appear, but the duration of flooding required to induce wilting depends on the plant species studied. Stem elongation in tomato slows down or stops completely after 24 hr of flooding.[22] In tobacco, Kramer[22] demonstrated that water permeability of the root system temporarily decreased after flooding, resulting in wilting of the plants. The development of adventitious roots and the epinasty of the leaves, however, could not be attributed to a reduced absorption of water. In sunflower and tomato plants, the development of adventitious roots occurs after 2-4 days of flooding, the ability to develop adventitious roots being genetically determined.[30] In sunflower plants, Kawase[17] observed roots emerging from the hypocotyl after 24 hr of flooding. The reductions both of growth and crop yield are very severe under conditions of prolonged flooding combined with high temperatures.[17,63]

Fluctuations in the ground water table do not always affect the whole root system. When part of the root system of a plant is damaged by flooding, shoot growth will be affected, especially during the early stages of vegetative growth, when the rate of shoot growth is maximal. Each reduction of the root surface will thus delay the moment when the foliage reaches a closed stage, thus influencing the final yield.

Yu et al.[6] reported that half-flooding treatments, in which only a part of the root system was flooded, did not affect dry matter production of corn and sunflower when compared with plants grown in drained cylinders. Obviously, growth activity of the roots can easily shift from one part of the root system to another, and the distribution of roots in the various regions of the soil is determined by local differences in soil conditions. Growth of the root system as a whole is mainly limited by the carbohydrate supply from the shoot. When extending roots meet a particular soil region of favorable external conditions, root growth in this specific region is promoted, and the carbohydrate transport is directed to the roots extending to this particular region. El Nadi et al.[64] showed that root growth could be shifted from one soil region to another by varying the water supply.

In conclusion, it appears that moderate shifts in the water table will not affect shoot growth extensively during vegetative development of the plant. Roots are probably able to change their distribution pattern due to the presence of many dormant primordia.

EFFECTS OF LIMITED OXYGEN SUPPLY TO THE ROOTS DURING THE GENERATIVE STAGE OF THE PLANT

During the flowering stage and during fruit set in determinate plants, root growth is reduced or completely stopped. This may be explained as a competition for carbohydrates between the roots and the generative parts of the plant, where the latter profit from being closer to the source of synthesis. Lack of oxygen in the root environment in this stage of plant development is therefore very harmful, because the root system is no longer able to shift its root growth to soil layers with a higher oxygen content or to form new adventitious roots. Under conditions of normal oxygen supply, senescence of many plants starts after flowering, and this senescence is attributed to a reduced cytokinin production by the root tips. Since the yield of harvested fruits and seeds will depend on the continuation of a photosynthesizing leaf canopy after fruit set, any measure that favors cytokinin production, such as a supply of nitrogen, will increase the yield of fruits and seeds. A factor like oxygen deficiency in the root environment will depress cytokinin production and thus reduce the yield.

REFERENCES

1. Cannon, W. A. "Physiological Features of Roots, with Especial Reference to the Relation of Roots to Aeration of the Soil," *Carnegie Inst. Wash. Publ.* 368:1-168 (1925).
2. Adreeva, I. N., G. I. Kozlova and B. B. Vartepetian. "Formation of the Mitochondria of Rice Coleoptiles During the Process of

Germination under Aerobic and Anaerobic Conditions," *Sov. Plant Physiol.* 23:89-98 (1976).

3. Sifton, H. B. "Air Space Tissue in Plants," *Bot. Rev.* 11:108-143 (1945).

4. Dunn, G. A. "Note on the Histology of Grain Roots," *Am. J. Bot.* 8:207-211 (1921).

5. Bryant, A. E. "Comparison of Anatomical and Histological Differences Between Roots and Barley Grown in Aerated and Non-aerated Culture Solutions," *Plant Physiol.* 9:389-391 (1934).

6. Yu, P. T., L. H. Stolzy and J. Letey. "Survival of Plants Under Prolonged Flooded Conditions," *Agron. J.* 61:844-847 (1969).

7. Pitman, M. G. "Adaptation of Barley Roots to Low Oxygen Supply and its Relation to Potassium and Sodium Uptake," *Plant Physiol.* 44:1233-1240 (1969).

8. McPherson, D. C. "Cortical Air Spaces in the Roots of *Zea Mays* L.," *New Phytol.* 38:190-202 (1939).

9. Schramm, R. J., Jr. "Anatomical and Physiological Development of Roots in Relation to Aeration of the Substrate," PhD Thesis, Duke University (Ann Arbor, Michigan: University Microfilms, 1960).

10. van der Heide, H., B. M. de Boer-Bolt and M. H. van Raalte. "The Effect of a Low Oxygen Content of the Medium on the Roots of Barley Seedlings," *Acta Bot. Neerl.* 12:231-247 (1963).

11. de Wit, M. C. J. "Reakties van Gerstwortels op een Tekort aan Zuurstof," PhD Thesis, Groningen 1974.

12. Luxmoore, R. J. and L. H. Stolzy. "Root Porosity and Growth Responses of Rice and Maize to Oxygen Supply," *Agron. J.* 61:202-204 (1969).

13. Grable, A. R. "Soil Aeration and Plant Growth," *Adv. Agron.* 18:57-105 (1966).

14. van Overbeek, J. "Auxins," *Bot. Rev.* 25:269-350 (1959).

15. Datko, A. H. and G. A. MacLachlan. "Indoleacetic Acid and the Synthesis of Glucanases and Pectic Enzymes," *Plant Physiol.* 43:735-742 (1968).

16. Kefford, N. P. "Auxin-Gibberellin Interaction," *Plant Physiol.* 37:380-386 (1962).

17. Kawase, M. "Role of Ethylene in Induction of Flooding Damage in Sunflower," *Physiol. Plant.* 31:29-38 (1974).

18. Kleinendorst, A. and R. Brouwer. "Responses of Two Different Clones of Perennial Ryegrass to Aeration of the Nutrient Solution," *Jaarb. I. B. S.* 29-39 (1967).

19. Vose, P. B. "Nutritional Response and Shoot/Root Ratio as Factors in the Composition and Yield Genotypes of Perennial Ryegrass," *Ann. Bot.* (N.S.) 26:425-437 (1962).

20. Brouwer, R. "Ion Absorption and Transport in Plants," *Ann. Rev. Plant Physiol.* 16:241-266 (1965).

21. Geisler, G. "The Morphogenetic Effect of Oxygen on Roots," *Plant Physiol.* 40:85-88 (1965).

22. Kramer, P. J. "Causes of Injury to Plants from Flooding of Soil," *Plant Physiol.* 26:722-736 (1951).

23. Alberda, Th. "Growth and Root Development of Lowland Rice and its Relation to Oxygen Supply," *Plant Soil* 5:1-28 (1953).

24. Arikado, H. and Y. Adachi. "Anatomical and Ecological Responses of Barley and some Forage Crops to the Flooding Treatment," *Bull. Fac. Agric. Mie Univ.* 11:1-29 (1955).

25. Valoras, N. and J. Letey. "Soil Oxygen and Relationship to Rice Growth," *Soil. Sci.* 101:210-215 (1966).

26. Greenwood, D. J. and D. Goodman. "Studies on the Supply of Oxygen to the Roots of Mustard Seedlings (*Sinapis alba L.*)," *New Phytol.* 70:85-96 (1971).

27. Kordan, H. A. "Mitonic Activity in Rice Seedlings Germinating under Oxygen Deficiency," *J. Cell. Sci.* 20:57-59 (1976).

28. Kramer, P. J. *Plant and Soil Water Relationships: A Modern Synthesis* (New York: McGraw-Hill Book Co., 1969).

29. Brouwer, R. "Some Physiological Aspects of the Influence of Growth Factors in the Root Medium on Growth and Dry Matter Production," *Jaarb. I. B. S.* 11-30 (1963).

30. Kramer, P. J. and W. T. Jackson. "Causes of Injury to Flooded Tobacco Plants," *Plant Physiol.* 29:241-245 (1954).

31. Jackson, W. T. "The Relative Importance of Factors Causing Injury to Shoots of Flooded Tomato Plants," *Am. J. Bot.* 43:637-639 (1956).

32. Rhoades, E. D. "Inundation Tolerance of Grasses in Flooded Areas," *Trans. Am. Soc. Agric. Eng.* 7:164-166 (1964).

33. Grable, A. R. and R. E. Danielson. "Influence of CO_2 Growth of Corn and Soybean Seedlings," *Soil Sci. Soc. Am. Proc.* 29:233-238 (1966).

34. Williamson, R. E. "Effect of Soil Gas Composition and Flooding on Growth of Nicotana Tabacum," *Agron. J.* 62:80-82 (1970).

35. Ponnamperuma, F. N., E. Martinez and T. Loy. "Influence of Redox Potential and Partial Pressure of Carbon Dioxide on pH Values and the Suspension of Flooded Soils," *Soil Sci.* 101:421-431 (1966).

36. Brouwer, R. "Water Absorption by the Roots of *Vicia faba* at Various Transpiration Strength," *Proc. Kon. Ned. Acad. Wet.* C57: 68-80 (1954).

37. Mohammed, S. A. and R. E. Williamson. "Interaction of Gas Composition and Salinity upon Root Cell Division of *Vicia faba L.*," *Agron. J.* 62:18-20 (1970).

38. Turkova, N. S. "Growth Reactions in Plants Under Excessive Watering," *Akad. Nauk. SSSR Leningrad C. R.* 42:87-90 (1944).

39. Williamson, C. E. "Ethylene, a Metabolic Product of Diseased or Injured Plants," *Phytopathol.* 40:205-208 (1950).

40. Phillips, I. D. J. "Root-Shoot Hormone Relations. I. The Importance of an Aerated Root System in the Regulation of Growth Hormone Levels in the Shoot of *Helianthus annuus,*" *Ann. Bot. N. S.* 28:17-35 (1964).

41. Yamasaki, T. *Nat. Inst. Agr. Sci. Bul.* 1:1-92 (1952) (English Summary).

42. Phillips, I. D. J. "Root-Shoot Hormone Relations. II. Changes in Endogenous Auxin Concentration Produced by Flooding of the Root System in *Helianthus annuus*," *Ann. Bot. N. S.* 28:37-45 (1964).

43. Crozier, A. and D. M. Reid. "Do Roots Synthesize Gibberellins?" *Can. J. Bot.* 49:967-975 (1971).

44. Kende, H. "Kinetin-Like Factors in the Root Exudate of Sunflowers," *Proc. Nat. Acad. Sci.* 53:1302-1307 (1965).

45. Burrows, W. J., and D. J. Carr. "Effect of Flooding the Root System of Sunflower Plants on the Cytokinin Content in the Xylem Sap," *Physiol. Plant.* 22:1105-1112 (1969).

46. Reid, D. M. and A. Crozier. "The Effects of Depressed Oxygen Concentrations," *J. Exp. Bot.* 6:80-93 (1955).

47. Chibnall, A. C. "Protein Metabolism in Rooted Runner Bean Leaves," *New Phytol.* 53:31-38 (1954).

48. Richmond, A. E. and A. Lang. "Effect of Kinetin on Protein Content and Survival of Detached *Xanthium* Leaves," *Science* 125(1): 650-651 (1957).

49. Smith, K. A. and S. W. F. Restall. "The Occurrence of Ethylene in Anaerobic Soil," *J. Soil Sci.* 22:430-443 (1971).

50. Crossett, R. N. and J. D. Campbell. "The Effect of Ethylene in the Root Environment upon the Development of Barley," *Plant Soil* 42:453-464 (1975).

51. Jackson, M. B. and D. J. Campbell. "Waterlogging and Petiole Epinasty in Tomato, the Role of Ethylene and Low Oxygen," *New Phytol.* 76:21-31 (1976).

52. Mapson, L. W. "Biogenesis of Ethylene," *Biol. Rev.* 44:155-187 (1969).

53. Aung, L. H. In *The Plant Root and Its Environment*, E. W. Carson, Ed. (Charlottesville, Virginia: University Press of Virginia,1974), p 29.

54. Brouwer, R. "Some Aspects of the Equilibrium Between Overground and Underground Plant Parts," *Jaarb. I. B. S.* 31-39 (1963).

55. Brouwer, R. "De Invloed van Aeratie van het Wortelmilieu op de Groei van Bruine Bonen," *Jaarb. I. B. S.* 11-21 (1960) (English Summary).

56. Ranson, S. L. and B. Parija. "Experiments on Growth in Length of Plant Organs. II. Some Effects of Depressed Oxygen Concentrations," *J. Exp. Bot.* 6:80-93 (1955).

57. Troughton, A. "The Effect of Aeration of the Nutrient Solution on the Growth of *Lolium perenne*," *Plant Soil* 36:93-108 (1972).

58. Luxmoore, R. J., R. E. Sojka and L. H. Stolzy. "Root Porosity and Growth Responses of Wheat to Aeration and Light Intensity," *Soil Sci.* 113:354-357 (1972).

59. Sojka, R. E., L. H. Stolzy, and M. R. Kaufmann. "Wheat Growth Related to Rizosphere Temperature and Oxygen Levels," *Agron. J.* 67:591-596 (1975).

60. Watson, D. J. "Physiological Basis of Variation in Yield," *Adv. Agron.* 4:101-145 (1952).

61. Beckel, D. K. B. "Cortical Disintegration in the Roots of *Bouteloua gracilis* (H. Bik) lag.," *New Phytol.* 55:183-190 (1956).

62. Hook, D. D., C. L. Brown and P. P. Kormanik. "Inductive Flood Tolerance in Swamp Tupelo (*Nyssa sylvatica* var. *biflora* (Walt) Sarg.)," *J. Exp. Bot.* 22:78-89 (1971).

63. Cameron, D. G. "Lucerne in Wet Soils; the Effect of Stage Regrowth, Cultivar, Air Temperature and Root Temperature," *Aust. J. Agric. Res.* 24:851-860 (1973).

64. El Nadi, E. H., R. Brouwer and J. Th. Locher. "Some Responses of the Root and the Shoot of *Vicia faba* plants to Water Stress," *Neth. J. Agric. Sci.* 17:133-142 (1969).

12

SHORT-TERM FLOODING*

B. D. Meek

Soil Scientist
Imperial Valley Conservation
    Research Center
Brawley, California 92227

L. H. Stolzy

Professor of Soil Physics
University of California
Riverside, California 92521

INTRODUCTION

Gas exchange is necessary to aerate plant roots, with most of the exchange occurring through the soil under well drained conditions, but in saturated soils, the gas exchange may be through the plant. Soil aeration is the exchange of carbon dioxide (CO_2) and oxygen (O_2) gases between the atmosphere and the soil pore spaces. The restriction of soil aeration for even one day may reduce plant growth, and longer periods of restriction may result in plant death.

Although soil aeration is usually adequate in agricultural soils, it may be restricted for short periods of time because of rainfall or application of irrigation water. If the soil moisture is increased above a critical value, narrow pores between soil voids will be blocked by water films, and the gaseous phase will become discontinuous. The result will be little, if any, gas exchange between the atmosphere and the soil voids. At this time, soil O_2 will decrease proportionately with O_2 consumption. This decrease

---

*Contribution of the Western Region, Agr. Res. Ser. USDA, Brawley, California and Dept. of Soil and Environmental Sciences, University of California, Riverside.

351

will continue until the soil moisture decreases sufficiently to again allow gaseous exchange.

Plants vary widely in their response to low-aeration status. Generally, critical O_2 levels begin in the range of 5-10% O_2 by volume. For most plants, root growth will be limited in soils with fewer than 10% air-filled pore spaces, which may have oxygen diffusion rates (ODR) of less than 0.2 $\mu g/cm^2/min$. Crops like rice can grow with no soil O_2, while other plants are very sensitive to low O_2 levels.

Measurement of soil aeration under short-term flooding is very difficult because the system varies rapidly with location in the soil and with time. An index of soil aeration can be obtained by several methods. However, each method may be best for only a particular condition and should be viewed with consideration for the complete system.

Several good reviews have dealt with soil aeration. Russell[1] and Grable[2] made extensive studies of soil aeration and plant growth, and Wesseling[3] discussed the influence of drainage on soil aeration.

In this chapter we will emphasize the effects of changes in soil aeration that occur in the field as the result of short-term flooding. We will cover: (1) methods for evaluating soil aeration, (2) factors influencing soil aeration, (3) aeration effects on plant growth, and (4) effects of water management.

Variation in Soil Gaseous Phase

Measurement of soil aeration is complicated because the soil system is very heterogeneous. The soil gaseous phase exists within two types of pores: the smaller ones within the aggregates or crumbs and the larger ones between the aggregates. Pore types are illustrated by data collected on sand, clay, and Fe-zeolite (Table I). When soil aeration is restricted, the smaller pores within the aggregates will have a lower O_2 level and may have no O_2, while the larger pores may be well aerated. For example, a soil crumb may be completely aerobic, while one twice its radius would have 30% of its volume anaerobic.[4] A crumb 10 times the radius of the aerobic crumb would have 84% of its volume anaerobic. The plant must integrate the different O_2 levels encountered in the soil by its root system. This variation makes it difficult to use average values to characterize soil aeration; therefore it may be better to define the part of the soil that has restricted aeration. Fluhler et al.[5] suggested that soil aeration may be understood by a statistical expression of microsite heterogeneity. Because of the large changes in soil O_2 over short distances, aeration sensors may give different values for the same system. The different values will be caused by the variation in the sphere of influence of different types of aeration sensors. For example, the sphere of influence for the O_2 membrane electrode will be much larger than for the ODR electrode.

Table I. Intercrumb and Crumb Porosities of Various Materials[29]

Separate	Total Porosity at Saturation % by Volume	Intercrumb Porosity % by Volume	Crumb Porosity % by Volume
Quartz Sand	37.0	30.0	7.0[a]
Clyde Clay	63.5	31.5	32.0
Brookstone Clay	68.0	31.0	37.0
Fe-zeolite[b]	70.0	30.0	40.0

[a]The 7.0% water undoubtedly represents film water around the sand grains. The intercrumb porosity is interparticle porosity.

[b]250-420 μm separates.

Dynamics of Soil Aeration

The location of a site in the soil profile with respect to its macro and micro position will influence the rate of change of O_2 at that site. There are rapid changes in soil aeration level when the soil surface is initially sealed by flooding. The O_2 level can decrease from 21 to 0% by volume in less than 24 hr, if O_2 consumption is high. Stolzy and van Gundy[6] found low ODR levels one day after irrigation but values nearly doubled within 4 days (Figure 1).

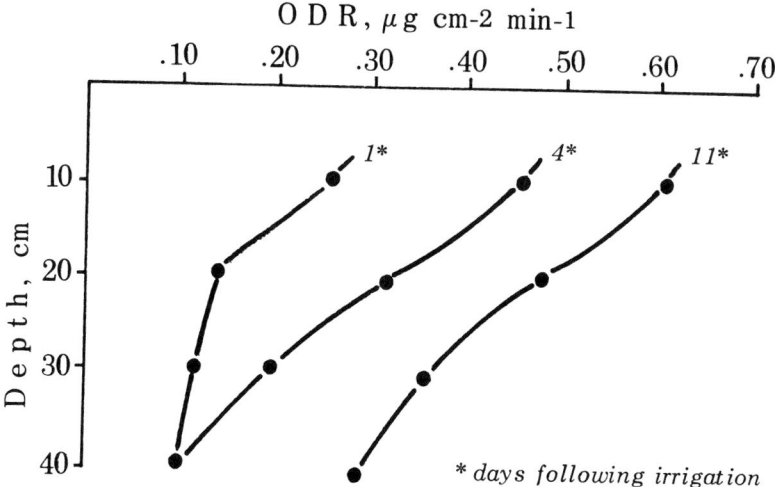

Figure 1.   Oxygen diffusion rate (ODR) as a function of soil depth on various days after irrigation in cotton field.

Soil profile characteristics exert a great influence on the rate of change of soil moisture tension and soil aeration status. After irrigation, there are major changes in hydraulic gradient in the soil surface because of drying, with very little or no change at greater depth. A study on clay loam soil planted to wheat showed the hydraulic gradient was -23 at the 15-30-cm depth, 15 days after irrigation, but only -3 at the 30-60-cm depth, while a positive gradient was present at the 60-90-cm depth (Figure 2). For the same soil, Figure 3 shows the change in ODR at different depths for an irrigation on March 2 as a function of time. There are rapid changes in ODR at the 2.5-, 5.0- and 10.0-cm depths, but almost no changes at deeper depths during a 10-day period.

FACTORS TO EVALUATE SOIL AERATION

The lack of simple accurate methods to characterize soil aeration has complicated characterization of the various factors that influence soil aeration. Many soil aeration methods are available to evaluate the soil aeration status. Some of the more common methods are to measure (1) soil O_2 content, (2) redox potential Eh, (3) O_2 diffusion rate, (4) chemical ion forms in solution and (5) trace gases in the soil and organic compounds in plants. Since the different methods monitor various aspects of soil aeration, the use of the findings will determine the selection of the most effective method. For example, ODR is useful for evaluating plant growth, while redox potential data is more useful in defining potential for denitrification.

Gaseous Oxygen

Measuring gaseous O_2 content was one of the first methods used to evaluate soil aeration. Most studies have analyzed gas samples that have been extracted from the soil. It has been difficult to extract these samples without contamination from atmospheric oxygen. Contamination can occur from many sources but the most probable cause is the flow of O_2 down the side of the sampling tube. Because of these problems, much of these data are questionable. However, use of a diffusion chamber for taking O_2 samples has improved the accuracy of this method.

Polarographic electrodes to measure O_2 content in place have been of limited use because of problems with temperature compensation, lack of a commercial source for the electrodes, need for frequent calibrations, and unreliability of the electrodes.

Willey and Tanner[7] published details of a membrane-covered electrode which is temperature-compensated and has a probe design that permits its

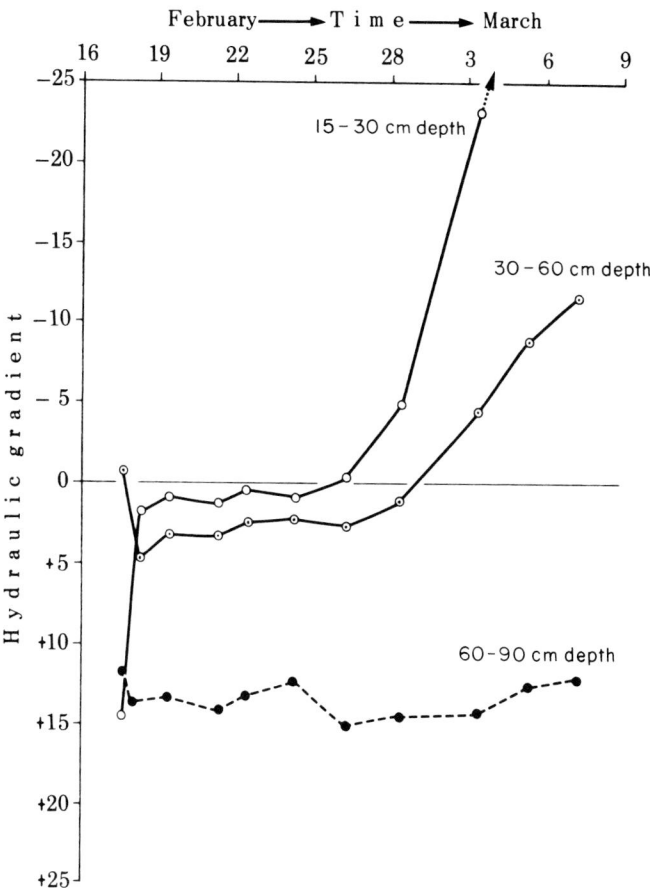

Figure 2. Average hydraulic gradients between various soil depths in the profile of a fine-textured soil profile following an irrigation.

installation in an access tube. Electrodes are now being produced commercially (Jensen Instruments, P.O. Box 44021, Tacoma, Washington, 98444) that are temperature-compensated and will fit into an access tube so they can be removed easily for calibration and service.*

The double membrane-covered electrode has some advantages over the electrodes used to measure ODR. Membrane electrodes are not

---

*Mention of a trademark of proprietary product does not constitute a guarantee or warranty of the product by the U. S. Department of Agriculture, and does not imply its approval to the exclusion of other products that may also be suitable.

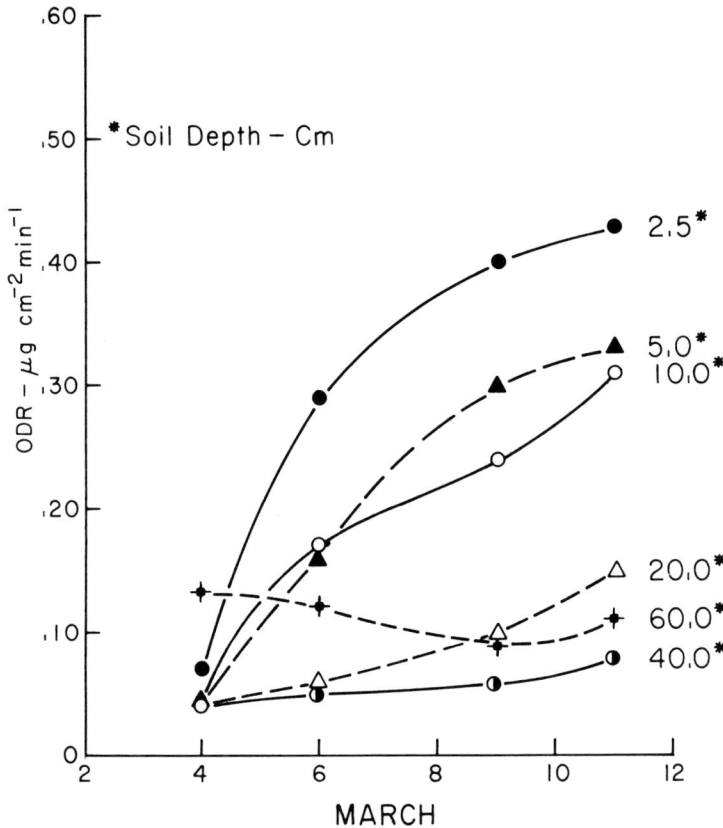

Figure 3. Average ODR as a function of time at different soil depths in the profile of a fine-textured soil profile. Reproduced by permission from *The Israel Journal of Agricultural Research* 23:167(1969).

as variable as ODR electrodes in output because they are influenced by a greater volume of soil. It is not necessary to reinsert the membrane electrode each time readings are taken.

Redox Potential (Eh)

Redox potential, which measures the tendency of the soil solution to receive or supply electrons, is used in research studies as an index of soil aeration. Changes in Eh with time will be small when there is sufficient gaseous O_2 present in the soil. Redox potential shows significant changes only when gaseous O_2 has been depleted by soil microorganisms. After atmospheric O_2 is depleted, the soil microorganism will use nitrate (NO_3), manganese dioxide (MnO_2), or some other substance for an electron acceptor. Meek and Grass[8] studied the response of redox electrodes to flooding in a tank of Holtville silty clay soil that was flooded and then allowed to air dry in the laboratory. The Eh decreased for 12 days, after which it increased rapidly as O_2 flowed into the soil (Figure 4).

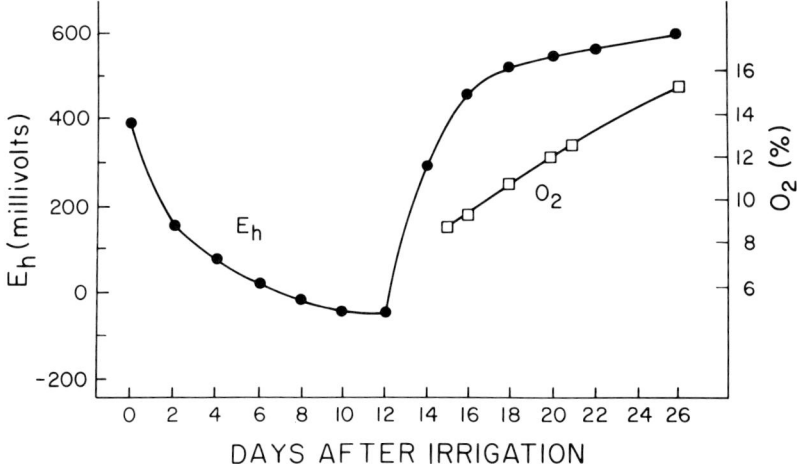

Figure 4.   Redox potential and gaseous oxygen at the 28-cm depth of soil, which was flooded and allowed to dry.[8]

Because Eh shows only small changes in the region where there is gaseous O_2, its use for evaluating plant responses to soil aeration is limited. However, it is very useful for evaluating oxidation-reduction changes in soil and its effect on the amounts of reduced forms of manganese (Mn) and iron (Fe) found in the soil solution.

Some problems encountered in measuring Eh are the poisoning of electrodes, lack of standard methods, and variation in the system being measured in the field. Meek and Grass[8] found that in the field, 22 electrodes would be needed to obtain an average value with ± 40 mV of the true value under conditions of rapid changes.

The advantages of Eh are that the electrodes are inexpensive and can be inserted with only a minor change in the soil system in the field; also, many readings can be made in a short period of time. Redox potential is more useful under long-term flooding when the O_2 status is lower.

Oxygen Diffusion Rate (ODR)

The ODR may be the soil aeration index most closely related to plant growth. The platinum-electrode method measures the reduction of O_2 at the surface of an electrode when an electrical potential is applied. The reading is controlled initially by the amount of O_2 in the solution surrounding the electrode but, when that is reduced, the reading will be controlled by the diffusion of O_2 to the electrode.

The relationship between Eh and ODR is shown in Figure 5. The Eh changed only a small amount until ODR values decreased below 9.5 x 10^{-8} g/cm^2/min. ODR is more sensitive in the high range of soil aeration while Eh is more sensitive in the low range.

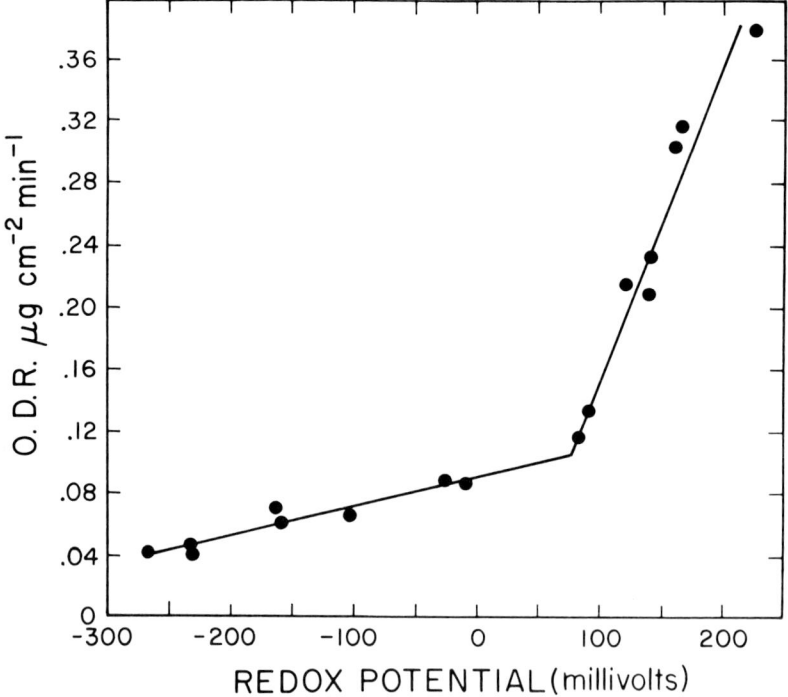

Figure 5. Relationship of oxygen diffusion rates to redox potentials.[25]

The diffusion of O_2 through water films to the electrode should resemble O_2 diffusion to roots. Therefore, ODR is influenced by the same factors that influence O_2 flow to roots, namely water-film thickness and diffusion path.

ODR values are easily made in soil, and the electrodes and instrumentation are sensitive within the range of plant growth where soil aeration may be limiting.

Disadvantages of ODR values are that they give good readings only over a limited soil moisture range and the readings are so variable that many measurements are needed to characterize soil aeration in the field. Dasberg and Bakker[9] reported that variability is so great that measurements should be interpreted with care.

Reduced Chemical Forms

The measurement of elements in the soil solution can be useful for evaluating soil aeration, especially at low O_2 levels. Species that may be useful are soluble forms of Mn, Fe and sulfur (S) which are brought into solution as a result of the oxide form being reduced and the O_2 being used as an electron acceptor when atmospheric O_2 is not available. Nitrate (NO_3) is reduced at about +300 mV, Mn^{+4} at +200 to +300 mV, Fe^{+3} at 0 to +100 mV, and SO_4^{-2} at almost -200 mV. Usually it is necessary to have long-term flooding, such as rice culture, to produce H_2S and Fe^{+2}. Concentrations of Fe, Mn and NO_3 in the soil solution as a function of Eh are presented in Table II. At Eh values of +295 to +315 mV, only trace amounts of Mn and Fe were found in the soil solution and large amounts of NO_3. When the soil was reducing (Eh +35 to +85 mV), amounts of Fe and Mn in the soil solution were substantial and NO_3 was low.

Under long-term flooding, toxic levels of Mn^{++}, Fe^{++} and S^{+2} may develop.

Table II. Elemental Composition and Redox Potential in a Sorghum Field at the 80-cm Depth for Two Irrigations[8]

Treatment	Eh mV	Mn ppm	Fe ppm	NO_3^-n ppm
Irrigation No. 4				
Check	315	0.04	0.00	16.7
Manure residual		0.43	0.01	80.0
Manure	35	7.88	3.47	0.3
Manure (double irrigation)	55	1.62	0.06	0.1
Irrigation No. 6				
Check	295	0.07	0.02	11.2
Manure residual	275	0.33	0.04	57.0
Manure	85	6.70	2.71	0.1
Manure (double irrigation)	65	1.25	0.05	1.0

Trace Gases in Soil and Organic Compounds in Plants

In the future, information correlating plant responses with trace gases in the soil atmosphere and organic compounds in plants may be sufficiently developed to help characterize soil aeration. Trace gases that may be useful to characterize soil aeration would be nitrous oxide (N_2O), ethylene (C_2H_4), hydrogen sulfide (H_2S) and methane (CH_4). Alcohol in tomato tissue has been suggested as a method of evaluating soil aeration. Because they are produced at a soil aeration level which is important to plant growth, C_2H_4 and N_2O levels can be used to characterize soil aeration and its effect on plant growth. Both H_2S and CH_4 are present only under extremely low soil aeration levels, so they would be of limited use in the evaluation of aeration effects on plant growth.

Roulier et al. (unpublished results) studied N_2O levels in the soil as an index of denitrification, and found that N_2O ranged from 0.5 to 10.0 ppm (Figure 6). Increases in N_2O correlated well with rainfall, irrigation, and manure applications. Figure 6 shows the N_2O levels in untreated soil and soil which had received manure at the rate of 45 metric ton/ha. The application of manure doubled the N_2O concentration in the surface soil. Dowdell and Smith[10] found that the N_2O level was inversely related to the O_2 content of the soil atmosphere.

Sheard and Leyshon[11] measured C_2H_4 levels as high as 16.7 ppm, 13 days after flooding. Smith and Dowdell[12] reported that the concentration of C_2H_4 was influenced by temperature, decrease in O_2 content, availability of substrates, and soil moisture content.

Fulton and Erickson[13] presented data defining the relationship between ODR and ethanol concentration in xylem exudate. They found that ethanol concentrations were present when ODR values were lower than 0.38 $\mu g/cm^2/min$.

FACTORS CONTROLLING SOIL AERATION

Soil Moisture Tension

Soil moisture tension is the most important factor controlling soil aeration status. When soil moisture tension is lowered, changes in soil aeration will be small until a critical value is reached (usually about 20 cb of water tension). Below this point, changes will be large. The critical value of soil moisture tension is that point at which the voids become so restricted they are no longer continuous. At this value, there is little exchange of gas between the soil and atmosphere.

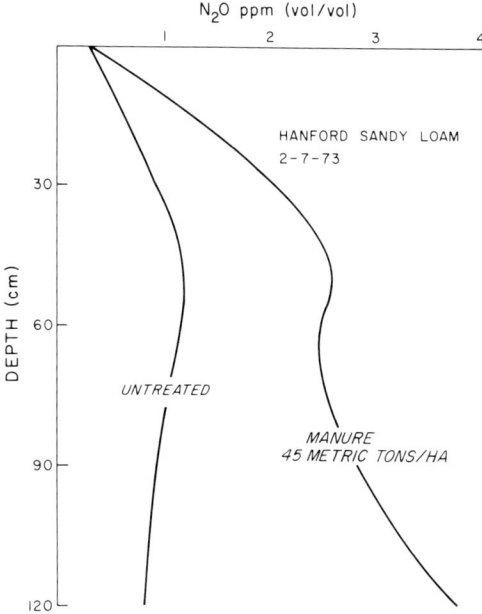

Figure 6. Nitrous oxide concentration in the soil atmosphere of two plots of Hanford sandy loam with and without a manure application.

The soil moisture tension will determine the air content of a soil, and either soil moisture tension or soil air content can be used to evaluate soil aeration. Dasberg and Bakker[9] found a high correlation ($r = 0.82$) between dry matter production and mean soil air content.

The critical soil moisture tension, above which there will be adequate soil aeration, will vary as a function of soil texture and aggregate size. Stolzy[14] presented data for a Yolo loam, which had been treated to obtain three different aggregate-size ranges (Figure 7). He found that the water release curve for the coarse (0.84-2.00 mm) aggregate size resembled that for sand.

Under some conditions, soil moisture tension can be used to evaluate soil aeration. Initial readings can be made with O_2 sensors to find the relationship between the sensor readings and soil moisture tension for a given soil. After this, soil aeration can be evaluated by using a certain critical soil moisture tension as an indication of restricted aeration.

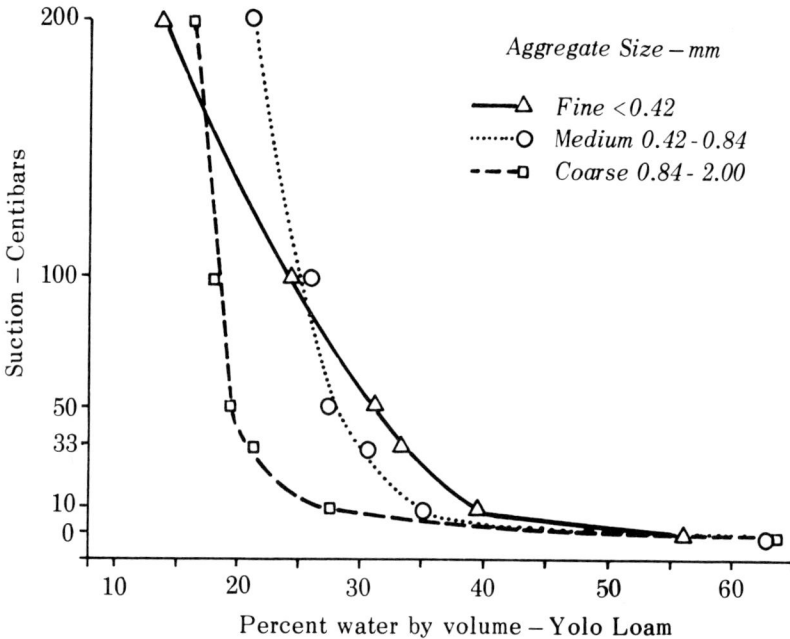

Figure 7.   Water release curves of Krilium-treated Yolo silt loam aggregates in three different size ranges.

Oxygen Consumption

When soil moisture tension is below the critical value, the change in O_2 level will depend on O_2 consumption. When flooded for short periods by irrigation or rain, soil may decrease below the critical soil moisture tension. Daily values of O_2 consumption of 2.2 liter/m^2 in a sandy soil and 1.8 liter/m^2 in a peat topsoil have been measured.

Oxygen consumption depends mainly on a source of energy, temperature and soil texture. Organic amendments can drastically affect soil aeration values. Meek and Grass[8] evaluated the effect of crop residue on soil aeration levels after flooding (Figure 8). Cotton residues were mixed with the top 15 cm of a Holtville silty clay loam soil. Location D was less saline, which resulted in 14.5 metric ton/ha of cotton residue being turned

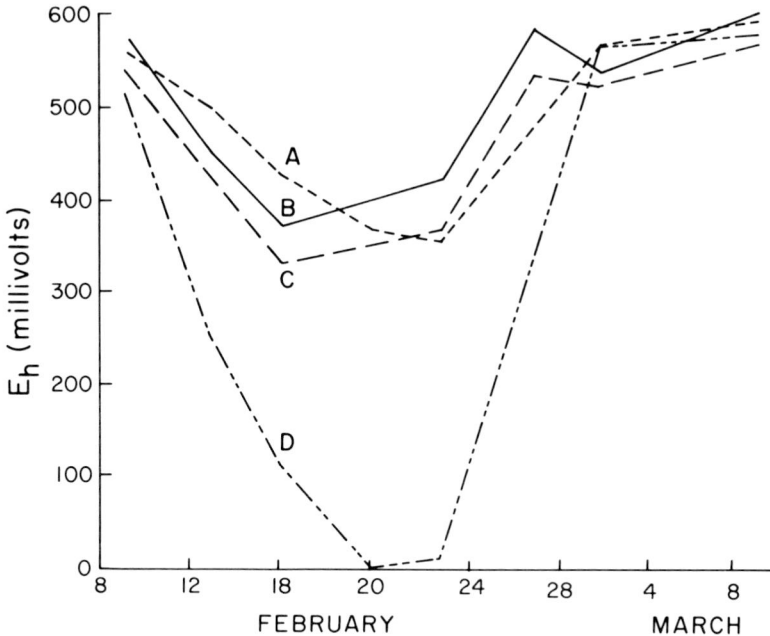

Figure 8. Eh values as a function of time under flood irrigation with D receiving double the amount of crop residue.[8]

under compared with less than one half this amount for other locations. The soil treatment receiving a large amount of crop residue was subjected to low soil aeration levels (Eh values as low as 0 mV) compared with high soil aeration levels for other treatments (Eh values above +350 mV).

Temperature

Temperature affects soil aeration indirectly by its effect on respiration of plants and microorganisms. The effect of temperature on diffusion and mass flow is minor. An increase of 15°C may result in a threefold increase in respiration rate; therefore, temperature will have a significant effect on soil O_2 level when there is a large active root system or microbial population. Luxmoore[15] calculated a Q_{10} value of 1.78 for 5-15°C and a value of 1.5 for 15-25°C for the increase in respiration rate of root tips. Meek et al.[16] studied the effect of organic matter (OM) and temperature on Eh after flooding (Figure 9), and found only a small difference in Eh between the high and medium temperature when no OM was added. When OM

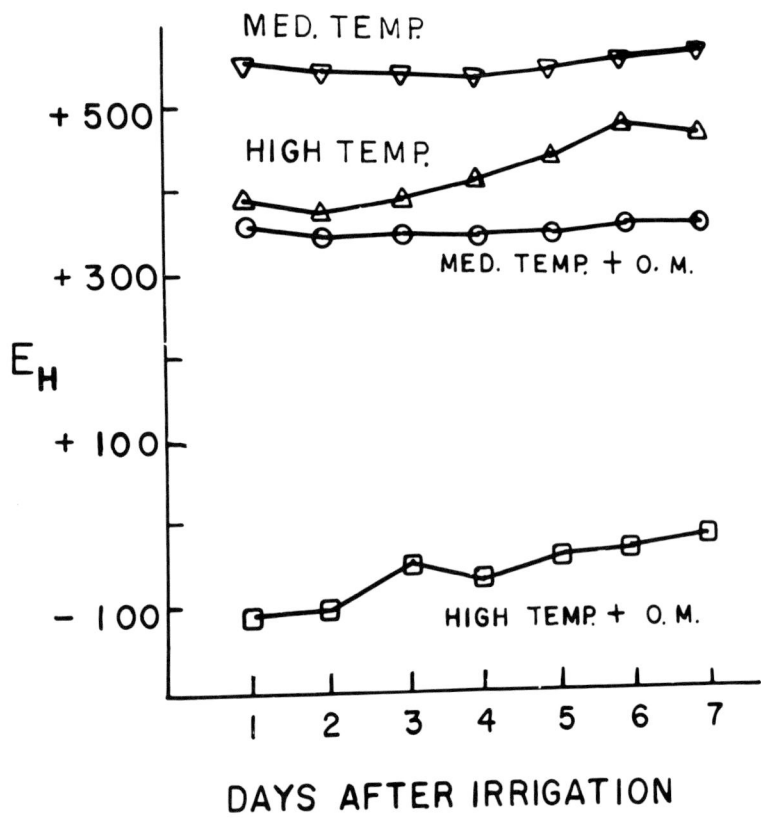

Figure 9.   Effect of organic matter and temperature on Eh (mV) measured at 20 cm for days after initial irrigation.[16]

was added, there were major differences in Eh between the two temperatures because of increased respiration rate of the microbial population under the high temperature.

Higher temperatures will result in higher values of evapotranspiration (Et), which will increase soil moisture tension and increase the soil aeration level. The effects of temperature on respiration and Et tend to counteract each other.

Effect of Plants on Soil Aeration

Plants affect the O_2 levels in the soil profile. The extracting of water by plants creates areas around roots where O_2 can flow into the soil. In some fine-textured clay soils, roots must extract moisture so that O_2 can flow into the soil, because drainage alone does not increase the soil moisture tension sufficiently over short periods of time for gaseous exchange with the atmosphere. Grable[17] reported that in larger aggregates, gravity alone provided sufficient drainage for germination, but in the smallest aggregates or densest soil, it required 48-68 cm of suction for germination.

EFFECT OF LOW OXYGEN STATUS ON PLANTS

Evaluating the effects of short-term flooding on plant growth is difficult because the soil aeration status is continually changing. Damage caused by temporary flooding will depend on plant species, growing stage, temperature and the duration of flooding. Some plants will die from short periods of low aeration, while others can grow under conditions of 0% soil O_2. Grasses, in general, will tolerate lower O_2 levels than other species. Mature plants will usually show less injury because of flooding than immature ones. Low O_2 status may cause soil fermentation, which may result in the buildup of toxic products. Long-term flooding is more easily characterized because, after the initial change, the soil O_2 status is fairly stable.

Effect on Respiration

The effect on plant respiration rate is the most important effect of low soil aeration level. This effect will vary depending on the plant and its physiological age. Flooding reduces water absorption directly by decreasing the permeability of roots to water and indirectly by reducing the size of the root system. Figure 10 presents data giving the cumulated water transpired at the indicated O_2 levels. As the O_2 level decreased, the amount of water transpired decreased by about 50%. Plants adapted to the low O_2 level and, after 35 days, the rate of water transpiration was similar for all treatments.

Plant Growth

Although low O_2 values will reduce both top and root growth, root growth is affected more.

Aceves-Navarro[18] found that top growth of wheat was reduced 50% by 0% oxygen in the root system, compared with adequate levels. In studies on the effect of short-term flooding, Leyshon and Sheard[19] found barley top growth was more sensitive to short-term flooding at an early growth stage (14 days) than at 21, 28 or 35 days; but the younger plants were capable of greater recovery than were the older plants. Barley grain yield was reduced 55% when flooded 28 days after seeding, but only 35%

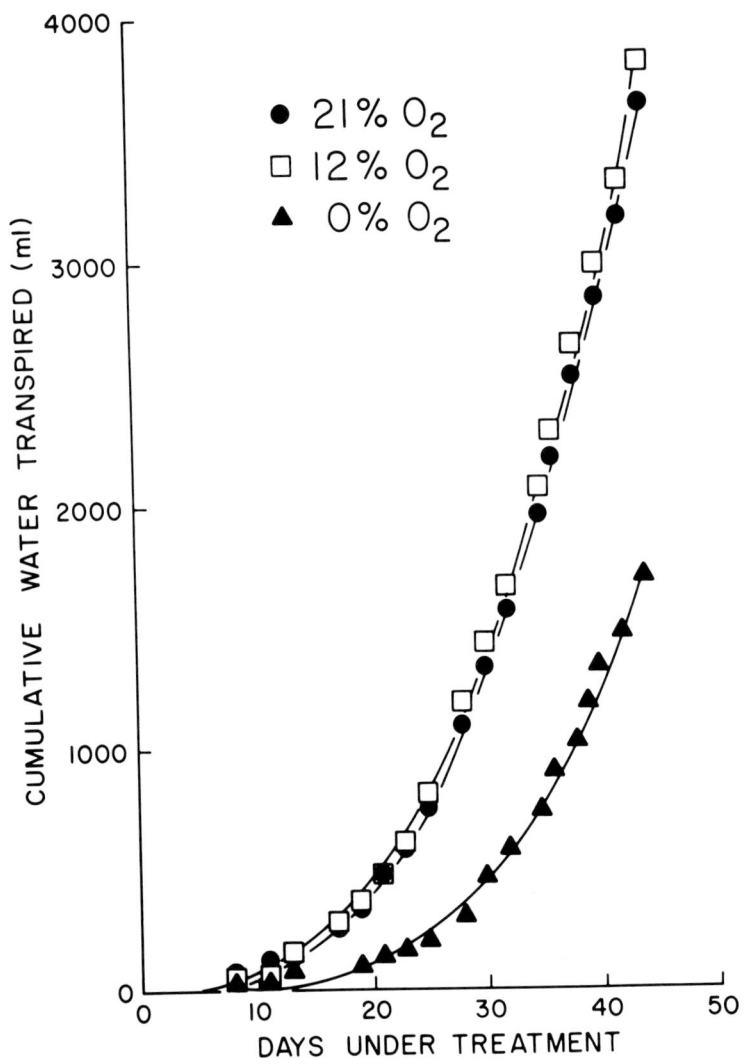

Figure 10. Cumulative water transpired by wheat variety Inia 66 grown for 45 days at three different oxygen levels in the rooting media.[18]

when flooded 35 days after seeding. Anaya and Stolzy[20] found that wheat yields were higher at 9.6% O_2 than at 21%. They suggested that excess O_2 supply to the roots may result in O_2 toxicity or luxury consumption of photosynthesis products.

Roots grown in areas of poor aeration are shorter, thicker and more branched, with fewer root hairs than those grown under adequate aeration.

Aceves-Navarro[18] found that when wheat was grown under low O_2 levels, the root systems were decreased by 72%. He suggested 5% as the critical O_2 content for wheat germination. Patrick *et al.*[21] found a good relationship between average soil O_2 content at the 90 cm depth and cotton root density at the 60-90-cm depth (Figure 11). The root tip is more sensitive to low O_2 than are other parts of the root. Anaya[22] reported that root elongation for wheat was 25, 32 and 36 cm for O_2 levels of 2, 6 and 10% respectively, after 21 days.

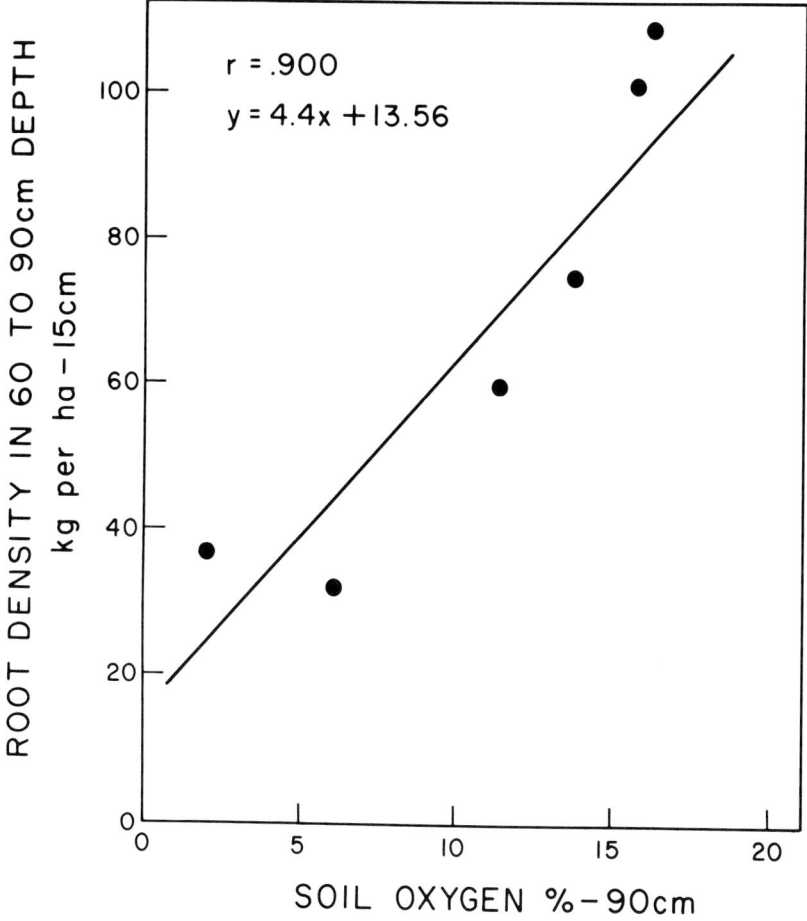

Figure 11. Relationship between average oxygen content at the 90-cm depth during June and amount of roots in the 60-90-cm depth (for a 15-cm increment) at the end of the growing season in 1970.[21]

Nutrient uptake is reduced by a low O_2 status. Leyshon and Sheard[19] found that low O_2 levels reduce N, P and K concentration by 51, 61 and 58%, respectively, and may account for the decrease in barley growth. Shoulders and Ralston[23] reported that low O_2 levels reduced the uptake of P, K, Ca and Mg by slash pine but increased NO_3 uptake. Nutrient interactions can occur physiologically inside the plant or occur chemically in the soil. Reduced O_2 transport from roots probably causes a decrease in mineral content of aboveground plant portions. Low O_2 levels may reduce plant growth because of the toxic effects of high levels of Fe and Mn in the soil solution.

Modification of Root Systems

Many plants can modify their root systems so that they can grow under low soil aeration levels. The degree of modification will depend on the plant species, with some plants showing no modification. Therefore, they grow only under high soil aeration levels. Plants like rice are able to modify their root and stem systems to such a degree that they can grow with 0% soil O_2.

Plant modification, which enables plants to survive under low aeration levels, includes development of a more porous root system, a shallow and more branching root system, and anaerobic respiration to conserve O_2. Under long-term flooding, plants have sufficient time to modify their root systems, but under short-term flooding, the changes in O_2 level are rapid, and the plants have less time for modification.

WATER MANAGEMENT

Water management is a practical method of controlling soil aeration levels especially under irrigation. Application of rain or irrigation water causes variations in soil mioisture tension and corresponding variations in soil aeration levels in the profile. The plant must integrate many levels of soil aeration in its root system, and this makes characterization more difficult. There have been few continuous measurements of soil aeration under field conditions, and the effect of having part of the root system of a plant under low aeration and the rest under high aeration is not well defined.[24]

Drainage

The main objective of drainage is to improve soil aeration. Under short-term flooding, the drain system should rapidly lower the water table so that the capillary fringe will not affect growth. How long plants may tolerate a high water table without injury will depend on many factors, but temperature is probably the most important, since it controls O_2 consumption.

Many studies have probed the effect of high water tables on crop growth. Shalhevet et al.[25] found that sugarbeet yields were 35% of maximum when compared with adequate drainage, when the water table was maintained at 20 cm. Tondreau[26] reported that the yield of ryegrass (water table 135 cm) was four times the yield obtained with a 50-cm water table. A water table of at least 90 cm is usually necessary for maximum crop growth.

The effects of a fluctuating water table on plant growth are not as well understood as the effects of a static water table. Shalhevet et al.[25] found that the effective depth of a fluctuating water table was the highest level attained during the growing season. When a water table rises above part of the root system, it will result in pruning of the root system, so this may explain their findings.

Frequency and Length of Flooding

Longer flooding times may result in soil aeration being limited for a longer period of time. Eh remained below +500 mV for 0, 20 and 45 hr for flooding times of 4, 8 and 24 hr, respectively.[8] Shorter flooding times will reduce the time period that plants are subjected to low O_2 levels, but will increase the number of flooding times because more irrigation will be necessary.

Short-term flooding lowers the O_2 levels, but usually there is sufficient time for O_2 to flow back into the soil before the next flooding period. Low O_2 levels will be present for long periods of time if the flooding interval is so short that gaseous O_2 does not flow into the soil profile between irrigations. Measurements made by Meek et al.[27] illustrated this point (Figure 12). Redox potential oscillated in phase with irrigations under long-term intervals. However, when irrigations were conducted at short intervals, the Eh decreased steadily, resulting in poor crop growth and loss of NO_3 by denitrification.

Soil Hydraulic Conductivity

Layers of restricted hydraulic conductivity may cause low soil O_2 levels since they cause perched water tables, which prevent the flow of O_2 through this zone. Tensiometer data are helpful in evaluating these layers, and if tension is below the critical values, soil aeration may be a problem. In a study on the effect of tillage on a Holtville silty clay loam soil that had been tilled to 22, 60 or 120 cm, O_2 levels were lower at the 60-cm

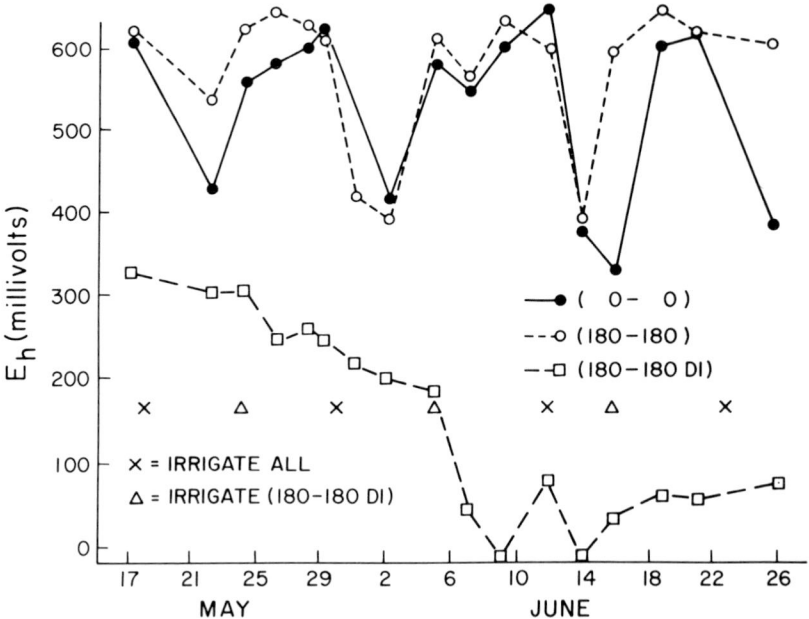

Figure 12. Redox potential (Eh) at 20 cm for three treatments (0-0), no manure and normal irrigation (180-180); 180 metric ton/ha of manure applied each year and normal irrigation (180-180DI); 180 metric ton/ha of manure applied each year and irrigation at one half the normal interval.[27]

depth for a longer time period than when soil was tilled only to 22 cm (Figure 13). At the 30-cm depth, there were no significant differences in O_2 levels as a result of tillage.

The effect of a certain aeration level may be different in the field than in pot studies, because flooding fills most of the pores. However, in the field, flooding may saturate only the first few centimeters of surface soil because of surface sealing and swelling.

Irrigation Methods

Drip or sprinkler irrigation may result in a higher soil O_2 level than that present under flood irrigation. Flood irrigation reduces the soil O_2 level because it seals the soil surface to flow of gaseous oxygen. Proper use of drip or sprinkler irrigation can maintain the soil moisture tension above the critical level necessary for adequate soil aeration. Soil aeration levels may be extremely variable under furrow irrigation, especially if the wetted zones of adjacent furrows do not overlap.

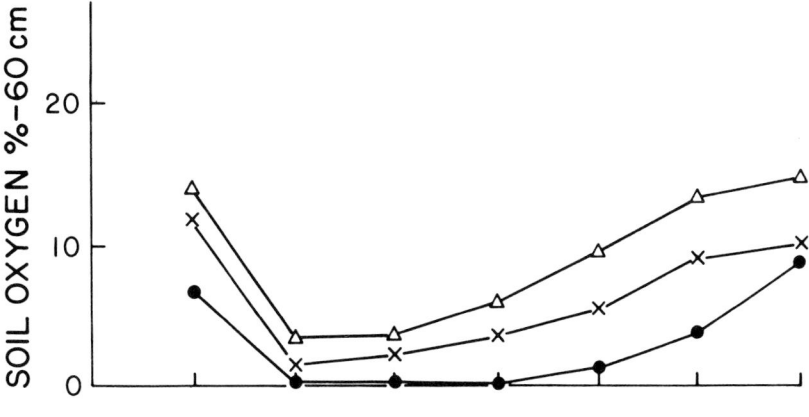

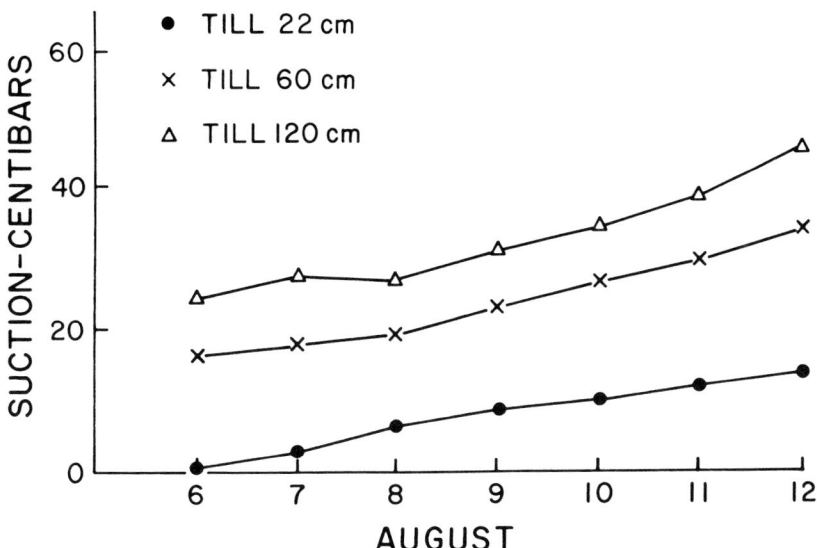

Figure 13.   Effect of tillage on soil oxygen levels and soil moisture tension at 60 cm for plots irrigated on August 5.

Meek and Grass[8] found that soil aeration levels in the field remained highest under drip irrigation, intermediate under sprinkler irrigation, and lowest under furrow irrigation. Soil aeration may be improved by reducing the intensity of water application by sprinklers, according to Gornat *et al.*[28], who found high ODR values (above 0.2 $\mu g/cm^{-2}/min$) when sprinkler intensity was low. There are few field measurements of the effect of irrigation method on soil aeration level, and more studies need to be conducted in this area.

REFERENCES

1. Russell, M. B. "Soil Aeration and Plant Growth," in *Soil: Physical Conditions and Plant Growth*, B. T. Shaw, Ed., *Agron*, 2:253-301. (New York: Academic Press, 1952).
2. Grable, A. R. "Soil Aeration and Plant Growth," in *Adv. in Agron.*, A. G. Norman, Ed., 18:57-106. (New York: Academic Press, 1966)
3. Wesseling, J. "Crop Growth and Wet Soils in Drainage for Agriculture," J. van Schilfgaarde, Ed., *Agron.* 17:7-37. Madison, Wisc: American Society of Agronomy, 1974).
4. Currie, J. A. "The Importance of Aeration in Providing the Right Conditions for Plant Growth," *J. Sci. Food Agri.* 7:380-385 (1962).
5. Fluhler, H., L. H. Stolzy, and M. S. Ardakani. "A Statistical Approach to Define Soil Aeration in Respect to Denitrification," *Soil Sci.* 122: 115-123 (1976).
6. Stolzy, L. H., and S. D. van Gundy. "The Soil as an Environment for Microflora and Microfauna," *Phytopathol.* 58(7):889-899 (1968).
7. Willey, C. R., and C. B. Tanner. "Membrane-Covered Electrode for Measurement of Oxygen Concentration in Soil," *SSSP* 27:511-515 (1963).
8. Meek, B. D., and L. B. Grass. "Redox Potential in Irrigated Desert Soils as an Indicator of Aeration Status," *SSSP* 39:870-875 (1975).
9. Dasberg, S., and J. W. Bakker. "Characterizing Soil Aeration Under Changing Soil Moisture Conditions for Bean Growth," *Agron J.* 62: 689-692 (1970).
10. Dowdell, R. J., and K. A. Smith. "Field Studies of the Soil Atmosphere II Occurrence of Nitrous Oxide," *J. Soil Sci.*, 25:231-238 (1974).
11. Sheard, R. W., and A. J. Leyshon. "Short-Term Flooding of Soil: Its Effect on the Composition of Gas and Water Phases of Soil and on Phosphorus Uptake of Corn," *Can. J. Soil Sci.*, 56:9-20 (1976).
12. Smith, K. A., and R. J. Dowdell. "Field Studies of the Soil Atmosphere I. Relationships Between Ethylene, Oxygen, Soil Moisture Content, and Temperature," *J. Soil Sci.*, 25:219-230 (1974).
13. Fulton, J. M., and A. E. Erickson. "Relation Between Soil Aeration and Ethyl Alcohol Accumulation in Xylem Exudate of Tomatoes," *SSSP* 28:610-614 (1964).
14. Stolzy, L. H. "Soil Atmosphere in The Plant Root and Its Environment," E. W. Carson, Ed. (Charlottesville, Virginia: University Press of Virginia, 1974), pp. 335-361.
15. Luxmoore, R. J., and L. H. Stolzy. "Oxygen Diffusion in the Soil-Plant System. V. Oxygen Concentration and Temperature Effects on Oxygen Relations Predicted for Maize Roots," *Agron. J.* 64:720-725 (1972).

16. Meek, B. D., A. J. MacKenzie, and L. B. Grass. "Effect of Organic Matter, Flooding Time, and Temperature on the Dissolution of Iron and Manganese from Soil in Sites," *SSSP* 32:634-648 (1968).

17. Grable, A. R. "Effects of Tillage on Soil Aeration in Tillage for Greater Crop Production," *Proc. Amer. Soc. of Agric. Eng.*, St. Joseph, Missouri, December 11, 12, 1967, pp. 44-46, 55.

18. Aceves-Navarro, E. "Effects of Soil Aeration, Water, and Nitrogen Fertilization on Nutrition, Yield, and Quality of Wheat," Unpublished dissertation, University of California, Riverside, 1974.

19. Leyshon, A. J., and R. W. Sheard. "Influence of Short-Term Flooding on the Growth and Plant Nutrient Composition of Barley," *Can J. Soil Sci.* 54:463-473 (1974).

20. Anaya, M. G., and L. H. Stolzy. "Wheat Response to Different Soil Water-Aeration Conditions," *SSSP* 36:485-489 (1972).

21. Patrick, W. H. Jr., R. D. Delaune, and R. M. Engler. "Soil Oxygen Content and Root Development of Cotton in Mississippi River Alluvial Soils," *Louisiana State Univ. Bull.* No. 673(1973).

22. Anaya, M. "Effects of Soil Aeration, Water, and Nitrogen Fertilization on Nutrition, Yield, and Quality of Wheat," Unpublished dissertation, University of California, Riverside, 1972.

23. Shoulders, E., and C. W. Ralston. "Temperature, Root Aeration, and Light Influence Slash Pine Nutrient Uptake Rate," *Forest Sci.* 21: 401-410(1975).

24. Letey, J., L. H. Stolzy, and W. D. Kemper. "Soil Aeration in Irrigation of Agricultural Lands," R. M. Hagan, H. R. Haise, and T. W. Edminster, Eds., *Agron.* 11:941-949 (Madison Wisc: American Society of Agronomy, 1967)

25. Shalhevet, J., H. Enoch, and S. Dasberg. "Response of Sugar Beet To Soil Drainage and Aeration," *Israel J. Agric. Res.* 19(4):161-170 (1969).

26. Tondreau, J. E. "Influence of a Fluctuating Water Table on Soil Aeration," Unpublished thesis, Utah State University, Logan, 1975.

27. Meek, B. D., A. J. MacKenzie, T. J. Donovan, and W. F. Spencer. "The Effect of Large Applications of Manure on Movement of Nitrate and Carbon in an Irrigated Desert Soil," *J. Environ. Qual.* 3:253-258 (1974).

28. Gornat, B., H. Enoch, and D. Goldberg. "The Effect of Sprinkler Intensity and Soil Type on Oxygen Flux During Irrigation and Drainage," *SSSP* 35:668-670 (1971).

29. Nelson, W. R., and L. D. Baver. "Movement of Water Through Soils in Relation to the Nature of the Pores," *SSSP* 5:69-76 (1940).

13

CHEMICAL AND MICROBIOLOGICAL PROPERTIES
OF ANAEROBIC SOILS AND SEDIMENTS

R. P. Gambrell and W. H. Patrick, Jr.
Center for Wetland Resources
Louisiana State University
Baton Rouge, Louisiana 70803

INTRODUCTION

Aerobic (oxidized) and anaerobic (reduced) soils are distinguished by the availability of oxygen for chemical and microbiological oxidative processes occurring in these soils. The primary factor limiting oxygen availability in anaerobic soils is the presence of free water in much or all of the internal pore space between individual soil particles and aggregates. Aqueous diffusion of dissolved oxygen in quiescent interstitial waters of flooded soils and sediments is approximately four orders of magnitude slower than gaseous diffusion through a porous medium. Thus, in many flooded soils, the chemical and microbial demand for oxygen greatly exceeds the rate of oxygen resupply, and anaerobiosis may develop. Water saturation contributing to temporary anaerobic conditions occurs in some upland soils in humid climates as a result of poor internal drainage. However, lowland soils that are continuously or seasonally flooded account for the greater area of economically and ecologically important anaerobic soils. Some of these wetland soils are flooded as a management practice, to enhance their agronomic productivity. Most rice production, especially in developing nations, utilizes flooded soils. One third of mankind reportedly depends on rice for half or more of its food, as rice is the primary or secondary food source for 90% of the low-income people in the most densely populated areas of the earth.[1] Anaerobic swamp and marsh soils are naturally flooded due to their low stratigraphy in relation to nearby rivers, lakes and coastal waters and occupy large areas of the earth's surface. These soils are important contributors to the abundant productivity by highly diversified flora and fauna populations character-istic of wetland ecosystems.[2,3]

In addition to the restrictions placed on plant populations, which can adapt to flooded soils because of limited oxygen for root respiration, the availability of biostimulants and toxicants is also strongly affected by the oxidation status of wetland soils. The objectives of this chapter are to discuss the chemical and microbiological processes characteristic of anaero-bic soils and the influence of these properties on the bioavailability of nutrients and potentially toxic substances.

INTERACTIONS OF SOIL AERATION

Oxygen Availability

In a well drained soil, most of the pore spaces surrounding individual soil particles and aggregates are gas-filled and interconnected with the atmosphere. This permits relatively rapid gaseous diffusion of oxygen throughout the plant rooting depth. Though there may be a reduction in gaseous oxygen content with depth in some soils,[4] there is sufficient molecular oxygen transport across the gas-liquid interface of the soil solu-tion to maintain some dissolved oxygen in this solution. As a result, the soil is maintained in an oxidized condition. The potential oxygen resupply rate by this process is usually more than sufficient to meet soil and root oxygen demand.

Excess water applied to a permeable upland soil by precipitation, irrigation, or temporary flooding will rapidly drain from the upper profile through the interconnected pore spaces. Much of this pore space is again filled with gas, which is continuous with the atmosphere, after draining for several hours. Thin, oxidized moisture films of one third bar tension or less will remain around individual soil particles and aggregates due to the hydrophilic nature of mineral and organic soil components. This relatively thin film of moisture on soil solids and root surfaces supports life by providing a medium for chemical reactions affecting nutrient avail-ability, serving as a nutrient and moisture reservoir and supporting transport of nutrients from the soil solid phase to the plant root system.

Upon flooding, gaseous diffusion of oxygen into the profile is not possible. As a result of prolonged flooding and continued oxygen demand for root and microbial respiration, as well as chemical oxidation of re-duced organic and inorganic components, the oxygen content of the soil solution begins an immediate decline and may be depleted within several hours to a few days.[5,6] Flooding an otherwise aerobic soil may result in a flush or surge of microbial activity which often results in rapid development of strong reducing conditions, particularly where a considerable organic energy source is available. No measurable oxygen is found beneath the surface of most continuously flooded soils and sediments, as indicated by the low oxidation-reduction (redox) potentials measured in agricultural, swamp and marsh soils as well as in ocean and lake sediments.[7-15] As a result of oxygen demand exceeding oxygen resupply in flooded soils and sediments, deep reduced horizons are found in these systems.

Oxidized and Reduced Zones in Flooded Soils and Sediments

Restricting gaseous transport of oxygen into a soil or sediment by flooding does not necessarily result in the development of a uniformly reduced profile. Dissolved oxygen from the overlying floodwater may diffuse across the surface water-soil or sediment interface, resulting in the presence of a thin, oxidized surface horizon overlying a deep, reduced horizon.[8,16-18] Shallow surface water flooding soils and sediments normally contain a uniform distribution of several $\mu g/ml$ of dissolved oxygen, often approaching saturation levels. This relatively high and uniform oxygen content is due to:

1. the rapid rate of oxygen transport across the atmosphere-surface water interface,
2. the small population of oxygen-consuming organisms present,
3. photosynthetic oxygen production by algae within the water column, and
4. surface water mixing by convection currents and wind action.

Thus, a small but continuous supply of dissolved oxygen is usually available at the soil or sediment-floodwater interface. Further oxygen transport into quiescent interstitial waters is by molecular diffusion of dissolved oxygen in response to an oxygen concentration gradient.

Dissolved oxygen diffusion is reported to occur at 1/10,000 the rate of gaseous diffusion.[19] As a result of the slow rate of oxygen transport through interstitial water and a comparatively high oxygen demand, the surface oxidized soil or sediment horizon is thin and ranges from a few millimeters to a few centimeters in depth, depending on the oxygen consumption capacity of the material. Within this thin, oxidized horizon, microbial metabolism and chemical transformations of nutrients and potential toxicants are similar to those in aerobic soils. Though this oxidized surface horizon is thin, biological and chemical processes occurring in this zone strongly influence the availability of both nutrients and toxins in flooded soils and sediment-water systems.

A thin, oxidized surface soil or sediment layer of fairly uniform thickness overlying a deep anaerobic horizon is characteristic of many swamp and marsh soils, shallow lake and ocean sediments, and agricultural soils flooded for crop production. However, this two-layered model does not apply to all flooded soils and sediments. Ponnamperuma[18] described thermally stratified deep water lakes in which dissolved oxygen cannot penetrate to the sediment surface before being depleted. During these periods, anaerobic conditions extend from deep within the sediment well into the surface water. Such conditions are reported to favor considerable exchange of nutrients from the sediment to the overlying water column.[10,11]

Also, heterogeneous oxidized and reduced zones may exist in soils subject to seasonal or short-term flooding. Greenland[20] suggested that a heterogeneous distribution of available organic carbon and active microbial populations may contribute to numerous anaerobic microzones in predominantly aerobic soils. Such zones would likely occur in the center of saturated soil aggregates.[19]

Daniels et al.[21] reported the dissolved oxygen content within fluctuating water tables of Aqult and Udult soils in the North Carolina Coastal Plain. Instead of a rather uniform reduction in dissolved oxygen with increasing depth, they found quite variable oxygen concentrations with depth and time beneath the water table. They emphasized that saturation does not ensure development of anaerobic conditions in these soils. An abundant energy source for biological activity must be available as well. They also suggested that sharp reductions in dissolved oxygen levels during the year were associated with large rainfalls, transporting soluble carbon from surface forest litter into the profile.

Numerous variations exist in the geometry and oxidation intensity of oxidized and reduced zones in flooded soils and sediments, as discussed above. However, this chapter will be primarily concerned with flooded agricultural and wetland soils and shallow sediments, which are usually characterized by a thin, oxidized surface layer of uniform thickness overlying deep anaerobic subsurface horizons (Figure 1).

Depth of Oxidized Surface Horizon

The depth of the oxidized layer depends on a balance between the rate of oxygen diffusion into the surface horizon and its consumption.[11] Oxygen consumption rates have long been thought to be a function of microbial respiration. Recently, however, Howeler and Bouldin[22] have experimentally demonstrated that oxygen consumption rates in some flooded soils can best be described by models including oxygen consumption for both biological respiration and for chemical oxidation of both

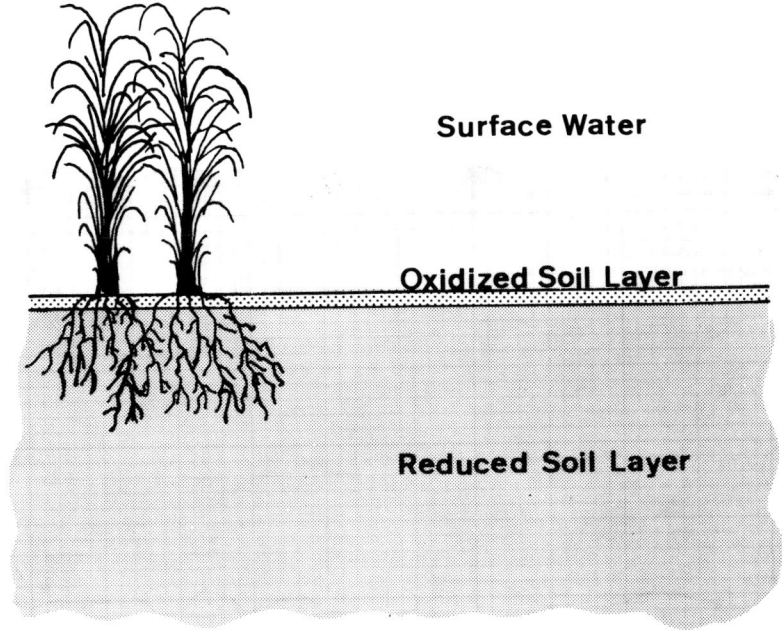

Figure 1. Surface water, oxidized and reduced soil layers characteristic of flooded agricultural and wetland soils.

mobile and nonmobile constituents.   Reduced iron and manganese ions were thought to represent the bulk of the mobile reductants that diffused upward from the reduced horizon into the oxidized zone.   Precipitated ferrous iron, manganous manganese and sulfide compounds, encountered as the oxidized zone increased in thickness, likely comprised much of the nonmobile constituents.   In the soils and sediments studied, chemical reductants such as ferrous iron accounted for about 50% of the oxygen consumption, though Howeler[23] pointed out that the ratio between biological and chemical oxygen consumption rates may vary widely depending on the organic matter content of the soil or sediment.

Redox Potential as a Measure of Oxidation-Reduction Intensity

Just as pH is a measure of hydrogen ion activity, redox potential is a measure of the electrochemical potential or electron availability in chemical and biological systems.   Electrons are essential to all chemical reactions. Chemical species that lose electrons become oxidized.   Conversely, reduction occurs as a chemical species gains electrons.   Thus, a measure of redox potential (electron availability) indicates the intensity of oxidation or reduction of a chemical or biological system. In an aqueous system, the intensity of oxidation is limited by the electrochemical potential at which water becomes unstable and releases molecular oxygen. Similarly, the potential at which molecular hydrogen is released from water represents the lower limit of reduction in aqueous systems.[24] Within the limits imposed by the stability of water, the oxidation states of hydrogen, carbon, nitrogen, oxygen, sulfur and several metals may be affected by the oxidation-reduction potential of a system, though the measured redox potential is dependent on the chemical activity of a few of the more abundant oxidized and reduced forms of these elements present.[25]

Use of Redox Potential Measurements

Redox potential measurements in soils and sediment-water systems are useful and have gained widespread acceptance. However, the value of these measurements depends, to a considerable extent, on the techniques used and their valid interpretation, as a number of theoretical and practical limitations exist. For example, electrodes used for measuring redox potential should be inert to the chemical species present. However, no electrode is completely inert. Though gold electrodes have been used, platinum has generally been found to be most satisfactory. Redox potential measurements can quantitatively describe the ionic distribution only between chemical species which interact with the transfer of electrons. Examples of such redox couples include ferrous and ferric iron, nitrite and nitrate nitrogen, and sulfide and sulfate sulfur. In soils and sediment-water systems, there are usually many redox couples present, and most are not chemically reactive with others. Unless a given redox couple is present at a relatively high concentration, electrodes used to measure redox potential are not specific for a single redox couple. Thus, the platinum electrode responds to the electrochemical potentials of all redox couples present. The resulting measurement represents a mixed potential, which reflects a weighted average of the potentials contributed by each of the redox couples present in the system.[26] Laitinen[27] described how several redox couples, each having greatly different potentials as separate redox systems, may be added together to produce a composite or mixed potential, which is several hundred millivolts from the potential of the individual couples. A redox equilibrium is almost never achieved in a natural system due to the almost continuous addition of organic matter which may be oxidized and thus serve as an electron donor.[25]

A system that is relatively resistant to changes in redox potential upon small additions of a reductant or oxidant is said to be well poised, a term analogous to buffering capacity in pH measurements. Poise is enhanced as the ratio of oxidant to reductant approaches unity and increases as the concentration of both the oxidant and reductant increases. Reduced soils and sediments are generally well poised as a result of the presence of relatively high concentrations of soluble iron and perhaps manganese. Redox potential measurements are generally not well poised and are poorly reproducible in many oxidized systems, because of the very low concentrations of soluble, chemically reactive species present. Thus, redox potential measurements are more satisfactory indicators of the intensity of reduction of anaerobic systems and are somewhat limited in estimating the degree of oxidation in aerobic soils, sediments and surface waters. Aerated systems are better characterized by measuring oxygen content or oxygen diffusion rates.[17]

Despite theoretical limitations and a number of procedural difficulties, redox potential measurements represent a rapid and convenient indicator of the intensity of reduction in soils and sediments. These measurements give a qualitative measure of oxidation-reduction conditions, enabling reasonable understanding and prediction of the chemical and biological transformations occurring in reduced systems affecting the availability of many nutrients and toxins.

Additional information on theoretical and practical considerations of redox potential measurements in complex natural systems is available in the literature.[18,25,26,28]

Relationship between pH and Redox Potential

Baas Becking et al.[24] described four general types of chemical reactions in natural systems:

1. reactions in which neither protons nor electrons are exchanged, i.e.,
   $Fe_2O_3 + H_2O \rightleftharpoons 2FeOOH$
2. reactions involving protons, i.e., $H_2CO_3 \rightleftharpoons H^+ + HCO_3^-$
3. reactions involving electrons, i.e., $Fe^{2+} \rightleftharpoons Fe^{3+} + e^-$
4. reactions in which both electrons and protons are transferred, i.e.,
   $FeSO_4 + 2H_2O \rightleftharpoons SO_4^{2-} + FeOOH + 3H^+ + e^-$

Reactions of the first type have no effect on the pH or redox potential of natural systems and similarly are not influenced by a change in the pH-redox potential environment. Reactions of the second type primarily involve the dissociation of acids, while reactions of the third type are characterized by loss or gain of electrons affecting the valence state of metals. Most reactions in the natural environment were reported to involve both protons and electrons, and a relationship exists between these parameters in most reactions.

Bass Becking et al.[24] expressed the relationship between pH and redox potential as follows:

$$Eh = E_O - 59(a/n)pH$$

where Eh is redox potential, E_O is the standard potential at equal activities of reduced and oxidized species, a is the number of protons transferred, and n the number of electrons involved in the reaction. This equation describes the slope or gradient of the stability boundary between the oxidized and reduced forms of a chemical species involved in redox reactions. When a is equated to n, as many researchers have generally assumed, a -59 mV per pH unit relationship is obtained. However, Baas Becking et al.[24] pointed out that the a/n ratio of natural systems may range from zero to infinity, and examples are given for four common reactions of iron in which the a/n ratio is 1, 2, 3 or 4. Bohn[29] pointed out that while the relationship between pH and redox potential is usually linear, it is not necessarily always so.

Ponnamperuma[18] also discussed the relationship between pH and redox potential for a general oxidation-reduction reaction in equilibrium:

$$(Ox + ne^- + mH^+ = Red)$$

In the absence of hydrogen ion transfer, redox potential is given by

$$Eh = E_O + \frac{RT}{nF} \ln \frac{(Ox)}{(Red)}$$

where E_O is the standard potential, n is the number of electrons, R is the gas constant, T represents temperature, F is the Faraday constant, and Ox and Red represent the chemical activities of the oxidized and reduced species, respectively. An additional term was included for reactions involving proton transfer:

$$Eh = E_O + \frac{RT}{nF} \ln \frac{(Ox)}{(Red)} + \frac{mRT}{nF} \ln H^+$$

where m is the number of protons and H^+ is hydrogen ion activity. From this equation, it is apparent that a change in pH during a redox reaction is not regulated solely by the number of hydrogen ions consumed or hydroxyl ions produced, but by the ratio of hydrogen ions consumed to electrons consumed.

In monoelemental chemical systems, theoretically derived redox potential/pH slopes have been found to coincide closely with experimentally determined slopes.[30,31] However, in soil and sediment systems containing many redox couples, which range widely in concentration, there is little basis for expecting pH/redox potential slopes to coincide with those determined theoretically or measured in simple aqueous systems. Bohn[29] explained that different systems may yield dissimilar pH/redox potential slopes because of the complexity of the reactions regulating changes in pH and redox potential and the variability in these reactions from one natural system to the next. He points out that many silicates, carbonates and soluble hydroxides, which buffer pH, are not sensitive to changes in redox potentials.

Thus, an interaction usually exists between pH and redox potential in chemical reactions of natural systems, but in many cases it is difficult to ascertain what the relationship is. Bohn[25] concluded that while adjustment of mixed potentials (several redox couples contributing to the measured redox potential) by -59 mV/pH unit has little experimental or theoretical justification, it is convenient and has been used with reasonable success in making comparisons between different mediums, though one should be aware of the limitations involved.

The pH of both acid and alkaline oxidized soils tends to converge toward pH 7 when these soils are flooded and become reduced.[32] Drainage and subsequent oxidation of sediments and submerged soils reverse these pH changes. Ponnamperuma[18] stated that the rate and magnitude of pH alteration of acid soils upon reduction is dependent on a number of soil properties, notably the soil organic matter and reactive iron contents. An initial reduction in pH may occur upon flooding even in acid soils due to accelerated carbon dioxide production by aerobic bacteria. However, acid soils eventually increase in pH due to the reduction of iron. Subsequent pH stability after prolonged flooding is again influenced by the partial pressure of carbon dioxide.[18] In acid coastal plain soils subjected to periodic flooding, van Breemen[33] concluded that pH is fundamentally related to the solubility, dissociation constant, and partial pressure of carbon dioxide, though oxidation and reduction of iron and sulfur compounds contributed to the observed pH changes. From his own work and the published literature, Ponnamperuma[18] suggested that the pH decrease observed as calcareous and sodic soils are submerged is also dependent on changes in the levels of carbon dioxide.

Chemical and Biological Transformations Indicative of Anaerobiosis

In oxidized soils containing molecular oxygen, redox potential is reported to range from about +400 to +700 mV.[17] This narrow range, lack of adequate poise, and poor reproducibility in oxidized systems limits the usefulness of redox potential measurements for characterizing aerobic soils and sediments. As previously discussed, oxygen content and oxygen diffusion rates are better indicators of oxidation intensity in these systems. In sediments and submerged soils, redox potential ranges from around -400 mV (strongly reduced) to +700 mV (well oxidized) and is better poised and fairly reproducible at the more reduced levels. Thus, redox potential measurements have gained increasing acceptance as a means of characterizing reduced soils and sediment-water systems.

As the oxidation condition of a soil changes from weakly oxidized to strongly reduced, a sequential reduction of redox-active inorganic constituents is noted. Turner and Patrick[6] reported only slight decreases in redox potential in 10 oxidized soils, as the oxygen content of the soil atmosphere declined from 21 to around 4%. Additional lowering of the oxygen content resulted in sharp decreases in redox potential. Redox potential at oxygen depletion ranged between 320 and 340 mV for all soils (corrected to pH 7 by -59 mV per unit pH change). Though oxygen may effectively poise redox potential at this level, the poising effect of oxygen upon submergence is temporary, as available oxygen is usually depleted within hours or a few days.

Nitrate is the next oxidant to be reduced following oxygen depletion. In laboratory studies using a range of carefully controlled redox potential levels, Patrick[34] reported 338 mV to be the critical potential for nitrate reduction in a Crowley soil at pH 5.1.

In near-neutral soils, or after appropriate correction to pH 7.0, nitrate reduction is reported to occur around 220 mV.[25] As with oxygen, the presence of nitrate in a submerged soil will tend to poise redox potential around this level until nitrate reduction is nearly complete.[35] Nitrate poising may persist for several days to several weeks, depending somewhat on the initial nitrate level. Nitrate reduction may begin before complete removal of oxygen, but complete nitrate reduction probably does not occur until all oxygen is depleted.[6]

Manganic compounds (Mn^{4+}) reduce to the more soluble manganous (Mn^{2+}) form as the redox potential decreases to approximately 200 mV.[6,36]

Ferric iron is reported to be stable until redox potential decreases to around 120 mV, assuming pH 7.0 conditions and a redox potential/pH slope of -60 mV/pH unit.[25,37]

Connell and Patrick[38] found that sulfate reduction was initiated when redox potential fell to approximately -150 mV, while Harter and McLean[39] reported sulfate-reducing organisms were active at -75 mV or less.

Cappenberg[40] reported the maximum population of methane-producing bacteria occurs between -250 to -300 mV. It is generally believed that most of the sulfate must be reduced to sulfide before methane production begins. The spatial and temporal separation between sulfide and methane production has been attributed to mutually exclusive metabolic processes occurring between the two bacterial populations,[41] the inhibitory effect of sulfide on methane production,[42,43] and possibly to redox potential difference as well as a dependence of the methogenic bacteria on substrates produced by sulfate-reducing bacteria.[43]

One redox component is not always completely reduced before reduction of the next most easily reduced component begins. Turner and Patrick[6] reported some nitrate may be reduced in the presence of low oxygen levels, manganese may reduce in the presence of nitrate, but probably not in the presence of oxygen, while some manganese may continue to reduce after ferric iron reduction begins. In other cases, relatively complete reduction of one component has been found to occur before a component that is more difficult to reduce is affected. Turner and Patrick[6] did not observe iron reduction in the presence of nitrate. Significant methane production is observed only after complete sulfate reduction.[41]

Despite some overlap of several of these reduction reactions, knowledge of the approximate critical reduction potentials for various inorganic components is useful for characterizing the intensity of reduction of a soil or sediment-water system and for predicting chemical transformations affecting the availability of a number of nutrients and potential toxins. The approximate critical threshold potential values for the various inorganic systems discussed are shown in Figure 2.

Patrick and DeLaune[16] described another useful approach for characterizing vertical boundaries of the oxidized and reduced horizons of flooded soils and sediments. The apparent thickness of the surface-oxidized layer was dependent on the oxidized and reduced components of the system, which were evaluated. The distribution of soluble manganese with depth indicated the thinnest oxidized surface layer, total sulfide content indicated

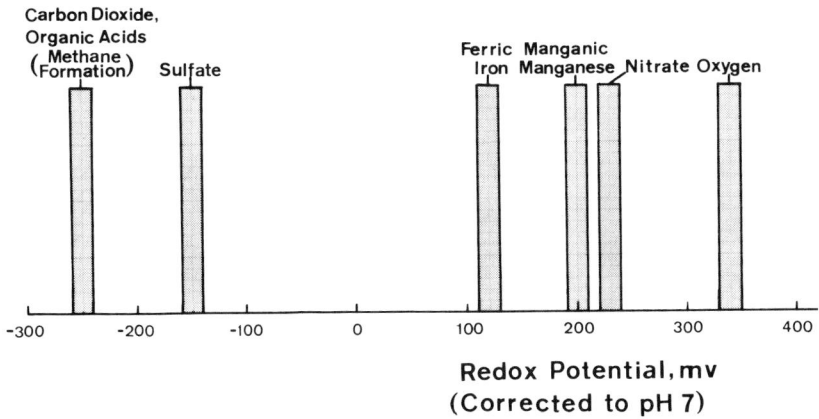

Figure 2. Approximate redox potentials below which several redox systems become unstable.

the thickest oxidized layer, and ferrous iron gave an intermediate thickness that corresponded closely with the redox potential profile (Figure 3).

Oxidation-reduction intensity also influences the total ion content or ionic strength of soil solutions and interstitial waters. Flooding and subsequent reduction of an oxidized soil causes an initial increase in the total ion content of the soil solution. With time, the ionic strength of the reduced soil will decrease and again approach prereduction levels.[32] In acid and near-neutral soils, the solubilization of iron and manganese, which forms from insoluble ferric and manganic oxides and hydroxides, contributes to the increase in ionic strength upon reduction. Gotoh and Patrick[36,44] have shown substantial increases in water-soluble manganese and iron in a Crowley silt loam as the redox potential of this soil was lowered. An increase in the relatively mobile forms of these metals was accompanied by a decrease in the reducible levels, indicating that some of the insoluble oxides and hydroxides became unstable and dissolved under reduced conditions. Redman and Patrick[45] flooded 26 Louisiana soils and found an increase in specific conductivity in 24 of these soils after 30 days which was closely correlated with the soil organic matter content. The apparent aberrant behavior of the remaining two soils was attributed to nitrate removal by denitrification, as these two soils initially contained high nitrate contents.

MICROBIOLOGICAL ACTIVITY

Aerobic and facultative anaerobic populations predominate in well aerated soils and utilize molecular oxygen as a terminal electron acceptor during respiration. Upon submergence, the respiratory activity of these organisms will rapidly deplete the available oxygen supply. Thereafter, the numbers and activity of aerobic organisms decrease markedly, while facultative and obligate anaerobic bacterial populations proliferate. These microbes may use organic degradation products or inorganic constituents other than oxygen as an electron acceptor. In the thinly oxidized surface horizon typically found overlying a deep, reduced horizon in flooded soils and sediments, aerobic microbial populations will function normally while facultative and obligate anaerobes are active in subsurface reduced horizons. The combined activity of these two groups of organisms on materials that diffuse across the aerobic-anaerobic boundary affects the bioavailability of a number of nutrients and potential toxins in soils and sediment-water systems, as will be discussed in later sections.

All microorganisms require various substrate materials to provide energy for metabolic processes and organic building blocks for increasing cellular biomass, and to serve as electron acceptors in energy-yielding respiration

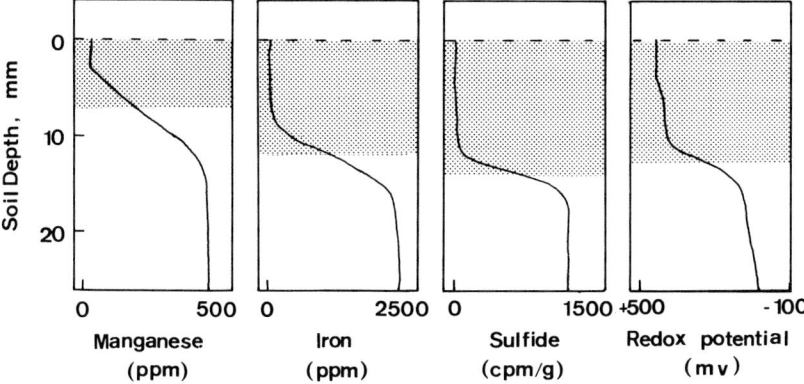

Figure 3. Depth of surface-oxidized layer as measured by sodium acetate-extractable manganese, ferrous iron, sulfide and redox potential profile after 13 weeks of flooding. (Reproduced from *Soil Sci. Soc. Am. Proc.* 36:573-576, 1972).[16]

processes. Oxidation of various organic and inorganic compounds provides the metabolic energy for most organisms, and these compounds release electrons in the process. Since oxidation must be accompanied by reduction, acceptors are required for electrons released from the oxidation of energy substrates and these electron acceptors become reduced. Where available, oxygen is the preferred electron acceptor by microorganisms because of the greater energy yield. Where oxygen is unavailable, facultative and obligate anaerobes utilize either inorganic (anaerobic respiration) or organic (fermentation) substrates as terminal electron acceptors.[46] Inorganic compounds, which may be utilized as electron acceptors, include nitrate and nitrite nitrogen, manganic manganese, ferric iron, and sulfate or elemental sulfur. Inorganic dissimilation products of organic matter degradation such as carbon dioxide, molecular nitrogen, and hydrogen ions may be reduced under some conditions. Some organic compounds may also serve as electron acceptors.

Aerobic and anaerobic respiration differ greatly in the amount of energy released by decomposition of organic energy substrates. Aerobic respiration yields considerably more energy than anaerobic respiration per unit of decomposition, resulting in more efficient assimilation of organic substrate into cellular material. Anaerobic bacteria assimilate carbon during decomposition with much less efficiency. Thus the rate of net organic matter loss is slowed in anaerobic soils and sediments,[47,48] and organic matter readily accumulates in reduced systems compared to oxidized environments. Reddy and Patrick[49] reported that a 128-day anaerobic incubation period resulted in the loss of approximately half the total indigenous carbon from a rice soil (Crowley silt loam) as compared with aerobic conditions. Equal periods of alternating aerobic and anerobic conditions resulted in intermediate levels of carbon loss. Parr and Reuszer[50] reported that total decomposition of wheat straw in anoxic soil was only 13% of that decomposed when the soil was aerated with 21% oxygen.

Another equally striking aspect of aerobic versus anaerobic degradation of organic matter is the nature of the end products produced. In aerobic soils, the primary end products of organic matter degradation are carbon dioxide, water , nitrate and sulfate, plus some residual humic materials. Anaerobic respiration results in the production of carbon dioxide, methane, hydrogen, ammonia, amines, mercaptans, hydrogen sulfide, numerous low-molecular-weight organic acids and additional residual humic materials.[18] Ponnamperuma[18] and Tusneem and Patrick[51] describe how breakdown metabolic pathways of carbohydrates and other organic compounds differs in aerobic and anaerobic soils contributing to the differences noted in end products of organic matter degradation.

Though little information is available on the nature of the residual humic materials resulting from aerobic and anaerobic respiration, it is likely that partial degradation under reduced conditions contributes to larger and more structurally complex humic materials than produced by aerobic degradation. The ecological and environmental implications of these differences will be discussed more fully in a later section on trace and toxic metal chemistry.

From the previous discussion, it is apparent that reduction of soils and sediment-water systems requires an absence of molecular oxygen, anaerobic microbial activity, and a readily available organic energy source. The rate of reduction of recently submerged soils, as well as intensity of reduction of continuously submerged wetland soils and sediments, are regulated by the amount and microbial availability of an organic energy supply, temperature, pH, and the kind and amount of electron acceptors present. Reduced organic and inorganic substances, which diffuse out of bacterial cells or are produced by extracellular enzymes, contribute to the low redox potentials associated with wetland soils and sediments. Thus, microbial activity and soil properties affecting microbial metabolism are important factors influencing chemical transformations of nutrients and toxicants in flooded soils and sediments.

NITROGEN

Nitrogen is the nutrient most limiting to agricultural production in both well drained and flooded agricultural soils throughout the world. Broome *et al.*,[52] Patrick and DeLaune,[53] and Sullivan and Daiber[54] reported that nitrogen availability is also apparently limiting primary productivity in estuarine marshes in Louisiana, North Carolina and Delaware, respectively, though the total quantity of major plant nutrients in some marsh soils is high.[55,56] Despite fertilizer applications to flooded rice soils and the high total nitrogen content of many marsh soils and sediments, nitrogen deficiencies are common in these systems. This is because soil and sediment reduction, as a result of submergence, favors physical, chemical and biological processes, which remove available nitrogen. These processes are strongly influenced by oxidation-reduction intensity and particularly by the presence of a well oxidized surface water or soil layer overlying a deep, reduced horizon.

Forms of Nitrogen

Biologically important forms of nitrogen include organically combined nitrogen in living organisms and organic detritus as well as numerous inorganic nitrogen compounds, which are derived primarily from the biological degradation of soil and sediment organics. Most of the nitrogen in soils and sediments is organically bound. The important inorganic nitrogen compounds involved with nitrogen transformations in soils and sediments include ammonium, nitrite, nitrate, nitrous oxide and molecular nitrogen. The primary sources of inorganic nitrogen in natural ecosystems are symbiotic and nonsymbiotic fixation of atmospheric molecular nitrogen by bacteria and blue-green algae. In recent decades, agricultural soils have been extensively amended with inorganic nitrogen manufactured by chemical nitrogen fixation to enhance crop production. Hutchinson[57] states that the quantity of nitrogen chemically fixed by man is now approaching the rate of biological fixation.

Nitrogen Mineralization

Mineralization refers to the biological transformation of organically combined nitrogen to ammonium nitrogen during organic matter degradation. Ammonium nitrogen is a nutrient form available for microbial immobilization or plant uptake. Mineralization occurs in both oxidized and reduced soils and sediments. Because the rate of organic matter degradation is considerably greater in oxidized soils, one would expect net nitrogen mineralization and subsequent availability to be greatest under these conditions. However, the nitrogen requirements for anaerobic metabolism are lower than for aerobic metabolism, resulting in a greater release or availability of ammonium nitrogen than would otherwise be expected in reduced systems. This has been demonstrated experimentally by a number of investigators.[47,58-60]

Inorganic Nitrogen Transformations

Ammonium nitrogen in anaerobic soils and sediments is subject to several possible fates. As a cation, it may be adsorbed by ion exchange mechanisms to colloidal mineral and organic components of the soil solid phase. Exchangeable ammonium is in equilibrium with the solution phase and may subsequently be released in a soluble form. Soluble and easily exchangeable ammonium are readily available for plant uptake or microbial immobilization. Ammonium is relatively immobile in soils and sediments. However, the low assimilatory demand by anaerobes may result in considerable quantities of ammonium nitrogen accumulating in reduced soils and sediments, which exceed the capacity for effective retention by cation exchange processes. This may contribute to the high levels of ammonium reported dissolved in interstitial waters relative to levels usually found in an oxidized surface water column.[61,62,63]

Except for immobilization and plant uptake, ammonium nitrogen is chemically stable in reduced soils and sediments. However, ammonium nitrogen in the thin, surface-oxidized soil or sediment layer is readily oxidized by chemoautotropic bacteria, first to nitrite, then to nitrate. As a result of ammonium oxidation and subsequent depletion in this surface layer, underlying ammonium may be mobilized by diffusion in response to a concentration gradient and be transported across the reduced-oxidized interface into the surface-oxidized horizon. In simulated studies in the laboratory on rice soils, Reddy et al.[64] have shown this process to be effective to a depth of approximately 12 cm during a 120-day incubation. In this study, 50% of the ammonium lost from the anaerobic soil was attributed to upward diffusion and subsequent oxidation of the ammonium (Figure 4). In a Lake Ontario sediment core, Kemp and Mudrochova[65] reported this process to be active to a depth of approximately 150 cm, while Chen et al.[66] reported considerable exchange of ammonium from the top 4 cm of a Wisconsin lake sediment into the overlying water column, but negligible exchange beneath 8 cm. Graetz et al.[61] suggested from laboratory studies that nitrogen mineralization and upward diffusion contributed approximately 30% of the nitrogen entering Lake Mendota, Wisconsin, from external sources.

In the thinly oxidized soil and sediment horizon located just beneath the floodwater column, the nitrate formed from ammonium oxidation is also subject to several possible fates, including assimilative reduction by plants or microbes, leaching, or dissimilative reduction (reduction to inorganic compounds) if transported into the reduced zone. In weakly acid and near-neutral soils, retention of nitrate by the soil solid phase is negligible, resulting in almost all the nitrate being in soluble form. Thus,

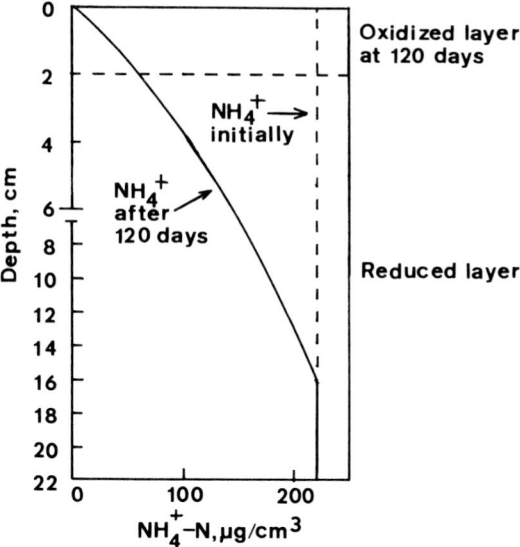

Figure 4. Removal of ammonium nitrogen from a flooded rice soil by nitrification and upward ammonium diffusion.[64]

nitrate is considerably more mobile in soils and sediments than ammonium, and is subject to mass flow with the movement of water and diffusion in response to a concentration gradient. Nitrate is not stable and is rapidly depleted from subsurface reduced horizons. Thus nitrate in the aerobic surface horizon tends to diffuse back down into the reduced layer, again in response to a concentration gradient.[67] Since nitrate reduction or denitrification readily occurs in the reduced horizon, the concentration gradient is maintained.

Denitrification is the reduction of nitrate and/or nitrite nitrogen to volatile gases that may escape into the atmosphere. Nitrous oxide and particularly molecular nitrogen are the predominant products of biological denitrification, and this process is highly dependent on oxidation-reduction intensity.[34] Although many factors influence denitrification, this process generally occurs in a sediment or flooded soil where an ample organic energy supply is supporting considerable microbial activity such that the oxygen resupply rate is inadequate to meet the demand for a terminal electron acceptor by microbial respiration. Nitrate may also be assimilated into microbial tissue in flooded soils. However, little nitrate is assimilated in reduced soils because:

1. the nutritional requirement for nitrogen by obligate and facultative anaerobic organisms in reduced soils is usually less than the requirement for electron acceptors, and

   2. reduced soils and sediments usually have an adequate supply of
   ammonium nitrogen, which is the preferred nitrogen source for
   nutritional purposes.

As a result, very little of the nitrate nitrogen that moves into reduced
soils or sediments is transformed to the organic fraction.

The presence of an oxidized surface horizon overlying a subsurface
reduced horizon is critical to the overall loss of nitrogen from flooded
soils and sediments. The sequential processes involved are mineralization,
upward diffusion of ammonium, nitrification, downward diffusion of
nitrate, and denitrification (Figure 5). Reddy et al.[64] suggested that the
rate-limiting step may be nitrification in many instances. Substantial
losses of nitrogen have been attributed to the nitrification-denitrification
sequence in flooded soils.[51,68,69] Recent studies have shown that more
ammonium was lost than was initially present in the surface-oxidized layer
of wetland soils.[67,70,71] These studies again point to the importance of
ammonium diffusion from the reduced soil layer to the surface-oxidized
layer, where nitrification, diffusion and subsequent denitrification occur.
Abichandani and Patnaik[72] thought 20-40% of the ammonium fertilizers
applied to rice was lost, and in Japan, Mitsui[73] reported 30-50% losses of
applied nitrogen.

In recent years, tracer studies utilizing Nitrogen-15 have been useful
in determining the fate of fertilizer nitrogen applied to agricultural
soils.[68,74-76] Where laboratory and field tracer studies have been conducted on flooded soils, they have confirmed earlier reports of substantial
nitrogen losses from these soils. Patnaik[77] could not account for 23%
of the labeled nitrogen incubated with a waterlogged soil. In a field
study in which labeled ammonium sulfate was applied to rice by deep
placement, Patrick and Reddy[70] determined that 49% of the applied
nitrogen was recovered by the rice, 26% remained in the soil (either in
root or as soil organic nitrogen) and 25% was lost, presumably by the
nitrification-denitrification sequence. Though such losses from rice soils
cannot be eliminated, they may be reduced by deep placement of ammonium nitrogen.[78,79]

The presence of plants will influence nitrogen losses from flooded soils
and shallow sediments. Inorganic nitrogen assimilation into plant tissue
by uptake reduces nitrogen losses. In addition, root activity may influence
the nitrification-denitrification sequence. Rice and marsh plants are known
to transport oxygen from aboveground tissue to the roots, where some of
this oxygen may diffuse out of the root, resulting in thinly oxidized rhizospheres in a predominantly reduced soil. Such a process will increase the
total volume of oxidized soil and may increase the nitrification rate,
which has been determined to be rate-limiting in the overall nitrification-
denitrification sequence. On the other hand, Woldendorp[80] and Stefanson[81]

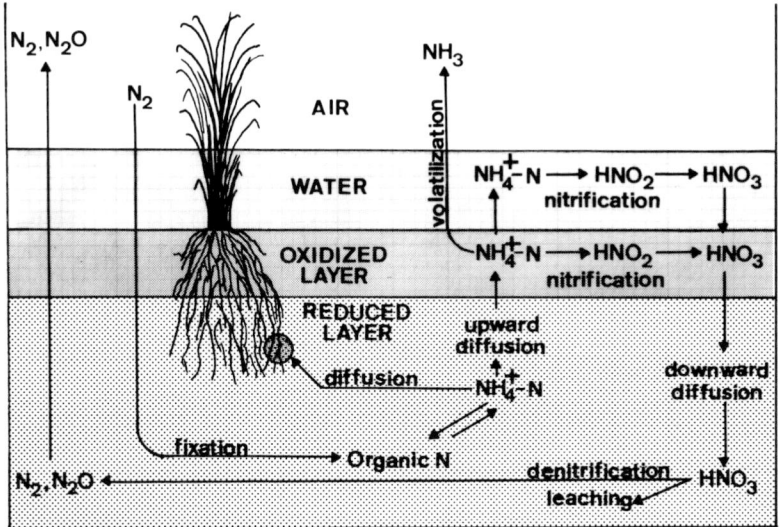

Figure 5. Nitrogen transformations in a flooded soil-plant system.

have suggested that under some conditions plant roots may favor denitrification by increasing oxygen depletion from the rhizosphere and secreting organic compounds, which may be used as an energy source by the denitrifying organisms. Stefanson[82] has found the presence of plants may influence which gaseous denitrification products are formed.

Nitrogen losses by denitrification are not always undesirable. Removal of excess inorganic nitrogen from agricultural drainage water and some surface waters is important in maintaining or enhancing the quality of ground and surface waters. This topic will be discussed briefly later in this chapter.

PHOSPHORUS

Studies have shown little or inconsistent responses by lowland rice to phosphorus fertilization, while upland crops grown on the same soils respond markedly to applied phosphorus.[17] Since rice has about the same phosphorus requirement as other crops, it is apparent that soil factors regulating transformations affecting the availability of phosphorus to plants are strongly influenced by the reducing conditions that develop upon waterlogging. A substantial increase in the availability of both native and applied phosphate in flooded soil (reduced) compared to well drained soils has been well established.[83-86]

Redman and Patrick[45] reported that extractable phosphorus in 26 Louisiana soils increased an average of 21% due to submergence. Poorly drained soils released the greatest amount of phosphorus upon submergence while better-drained "recent" alluvial soils released little or no phosphorus. Khalid et al.[87] reported an increase in soluble phosphorus in 13 of 20 rice soils when incubated under reduced vs oxidized conditions.

Mortimer[10] reported that soluble phosphorus increased more than 100 times in the hypolimnion of a lake following reduction at the sediment-water interface. Apparently, reduction at the interface allowed phosphate to diffuse upward from the interstitial waters of reduced sediments which have been reported to be enriched in soluble phosphorus compared to overlying oxidized surface waters.[88] A thin, oxidized surface layer likely serves as a barrier or trap for soluble phosphorus diffusing upward from reduced, subsurface horizons.

Phosphorus is present in soils and sediments in both organic and inorganic combinations, and the relative proportions are reported to vary widely. Hesse[89] found that organic phosphorus comprised 87% of the total phosphorus in swamp mud. In lake sediments, organic phosphorus has been reported to constitute 10-70% of the total phosphorus.[90,91] From a survey of the literature, Syers et al.[92] concluded that organic phosphorus appears to correlate with organic matter in some sediments. Soil organic phosphates, like nitrogen, are not available to plants until released by organic matter decomposition. Phosphorus availability from organic sources is not usually considered important because of the slow rate of organic matter mineralization, especially in flooded soils. Thus, most soil and sediment research on phosphorus availability has focused on inorganic phosphorus forms, though organic phosphorus may comprise a substantial reservoir.

Chang and Jackson[93] developed a method of fractionating inorganic soil phosphate into its major forms. Though a number of limitations have been identified,[94-98] this procedure and numerous modifications have been useful for studying the effect of various soil conditions on chemical and biological transformations affecting the bioavailability of soil and sediment phosphates. The procedure estimates aluminum-, iron-, and calcium-bound phosphates, and the amount of inorganic phosphate released by reduction and solubilization of ferric oxides in soils and sediments. Of these forms, transformations affecting the availability of reductant soluble phosphates are considered economically important to lowland rice production and ecologically significant in sediment-water systems. This phosphate is thought to be precipitated with ferric iron or to exist as occluded aluminum and iron precipitates within colloidal, amorphous ferric oxyhydroxide particulates or coatings on clay minerals.[99] Phosphate bound in this form is unavailable in upland oxidized soils. However, the instability of these hydrous ferric compounds under reduced soil or sediment conditions favors the release of reductant soluble phosphate.

Ponnamperuma[18] attributed the increase in soluble phosphorus upon flooding acid soils to:

1. hydrolysis of ferric and aluminum phosphates,
2. release of phosphorus adsorbed to clays and hydrous oxides by anion exchange processes, and
3. reduction of ferric compounds containing phosphate to the more soluble ferrous compounds.

It was suggested an increase in pH accompanying soil reduction may contribute to processes 1 and 2 above. Calcium phosphate compounds which predominate in alkaline soils become more soluble as pH approaches 7.0 upon reduction.

Factors Affecting Phosphorus Availability

A number of important factors have been identified in regulating phosphate fixation and release in soils and sediments. These include the types and amounts of clay minerals, the quantities of calcium, aluminum, iron and magnesium, pH, organic matter content, and the oxidation-reduction intensity of the soil. In flooded soils and sediments, oxidation-reduction conditions are the key to several processes affecting phosphorus availability. The redox potential mediated influence on pH and subsequent phosphorus release as a system becomes reduced has been previously mentioned. The presence of organic matter affects the availability of soil phosphate by reduction and chelation.[86] The presence of an ample supply of organic matter encourages development of strongly reducing conditions in flooded soils and sediments, which tend to solubilize ferric oxyhydroxide particulates and coatings, increasing phosphorus availability.

These increased total levels and structural complexity of organic compounds in anaerobic systems likely favor chelation of the orthosphosphate ions as both soluble and insoluble complexes that compete with precipitation reactions. The formation of soluble complexes would enhance phosphorus availability. Transformations of inorganic phosphate to insoluble organic combinations in flooded soils have been reported to reduce phosphate availability.[100,101] Aluminum may stabilize insoluble organic phosphorus complexes in anaerobic soils, reducing phosphorus availability,[102] though this process is difficult to substantiate from studies of proposed mechanisms.[103] Vijayachandran and Harter[104] suggested that while the role of organics in phosphorus adsorption has been primarily attributed to the iron and aluminum chelated by the organic matter, their studies of a number of soils varying widely in properties from a variety of locations in the United States indicated anion adsorption sites on organic matter were responsible for the correlations between phosphorus adsorption and organic carbon content rather than chelated iron or aluminum.

Most of the research concerned with oxidation-reduction effects on phosphorus chemistry in soils and sediments has focused on reductant soluble phosphorus, that is, inorganic phosphorus associated with ferric iron. In a laboratory study on the effects of controlled pH and redox potential on extractable phosphate from a flooded soil, Patrick et al.[105] found that both pH and redox potential strongly influenced the dissolution of strengite ($FePO_4 \cdot 2H_2O$). At all pH levels, a decrease in redox potential enhanced the quantities of iron and phosphorus dissolved. Maximum dissolution occurred under conditions of low pH and low redox potential. Though Tanaka et al.[106] attributed most of the increase in soluble levels of phosphate after soil submergence to the increase in pH usually accompanying soil reduction which favors hydrolysis of iron and aluminum phosphates, the results of Patrick et al.[105] showed that a decrease in redox potential at a constant pH markedly increased phosphate release.

In an earlier study, Patrick[37] found that extractable phosphorus increased by a factor of three as redox potential was reduced from +200 to -200 mV. A sharp increase in phosphate release was noted at the same potential at which ferric iron becomes unstable, indicating the increase in extractable phosphate was dependent on the reduction of ferric iron compounds.

Using the Chang and Jackson[93] fractionation procedure, Mahapatra and Patrick[107] studied the effect of waterlogging on transformations of inorganic phosphate in 16 soils used for lowland rice production. A decrease in reductant soluble phosphate as a result of waterlogging was generally accompanied by an increase in iron phosphate, while aluminum and calcium phosphates were little affected (Figure 6).

Nature of Reductant Soluble Phosphate

Syers et al.[92] concluded from a review of the literature that fixation of inorganic phosphorus added to lake sediments occurred by sorption processes rather than by precipitation. The nature of the sorbing substance has been the subject of much research which has further emphasized the role of poorly crystalline ferric iron compounds. McKeague and Day[108] have reported that amorphous and poorly crystalline iron oxides and hydroxides are extracted by an oxalate reagent, which has little effect on better crystalline iron oxides and iron contained in primary silicates. Poorly crystalline iron compounds are the reductant soluble ferric compounds that are highly stable in oxidized environments, but become

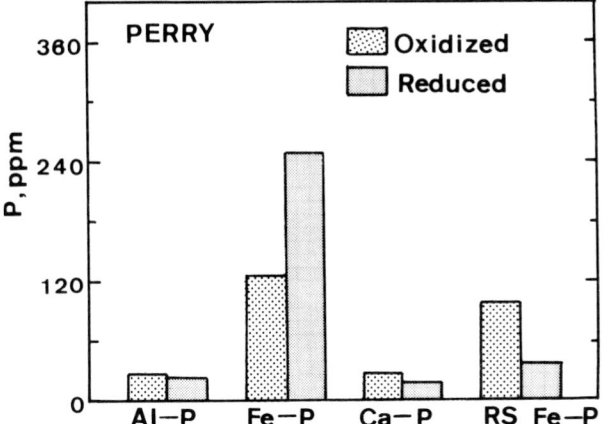

Figure 6. The levels of aluminum, iron, calcium and reductant soluble phosphorus, as affected by oxidized and reduced soil conditions. (Reproduced from *Soil Sci.* 107:281-288, 1969).[107]

unstable in reduced soils and sediments, releasing adsorbed and occluded inorganic phosphates. Numerous investigators have found that oxalate-extractable iron is associated with phosphorus adsorption.

Williams *et al.*[109] reported that inorganic phosphorus in both calcareous and noncalcareous Wisconsin surface lake sediments was closely related to poorly crystalline hydrous oxides of iron extracted by oxalate. Saunders[110] also obtained a close relationship between phosphate retention and oxalate-extractable iron and aluminum in a number of New Zealand soils. Shukla *et al.*[111] found that oxalate extraction removed sediment components active in inorganic phosphate sorption. Upchurch *et al.*[112] reported a high correlation ($r = .99$) between what Wentz and Lee[113] termed available phosphorus and oxalate-extractable iron in the Pamlico Estuary of North Carolina.

Patrick and Khalid[99] and Khalid *et al.*[87] have recently reported that reduced soils released more phosphate to soil solutions low in soluble phosphate and sorbed more phosphate from solutions containing high levels of soluble phosphate than did the same soils under oxidized conditions. It was suggested that the reduction of insoluble ferric oxyhydroxides to more highly dispersed ferrous forms increases the surface area of the reactive iron compounds responsible for phosphate retention. This was supported by the recovery of more oxalate-extractable (poorly crystalline) iron from the reduced soils.

Soil Testing for Phosphorus in Reduced Soils

A few studies are available in the literature in which soil test results have enabled researchers to make adequate predictions of the response of lowland rice to phosphate fertilizers. However, Patrick and Mahapatra[17] pointed out that most soil test methods commonly used to estimate available phosphate for upland crops have not been satisfactory indicators of soil phosphorus availability to rice. These tests have generally been applied to air-dry soil samples and do not account for the increase in phosphate availability to lowland rice caused by soil submergence and subsequent reduction. Mahapatra and Patrick[114] later reported extractable iron, aluminum and calcium phosphates of air-dry soils were highly correlated with the Bray-extractable phosphorus of reduced rice soils in Louisiana. It was suggested that identification and measurement of phosphorus forms in air-dry soils, which become available upon reduction, would permit phosphate fertility evaluation of flooded soils by chemical testing of air-dry samples.

More recently, Khalid et al.[84] found that phosphorus adsorbed from 0.1 and 0.2 mg/ml phosphorus solutions under reduced conditions were highly correlated with rice yields. The currently used soil testing procedure on air-dry samples was of less value in predicting fertilizer phosphorus requirements for lowland rice in this study. They concluded that soil test correlations for phosphorus with rice yields may be improved by modifying soil testing procedures to account for the influence of soil reduction processes on phosphorus availability.

SULFUR

Sulfur is found in wetland soils and sediments in both organic and inorganic combinations. Sulfur is subject to oxidation-reduction reactions and its most common valences are -2 (sulfide), 0 (elemental sulfur) and +6 (sulfate). Principal chemical transformations in oxidized soils and sediments are assimilatory sulfate reduction into organic sulfur compounds of plant and microbial tissue, oxidation of inorganic sulfide and elemental sulfur, and mineralization of organic sulfur compounds to inorganic sulfate.[48] In anaerobic soils and sediment-water systems, inorganic sulfate is reduced biologically to sulfide by respiratory sulfate reduction in which certain bacteria utilize sulfate as a terminal electron acceptor. Anaerobic decomposition of organic matter also contributes to an accumulation of sulfide. Postgate[115] stated that bacteria capable of sulfur transformations may be considered ubiquitous in terrestrial environments, tolerating an unusually wide range of salinity, temperature and pressure.

Several investigators have studied the influence of pH and redox potential on dissimilatory sulfate reduction. Harter and McLean[39] reported rapid sulfide accumulation in soils with redox potentials of -75 mV or less.

Connell and Patrick[38] reported -150 mV to be the critical potential below which sulfate was readily reduced to sulfide. Soil pH has also been found to regulate sulfide formation. Connell and Patrick[38] found little sulfide accumulation from reduced Texas and Louisiana rice soils when pH was outside the range of 6.5-8.5. Maximum sulfide accumulation was found over the relatively narrow pH range of 6.8-7.0. Baas Becking and Moore[116] reported that *Desulfovibrio* could reduce sulfate over the pH range of 4.2 to over 10 within a redox potential range of +110 to -500 mV.

Engler and Patrick[117] demonstrated that a number of chemical oxidants are effective in promoting sulfide oxidation or retarding sulfide formation. The oxidants studied included oxygen, potassium nitrate, manganese dioxide, ferricitrophosphate and ferric phosphate. It was found that the more soluble oxidants (potassium nitrate and ferricitrophosphate) were most effective in retarding sulfate reduction over short time intervals, while the less-soluble compounds were ineffective in delaying sulfide production, but did prevent maximum sulfide buildup. It was suggested that effects on sulfide production were due to the oxidants' influence on redox potential.

The toxicological importance of sulfur and its role in regulating the availability of other toxins will be discussed in later sections. Postgate[115] reviews the growth, nutrition, biochemistry and economic importance of sulfate-reducing bacteria.

TRACE METALLIC NUTRIENTS

The plant availability of trace metallic nutrients is also influenced by oxidation-reduction intensity. Some of these metals are directly subject to redox reactions affecting their chemical valence state and, thus, speciation. All are influenced indirectly by a number of regulatory processes influenced by the oxidation-reduction status of a soil or sediment.

Chemically reactive iron and manganese of biological importance are present in two oxidation states. In reduced soils, ferrous (Fe^{2+}) and manganous (Mn^{2+}) compounds predominate. These forms are considered relatively soluble and available to organisms. Their abundance and stability are favored by increases in acidity and decreases in oxidation intensity. In oxidized environments, sparingly soluble ferric (Fe^{3+}) and manganic (Mn^{4+}) compounds predominate and are highly stable. Upon waterlogging of several rice soils, Turner and Patrick[6] noted an increase in the quantity of reduced iron and manganese with time which corresponded with oxygen and nitrate depletion and a reduction in redox potential.

Gotoh and Patrick[36,44] studied the effects of controlled redox potential and pH on the transformations of iron and manganese between the water-soluble, exchangeable and reducible forms in a soil suspension. Water-soluble and exchangeable iron levels increased sequentially as both pH and redox potential decreased. Soluble iron ranged from more than 3600 $\mu g/g$ under strongly reduced, acid conditions to undetectable levels in weakly alkaline, oxidized suspensions. Levels of reducible iron, thought to consist of ferric oxides and hydroxides, were also influenced by pH and redox potential. Decreased levels of this form, as pH and redox potential were lowered, were accompanied by equivalent increases in soluble plus exchangeable levels. Dissolved and exchangeable manganese responded similarly to changes in pH and redox potential.

Mandal[118] reported on yield and the nitrogen, iron and manganese content of rice as a function of the oxygen content of the experimentally controlled atmosphere overlying the floodwater. The anaerobic and semi-anaerobic (0 and 10% oxygen) treatments resulted in greater accumulation of both metals in the shoot tissue than was found in better-oxidized treatments. The aerobic treatments were thought to favor the stability of the sparingly soluble oxides and hydroxides of these metals, which reduced availability to plants.

Weeraratna[119] reported that manganese uptake by rice from wetland soils was greater under flooded conditions than under nonflooded conditions. The increased availability upon flooding these soils was attributed to the increased levels of soluble manganese in reduced (flooded) soils. Flooding decreased manganese uptake from an upland soil compared to nonflooded conditions, but this was thought to result from increased iron availability and subsequent iron toxicity to plants grown in the flooded, upland soil. Clark, Nearpass and Specht[120] also reported that manganese availability to rice was enhanced by submerged soil conditions.

Jugsujinda[121] found that uptake of iron, manganese and zinc was strongly influenced by redox potential and pH in a laboratory study in which these parameters were controlled. Labeled iron uptake was high under reduced conditions at low pH. Labeled manganese uptake was considerably greater under reduced conditions at both pH 5.5 and 7.5.

Jones and Etherington[122] reported the effects of waterlogging on the uptake of iron and manganese by *Erica cinerea* and *Erica tetralix* in a study to determine factors favoring the establishment of *E. tetralix* on wetter sites. Both species accumulated significantly more iron from waterlogged soils than from the control treatment in both shoot and root tissue. Waterlogging enhanced the uptake of manganese by the leaves of *E. cinerea*.

Waid[123] reviewed the role of hydroxamic acids produced by plants and microbes in soils. It was suggested that oxidized soils may stimulate biological production and exudation of iron-solubilizing organic chelates in response to iron deficiency in oxidized soils. This process may oppose to some extent, the reduced availability of iron in oxidized soils where sparingly soluble ferric compounds are stable.

Although soil zinc and several other metallic nutrients are not subject to redox transformations as are iron and manganese, the soluble levels of these metals are influenced by a number of processes that are regulated by oxidation intensity. Most of the literature available on the effects of oxidation-reduction conditions on chemical forms of trace metals is concerned with these metals from a geochemical or environmental point of view. However, the rather limited information available on plant uptake of metallic nutrients under different aeration conditions indicates the importance of oxidation-reduction intensity.

Citing other published literature, Lucas and Knezek[124] reported mixed effects of poor aeration on plant zinc availability. However, much of the available literature indicates that zinc is less available to crops in more reduced soils.[125,126] In studies of zinc uptake by rice grown in soil suspensions of controlled pH and oxidation intensity, reduced soils decreased zinc uptake over oxidized soil at each of four pH levels studied.[121]

Regulatory processes affecting the chemical and biological availability of trace metals will be discussed in a later section.

PLANT TOXINS AND ENVIRONMENTAL CONTAMINANTS IN WETLAND SOILS AND SEDIMENTS

Iron

Increased soluble levels of iron in some reduced soils have been reported to result in iron toxicity to crops.[127,128] The adverse effects of high levels of available iron in reduced soils may result from direct toxicity. However, several reports in recent years suggest the mechanism of iron toxicity may involve indirect effects of excess iron. Howeler[127] postulated that excess soluble iron in flooded Oxisols in Columbia may coat roots with an iron oxide barrier layer, thereby reducing nutrient transport from the soil into the plant. As a result, affected rice plants were thought to be deficient in phosphorus, potassium, calcium and magnesium. Jones[129] also suggested that reduced phosphorus uptake by some dune and dune slack grass species may be attributed to phosphorus immobilization at the root surface due to the high levels of iron associated with roots in waterlogged soils.

Armstrong and Boatman[130] suggested that a high available iron level served as a sink for oxygen transported through stems to roots in rice. A relatively thick sheath of hydrous ferric oxide, generally found around the roots of swamp and marsh plants, supports this hypothesis and may contribute to the barrier effect reported in more recent studies cited above. It was also suggested this iron coating may sometimes be beneficial by protecting the root from sulfide toxicity as a consequence of sulfide precipitation with iron just outside the root.

Howeler[127] suggested the detrimental effects of either direct or indirect iron toxicity may be minimized by liming and various management practices to control the excess soluble iron levels resulting from prolonged intensive soil reduction.

Sulfide

Waterlogged soils and sediments containing appreciable organic carbon may become strongly reduced, resulting in sulfate reduction and sulfide accumulation. The detrimental effect of hydrogen sulfide on root function and plant growth is well established in the literature.[131-133] Ponnamperuma[18] pointed out that sulfide formation may adversely affect plant growth because of:

1. toxicity of free sulfide to plant roots,
2. insufficient nutrient availability of sulfur due to its precipitation with trace metals, and
3. immobilization of zinc and copper by sulfide precipitation.

The capacity for rice and other marsh plants to oxidize their rhizospheres likely counters, to some extent, sulfur deficiencies resulting from sulfide precipitation with trace metals. Several workers have suggested that rice roots may oxidize metal sulfides next to the root, then adsorb the sulfate.[12,134] Additionally, others have shown that the varietal susceptibility of rice plants to diseases associated with hydrogen sulfide is related to the oxidizing capacity of the roots.[131,135,136]

Several studies on the mechanism of toxicity have indicated the presence of hydrogen sulfide may limit nutrient uptake.[137,138] Ford[139] and Mitsui and Kumazawa[140] suggested the mechanism involved is an adverse effect of hydrogen sulfide on enzymatic reactions.

Where sulfide toxicity has been a problem in highly reduced soils, amendments with manganese dioxide have been beneficial.[141,142] Engler and Patrick[117] studied the effects of several chemical oxidants on inhibition of sulfide formation as a soil becomes reduced in the laboratory. They reported soluble oxidants such as potassium nitrate and ferricitrophosphate were very effective in delaying sulfide production, but were not effective over a long period. The less-soluble oxidants of manganese dioxide and ferric phosphate were not as effective in delaying sulfide formation, but did retard maximum sulfide formation for an extended period.

Though some free hydrogen sulfide may be produced in reduced soils,[38,131] and sulfide toxicity has been reported to occur on iron-excess soils,[137] most researchers believe that the formation of insoluble ferrous sulfide effectively regulates sulfide toxicity in most soils.[143,144] Exceptions include reduced, alkaline soils where more free hydrogen sulfide may be generated due to the decreased soluble levels and mobility of iron[145] and old degraded rice soils where reactive iron is apparently limiting.[73]

Patrick and Mikkelsen[32] pointed out that the sequential reduction of oxidized soil components is important to the regulation of sulfide toxicity. Reducible iron, if present, is always transformed to the relatively mobile ferrous form prior to initiation of sulfate reduction. In addition, the several oxidation-reduction systems between the sulfate-sulfide couple and the irreversible oxygen system function to minimize sulfide toxicity by buffering redox potential at an intermediate level above the critical potential for sulfide formation.

Nitrogen

The physical, chemical and microbial processes affecting nitrogen transformations in flooded soils and sediments have been covered previously in this chapter. Now we will discuss briefly the role of reduced soils and sediments in removing excess nitrogen from the environment.

Wastes containing nitrogenous materials and unutilized fertilizer nitrogen may have adverse effects on ground and surface water quality. Nitrogen has been found to be the critical limiting factor for undesirable eutrophication in many coastal waters[146-148] and for some freshwater lakes.[149,150] The chemical and physical properties of many soils do not favor removal of unutilized fertilizer or waste nitrogen, and substantial accumulations or leaching losses of nitrate may occur.[151-153] However, factors may be favorable for denitrification in subsurface horizons of many other soils, especially if these soils are poorly drained or have restrictive or saturated horizons.[154-156]

Water-saturated horizons beneath agricultural soils, sediments and swamp and marsh soils are important in regulating excess nitrogen levels for nitrogen-sensitive ecosystems. In these soils and sediments, substantial amounts of excess nitrogen may be removed *in situ* by denitrification.[66,157-159]

In recent years, considerable effort has been directed to managing soils possessing the potential for denitrification to maximize the gaseous removal of residual fertilizer nitrate in the field, or the nitrogen in wastes applied to soil for disposal.[160,161,162] The nitrogen cycle in lake sediment-water systems has also been shown to be important to water quality. Though the overlying water column may receive substantial quantities of inorganic nitrogen from sediments,[61] as well as from external sources, sediments play an important role in removing excess nitrogen entering lakes from surface and subsurface drainage.[61,65,66,163]

Engler and Patrick[164] and Engler et al.[165] have shown in laboratory studies that swamp and marsh soils have a large capacity to remove nitrate nitrogen, which may enter these wetlands from local agricultural drainage or other sources. A simulated saltwater marsh was reported to remove the equivalent of 7.4 kg nitrogen/ha/day by denitrification. The nitrate removal rate of a freshwater swamp was approximately half as great, but still substantial in terms of overall denitrifying capacity.

In field studies of nutrient utilization by *Spartina alterniflora* in a Louisiana salt marsh, Patrick and DeLaune[53] found that nitrogen may limit the growth of this species despite high levels of total sediment nitrogen and comparatively high nitrogen mineralization rates in these marshes.[55] They concluded that the mineralization-nitrification-denitrification sequence resulted in considerable inorganic nitrogen loss before it could be utilized by the marsh plants. These findings also suggest that wetlands have a high capacity for removing excess nitrogen.

Keeney[166] recently reviewed the factors affecting nitrogen loss, mechanisms of loss and the environmental importance of the nitrogen cycle in sediment-water systems.

ORGANIC PESTICIDES

The movement, persistence and fate of synthetic organic pesticides in soils and sediments are subjects of scientific and ecological interest. Extensive research on the fate of pesticides has shown that a small proportion of that applied often moves with drainage water, and particularly with suspended solids in surface drainage, into waterways and wetlands, where these materials may accumulate to varying extents in receiving waters, sediments and biota. Soil and sediment oxidation-reduction intensity has been found to affect the persistence of residual pesticides, and some recent investigations have considered the possibility of managing soil oxidation conditions to minimize the adverse environmental impact of these materials. Though most agricultural pesticides are applied to upland soil-crop systems in which the soils are aerobic, the soils and sediments of wetlands receiving drainage are usually anaerobic. Studies have shown that a number of different classes of pesticides degrade more rapidly in anaerobic systems than in aerobic environments. This is an interesting contrast to the generally slower degradation reported for detrital plant and animal material in anaerobic soils and sediments.

A number of studies and reviews have shown that DDT is less persistent in anaerobic soils than aerobic soils.[167-170]

Spencer et al.,[171] citing the literature, indicated that DDE is the principal degradation product of DDT in oxidized soils, while DDD is the chief residual form under reduced conditions. To minimize volatilization of DDE, the more hazardous of the two breakdown products, they suggested management practices should focus on flooding, when feasible, to encourage the DDD breakdown pathway since, once formed, DDE is quite persistent even in reduced soils.[172]

MacRae et al.[173] reported that benzene hexachloride is slowly degraded to carbon dioxide in flooded soils, but is more persistent in oxidized soils. In a review on dinitroaniline herbicides in soils, Helling[174] reported that several of these compounds are known to be lost more quickly from reduced soils than from better-oxidized soils. Parr and Smith[175] reported toxaphene degradation was more rapid as redox potential decreased from 0 to -100 mV. Sethunathan[176] found that diazinon persisted for 180 days in unflooded soils, but disappeared within 60 days in flooded soils. Kearney et al.[177] noted from the literature that atrazine and trifluralin were lost more quickly under reduced conditions. Willis et al.[170] reported that oxygen exclusion alone did not significantly enhance trifluralin degradation, but rapid degradation was initiated when redox potential dropped below some level between +150 and +50 mV. Guenzi et al.[168] concluded that if the redox potential of field soils can be decreased to about 250 mV (a mildly reduced condition), residues of heptachlor, DDT, lindane and endrin can be substantially reduced.

The available literature suggests that compared to oxidized conditions, reduced soils and sediments result in different degradative microbial populations and breakdown pathways,[172,174,178,179] as well as differences in pesticide adsorption to soil and sediment solids. In a review of pesticide-sediment-water interactions, Pionke and Chesters[180] indicated that adsorption of many nonionic pesticides is strongly correlated with soil organic matter content. It was suggested that the greater accumulation of organic matter in reduced sediment layers, as well as differences in structural composition of the organics, is likely to increase the pesticide adsorptive capacity of reduced soils and sediments relative to oxidized systems. They also suggested that colloidal hydrous ferric oxides, whose formation and stability are favored in oxidized environments, are also thought to be effective adsorbents for some pesticides. Thus, suspended colloidal iron could scavenge some pesticides from the water column. The presence of a thin, oxidized sediment horizon containing ferric iron may also be important to the fate of pesticides in sediment-water systems scavenging pesticides from the surface water column or serving as a barrier to pesticides diffusing upward from subsurface, reduced horizons.

TRACE METALS

Waterways have historically been used for waste disposal in industrial societies and, more recently, waste application to soils for treatment is receiving much attention. Many industrial and most municipal waste discharges are known to contain high levels of one or several potentially toxic trace metals.[181,182] Though efforts are being initiated to minimize this type of pollution, substantial quantities of metals have accumulated in sediments. Sediments, and reduced sediments in particular, are effective sinks for most metal contaminants. There is some concern that dredging of contaminated sediments may result in metal release and possibly adverse environmental effects in aquatic and wetland ecosystems.[183,184] Perkins[3] pointed out that many wetlands of the world have been used directly as waste disposal sites since, until recently, they were considered wastelands rather than valuable resources.

The application of secondary sewage effluents to agricultural soil-crop systems has gained acceptance in recent years, both for productive nutrient and water utilization as well as for waste treatment purposes. Extensive research has been and is being conducted to determine if this practice may contaminate human food supplies with toxic metals. Well drained, upland soils are usually selected for land treatment systems. However, the application of large quantities of treated effluent in aqueous slurries may, in some cases, result in development of strongly reduced conditions within the soil profile.[185,186] Thus, anaerobic chemical and microbiological transformations affecting the availability of toxic metals applied with sludge may play an important regulatory role in some upland soils.

Metals are present in soils and sediments in many chemical forms that differ greatly in their bioavailability. The mineral fractions of soils and sediments contain some toxic metals bound within the crystalline structural lattice as a primary constitutent. Metals in this form are essentially unavailable to biota. Metals dissolved in soil solution, or interstitial or surface waters are considered readily available to biota. Metals weakly adsorbed to the solid mineral or organic colloidal phase by ion exchange mechanisms are also considered readily available, as this form may equilibrate with the aqueous phase. Between the unavailable and readily available metals are a number of chemical forms that are potentially available. The regulatory processes binding metals in these forms are strongly influenced by oxidation intensity and include metal precipitation as insoluble sulfides in reducing environments,[134,187] metal coprecipitation or adsorption with colloidal hydrous oxides of iron and manganese, primarily in oxidized environments,[188-191] and metal complex formations with insoluble humic materials.[192-195]

In reduced soils and sediments, the formation of stable, insoluble metal sulfide precipitates is important in limiting the mobility and bioavailability of most metals.[134,187,196] Engler and Patrick[134] reported that metal sulfides of manganese, iron, zinc, copper and mercury were stable in a flooded (reduced) soil as determined by the solubility of labeled sulfur. In an oxidized soil, the stabilities of these metal sulfides were considerably lessened, and the reduction in stability was related to their solubility product constants. Rice plants grown in strongly reduced soils amended with the metal sulfides were found to take up tagged sulfur, indicating oxidation in the rhizosphere and increased availability of the metals and sulfur to the growing rice roots.

Trace metals associated with colloidal hydrous oxides of iron and manganese are probably not in equilibrium with the solution phase in neutral or alkaline oxidized soils and sediment-water systems favoring the stability of these compounds. Development of reduced conditions or transport of these materials to reduced environments will result in instability of colloidal hydrous oxides, likely causing release of some adsorbed or coprecipitated trace metals. However, in reduced environments, other regulatory mechanisms such as complexation with insoluble organics or sulfide precipitation will tend to immobilize metals which may be released by dissolution of hydrous oxides. For example, Sanchez and Lee,[197] working with copper in Lake Monona, Wisconsin, reported that as hydrous metal oxides containing copper became reduced, the copper was immobilized by sulfide precipitation.

Metal complex formation with insoluble organics is reported to be an important regulatory process affecting immobilization of metals. Changes in oxidation-reduction intensity likely contribute to quantitative as well as qualitative changes in humic substances affecting the fixation of trace metals. The greater accumulation of humic materials in reduced environments has already been discussed. In addition, compared to oxidized systems, it is probable that humic materials in reduced environments are characterized by large molecular weights and greater structural complexity, which may increase the metal retention capacity and the metal bonding stability of insoluble humic materials. In studies of the effect of controlled pH and redox potential levels on chemical transformations of trace metals, Gambrell et al.[198] found a decrease in stability of copper, lead and cadmium complexes with insoluble organics as reduced sediment materials were subjected to an oxidized environment. In the case of cadmium, as redox potential was increased from strongly reducing to well oxidized levels, insoluble organically bound cadmium (Figure 7) was transformed to more available soluble and exchangeable forms. Stevenson and Ardakani[195] suggested a reduction in metal availability due to insoluble organic complex formation in reduced systems may be offset to

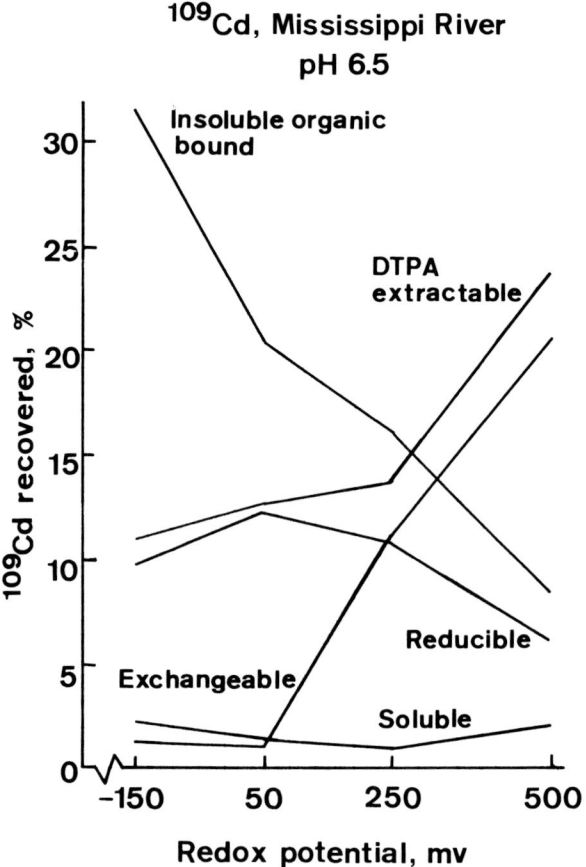

Figure 7. The effect of redox potential on the distribution of cadmium-109 among selected chemical forms in Mississippi River sediment suspensions incubated at pH 6.5.

some extent by an increase in soluble organic acids, which maintain some metals in solution as soluble organic complexes.

From the literature, it is apparent that several metal regulatory factors are influenced by oxidation-reduction intensity. Fixation or release of metals induced by a change in oxidation status of a soil, sediment, or dredged material may be difficult to predict, as a complex interaction between several regulatory factors may be involved.

In studies using controlled pH and oxidation levels, Jugsujinda[121] found uptake of native and added radiotracer zinc by rice seedlings was greater from aerobic than anaerobic soil suspensions. It was suggested that insoluble zinc sulfide formation and possible greater zinc complexation with insoluble humic materials contributed to the decrease in zinc availability under reduced conditions. It has also been reported elsewhere that flooding depresses zinc uptake by rice plants.[126]

Reddy and Patrick[199] have shown that uptake of indigenous cadmium by rice is enhanced by sequential increases in redox potential from strongly reduced to moderately oxidized conditions, especially in weakly-to-moderately acid soils (Figure 8). Lead uptake was enhanced by low pH levels and little affected by changes in oxidation-reduction intensity. Gambrell et al.[200] have reported that acid-oxidizing soil environments increase cadmium uptake by several marsh plants also.

SUMMARY

Anaerobic agricultural soils play a major role in providing food to a large segment of the earth's human population. In addition, swamps and marshes occupy large areas of the earth's surface, and these wetland soils

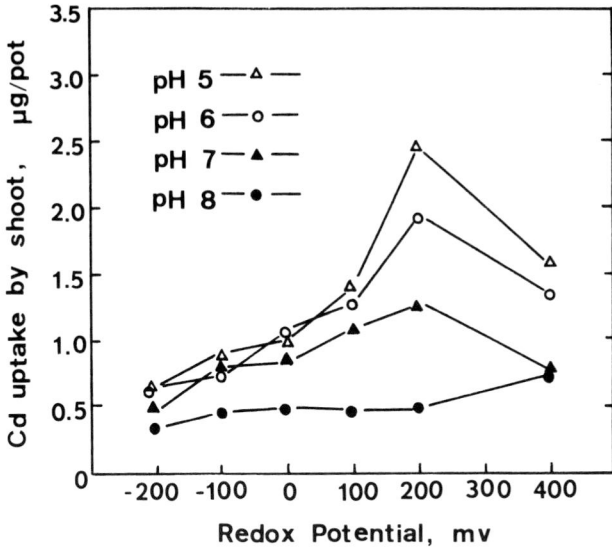

Figure 8. The effect of pH and redox potential on the accumulation of cadmium by lowland rice (*Oryza sativa*) shoots. (Reproduced from *J. Environ. Qual.* 6:260, 1977).[199]

represent an important resource contributing to the abundant productivity by the diverse flora and fauna populations of wetland ecosystems.

The chemistry and microbiology of sediments are similar to anaerobic soils. Wetland soils and sediments are thought to degrade, inactivate, or otherwise minimize adverse environmental impact from a number of materials that drain into waterways, such as excess plant nutrients, toxic metals and pesticides. However, sediments may release some of the harmful materials accumulated to bioavailable forms if the sediment is disturbed or its oxidation-reduction conditions altered, as sometimes occurs during dredged material disposal. Because of their capacity to accumulate biostimulants and environmental toxins, chemical and biological processes affecting the availability of these substances in sediments are important to environmental quality.

Anaerobic soils and sediments are generally characterized by water saturation, which limits oxygen transport into these systems, and an active microbial population feeding on the usually abundant organic energy supply of flooded soils and sediments. Since oxygen is the preferred electron acceptor for microbial respiration, oxygen demand for both chemical and biological processes usually exceeds the rate of oxygen resupply in flooded soils. As a result, microbiological populations, which may use electron acceptors other than molecular oxygen, flourish, and their predominant metabolic pathways affecting both inorganic and organic constituents of wetland soils differ markedly from oxidized environments.

The bioavailability of most nutrients and toxins respond to the oxidation-reduction conditions of soils and sediment-water systems. In many cases, the regulatory processes may be considered a direct response to oxidation intensity. For example, ferric iron and nitrate nitrogen are reduced as oxygen becomes limiting for biological respiration. Phosphorus and trace metal occlusion and/or adsorption with colloidal hydrous oxides of iron in oxidized environments represent examples of an indirect availability response to changes in oxidation intensity. Also, the availability of many trace metals that do not participate directly in oxidation-reduction reactions in the natural environment may be strongly influenced by the bonding stabilities between metals and naturally occurring organic chelates. Soil and sediment organic matter is thought to be quantitatively and qualitatively affected by oxidation intensity.

Anaerobic soils and sediments under most shallow water columns are characterized by a thin, surface-oxidized horizon overlying a thick, reduced horizon. Nutrient and toxin transformations in this zone are similar to those in oxidized soils. Though the surface-oxidized horizon is thin, typically a few millimeters to a few centimeters in depth, it is nevertheless an important component of most anaerobic soil systems. Diffusion of mobile constituents from one horizon to another where different chemical and biological processes are occurring affects the loss or retention of these substances. For example, substantial losses of ammonium nitrogen may occur by the nitrification-denitrification sequence in flooded soils and sediments having oxidized and reduced horizons. Also, there is evidence that a thin, surface-oxidized horizon acts as a barrier to phosphorus migration from within sediments to the overlying water column and may reduce levels of dissolved pesticides and trace metals in surface waters.

Despite a number of theoretical limitations and procedural difficulties, redox potential measurements have been found to be useful indicators of the intensity of oxidation or reduction in flooded soils and sediments. Redox potentials have been used successfully to predict the occurrence of chemical and biological transformations affecting the availability of a number of biostimulants and toxins. Research to increase our understanding of the chemical and biological processes occurring in anaerobic soils should improve management practices for these systems to increase productivity of flooded agricultural soils and minimize adverse environmental effects in other wetland ecosystems influenced by man's activities.

ACKNOWLEDGMENTS

The authors wish to express appreciation to R. A. Khalid, K. R. Reddy, and R. D. DeLaune for their review and suggested improvements to this manuscript and to Ms. Judy Henderson for typing and editorial assistance.

REFERENCES

1. International Rice Research Institute. "Research Highlights," in *The International Rice Research Institute Annual Report for 1973*, IRRI, Los Banos, Philippines (1974).
2. Day, J. W., Jr., W. G. Smith, P. R. Wagner and W. C. Stowe. "Community Structure and Carbon Budget of a Salt Marsh and Shallow Bay Estuarine System in Louisiana," Center for Wetland Resources, Louisiana State University, Baton Rouge, Louisiana, Publ. No. LSU-SG-72-04 (1973).
3. Perkins, E. J. *The Biology of Estuaries and Coastal Waters* (London: Academic Press, 1974).
4. Russell, J. E. *Soil Conditions and Plant Growth* (New York: John Wiley and Sons, Inc., 1961).
5. Evans, D. D. and A. D. Scott. "A Polarographic Method of Measuring Dissolved Oxygen in Saturated Soil," *Soil Sci. Soc. Am. Proc.* 19:12 (1955).
6. Turner, F. T. and W. H. Patrick, Jr. "Chemical Changes in Waterlogged Soils as a Result of Oxygen Depletion," *Trans. 9th Internat. Cong. Soil Sci.* 4:53 (1968).

7. Berryhill, H. L., Jr., V. E. Swanson and A. H. Love. "Organic and Trace-Element Content of Halocene Sediments in Two Estuarine Bays, Pamlico Sound Area, North Carolina," Geological Survey Bulletin 1314-E,E1-E32 (1972).

8. Hallberg, R. O. "Metal Distribution Along a Profile of an Intertidal Area," *Est. Coast. Mar. Sci.* 2:153 (1974).

9. Ho, C. L. and J. Lane. "Interstitial Water Composition in Barataria Bay (Louisiana) Sediment," *Est. Coast. Mar. Sci.* 1:125 (1973).

10. Mortimer, C. H. "The Exchange of Dissolved Substances between Mud and Water in Lakes," *J. Ecol.* 29:280 (1941).

11. Mortimer, C. H. "The Exchange of Dissolved Substances between Mud and Water in Lakes," *J. Ecol.* 30:147 (1942).

12. Ponnamperuma, F. N. "Dynamic Aspects of Flooded Soils and the Nutrition of the Rice Plant," in *The Mineral Nutrition of the Rice Plant* (Baltimore, Maryland: John Hopkins Press, 1965), p. 295.

13. Vanderpost, J. M. "Bacterial and Physical Characteristics of Lake Ontario Sediment During Several Months," *Proc. 15th Conf. Great Lakes Res.* 15:198 (1972).

14. Weiler, R. R. "The Interstitial Water Composition in the Sediments of the Great Lakes—I. Western Lake Ontario," *Limnol. Oceanogr.* 18:918 (1973).

15. Windom, H. L. "Environmental Aspects of Dredging in Estuaries," *J. Am. Soc. Civil Eng., Coast. Eng. Div.* 98:475 (1972).

16. Patrick, W. H., Jr. and R. D. DeLaune. "Characterization of the Oxidized and Reduced Zones in Flooded Soil," *Soil Sci. Soc. Am. Proc.* 36:573 (1972).

17. Patrick, W. H., Jr. and I. C. Mahapatra. "Transformation and Availability to Rice of Nitrogen and Phosphorus in Waterlogged Soils," *Adv. Agron.* 20:323 (1968).

18. Ponnamperuma, F. N. "The Chemistry of Submerged Soils," *Adv. Agron.* 24:29 (1972).

19. Greenwood, D. J. "The Effect of Oxygen Concentration on the Decomposition of Organic Materials in Soil," *Plant Soil* 14:360 (1961).

20. Greenland, D. J. "Denitrification in Some Tropical Soils," *J. Agric. Sci.* 58:227 (1962).

21. Daniels, R. B., E. E. Gamble and S. W. Buol. "Oxygen Content in the Ground Water of Some North Carolina Aqults and Udults," in *Field Soil Water Regime*, M. Stelly, R. C. Dinauer and J. M. Hach, Eds. (Madison, Wisconsin: Soil Science Society of America, Inc., 1973), p. 153.

22. Howeler, R. H. and D. R. Bouldin. "The Diffusion and Consumption of Oxygen in Submerged Soils," *Soil Sci. Soc. Am. Proc.* 35:202 (1971).

23. Howeler, R. H. "The Oxygen Status of Lake Sediments," *J. Environ. Qual.* 1:366 (1972).

24. Baas Becking, L. G. M., I. R. Kaplan and D. Moore. "Limits of the Natural Environment in Terms of pH and Oxidation-Reduction Potentials," *J. Geol.* 68:243 (1960).

25. Bohn, H. L. "Redox Potentials," *Soil Sci.* 112:39 (1971).

26. Bohn, H. L. "Electromotive Force of Inert Electrodes in Soil Suspensions," *Soil Sci. Soc. Am. Proc.* 32:211 (1968).

27. Laitinen, H. A. *Chemical Analysis* (New York: McGraw-Hill Book Co., 1960).

28. Stumm, W. and J. J. Morgan. *Aquatic Chemistry; An Introduction Emphasizing Chemical Equilibria in Natural Waters* (New York: Wiley-Interscience, 1970), p. 300.

29. Bohn, H. L. "The EMF of Platinum Electrodes in Dilute Solutions and Its Relation to Soil pH," *Soil Sci. Soc. Am. Proc.* 33:639 (1969).

30. Collins, J. F. and S. W. Buol. "Effects of Fluctuations in the Eh-pH Environment on Iron and/or Manganese Equilibria," *Soil Sci.* 110:111 (1970).

31. Hem, J. D. "Chemical Equilibria and Rates of Manganese Oxidation," U.S. Geological Survey Water Supply Paper 1167-A, 1 (1973).

32. Patrick, W. H., Jr. and D. S. Mikkelsen. "Plant Nutrient Behavior in Flooded Soil," in *Fertilizer Technology and Use*, R. C. Dinauer, Ed., 2nd ed. (Madison, Wisconsin: Soil Science Society of America, Inc., 1971), p. 187.

33. van Breemen, N. "Acidification and Deacidification of Coastal Plain Soils as a Result of Periodic Flooding," *Soil Sci. Soc. Am. Proc.* 39:1153 (1975).

34. Patrick, W. H., Jr. "Nitrate Reduction Rates in a Submerged Soil as Affected by Redox Potential," *Trans. 7th Internat. Cong. Soil Sci.* 2:494 (1960).

35. Bailey, L. D. and E. G. Beauchamp. "Nitrate Reduction, and Redox Potentials Measured with Permanently and Temporarily Placed Platinum Electrodes in Saturated Soils," *Can. J. Soil Sci.* 51:51 (1971)

36. Gotoh, S. and W. H. Patrick, Jr. "Transformation of Manganese in a Waterlogged Soil as Affected by Redox Potential and pH," *Soil Sci. Soc. Am. Proc.* 36:738 (1972).

37. Patrick, W. H., Jr. "Extractable Iron and Phosphorus in a Submerged Soil at Controlled Redox Potential," *Trans. 8th Internat. Cong. Soil Sci.* 3:605 (1964).

38. Connell, W. E. and W. H. Patrick, Jr. "Sulfate Reduction in Soil: Effects of Redox Potential and pH," *Science* 159:86 (1968).

39. Harter, R. D. and E. O. McLean. "The Effect of Moisture Level and Incubation Time on the Chemical Equilibria of a Toledo Clay Loam Soil," *Agron. J.* 57:583 (1965).

40. Cappenberg, T. E. "Interrelations between Sulfate-Reducing and Methane-Producing Bacteria in Bottom Deposits of a Fresh-Water Lake I. Field Observations," *Antonie van Leeuwenhoek J. Microbiol. Serol.* 40:285 (1974).

41. Martens, C. S. and R. A. Berner. "Methane Production in the Interstitial Waters of Sulfate Depleted Marine Sediments," *Science* 185:1167 (1974).

42. Butlin, K. R., S. C. Selwyn and D. S. Wakerley. "Sulphide Production from Sulphate-Enriched Sewage Sludges," *J. Appl. Bacteriol.* 19:3 (1956).

43. Cappenberg, T. E. "Relationships Between Sulfate-Reducing and Methane Producing Bacteria," *Plant Soil* 43:125 (1975).

44. Gotoh, S. and W. H. Patrick, Jr. "Transformation of Iron in a Waterlogged Soil as Influenced by Redox Potential and pH," *Soil Sci. Soc. Am. Proc.* 38:66 (1974).

45. Redman, F. H. and W. H. Patrick, Jr. "Effect of Submergence on Several Biological and Chemical Soil Properties," Louisiana Agricultural Experiment Station, Louisiana State University, Baton Rouge, Louisiana, Bull. No. 592 (1965).

46. Doelle, J. H. *Bacterial Metabolism* (New York: Academic Press, 1969).

47. Acharya, C. N. "Studies on the Anaerobic Decomposition of Plant Materials. III. Comparison of the Course of Decomposition Under Anaerobic, Aerobic, and Partially Aerobic Conditions," *Biochem. J.* 29:1116 (1935).

48. Alexander, M. *Introduction to Soil Microbiology* (New York: John Wiley and Sons, Inc., 1961).

49. Reddy, K. R. and W. H. Patrick, Jr. "Effect of Alternate Aerobic and Anaerobic Conditions on Redox Potential, Organic Matter Decomposition, and Nitrogen Loss in a Flooded Soil," *Soil Biol. Biochem.* 7:87 (1975).

50. Parr, J. F. and H. W. Reuszer. "Organic Matter Decomposition as Influenced by Oxygen Level and Method of Application to Soil," *Soil Sci. Soc. Am. Proc.* 23:214 (1959).

51. Tusneem, M. E. and W. H. Patrick, Jr. "Nitrogen Transformations in Waterlogged Soil," Louisiana Agricultural Experiment Station, Louisiana State University, Baton Rouge, Louisiana, Bull. No. 657 (1971).

52. Broome, S. W., W. W. Woodhouse and E. D. Seneca. "An Investigation of Propagation and the Mineral Nutrition of *Spartina alterniflora*," Sea Grant Publication UNC-SG-73-14, North Carolina State University, Raleigh, North Carolina (1973).

53. Patrick, W. H., Jr. and R. D. DeLaune. "Nitrogen and Phosphorus Utilization by *Spartina alterniflora* in a Salt Marsh in Barataria Bay, Louisiana," *Est. Coast. Mar. Sci.* 4:59 (1976).

54. Sullivan, M. J. and F. C. Daiber. "Response in Production of Cord Grass *Spartina alterniflora* to Inorganic Nitrogen and Phosphorus Fertilizer," *Chesapeake Sci.* 15:121 (1974).

55. Brannon, J. M. "Seasonal Variation of Nutrients and Physico-Chemical Properties in the Salt Marsh Soils of Barataria Bay, Louisiana," unpublished M.S. Thesis, Louisiana State University, Baton Rouge, Louisiana, 1973.

56. Brupbacher, R. H., J. E. Sedberry and W. H. Willis. "The Coastal Marshlands of Louisiana. Chemical Properties of the Soil Materials," Louisiana Agricultural Experiment Station, Louisiana State University, Baton Rouge, Louisiana, Bulletin No. 672 (1973).

57. Hutchinson, G. E. "Eutrophication," *Am. Scientist* 61:269 (1973).

58. Acharya, C. N. "Studies on the Anaerobic Decomposition of Plant Materials. I. Anaerobic Decomposition of Rice Straw," *Biochem. J.* 29:528 (1935).

59. Patrick, W. H., Jr. and R. Wyatt. "Soil Nitrogen Loss as a Result of Alternate Submergence and Drying," *Soil Sci. Soc. Am. Proc.* 28:647 (1964).

60. Waring, S. A. and J. M. Bremner. "Ammonium Production in Soil Under Waterlogged Conditions as an Index of Nitrogen Availability," *Nature* 201:951 (1964).

61. Graetz, R. A., D. R. Keeney and R. B. Aspiras. "Eh Status of Lake Sediment-Water Systems in Relation to Nitrogen Transformation," *Limnol. Oceanogr.* 18:908 (1973).

62. Nissenbaum, A., B. J. Presley and I. R. Kaplan. "Early Diagenesis in a Reducing Fjord, Saanich Inlet, British Columbia—I. Chemical and Isotopic Changes in Major Components of Interstitial Water," *Geochim. Cosmochim. Acta.* 36:1007 (1972).

63. Sholkovitz, E. "Interstitial Water Chemistry of the Santa Barbara Basin Sediments," *Geochim. Cosmochim. Acta.* 37:2043 (1973).

64. Reddy, K. R., W. H. Patrick, Jr. and R. E. Phillips. "Ammonium Diffusion as a Factor in Nitrogen Loss from Flooded Soils," *Soil Sci. Soc. Am. J.* 40:528 (1976).

65. Kemp, A. L. W. and A. Mudrochova. "Distribution and Forms of Nitrogen in a Lake Ontario Sediment Core," *Limnol. Oceanogr.* 17:855 (1972).

66. Chen, R. L., D. R. Keeney, D. A. Graetz and A. J. Holding. "Denitrification and Nitrate Reduction in Wisconsin Lake Sediments," *J. Environ. Qual.* 1:158 (1972).

67. Patrick, W. H., Jr. and K. R. Reddy. "Nitrification-Denitrification Reactions in Flooded Soils and Water Bottoms: Dependence on Oxygen Supply and Ammonium Diffusion," *J. Environ. Qual.* 5:469 (1976).

68. Broadbent, F. E. and M. E. Tusneem. "Losses of Nitrogen from Some Flooded Soils in Tracer Experiments," *Soil Sci. Soc. Am. Proc.* 35:922 (1971).

69. International Atomic Energy Agency. "Rice Fertilization. A Six-Year Study on Nitrogen and Phosphorus Fertilizer Utilization," *Fert. Abstr.* 4:24 (1966).

70. Patrick, W. H., Jr. and K. R. Reddy. "Fate of Fertilizer Nitrogen in Flooded Rice Soil," *Soil Sci. Soc. Am. J.* 40:678 (1976).

71. Patrick, W. H., Jr. and S. Gotoh. "The Role of Oxygen in Nitrogen Loss From Flooded Soils," *Soil Sci.* 118:78 (1974).

72. Abichandani, C. T. and S. Patnaik. "Mineralizing Action of Lime on Soil Nitrogen in Waterlogged Rice Soils," *Internat. Rice Comm. Newsletter* 13:11 (1955).

73. Mitsui, S. *Inorganic Nutrition, Fertilization, and Amelioration for Lowland Rice* (Tokyo, Japan: Yokendo, Ltd., 1954).

74. Patrick, W. H., Jr. and M. E. Tusneem. "Nitrogen Loss From Flooded Soil," *Ecology* 53:735 (1972).

75. Westerman, R. L. and L. T. Kurtz. "Priming Effect of ^{15}N-Labeled Fertilizers on Soil Nitrogen in Field Experiments," *Soil Sci. Soc. Am. Proc.* 37:725 (1973).

76. Zamyatina, V. B., N. M. Varyushkina, L. I. Kirpaneva and V. I. Porshneva. "Transformation of Nitrogen Fertilizers in Soil," *Pochvovedeniye* 1:50 (1972) (In Russian).

77. Patnaik, S. "Nitrogen-15 Tracer Studies on the Transformation of Applied Nitrogen in Submerged Rice Soils," *Proc. Ind. Acad. Sci. Soc.* B61:25 (1965).

78. Mikkelsen, D. S. and D. C. Finfrock. "Availability of Ammonical Nitrogen to Lowland Rice as Influenced by Fertilizer Placement," *Agron. J.* 49:296 (1957).

79. Patrick, W. H., Jr., F. J. Peterson, J. E. Seaholm, M. D. Faulkner and R. J. Miears. "Placement of Nitrogen Fertilizers for Rice," Louisiana Agricultural Experiment Station, Louisiana State University, Baton Rouge, Louisiana, Bull. No. 619, 3 (1967).

80. Woldendorp, J. W. "Losses of Soil Nitrogen," *Dutch Nitrogenous Fert. Rev.* 12:32 (1968).

81. Stefanson, R. C. "Soil Denitrification in Sealed Soil-Plant Systems II. Effect of Soil Water Content and Form of Applied Nitrogen," *Plant Soil* 37:129 (1972).

82. Stefanson, R. C. "Soil Denitrification in Sealed Soil-Plant Systems," *Plant Soil* 37:113 (1972).

83. Aoki, M. "Studies on the Behavior of Phosphoric Acid Under Paddy Field Conditions. Part I," *J. Sci. Soil Manure, Tokyo* 15:182 (1941).

84. Khalid, R. A., W. H. Patrick, Jr. and F. J. Peterson. "Relationships between Soil Phosphorus and Rice Yield as Affected by Aerobic and Anaerobic Soil Conditions," *Proc. 16th Rice Tech.* Working Group, 103-104, Lake Charles, Louisiana (1976).

85. Shapiro, R. E. "Effect of Flooding on Availability of Phophorus and Nitrogen," *Soil Sci.* 85:190 (1958).

86. Shapiro, R. E. "Effect of Organic Matter and Flooding on Availability of Soil and Synthetic Phosphates," *Soil Sci.* 85:267 (1958).

87. Khalid, R. A., W. H. Patrick, Jr. and R. D. DeLaune. "Phosphorus Sorption Characterisitcs of Flooded Soils," *Soil Sci. Soc. Am. J.* 41:305 (1977).

88. Kaplan, I. R. and S. C. Rittenberg. "Basin Sedimentation and Diagenesis," in *The Sea*, M. N. Hill, Ed., Vol. 3 (New York: Interscience Publishers, 1963), pp. 583-619.

89. Hesse, P. R. "Phosphorus Fixation in Mangrove Swamp Muds," *Nature* 193:295 (1962).

90. Frink, C. R. "Chemical and Mineralogical Characteristics of Eutrophic Lake Sediments," *Soil Sci. Soc. Am. Proc.* 33:369 (1969).

91. Sommers, L. E., R. F. Harris, J. D. H. Williams, D. E. Armstrong and J. K. Syers. "Determination of Total Organic Phosphorus in Lake Sediments," *Limnol. Oceanogr.* 15:301 (1970).

92. Syers, J. K., R. F. Harris and D. E. Armstrong. "Phosphate Chemistry in Lake Sediments," *J. Environ. Qual.* 2:1 (1973).

93. Chang, S. C. and M. L. Jackson. "Fractionation of Soil Phosphorus," *Soil Sci.* 84:133 (1957).

94. Bromfield, S. M. "Phosphate Sorbing Sites in Acid Soils. I. An Examination of the Use of Ammonium Fluoride as a Selective Extractant for Aluminum-Bound Phosphate in Soils," *Aust. J. Soil Res.* 5:93 (1967).

95. Khin, A. and G. W. Leeper. "Modifications in Chang and Jackson's Procedure for Fractionating Soil Phosphorus," *Agrochim.* 4:246 (1960).

96. Smith, A. N. "Distinction between Iron and Aluminum Phosphate in Chang and Jackson's Procedure for Fractionating Inorganic Soil Phosphorus," *Agrochim.* 9:162 (1965).

97. Williams, J. D. H., J. K. Syers, D. E. Armstrong and R. F. Harris. "Characterization of Inorganic Phosphate in Noncalcareous Lake Sediments," *Soil Sci. Soc. Am. Proc.* 35:250 (1971).

98. Williams, J. D. H., J. K. Syers, R. F. Harris and D. E. Armstrong. "Fractionation of Inorganic Phosphate in Calcareous Lake Sediments," *Soil Sci. Soc. Am. Proc.* 35:250 (1971).

99. Patrick, W. H., Jr. and R. A. Khalid. "Phosphate Release and Sorption by Soils and Sediments: Effect of Aerobic and Anaerobic Conditions," *Science* 186:53 (1974).

100. Bartholomew, R. P. "Changes in the Availability of Phosphorus in Irrigated Rice Soils," *Soil Sci.* 31:209 (1931).

101. Paul, H. and W. H. DeLong. "Phosphorus Studies. I. Effects of Flooding on Soil Phosphorus," *Sci. Agric.* 29:137 (1949).

102. Gasser, J. K. R. "The Effects of Anaerobically Fermenting Rice Straw (*Oryza sativa*) on the Mobilization of Aluminum, Calcium, and Iron Phosphates," *Proc. 6th Internat. Cong. Soil Sci.* C:479 (1956).

103. Jackman, R. H. "Accumulation of Organic Matter in Some New Zealand Soils under Permanent Pasture," *N.Z. J. Agric. Res.* 7:472 (1964).

104. Vijayachandran, P. K. and R. D. Harter. "Evaluation of Phosphorus Adsorption by a Cross Section of Soil Types," *Soil Sci.* 119:119 (1975).

105. Patrick, W. H., Jr., S. Gotoh and B. G. Williams. "Strengite Dissolution in Flooded Soils and Sediments," *Science* 179:564 (1973).

106. Tanaka, A., N. Watanabe and Y. Ishizuba. "A Critical Study on the Phosphate Solution of Submerged Soils," *Soil Sci. Plant Nutr.* 16: 182 (Abstracts) (1970).

107. Mahapatra, I. C. and W. H. Patrick, Jr. "Inorganic Phosphate Transformation in Waterlogged Soils," *Soil Sci.* 107:281 (1969).

108. McKeague, J. A. and J. H. Day. "Dithionite- and Oxalate-Extractable Fe and Al as Aids in Differentiating Various Classes of Soils," *Can. J. Soil Sci.* 46:13 (1966).

109. Williams, J. D. H., J. K. Syers, S. S. Shukla, R. E. Harris and D. E. Armstrong. "Levels of Inorganic and Total Phosphorus in Lake Sediments as Related to Other Sediment Parameters," *Environ. Sci. Technol.* 5:1113 (1971).

110. Saunders, W. M. H. "Phosphate Detention by New Zealand Soils and Its Relationship to Free Sesquioxides, Organic Matter, and Other Soil Properties," *N.Z. Agric. Res.* 8:30 (1965).

111. Shukla, S. S., J. K. Syers, J. D. H. Williams, D. E. Armstrong and R. F. Harris. "Sorption of Inorganic Phosphate by Lake Sediments," *Soil Sci. Soc. Am. Proc.* 35:244 (1971).

112. Upchurch, J. B., J. K. Edzwald and C. R. O'Melia. "Phosphate in Sediments of Pamlico Estuary," *Environ. Sci. Technol.* 8:56 (1974).

113. Wentz, D. A. and G. F. Lee. "Sedimentary Phosphorus in Lake Cores—Observations on Depositional Pattern in Lake Mendota," *Environ. Sci. Technol.* 3:754 (1969).

114. Mahapatra, I. C. and W. H. Patrick, Jr. "Evaluation of Phosphate Fertility in Waterlogged Soils," *Proc. Internat. Symp. Soil Fert. Eval.*, New Delhi 1:53 (1971).

115. Postgate, J. R. "Recent Advances in the Study of the Sulfate-Reducing Bacteria," *Bacteriol. Rev.* 29:425 (1965).
116. Baas Becking, L. G. M. and D. Moore. "Biogenic Sulfides," *Econ. Geol.* 56:243 (1961).
117. Engler, R. M. and W. H. Patrick, Jr. "Sulfate Reduction and Sulfide Oxidation in Flooded Soil as Affected by Chemical Oxidants," *Soil Sci. Soc. Am. Proc.* 37:685 (1973).
118. Mandal, L. N. "Levels of Iron and Manganese in Soil Solution and the Growth of Rice in Waterlogged Soils in Relation to the Oxygen Status of Soil Solution," *Soil Sci.* 94:387 (1962).
119. Weeraratna, C. S. "Absorption of Manganese by Rice Under Flooded and Unflooded Conditions," *Plant Soil* 30:121 (1969).
120. Clark, F., D. C. Nearpass and A. W. Specht. "Influence of Organic Additions and Flooding on Iron and Manganese Uptake by Rice," *Agron. J.* 49:586 (1957).
121. Jugsujinda, Aroon. "Growth and Nutrient Uptake by Rice Under Controlled Oxidation-Reduction and pH Conditions in a Flooded Soil," Ph.D. Thesis, Louisiana State University, Baton Rouge, Louisiana (1975).
122. Jones, H. E. and J. R. Etherington. "Comparative Studies of Plant Growth and Distribution in Relation to Waterlogging. 1. The Survival of *Erica cinerea* L. and *E. tetralix* L. and its Apparent Relationship to Iron and Manganese Uptake in Waterlogged Soil," *J. Ecol.* 58:487 (1970).
123. Waid, J. S. "Hydroxamic Acids in Soil Systems," in *Soil Biochemistry*, E. A. Paul and A. D. McLaren, Eds., Vol 4 (New York: Marcel Dekker, 1975), p. 65.
124. Lucas, R. E. and B. D. Knezek. "Climatic and Soil Conditions Promoting Micronutrient Deficiencies in Plants," in *Micronutrients in Agriculture*, R. C. Dinauer, Ed. (Madison, Wisconsin: Soil Science Society of America, Inc., 1972), p. 265.
125. Mikkelsen, D. S. and D. M. Brandon. "Zinc Deficiency in California Rice," *Calif. Agric.* (September 8, 1975).
126. International Rice Research Institute. "Annual Report," IRRI, Los Banos, Philippines (1970).
127. Howeler, R. H. "Iron-Induced Oranging Disease of Rice in Relation to Physico-Chemical Changes in a Flooded Oxisol," *Soil Sci. Soc. Am. Proc.* 37:898 (1973).
128. Ponnamperuma, F. N., R. Bradfield and M. Peech. "Physiological Disease of Rice Attributable to Iron Toxicity," *Nature* 175:265 (1955).
129. Jones, R. "Comparative Studies of Plant Growth and Distribution in Relation to Waterlogging. VIII. The Uptake of Phosphorus by Dune and Dune Slack Plants," *J. Ecol.* 63:109 (1975).
130. Armstrong, W. and D. J. Boatman. "Some Field Observations Relating the Growth of Bog Plants to Conditions of Soil Aeration," *J. Ecol.* 55:101 (1967).
131. Hollis, J. P. "Toxicant Diseases of Rice," Louisiana Agricultural Experiment Station, Louisiana State University, Baton Rouge, Louisiana, Bulletin No. 614 (1967).

132. Ponnamperuma, F. N. "The Chemistry of Submerged Soils in Relation to the Growth and Yield of Rice," Ph.D. Thesis, Cornell University, Ithaca, New York (1955).
133. Vamos, R. "H_2S, the Cause of the Bruzone (Akiochi) Disease of Rice," *Soil Plant Food* 4:37 (1958).
134. Engler, R. M. and W. H. Patrick, Jr. "Stability of Sulfides of Manganese, Iron, Zinc, Copper, and Mercury in Flooded and Non-flooded Soil," *Soil Sci.* 119:217 (1975).
135. Armstrong, W. "Rhizosphere Oxidation in Rice: An Analysis of Intervarietal Differences in Oxygen Flux from the Roots," *Physiol. Plant.* 22:296 (1969).
136. Goto, Y. and K. Tai. "Studies on Oxidizing Power of Roots of Paddy Rice Plant. Part 1. On Difference Among Varieties of Seedling Roots," *J. Sci. Soil Manure* 26:403 (1956).
137. Hollis, J. P., A. I. Allam, G. Pitts, M. M. Joshi and I. K. A. Ibrahim. "Sulfide Diseases of Rice on Iron-Excess Soils," *Acta Phytopathol. Acad. Scient. Hung.* 10:329 (1975).
138. Joshi, M. M., I. K. A. Ibrahim and J. P. Hollis. "Hydrogen Sulfide: Effects on the Physiology of Rice Plants and Relation to Straighthead Disease," *Phytophathol.* 65:1165 (1975).
139. Ford, H. W. "Bacterial Metabolites that Affect Citrus Root Survival in Soils Subject to Flooding," *Proc. Amer. Soc. Hort. Sci.* 86:205 (1965).
140. Mitsui, S. and K. Kumazawa. "Dynamic Studies on the Nutrients Uptake by Crop Plants. Part 41. Nutrient and Redox Conditions," *Soil Sci. Plant Nutr.* 10:227 (Abstracts) (1964).
141. Ponnamperuma, F. N., W. L. Yuan and M. T. M. Nhung. "Manganese Dioxide as a Remedy for a Physiological Disease of Rice Associated with Reduction of the Soil," *Nature* 207:1103 (1965).
142. Yuan, W. L. and F. N. Ponnamperuma. "Chemical Retardation of the Reduction of Flooded Soils and the Growth of Rice," *Plant Soil* 25:347 (1966).
143. Connell, W. E. and W. H. Patrick, Jr. "Reduction of Sulfate to Sulfide in Waterlogged Soil," *Soil Sci. Soc. Am. Proc.* 33:711 (1969).
144. Mitsui, S. "Dynamic Aspects of Nutrient Uptake," in *The Mineral Nutrition of the Rice Plant* (Baltimore, Maryland: John Hopkins Press, 1965), p. 53.
145. Bloomfield, C. "Sulphate Reduction in Waterlogged Soils," *J. Soil Sci.* 20:207 (1969).
146. Goldman, J. C., K. R. Tenore and H. I. Stanley. "Inorganic Nitrogen Removal from Wastewater: Effect on Phytoplankton Growth in Coastal Marine Waters," *Science* 180:955 (1973).
147. Harrison, W. G. and J. E. Hobbie. "Nitrogen Budget of a North Carolina Estuary," Water Resources Research Institute of University of North Carolina, Raleigh, North Carolina, Report No. 86 (1974).
148. Ryther, J. H. and W. M. Dunstan. "Nitrogen, Phosphorus and Eutrophication in the Coastal Marine Environment," *Science* 171:1008 (1971).
149. Weiss, C. M. "The Relative Significance of Phosphorus and Nitrogen as Algae Nutrients," Water Resources Research Institute of University of North Carolina, Chapel Hill, North Carolina, Report No. 34 (1970).

150. Gerloff, G. C. and F. Skoog. "Nitrogen as a Limiting Factor for the Growth of *Microcystis aeruginosa* in Southern Wisconsin Lakes," *Ecology* 38:556 (1957).
151. Mielke, L. N. and J. R. Ellis. "Nitrogen in Soil Cores and Ground Water under Abandoned Cattle Feedlots," *J. Environ. Qual.* 5:71 (1976).
152. Ludwick, A. E., J. O. Reuss and E. J. Langin. "Soil Nitrates Following Four Years Continuous Corn in Irrigated Farm Fields of Central and Eastern Colorado," *J. Environ. Qual.* 5:82 (1976).
153. Bingham, F. T., S. Davis and E. Shade. "Water Relations, Salt Balance and Nitrate Leaching Losses of a 960-Acre Citrus Watershed," *Soil Sci.* 112:410 (1971).
154. Gambrell, R. P., J. W. Gilliam and S. B. Weed. "Denitrification in Subsoils of the North Carolina Coastal Plain as Affected by Soil Drainage," *J. Environ. Qual.* 4:311 (1975).
155. McGarity, J. W. "Denitrification Studies on Some South Australian Soils," *Plant Soil* 14:1 (1961).
156. Pratt, P. F., W. W. Jones and V. E. Hunsaker. "Nitrate in Deep Soil Profiles in Relation to Fertilizer Rates and Leaching Volume," *J. Environ. Qual.* 1:97 (1972).
157. Gambrell, R. P., J. W. Gilliam and S. B. Weed. "Nitrogen Losses from Soils of the North Carolina Coastal Plain," *J. Environ. Qual.* 4:317 (1975).
158. Meek, B. D., L. B. Grass and A. J. MacKenzie. "Applied Nitrogen Losses in Relation to Oxygen Status of Soils," *Soil Sci. Soc. Am. Proc.* 33:575 (1969).
159. Meek, B. D., L. B. Grass, L. S. Willardson and A. J. MacKenzie. "Nitrate Transformations in a Column with a Controlled Water Table," *Soil Sci. Soc. Am. Proc.* 34:235 (1970).
160. Jones, G. D. and P. J. Zwerman. "Rates and Timing of Nitrogen Fertilization in Relation to Nitrate-Nitrogen Outputs and Concentrations in the Water from Intercepted Tile Drains," *Search Agric.* Cornell University Agricultural Experiment Station, 2:1 (1972).
161. Lance, J. C., F. D. Whisler and R. C. Rice. "Maximizing Denitrification during Soil Filtration of Sewage Water," *J. Environ. Qual.* 5:102 (1976).
162. Willardson, L. S., B. D. Meek, L. B. Grass, G. L. Dickey and J. W. Bailey. "Nitrate Reduction with Submerged Drains," *Trans. Am. Soc. Agric. Eng.* 15:84 (1972).
163. Macgregor, A. N. and D. R. Keeney. "Denitrification in Lake Sediments," *Environ. Letters* 5:175 (1973).
164. Engler, R. M. and W. H. Patrick, Jr. "Nitrate Removal from Floodwater Overlying Flooded Soils and Sediments," *J. Environ. Qual.* 3:409 (1974).
165. Engler, R. M., D. A. Antie and W. H. Patrick, Jr. "Effect of Dissolved Oxygen on Redox Potential and Nitrate Removal in Flooded Swamp and Marsh Soils," *J. Environ. Qual.* 5:230 (1976).
166. Keeney, D. R. "The Nitrogen Cycle in Sediment-Water Systems," *J. Environ. Qual.* 2:15 (1973).

167. Guenzi, W. D. and W. E. Beard. "Anaerobic Conversion of DDT to DDD and Aerobic Stability of DDT in Soil," *Soil Sci. Soc. Am. Proc.* 32:522 (1968).

168. Guenzi, W. D., W. E. Beard and F. G. Viets, Jr. "Influence of Soil Treatment on Persistence of Six Chlorinated Hydrocarbon Insecticides in the Field," *Soil Sci. Soc. Am. Proc.* 35:910 (1971).

169. Parr, J. F., G. H. Willis and S. Smith. "Soil Anaerobiosis: II. Effect of Selected Environments and Energy Sources on the Degradation of DDT," *Soil Sci.* 110:306 (1970).

170. Willis, G. H., R. C. Wander and L. M. Southwick. "Degradation of Trifluralin in Soil Suspensions as Related to Redox Potential," *J. Environ. Qual.* 3:262 (1974).

171. Spencer, W. F., M. M. Cliath, W. J. Farmer and R. A. Shepherd. "Volatility of DDT Residues in Soil as Affected by Flooding and Organic Matter Applications," *J. Environ. Qual.* 3:126 (1974).

172. Farmer, W. J., W. F. Spencer, R. A. Shepherd and M. M. Cliath. "Effect of Flooding and Organic Matter Applications on DDT Residues in Soil," *J. Environ. Qual.* 3:126 (1974).

173. MacRae, I. C., K. Raghu and T. F. Castro. "Persistence and Biodegradation of Four Common Isomers of Benzene Hexachloride in Submerged Soils," *J. Agric. Food Chem.* 15:911 (1967).

174. Helling, C. S. "Dinitroaniline Herbicides in Soils," *J. Environ. Qual.* 5:1 (1976).

175. Parr, J. F. and S. Smith. "Degradation of Toxaphene in Selected Anaerobic Soil Environments," *Soil Sci.* 121:52 (1976).

176. Sethunathan, N. "Diazinon Degradation in Submerged Soil and Rice-Paddy Water," in *Fate of Pesticides in the Aquatic Environments,* Adv. Chem. Series No. 111, R. F. Gould, Ed. (Washington, D. C.: American Chemical Society, 1972), p. 244.

177. Kearney, P. C., D. D. Kaufman and M. Alexander. "Biochemistry of Herbicide Decomposition in Soils," in *Soil Biochemistry*, A. D. McLaren and G. H. Peterson, Eds. (New York: Marcel Dekker, 1967), p. 318.

178. Fries, G. F. "Degradation of Chlorinated Hydrocarbons under Anaerobic Conditions," in *Fate of Pesticides in the Aquatic Environment,* Adv. Chem. Series No. 111, R. F. Gould, Ed. (Washington, D. C.: American Chemical Society, 1972), p. 256.

179. Probst, G. W., G. Tomasz, R. J. Herberg, F. J. Holzer, S. J. Parka, C. van der Schans and J. B. Tepe. "Fate of Trifluralin in Soils and Plants," *J. Agric. Food Chem.* 15:592 (1967).

180. Pionke, H. B. and G. Chesters. "Pesticide-Sediment-Water Interactions," *J. Environ. Qual.* 2:29 (1973).

181. Furr, A. K., A. W. Lawrence, S. S. C. Tong, M. C. Grandolfo, R. A. Hofstader, C. A. Bache, W. H. Gutenmann and D. J. Lisk. "Multielement and Chlorinated Hydrocarbon Analysis of Municipal Sewage Sludges of American Cities," *Environ. Sci. Technol.* 10:683 (1976).

182. Page, A. L. "Fate and Effects of Trace Elements in Sewage Sludge when Applied to Agricultural Lands—A Literature Review Study," U.S. Environmental Protection Agency, National Environmental Research Center, Cincinnati, Ohio, EPA-670/2-74-005 (1974).

183. Lee, G. F. "Dredged Materials Research Problems and Progress," *Environ. Sci. Technol.* 10:334 (1976).
184. Smith, D. D. "New Federal Regulations for Dredged and Fill Material," *Environ. Sci. Technol.* 10:328 (1976).
185. Lance, J. C., F. D. Whisler and H. Bouwer. "Oxygen Utilization in Soils Flooded with Sewage Water," *J. Environ. Qual.* 2:345 (1973).
186. Whisler, F. D., J. C. Lance and R. S. Linebarger. "Redox Potentials in Soil Columns Intermittently Flooded with Sewage Water," *J. Environ. Qual.* 3:68 (1974).
187. Krauskopf, K. P. "Factors Controlling the Concentration of Thirteen Rare Metals in Sea-Water," *Geochim. Cosmochim. Acta.* 9:1 (1956).
188. Lee, G. F. "Role of Hydrous Metal Oxides in the Transport of Heavy Metals in the Environment," *Prog. Water Technol.* 17:137 (1975).
189. Lockwood, R. A. and K. Y. Chen. "Adsorption of Hg (II) by Hydrous Manganese Oxides," *Environ. Sci. Technol.* 7:1028 (1973).
190. Jenne, E. A. "Controls of Mn, Fe, Co, Ni, Cu, and Zn Concentrations in Soils and Water: The Significant Role of Hydrous Mn and Fe Oxides," in *Trace Inorganics in Water,* Adv. Chem. Series 73, R. A. Baker, Ed. (Washington, D. C.: American Chemical Society, 1968), p. 337.
191. Taylor, R. M. and R. M. McKenzie. "The Association of Trace Elements with Manganese Minerals in Australian Soils," *Aust. J. Soil Res.* 4:29 (1966).
192. Leland, H. V., S. S. Shukla and N. F. Shimp. "Factors Affecting Distribution of Lead and Other Trace Elements in Sediments of Southern Lake Michigan," in *Trace Metals and Metal-Organic Interactions in Natural Waters,* P. C. Singer, Ed. (Ann Arbor, Michigan: Ann Arbor Science Publishers, Inc., 1973), p. 89.
193. Schnitzer, M. and S. I. M. Skinner. "Organo-Metallic Interactions in Soils: 5. Stability Constants of Cu^{++}-, Fe^{++}-, and Zn^{++}-Fulvic Acid Complexes," *Soil Sci.* 102:361 (1966).
194. Schnitzer, M. and S. I. M. Skinner. "Organo-Metallic Interactions in Soils: 7. Stability Constants of Pb^{++}-, Ni^{++}-, Mn^{++}-, Co^{++}-, Ca^{++}-, and Mg^{++}-Fulvic Acid Complexes," *Soil Sci.* 103:247 (1967).
195. Stevenson, F. J. and M. S. Ardakani. "Organic Matter Reactions Involving Micronutrients in Soils," in *Micronutrients in Agriculture,* R. C. Dinauer, Ed. (Madison, Wisconsin: Soil Science Society of America Inc., 1972), p. 79
196. Morel, F. M. M., J. C. Westall, C. R. O'Melia and J. J. Morgan. "Fate of Trace Metals in Los Angeles County Wastewater Discharge," *Environ. Sci. Technol.* 9:756 (1975).
197. Sanchez, I. and G. F. Lee. "Sorption of Copper on Lake Monona Sediments—Effect of NTA on Copper Release from Sediments," *Water Resources Res.* 7:587 (1973).
198. Gambrell, R. P., R. A. Khalid, M. G. Verloo and W. H. Patrick, Jr. "Transformation of Heavy Metals and Plant Nutrients in Dredged Sediments as Affected by Oxidation-Reduction Potential and pH. Part II. Materials and Methods, Results and Discussion," Contract Report #DACW-39-74-C-0076, Office of Dredged Material Research, U.S. Army Engineer Waterways Experiment Station, Vicksburg, Mississippi (1977).

199. Reddy, C. N. and W. H. Patrick, Jr. "Effect of Redox Potential and pH on the Uptake of Cd and Pb by Rice Plants," *J. Environ. Qual.* 6:260 (1977).
200. Gambrell, R. P., R. A. Khalid, V. R. Collard, C. N. Reddy and W. H. Patrick, Jr. "The Effect of pH and Redox Potential on Heavy Metal Chemistry in Sediment-Water Systems Affecting Toxic Metal Bioavailability," in *Dredging, Environmental Effects and Technology* (San Pedro, California: World Dredging Conference, 1976), p. 579.

14

ISOLATION AND IDENTIFICATION OF ALCOHOL DEHYDROGENASES (ADH) FROM GERMINATING SEEDS

Sylva Leblová

Department of Biochemistry
Charles University
Prague, Czechoslovakia

ALCOHOL DEHYDROGENASE (ADH)

Alcohol dehydrogenase was determined in a number of germinating seeds of mono- and dicotyledonous plants with glycides, fats and proteins as reserve substances (Figure 1). The ADH activity increased during the first hours of germination in dependence on the cultivation conditions (the amount of water in the medium, temperature, number of germinating seeds, volume of the cultivation vessel[1,2]), and attained a maximum characteristic of the plant as far as the time and activity value were concerned; then the enzyme activity decreased. It was interesting that the maximum activity was reached sooner than the maximum substrate, *i.e.*, ethanol concentration (see Chapter 6, Figures 1 and 5).

The ADH activity was measured as the increment in the absorbance at 340 nm in a medium of 0.1 M sodium phosphate buffer at pH 8.5 with 0.01 M mercaptoethanol, 100 mM ethanol, and 860 μm NAD at 20°C. The activity unit was the amount of enzyme causing an increase in the absorbance of 0.001 in 1 minute.

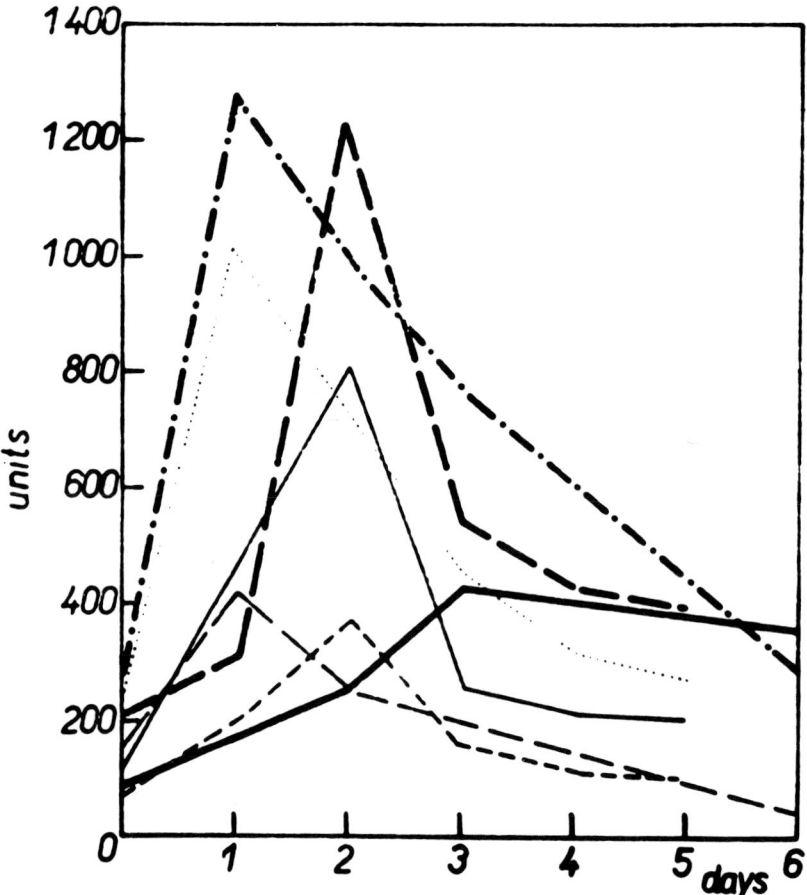

Figure 1.   The specific activity of alcohol dehydrogenases prepared from plants of different ages. Abscissa: age of plant in days; ordinate: specific activity of ADH in units. ——————— peas; ————— wheat; ---------- rape; - - - - - - - lentils; ················· rye; —·—·—·—·— barley; and — — — — — sunflowers.

ISOLATION OF ALCOHOL DEHYDROGENASE
FROM GERMINATING SEEDS

Alcohol dehydrogenase was isolated from plants with protein as reserve substances (broad bean, *Vicia faba*; lentil, *Lens esculenta* Moench; bean, *Phaseolus vulgaris* L.; pea, *Pisum sativum*; soya bean, *Glycine max.*), from plants with glycides in the seeds (maize, *Zea mays* L.; cucumber, *Cucumis sativus* L.; wheat, *Triticum aestivum* L. s.s.; rice, *Oryza sativa* L.), and from fatty seeds (rape, *Brassica napus* L.; sunflower, *Helianthus annuus* L.) during the first three days of germination after swelling time when enzyme activity was highest.

Isolation Procedure

Forty grams of germinating seeds were homogenized with 60 ml of 0.1 *M* sodium phosphate buffer at pH 8.5. After centrifugation at 10,000 g, the supernatant was fractionated with ammonium sulfate. The active precipitate was desalted with a G-25 Sephadex column using 0.01 *M* *tris*-acetate buffer at pH 6.4. This process was followed by chromatography on a column of DEAE cellulose, DE-32, using a *tris*-acetate buffer gradient in the 0.01-0.6 *M* range. After lyophilization, the five most active fractions were subjected to gel filtration on Sephadex G-150. All buffer solutions employed contained 0.01 *M* mercaptoethanol, and all processes were carried out at 4°C to avoid inactivation of the enzyme by polymerization. The percentage saturation with ammonium sulfate was between 30 and 60 and differed for each kind of plant. The results obtained in the isolation of several ADH's are given in Table I.

Table I. Specific Activity of ADH Isolated from Germinating Seeds of Several Plant Species

	Species						
Fraction	Pea	Kidney Bean	Broad Bean	Lentil	Rape	Maize	Rice
	Specific Activity $(10^{-3}$ units mg$^{-1})$						
NaPi Extract	0.92	0.36	0.15	0.22	0.11	0.54	1.65
Sulfate Fraction	5.21	3.75	4.75	2.50	0.49	3.40	8.15
Desalted Sulfate Fraction	5.20	–	4.65	2.48	0.55	10.77	11.42
Fraction after DEAE-Cellulose	95.50	16.70	33.30	16.60	4.71	31.63	98.38
Fraction after Gel Filtration	160.00	27.90	50.00	27.00	9.42	27.11	98.45

Activity of Isolated Plant Alcohol Dehydrogenases

The activities of all 11 isolated ADH's are given in Table II. By the purification procedure, the specific activity is increased compared with the initial extract, *e.g.*, 334 times for broad bean, 122 times for lentil, 77 times for bean, 80 times for rape, and 90 times for sunflower and rice. ADH was isolated from seedlings, *e.g.*, of pea, by Cossins *et al.*[3]. The activity increased 25 times compared with the initial extract. Pattee and Swaisgood[4] purified ADH about 20 times from peanut, and Davies[5,6] about 88 times from potato tubers. The most successful isolation so far is that from tea seeds, where the activity increased 1500 times on introduction of electrophoresis.[7,8]

Table II. Properties of Isolated ADH's

ADH from	Specific Activity (units/mg)	MV ±5000	Optimum pH	K_m(NAD) (0.10^{-4} M)	K_m(ethanol) (0.10^{-2} M)
			Parameters		
Broad Bean	50,000	60,000	8.7	1.1	1.3
Lentil	27,000	70,000	8.2	0.8	1.0
Bean	27,900	63,000	8.2	1.0	0.7
Pea	160,000	60,000	8.7	1.5	2.0
Soybean	43,300	53,000	8.7	1.1	1.3
Maize	31,625	62,000	8.7	1.1	1.8
Cucumber	38,000	57,000	8.7	1.1	1.1
Wheat	25,400	66,000	8.5	1.5	2.2
Rice	98,450	80,000	8.5	1.1	0.8
Rape	9,420	66,000	8.5	2.0	2.5
Sunflower	18,400	60,000	8.8	0.8	0.8
Yeast	–	150,000	7.9	1.4	1.7
Horse Liver	–	80,000	8.0	0.1	0.5

Stability of Enzyme Preparations

Effect of ADH Substrates and of SH-Substances on the Stability

The stability of the enzyme preparation in the incubation at 0°C was not increased by addition of 1% methanol or ethanol. Mercaptoethanol in a concentration of 5 mM and 10 mM cysteine exerted a protective effect. High concentrations of SH-substances were required for maintaining essential SH-groups in the native state.[9]

ADH Stability During Purification

During ADH isolation the sulfate fraction was inactivated relatively rapidly, but the enzyme obtained by the isolation procedure, including gel filtration and made more viscous by lyophilization, was stable for several weeks. The ADH stabilities during isolation were similar to various plant ADH's.

Thermostability

ADH's are thermolabile.[10] The enzyme was inactivated at termperatures above 60°C, and the thermal stability was not improved by addition of 0.1 mM or 0.25 mM NAD; but 0.4 mM, 0.8 mM and 1.5 mM NAD has a strong protective effect. Ethanol exhibited a mild protective effect. Plant ADH's behave differently from the liver enzyme [11-13] as far as the effect of NAD and ethanol on the thermal stability was concerned. The loss of activity was governed by first-order kinetics.

KINETIC DATA ON PLANT ADH's

Effect of pH on the Activity

Rate of Oxidation

The rate of oxidation of alcohols was monitored in a pH range of 4-10 in a *tris*-acetate or phosphate buffer, and it was found that the optimum value was located between 8.2 and 8.8 (Table II). The optimum pH was identical for ethanol and allyl alcohol.

Rate of Reduction

The rate of the reduction of aldehydes was highest in a pH region around 7. According to Eriksson,[14,15] the optimum pH for pea was 6.9, according to Cossins *et al.*[3] 6.3, and according to Suzuki,[16,17] 7-7.3; different workers employed different buffering systems.

Michaelis Constants

K_m Values for Oxidation

It was found that the K_m values are of the same order of magnitude: 10^{-2} M for the oxidation of ethanol and 10^{-4} M for the reduction of NAD (Table II). Similar values were obtained by Cossins *et al.*[3] for the pea enzyme.

K_m Values for Reduction

The K_m value for the reduction of aldehyde was measured for acetaldehyde ($10^{-3}$$M$), and for NADH ($10^{-5}$$M$), as substrates. Except for rice ADH, the Michaelis constants were lower for the substrate reduction than those for its oxidation. Suzuki[16,17] determined a K_m value lower than shown in Table III for pea ADH, namely, $4.3 \times 10^{-4}$$M$. It appears that the pH at which the determination was carried out and the temperature may play a role with different K_m values (Tables III and IV).

Table III. The Michaelis Constant-Temperature Dependence for Pea ADH (pH 8.5)

t (°C)	K_m(NAD) (mM)	K_m(ethanol) (mM)
21	0.15	30
32	0.21	42
41	0.29	51

Table IV. The pH Dependence of the Michaelis Constant for Pea ADH

	pH 7.4	pH 8.5
K_m(NAD) μM	200	150
K_m(NADH) μM	160	200
K_m(ethanol) mM	40	30
K_m(acetaldehyde) mM	4	6

Effect of Temperature and pH on K_m

The magnitudes of the Michaelis constants were determined in relation to the temperature over the range of 21-41°C, and it was found that they increased with temperature (Table III). The K_m versus pH dependence is given in Table IV.

Substrate Specificity

Oxidation of Alcohols

Plant ADH oxidizes many other alcohols in addition to ethanol.[2,3,10,14,15,18,19] Eriksson's experimental results for pea ADH are given in Table V. Differences in the substrate specificity for ADH's from broad bean, lentil, bean, soybean, maize, cucumber, wheat, rice, rape and sunflower are evident from Table VI.[20-27] The purity of the substrates used was controlled by gas chromatography. The individual ADH's differed in their relative oxidation rates for most substrates and sometimes qualitative differences were encountered also. Nonetheless, it generally holds that:

1. The rate of substrate oxidation decreased with an increasing number of carbons in the chain.

Table V. Relative Reaction Rates Obtained with Alcohol: NAD Oxidoreductase from Pea Seeds and Alcohols at Two Different Concentrations (NAD) = 5×10^{-4} M, pH = 8.5. Specific Activity of Enzyme (Ethanol Oxidation) = 30 μmol min^{-4} g Dry Weight (Ethanol) = 0.001 M

Alcohol	Alcohol (0.01 M)	Concentration (0.001 M)	Saturated[a]
Methanol	0.00	0.00	
Ethanol	1.00	1.00	
Propan-1-ol	0.18		
2-propen-1-ol (allyl alcohol)	1.41		
2-propyn-1-ol (propargyl alcohol)	0.00		
Butan-1-ol	0.27	0.22	
Trans 2-buten-1-ol	0.55	0.49	
Cis-2-buten-1-ol	0.20		
2-methyl-2-propen-1-ol	0.02		
Pentan-1-ol	0.06		
3-methyl-butan-1-ol	0.01		
3-methyl-2-buten-1-ol	0.46		
Hexan-1-ol		0.12	
4-methyl-pentan-1-ol		0.00	
Trans-2-hexen-1-ol		2.93	
Trans-3-hexen-1-ol		low	
Cis-3-hexen-1-ol		0.15	
5-hexen-1-ol		low	
Trans, trans-2,4-hexadien-1-ol		2.43	
Heptan-1-ol			0.33
Octan-1-ol			0.21
Nonan-1-ol			0.18
3-phenyl-2-propen-1-ol (cinamic alcohol)		0.22	

[a]Compared with the rate of ethanol at 0.001 M.

2. Unsaturated analogues were oxidized more rapidly than saturated; however, propargyl alcohol was not oxidized. An unsaturated bond far from a functional group decreased the reaction rate nonsubstantially, as can be seen for the hexacarbon alcohols in Eriksson's table.

3. Methyl branching caused a decrease in the reaction rate (3-methylbutan-1-ol, 4-methylpentan-1-ol, 2-methyl-2-propen-1-ol, where the methyl removes the effect of the double bond). On the other hand, the double bond had a positive effect with the couple 3-methylbutan-1-ol and 3-methyl-2-buten-1-ol, because the configuration on the C2 carbon was important; this was also verified by the finding that amino- and methoxy-derivatives of an alcohol in position 2 were not oxidized by alcohol dehydrogenase.

4. Secondary alcohols, diols, and terpenic, carbohydrate, cyclic and aromatic alcohols were not ADH substrates. An exception was formed by cinnamyl alcohol, in which the allyl alcohol chain apparently plays a role. Potato tubers ADH were also an exception, as

Table VI. Substrate Specificity of Plant Alcohol Dehydrogenases (the numbers in the table denote the relative oxidation rates. Alcohol concentration was 0.01 M, except for hexanol, isooctanol, cyclohexanol and cinnamyl alcohol, which were used as saturated aqueous solutions)

Relative Rate of Oxidation of ADH

Substrate	Beans	Broad Beans	Peas	Lentils	Rape	Sunflowers	Cucumbers	Maize	Rice	Wheat
Methanol	0	0	2	0	0	0	0	0	0	0
Ethanol	100	100	100	100	100	100	100	100	100	100
n-Propanol	39	46	44	55	60	90	52	8	19	40
2-Propen-1-ol	171	156	160	152	115	130	170	119	133	290
n-Butanol	19	32	30	30	54	75	26	6	–	30
2-Buten-1-ol	34	49	49	54	62	82	45	–	11	43
Isobutanol	0	6	10	0	–	–	–	–	16	–
Isoamyl alcohol	0	2	15	14	0	5	10	0	24	–
4-Penten-1-ol	3	8	10	8	–	0	8	5	19	9
n-Hexanol	5	16	12	20	25	12	20	0	0	0
Cyclohexanol	0	0	0	0	0	0	0	0	0	0
Isooctanol	0	0	0	0	0	0	0	0	0	0
Cinnamyl alcohol	10	13	16	16	12	10	10	10	–	21

2-butanol,1,3-butandiol,2-buten-1,4-diol,cyclohexanol,isooctanol,mercaptoethanol,2-methoxyethanol,2-aminoethanol,carbohydrate and terpenic alcohols were not attached at all, as well as propargyl alcohol and benzyl alcohol.

they were active toward terpenic and aromatic alcohols;[5,6] however, our review of alcohol dehydrogenases dealt only with the enzymes present in seedlings.

Reduction of Aldehydes

Similar to alcohols, ADH did not exhibit clear-cut specificity toward aldehydes. So far, we have not had at our disposal as wide a range of substrates as was tested by Eriksson[15] for pea ADH. We found that alcohol dehydrogenases reduced acetaldehyde, propanal, and butanol with similar Michaelis constants (*e.g.*, for the rape enzyme this constant had the values 1.3, 3.3 and 1.0, respectively, for the pea enzyme, 4.0, 4.7 and 4.8 x 10^{-3} M, respectively.[28,29] Isobutonal was oxidized more slowly; pivalaldehyde, cinnamylaldehyde, and benzaldehyde were not substrates of the studied ADH's.

ADH STRUCTURAL PROPERTIES

Molecular Weight of Plant ADH's

We determined the molecular weight by means of:

1. Gel filtration on a column of Sephadex G-200, with dimensions of 2.6 x 18 cm. The standards employed were albumin (M.W.67,000), ovalbumin (M.W. 45,000), myoglobin (M.W. 17,000), and gamma globulin (M.W. 157,000).
2. By electrophoresis on polyacrylamide gels with various concentrations of BIS and polyacrylamide. We applied the Hedrick and Smith method.[30] Myoglobin (17,000), albumin (67,000), ghymotrypsinogen A (25,000) and aldolase (147,000) were used as standards.

The molecular weights of the 11 ADH's studied were determined with a precision of ±5,000 and varied within 53,000-80,000 (Table II). Among plant ADH's, the molecular weight has so far been determined only for the pea enzyme[3] –60,000, in agreement with our result, and for the ADH from tea seeds and peanut seeds; the molecular weights of the latter were above 100,000.[4,7,8]

Zinc in the Plant ADH Molecule

So far, zinc has been determined only in the enzyme isolated from peanuts by Pattee and Swaisgood[4] (1.5 g-atoms per 112,000 g of proteins),and for the pea enzyme by Cossins *et al.*[3] (10 μg per mg of proteins). According to Fischer and Schwartz,[31] zinc is the metal in the maize ADH molecule: the addition of Zn during dialysis reassociation caused a significant increase in the enzyme activity.

Zinc ions apparently participate in the catalysis, similar to YADH and LADH.

The enzyme activity is substantially decreased by chelating agents. We tested the effect of α,α'-dipyridyl, o-phenanthroline, sodium azide, cupral, ferron, and salicylaldoxime on all the ADH's studied. We followed the inhibitive effect on the enzyme alone, the effect after preincubation of the enzyme with the inhibitor and the effect of inhibitors on binary enzyme-coenzyme and enzyme-ethanol complexes. While preincubation of the enzyme with the inhibitor mostly increased the inhibitive effect, binding of the enzyme with the coenzyme protected against inhibition, rather than enzyme combination with the substrate. Sensitivities of the individual ADH's toward inhibitors were not identical, and there were certain differences in the behavior during preincubation of the enzyme with the inhibitor and in the effect of inhibitors on the binary complexes, as follows from Table VII.

Table VII. The Effect of Some Inhibitors on the Oxidation of Ethanol by Plant ADH (the values in the table indicate % inhibition. I+S+NAD: effect of inhibitor with preceding incubation of the substrate or with NAD; I+E: enzyme preincubated 5 min with the inhibitor; E+NAD: enzyme preincubated 5 min with NAD; E+S: enzyme preincubated 5 min with ethanol)

Enzyme from	Inhibitor	Conc.I (M)	I+S+NAD	I+E	E+NAD	E+S
Bean	Sodium azide	1×10^{-3}	53	60	51	51
	α,α'-dipyridyl	5×10^{-4}	30	45	8	20.5
	o-phenanthroline	1×10^{-3}	58	48	40	50
	Cupral	4×10^{-1}	43	48.3	39	42.1
	Ferron	4×10^{-4}	19.8	24	15	20
	Salicylaldoxime	4×10^{-2}	64.2	67	58.3	61
Rape	Sodium azide	1×10^{-3}	48.3	52	30	31
	α,α'-dipyridyl	5×10^{-4}	44	15	6	44
	o-phenanthroline	1×10^{-3}	60	53	44	45.3
	Cupral	4×10^{-1}	51.2	53	45.1	50.8
	Ferron	4×10^{-4}	18.3	21.5	17	20
	Salicylaldoxime	4×10^{-2}	65.8	66	61	64.7
Wheat	Sodium azide	1×10^{-3}	30.2	48	20	20
	α,α'-dipyridyl	5×10^{-4}	20.3	21	11	12
	o-phenanthroline	1×10^{-3}	55	66	51.2	53
	Cupral	4×10^{-1}	50	55	26	30
	Ferron	4×10^{-4}	16	18	12	16
	Salicylaldoxime	4×10^{-2}	58	61	50	57

Sulfhydryl Groups in the Plant ADH Molecule

SH-groups were determined in ADH's isolated from germinating seeds using the Ellman reagent (5,5'-dithiobis-2-nitrobenoic acid), yielding a yellow reaction product with SH-groups.[32] The number of SH-groups determined by this method was low (5-7 per ADH molecule), because of the low degree of purification; this may be considered only as a qualitative detection.

SH-groups participate in process catalysis with plant alcohol dehydrogenases, as substances known as SH-poisons decrease the activity. The effect was found with pea, maize, wheat, rice, bean, barley and tea.[3,7,8,10,17,20-27,33-36]

MECHANISM

Inhibition Studies

The Effect of Chelating Agents and Substances
Bound to SH-Groups

The effect of chelating agents is depicted in Table VII, which indicates that ADH's apparently were metalloproteins. Substances known as SH-poisons also inhibited all the ADH's studied (Table VIII). Chloromercuribenzoate, iodacetamide, N-ethylmaleinimide and Na_3AsO_3 were tested. The inhibition effect was alleviated when the effect of these substances was again followed with the binary enzyme-coenzyme or enzyme-substrate complexes, similar to the chelating agents. This protection was more pronounced with ethylmaleinimide or iodoacetate. As the latter two substances are considered to react primarily with structural SH-groups, while chloromercuribenzoate also reacts with active center SH-groups, this suggests participation of SH-groups in the catalysis. The inhibitive effect always increased on preincubation of the inhibitor with the enzyme. Analogously to the effect of SH-poisons, protection against inhibition by these substances was also weakly qualitatively different with individual ADH's.

Table VIII. Effect of Substances Bound to SH-Groups of Rape ADH
(The numbers in the table give percentage inhibition. For symbols, see Table VII.)

Inhibitor	Conc. I (M)	I+NAD+S	I+E	E+NAD	E+S
Iodoacetamide	4×10^{-2}	50	61	33	39
Chloromercuribenzoate	1×10^{-3}	33	40	17	17
n-Ethylmaleimide	10^{-3}	33	40	19	30
Na_3AsO_3	2×10^{-4}	23	47	17	17

Iodoacetate Alkylates SH-Groups

We also found that the alkylation can be prevented by nucleotides. As follows from the inhibition constants given in Table IX, the protective effect increased in the series ATP, ADP, AMP.[37]

Table IX.   Inhibition Constants[a] of Nucleotides for Pea ADH Modified by Iodoacetate

Ligand	K_i μM
ATP	1250
ADP	250
AMP	70
NAD	100

[a]Inhibition constant K_i is the dissociation constant for the enzyme-inhibitor complex.

Phenanthroline protects plant ADH against alkylation, similar to LADH. Phenanthroline is a noncompetitive inhibitor, *i.e.*, it does not prevent ADH interaction with iodoacetate, but it protects ADH against alkylation of its sulfhydryl groups by another mechanism.   The inhibition constant for ADH-phenanthroline is 8 μM.

Chloride ions compete with iodoacetate, with an inhibition constant of 11 mM.   It can be assumed that this interaction is of electrostatic origin, as the protective effect of chloride ions on LADH could be decreased by increasing the ionic strength.[37]

Chloride ions and AMP are competitive inhibitors with respect to iodo-acetate.   If they act simultaneously, with variable AMP concentration, competition with a high interaction constant α is encountered.[38]   The value of interaction constant α equals 10, so that a) the two inhibitors are repulsed, which can be explained by the negative charge on the two ligands; and b) each inhibitor is bound to a different site on ADH.

In the simultaneous effect of AMP and phenanthroline, the former being an inhibitor competitive with respect to iodoacetate and the latter being noncompetitive, it was found, by following the inhibition with variable amounts of AMP and a constant amount of phenanthroline, that both ligands were noncompetitive.[37]

Intermediates of the Carbohydrate Metabolism

From carbohydrate metabolism intermediates, pyruvate, lactate, isocitrate, malate, succinate, and acetate were tested (Table X); their effect on plant ADH is interesting for two reasons:

Table X. The Effect of Carbohydrate Metabolism Intermediates on the Rate of Ethanol Oxidation by Alcohol Dehydrogenase (The numbers in the table are relative rates for 0.1 M ethanol and 0.1 M inhibitor concentrations.)

ADH from	Ethanol	Ethanol + Lactate	Ethanol + Malate	Ethanol + Pyruvate	Ethanol + Acetate	Ethanol + Succinate	Ethanol + Isocitrate
Broad bean	100	48	31	61	42	–	–
Lentil	100	75	11	100	14	63	–
Bean	100	39	22	92	40	29	105
Pea	100	16	5	28	17	31	105
Soya	100	62	34	68	21	52	97
Maize	100	92	15	110	19	37	–
Cucumber	100	51	25	89	35	47	81
Wheat	100	47	31	63	46	–	–
Rice	100	19	6	73	13	29	102
Rape	100	38	26	51	36	25	–
Sunflower	100	42	19	85	37	32	–
Liver[44]	100	99	0	0	100	125	–

1. It is well known that the concentrations of many acids change during anaerobiosis.[39-42] If these acids affect the ADH activity, they may carry out a regulation function in the anaerobic metabolism of plant cells.
2. Some of the acids may function as shuttles in the transfer of hydrogen from extramitochondrial NADH into mitochondria.[43]

All the substances studied, except for isocitrate, were found to inhibit plant ADH's. The percentage inhibition is given in Table X. The ethanol oxidation rate was most decreased by malate, less by acetate and lactate, and least by pyruvate, with the exception of pea and rice ADH's. All tested carbohydrate metabolism intermediates were noncompetitive inhibitors with respect to ethanol.

Amides and Oximes

The effect of two amides and two oximes on the rates of ethanol oxidation and acetaldehyde reduction by plant alcohol dehydrogenases was measured. Oximes decreased the plant ADH activity more than amides, and the reduction of acetaldehyde was inhibited more effectively than the oxidation of ethanol. Again the effect of preincubation of the inhibitor with the enzyme and the inhibitor effect on the binary complexes of the enzyme with the coenzyme or the substrate for the amides and oximes was followed, and it was found that the inhibitive effect increased for all 11 ADH's studied on enzyme preincubation with the inhibitor. The protective effect of the binary enzyme-coenzyme complex was more pronounced only in a few cases, and the enzyme-substrate complex did not affect the inhibition. Therefore, the formation of ternary complex E-NADH-I and competition for a binding site with acetaldehyde can be assumed for plants as with animals: the test substances, except for acetoxime, act as inhibitors noncompetitive with respect to ethanol and competitive with respect to acetaldehyde. The inhibition constants were of the order of 10^{-2} M for acetamide and butyrylamide and 10^{-3} M for acetoxime and cyclohexanone oxime.

Dimethylsulfoxide

Dimethylsulfoxide is used as an antialcoholic medicine.[44] This substance is an effective inhibitor of plant ADH's, noncompetitive with respect to ethanol and competitive with respect to acetaldehyde.

Heterocycles

Pyridine, pyrazole, and imidazole were found to inhibit plant alcohol dehydrogenase, similar to their inhibition of LADH. For the sake of comparison, the terminology from Theorell's works[45] for classification of inhibition types was employed.

Two inhibition types are generally differentiated in enzyme kinetics, competitive and noncompetitive, as has already been mentioned. Competitive inhibitors are bound at the enzyme active center and compete with the true substrate; they decrease the enzyme affinity toward the substrate and thus increase the effective value of the Michaelis constant. Noncompetitive inhibitors are bound outside the active center and hence do not compete with the substrate for the binding site, but they decrease the reaction rate without altering the Michaelis constant. There can be more complex situations; *e.g.*, an inhibitor may be bound close to the active center and hence partially competes with the substrate for the binding site (is competitive), but simultaneously it also decreases the V_{max} value. The inhibitor is then simultaneously competitive and noncompetitive.

Theorell and co-workers[45] found that liver alcohol dehydrogenase (LADH) forms ternary complexes EOI and ERI with many inhibitors. In complex EOI, the inhibitor occupies the ethanol binding site; in complex ERI it occupies the aldehyde binding site. The inhibitor effect can be schematically represented by the following reactions:

1. $E \;+\; O \rightleftharpoons EO$ E = enzyme

2. $EC \rightarrow +$ ethanol $\rightleftharpoons ER$ + aldehyde + H^+ O = NAD
       $+$ $I \rightleftharpoons EOI$ R = NADH

3. $ER \longrightarrow E + R$
       $+$ $I \rightleftharpoons ERI \rightarrow E + R + I$ I = inhibitor

If an inhibitor forms both complexes and the rate of decomposition of complex ERI to E + R + I is greater than that for decomposition of binary complex ER, then from a certain ethanol concentration on the ethanol, oxidation is not inhibited, but activated, by substance I. This inhibition type is termed by Theorell[45] as CIS. If the decomposition rate for ER is the same as that for ERI, and the complex stability is unchanged or ERI is not formed, then the inhibitor is competitive with respect to ethanol. Inhibitors forming complexes ERI and EOI have inhibition type M (mixed-competitive) with respect to alcohol, N (noncompetitive), or U-N (un-noncompetitive), depending on the increasing stability of ERI; those not forming EOI, but only ERI, exhibit U-type (uncompetitive) inhibition.

The effect of the studied heterocycles on the four alcohol dehydrogenases, together with the data on LADH, is illustrated in Table XI. Imidazole exhibited CIS-type inhibition toward all plant ADH's. The inhibition constants for lentils and rape were somewhat lower than those for the other

Table XL. Inhibition Constants of Alcohol Dehydrogenases Measured at pH 8.5 with Either 40-100 mM Ethanol and 860 μM NAD (K_{iEtOH}) or 100 mM Ethanol and 250-860 μM NAD (K_{iNAD}) (Values in mM except for pyrazole (μM). Type of inhibition: CIS, competitive inhibition and stimulation; C, competitive inhibition; M, mixed-competitive; U-N, un-noncompetitive; U, uncompetitive.)

Inhibition	Inhibitor	Pea ADH		Bean ADH		Lentil ADH		Rape ADH		LADH[45]	
		K_{iNAD}	K_{iEtOH}	K_{iNAD}	K_{iEtOH}	K_{iNAD}	K_{iEtOH}	K_{iNAD}	K_{iEtOH}	K_{iNAD}	K_{iEtOH}
Constant	Imidazole	4	7	3	12	1	1.5	2	1	3.6	7.6
	Pyridine	5	3	5	2	2.5	2	2	1	0.9	2.1
	Pyrazole	–	17	–	16	–	23	–	10	–	0.2
	2-benzyl-4,5-dihydroimidazole · HCl	9	13	9	13	9	9	–	–	–	–
Type	Imidazole	CIS	CIS	CIS	CIS	CIS	CIS	CIS	CIS	CIS	CIS
	Pyridine	C	C	C	C	C	C	C	C	C	C
	Pyrazole	U-N	C	U-N	C	U-N	C	U-N	C	U	C
	2-benzyl-4,5-dihydroimidazole · HCl	M	M	M	M	M	M	–	–	–	–

enzymes. The imidazole effect was very similar to that on LADH; therefore, the plant ADH probably formed ternary complexes of the EOI and ERI types with the coenzyme and imidazole, and complex ERI was decomposed more rapidly to E + R + I than complex ER. If imidazole formed only complex EI, increasing the substrate concentration would cause suppression of the inhibitor effect but not activation of the reaction.

Pyridine is a competitive inhibitor with respect to NAD and ethanol for all four enzymes. However, the K_i value with respect to NAD is higher than that with respect to ethanol, while the reverse is true for LADH. The values of the inhibition constants for the studied ADH's were similar.

2-Benzyl-4,5-dihydroimidazole·HCl exhibited M-type inhibition. The inhibition constants for peas and broad bean were lower with respect to NAD than to ethanol, while the opposite was true for lentils.

Pyrazole exhibits U-N-type inhibition toward NAD, very similar to U-type, for all plant ADH's. Therefore, the inhibition constants could not be calculated. Pyrazole is an inhibitor competitive with respect to ethanol for all plant ADH's and is the strongest inhibitor among those tested on ADH. Pyrazole is also a competitive inhibitor with horse and human LADH, with K_i of 10^{-7} and $10^{-6} M$, respectively.

Hence, only complex EOI is probably formed with pyrazole and pyridine, analogously to LADH, and complexes EOI and ERI are formed with imidazole and benzylhydroimidazole·HCl. Our assumptions concerning the formation of ternary heterocyclic complexes with plant ADH's and the coenzyme were derived from the similarity in the behavior of plant ADH's and LADH toward inhibitors. No direct proof of the existence of these complexes is available, by contrast to the LADH complexes.[46]

Fatty Acids

Fatty acids were tested as inhibitors of LADH.[47-49] With plant alcohol dehydrogenases, the effect of propionic and butyric acids and their mercaptoderivatives was followed. Both fatty acids inhibited the ethanol oxidation competitively and with a well-defined dissociation constant, while toward acetaldehyde they behaved as uncompetitive inhibitors.

The dissociation constants of mercaptopropionic and mercaptobutyric acids are lower than those for the unsubstituted acids: the position of the sulfhydryl group has no effect on the inhibition.

The inhibition of the reduction of acetaldehyde by fatty acids is uncompetitive; hence, fatty acids do not compete with acetaldehyde and do not form the ternary E-NADH-I (ERI) complex, but react only with the product of acetaldehyde reduction, E-NAD.[9,28]

Adenosine Triphosphate

Nucleotides of adenine are inhibitors competing with the coenzyme for LADH.[39,50,51] The same holds for the plant enzymes (Table XII). The inhibitory effect of ATP indicated bonding of the adenosine part of the coenzyme to the enzyme. A similar inhibitive effect was also exhibited by ADP and AMP. The increasing bond strength between ATP and the enzyme due to increasing pH can be explained by the negative charge on ATP. Inhibition constants similar to that for the pea enzyme were also found for other ADH's, and the inhibition type was determined as competitive with respect to NAD and NADH.

Table XII. The Inhibition Constants for Pea ADH and ATP in Dependence on the pH

pH	K_i mM
7.4	12.2
8.5	7.5
10.5	6.2

Chloride Ions

LADH,[52,53] as well as plant ADH's, are inhibited by chloride ions. (Table XIII).

Table XIII. Inhibition Constants for Chloride Ions with Pea and Liver ADH[9]

Substrate	ADH from Pea (K_i mM)	ADH from Liver (K_i mM)
NADH	80	30
NAD	100	60
Ethanol	120	100
Acetaldehyde	100	—

Chloride ions were competitive inhibitors with respect to NAD, NADH and ethanol, and exhibited the mixed-inhibition type with respect to acetaldehyde.

Inhibition of acetaldehyde was mainly due to chloride bonding to the E-NAD complex formed in the reduction of acetaldehyde, thus inhibiting the release of NAD from the enzyme. The enzyme-NADH-acetaldehyde-Cl' complex can also be formed to a certain degree, so that the inhibition was noncompetitive rather than uncompetitive.[9,28]

Berberine Derivatives and Psychopharmaceuticals

The alkaloid berberine, its derivatives and psychopharmaceuticals such as the thioxanthene derivative, chloroprothixene, were tested as effective LADH inhibitors.[54–57] With LADH it was found that the inhibitive effect increased with the hydrophobic nature of the substituents on the carbon in position 13 and that it was highest with 13-ethylberberine. As follows from Table XIV, the substituent on berberine had the same inhibitive effect on the liver and pea enzymes, but the inhibition constants were almost one order of magnitude higher for the plant ADH's than for the animal enzyme. The same was true for chloroprothixene.[58]

Table XIV. Inhibition Constants for Berberine Derivatives and Pharmaceuticals[9]

Inhibitor	$K_{0.5}$ μM Pea ADH	$K_{0.5}$ μM LADH
Berberine	500	120
13-Methylberberine	70	10
13-Ethylberberine	30	1.3
Chloroprothixene	75	3.5

8-Anilino-1-naphthalenesulphonic Acid (ANS)
and Auramine

Conformational changes occur during bonding of the coenzyme to the enzyme: E-NAD crystallizes in another modification than as the enzyme alone.[59] There are proofs that the coenzyme is bound to LADH by its adenosine part and that a certain orientation of the adenosine part of NAD with respect to ribose is required for bond formation in the hydrophobic region of the enzyme. The hydrophobic bonding site of LADH for the adenosine part of NAD enables bonding of various organic substances, among which we tested aminonaphthalenesulphonic acid, that forms a fluorescing complex with LADH,[60] and auramine on plant ADH.

The values of the dissociation constant of ADH complexes with ANS were determined from four parallel titrations of plant ADH with ANS; for pea ADH, this value was 13 μM at pH 7.5.

ANS was displaced from the ANS-enzyme complex by some compounds, such as NAD and NADH and *o*-phenanthroline; pyrazole and mercuric ions also decreased the fluorescence of the binary complex. If the pure enzyme were available, it would be possible to determine the active site concentration using ANS. Auramine also forms a fluorescing complex with plant ADH. This is also analogous to LADH, although the site of the bond with the enzyme is not known.[9]

Phenanthroline

Phenanthroline associates with Zn metalloproteins by forming inactive complexes. We have found that 1.4×10^{-3} M phenanthroline is required for a 50% inhibition of rape ADH in the oxidation of ethanol and 1.5×10^{-3} M phenanthroline in the reduction of acetaldehyde. The inhibition is of the mixed type with respect to ethanol and noncompetitive with respect to acetaldehyde. NAD and NADH are inhibited competitively by phenanthroline, with K_i equal to 1.1×10^{-3} M. The metal apparently plays a role in the coenzyme-enzyme interaction.[61-63]

Kinetic Measurements in the Presence of
Two Inhibitors Competitive with the Coenzyme

If two competitive inhibitors, I_1 and I_2, react with the enzyme, complexes EI_1, EI_2 and EI_1I_2 can be formed. The interaction is described by interaction constant α. If the two inhibitors interact with the same site in the enzyme, then α equals infinity; if they interact with different sites, then it holds that $\infty > 1 > 0$. If $\alpha < 1$, the inhibitors positively affect one another, if $\alpha > 1$, repulsion occurs (Table XV). The interaction constants were determined graphically, according to Yonetani and Theorell.[81]

Table XV. Interaction Constants for Two Inhibitors Competitive
with the Coenzyme by Rape ADH

Inhibitors	Interaction Constant
Phenanthroline-AMP	1
Phenanthroline-ADP	3.3
Phenanthroline-ATP	6.2
Phenanthroline-Chloride Ion	∞
ATP-Chloride Ion	5.5
AMP-ADP	∞
AMP-ATP	∞
ADP-ATP	∞
AMP-Adenine	∞
AMP-Adenosine	∞
Phenanthroline-Nicotinamide	∞
ATP-Nicotinamide	1

Therefore, the bonding sites for Cl^- and phenanthroline are identical, but those for Cl^- and ATP, as well as for phenanthroline and nucleotides, are different. Hence the coenzyme should react with the protein part at two sites, namely, at the metal component of the enzyme and further through the ADP-ribosyl residue. As the bonding site for nicotinamide is identical with that for o-phenanthroline, it can be assumed that the nicotinamide part of the coenzyme reacts with the metal part of the protein molecule. It follows from the kinetic measurements in the presence of nicotinamide and ATP that the two ligands are bound to different sites, independently of one another; hence the nicotinamide part of NAD reacts with the protein part of alcohol dehydrogenase at a site different from that for the ADP-ribosyl part.[61-63]

Kinetics of the Enzyme Catalysis

The kinetics of the redox reaction of horse liver alcohol dehydrogenase and of yeast alcohol dehydrogenase with various alcohols and aldehydes have been studied in detail.[64] It has been shown that the oxidation of ethanol by NAD and the reduction of acetaldehyde by NADH follow the Theorell and Chance mechanism,[65] *i.e.*, NAD or NADH, is bound first to the enzyme and NADH or NAD, dissociates last from the enzyme after completion of the redox reaction. (Abbreviations used: alcohol dehydrogenase-ADH, enzyme-E, NAD-S_1, NADH-S_1', ethanol-S_2, acetaldehyde-S_2'.)

$$S_1 + E \underset{k_{-1}}{\overset{k_{+1}}{\rightleftarrows}} ES_1 \underset{k_{-2}}{\overset{k_{+2}}{\rightleftarrows}} ES_1 S_2$$

$$\underset{k'}{\overset{k}{\rightleftarrows}} ES_1' S_2' \underset{k_{+2}'}{\overset{k_{-2}'}{\rightleftarrows}} ES_1' \underset{k_{+1}'}{\overset{k_{-1}'}{\rightleftarrows}} E + S_1'$$

The validity of these equations, referred to as the Theorell and Chance model, for animal liver ADH, has been demonstrated experimentally.

Another feature of the Theorell and Chance mechanism is that the formation of a ternary ADH-NAD-ethanol complex and a ternary ADH-NADH-acetaldehyde complex, and their reciprocal conversions, proceed faster than the remaining processes; therefore none of these reactions represents a limiting step. If the initial concentration of the coenzyme and of the substrate is sufficiently high, the dissociation of the product, *i.e.* of ADH-coenzyme, is the limiting step of the whole reaction.

In accordance with such a simple mechanism obviously proceed the reduction of aldehydes and the oxidation of ethanol, whereas a partial dissociation of NAD from the ternary ADH-NAD-ethanol complex is supposed to be necessary for the oxidation of higher alcohols.[66]

The kinetics of a two-substrate reaction can be expressed by the following general formula:

$$\frac{e}{v_0} = \Phi_0 + \frac{\Phi_1}{[S_1]} + \frac{\Phi_2}{[S_2]} + \frac{\Phi_{12}}{[S_1][S_2]} \qquad (1)$$

where e is the concentration of active sites, $[S_1]$ and $[S_2]$ the concentration of the coenzyme and the substrate, respectively, and v_0 the initial reaction rate. The values of the kinetic coefficients Φ (for oxidation of alcohols) and Φ' (for reduction of aldehydes) are obtained as follows. When the reciprocal reaction rate is plotted versus reciprocal substrate concentrations, we obtain a series of lines. The slopes and intercepts of the latter are plotted versus reciprocal coenzyme concentration and we obtain two lines whose intercepts and slopes directly indicate the set of four kinetic coefficients. The necessary and sufficient conditions for the reaction to follow the Theorell and Chance mechanism are:

$$\Phi'_0 = \frac{\Phi_1 \Phi_2}{\Phi_{12}}, \quad \Phi_0 = \frac{\Phi'_1 \Phi'_2}{\Phi'_{12}} \qquad (2)$$

These equations guarantee that the rate-determining step of the entire reaction is the dissociation of the enzyme from the ADH-coenzyme complex.[64-68]

The kinetic coefficients, given in Table XVI and determined at two pH values, comply with the conditions of Equation 2, valid for the Theorell and Chance mechanism. We can calculate from the values of these kinetic coefficients not only the Michaelis constants for all the participating substrates (Table XVII), but we can also determine the values of the dissociation constants of binary complexes of the enzyme with the two coenzymes (Table XVIII).

Table XVI. Values of Kinetic Coefficients from Equation 1 for Oxidation of Ethanol by NAD and Reduction of Acetaldehyde by NADH, Catalyzed by Pea ADH at pH 8.5 and 7.4
(Experimental conditions: [ethanol] = 5-40 mM, NAD = 50-500 μM, [acetaldehyde] = 1-8 mM, [NADH] = 50-500 μM)

	a	b		c	d
Φ_0, S	5.5	7	Φ'_0, S	1.5	1
Φ_1, mM S	0.8	0.35	Φ'_1, mM S	0.3	0.15
Φ_2, mM S	170	280	Φ'_2, mM S	9	4
Φ_{12}, mM^2 S	90	90	Φ'_{12}, mM^2 S	0.5	0.1

Measured at pH: [a] 8.5; [b] 7.4; [c] 8.5; [d] 7.5.

Table XVII.   Michaelis Constants of Basic Substrates of Pea ADH Calculated
from Values of Kinetic Coefficients

	a	b
$K_m(NAD) = \Phi_1/\Phi_0,\ \mu M$	150	200
$K_m(NADH) = \Phi_1'/\Phi_0',\ \mu M$	200	160
$K_m(ethanol) = \Phi_2/\Phi_0,\ mM$	30	40
$K_m(acetaldehyde) = \Phi_2'/\Phi_0',\ mM$	6	4

Measured at pH: [a] 8.5; [b] 7.4.

Table XVIII. Dissociation Constants of Binary Complexes of Pea ADH with NAD
and NADH, Calculated from Varlues of Kinetic Coefficients

	a	b
$K_{ADH\text{-}NAD} = \Phi_1/\Phi_0',\ \mu M$	500	300
$K_{ADH\text{-}NADH} = \Phi_1'/\Phi_0,\ \mu M$	60	20

Measured at pH: [a] 8.5; [b] 7.4.

The equilibrium constant K for the reaction ethanol + NAD $\underset{}{\overset{k}{\rightleftharpoons}}$ acetaldehyde + NADH + H^+ can be calculated from the kinetic coefficients as:

$$K = \frac{\Phi_{12}'}{\Phi_{12}}\ [H^+]$$

The calculated value of K for pH is $1.7 \times 10^{-11}\ M$. The thermodynamic value of the constant at 25°C is $0.98 \times 10^{-11}\ M$.[67]

The results obtained show that the oxidation of ethanol by NAD and the reduction of acetaldehyde by NADH, catalyzed by pea ADH, follow the Theorell and Chance mechanism, *i.e.*, a mechanism supposed to be operative also in the reactions catalyzed by horse liver and yeast alcohol dehydrogenase. The Theorell and Chance mechanism points to the decisive role of binary ADH-coenzyme complexes whose dissociation constants can be determined. The interpretation of the formation, reciprocal conversion, and dissociation of ternary ADH-NAD-ethanol and ADH-NADH-acetaldehyde complexes rests on somewhat uncertain grounds since these reactions are faster than the dissociation of binary ADH-coenzyme complexes and hence not reflected in the kinetic schemes.[68]

COMPARISON OF ADH's FROM PLANTS, ANIMALS (LADH) AND YEAST (YADH)

ADH Isolation

A procedure commonly used for LADH or YADH could not be employed for the isolation of the plant enzyme, as the plant enzymes are inactivated, *e.g.*, by thermal denaturation, by ethanol or acetone fractionation, and during preparative high-voltage electrophoresis. The enzyme stability is not increased by the addition of ethanol or methanol. On the other hand, YADH and LADH were prepared in the crystalline form long ago, and considerable successes were attained in the study of higher-order structures.[69-72]

ADH Molecular Weights

The molecular weights of the isolated plant ADH's were similar and varied within a range from 53,000-80,000 when determined by gel filtration or disk electrophoresis. As far as the molecular weight was concerned, plant ADH's resembled LADH from rats, human beings and horses, where the molecular weights also lie within 60,000-80,000 range. The yeast enzyme has a molecular weight of 150,000 and resembles the peanut and tea seed enzymes.[44,73-75]

Michaelis Constants and Optimum pH

For ethanol oxidation, the Michaelis constants for plant ADH's were of the order of $10^{-2} M$ and were lower for the aldehyde reduction, except for rice. The optimum pH for the substrate oxidation was between 8 and 9, and around 7 for the aldehyde reduction. NAD and NADH were reduced or oxidized with a rate constant of 10^{-4}-$10^{-5} M$. With all plant ADH's, the catalysis rate was larger for NAD than for NADP.

In their K_m values, plant ADH's resembled the yeast enzyme rather than the liver enzyme, as is documented in Table II. The optimum pH values for the substrate oxidation or reduction were different with LADH and YADH, similar to plant ADH's.

Substrate Specificity

Similar to LADH and YADH, plant ADH's also exhibited a wide substrate specificity toward alcohols (Tables V and VI). In addition to ethanol, its homologs were also oxidized, though more slowly, the oxidation of unsaturated homologs being faster than that of saturated. However, plant ADH's did not oxidize diols or cyclic, aromatic, secondary aliphatic, carbohydrate or terpenic alcohols. In certain respects, they resembled LADH;

in other respects, YADH (Table XIX):

1. The decrease in the oxidation rate with increasing chain length was similar to YADH, as with LADH alcohols up to butanol are oxidized faster than ethanol.
2. Unsaturated analogs were oxidized faster than ethanol, similar to LADH and YADH.[76]
3. Cyclohexanol is an LADH substrate, but was not oxidized by plant ADH.
4. Diols, cyclic alcohols and aromatic alcohols are LADH substrates but not substrates for plant ADH's.

Table XIX. Substrate Specificity of Alcohol Dehydrogenases (Measured at pH 8.5; substrate concentration 10^{-2} M, except for hexanol, cyclohexanol, cinnamyl alcohol and phenylethanol, where saturated solutions were used (−). The rate in question was not measured.)

Substrate	Relative Oxidation Rate of ADH		
	Liver[77]	Yeast[5]	Pea[22]
Methanol	0	0	1.5
Ethanol	100	100	100
n-Propanol	110	44	44
n-Butanol	163	26	30
n-Pentanol	118	14	15
n-Hexanol	−	−	12.3
2-Propanol	−	27	−
2-Butanol	36	−	0
2-Propene-1-ol	142	−	160
2-Butene-1-ol	−	−	49
Isobutanol	−	17	9.8
Isoamyl alcohol	124	−	15
4-Pentene-1-ol	−	−	6.4
Cinnamyl alcohol	−	−	16.4
2-Mercaptoethanol	−	−	3.3
Cyclohexanol	100	−	0
Benzyl alcohol	88	−	0
Tert. Butanol	0	−	0
Tert. Amyl alcohol	0	−	0

The specificity toward aldehydes was also quite broad. It is difficult to draw conclusions on this specificity on the basis of the results obtained so far with plant alcohol dehydrogenases. The magnitude of K_m for the first three members of the series, plant ADH's are more similar to YADH than to LADH, for which the Michaelis constant values substantially decrease with increasing chain length. It is interesting that unsaturated aldehydes are oxidized by the liver enzyme more slowly than the saturated analogs, in contrast to alcohols, where the situation is reversed.[76,77]

Enzyme Reaction Kinetics

Analysis of the kinetics of alcohol oxidation and acetaldehyde reduction at two pH values indicated that the reaction first involves binding of NADH or NAD to the enzyme. For the ethanol oxidation and acetaldehyde reduction, the last oxidation step was apparently rate-determining for the overall reaction rate. The LADH-catalyzed redox reaction obeyed a similar mechanism. The Theorell and Chance mechanism was verified only for the two basic substrates and, therefore, the oxidation and reduction mechanisms for higher alcohols and aldehydes cannot yet be specified.[63,68]

Catalysis Mechanism

Function of Zinc Atoms

Zinc atoms are present in YADH and LADH molecules: there are four in the LADH molecule, which consists of two subunits; in the YADH molecule there are also four, but the molecule is a tetramer. The function of these atoms has been studied intensively.[71,78-80]

Zinc atoms also play a role in catalysis by plant ADH's. The ADH is inhibited by chelating agents, the degree of inhibition being different for enzymes isolated from germinating seeds of various plants. The inhibitive effect of phenanthroline, for example, is decreased by ADH preincubation with the coenzyme, similar to LADH.

Sulfhydryl Groups

LADH contains 28 and YADH 36 sulfhydryl groups.[81] However, only two SH-groups in each ADH subunit are important for the enzyme catalytic function. These cysteinyl residues are the ligands for zinc ions in the catalytic center. Crystallographic studies support the assumption of short distance between the sulfhydryl groups and the nicotinamide part of the coenzymes.[70,72,82]

On alkylation of the SH-groups by iodoacetate or iodacetateamide, the enzyme ability to bind the coenzyme is not decreased substantially; the nativeness of the essential SH-groups is, however, a condition for the formation of ternary enzyme-coenzyme-substrate complexes, without which the redox reaction cannot take place.[38]

Sulfhydryl groups are also present in plant ADH molecules, although their number cannot be determined because of a low degree of enzyme purification. For preservation of the enzyme activity, the presence of mercaptoethanol, cysteine, or dithreitol is required; substances known as SH-poisons inhibit the enzymes. Alkylation of the essential SH-groups by iodoacetate is blocked by substances bound close to the active center. These are ATP, ADP and AMP with inhibition constants of 1250 μM, 250 μM, and 70 μM, respectively, NAD with an inhibition constant of 100 μM, as well as phenanthroline (8 μM) and chloride ions (11 μM). All the ligands tested are competitive with respect to iodoacetate, except for phenanthroline.[9,37]

The ADH Active Center

The high-resolution X-ray structure of horse liver ADH has revealed that the active zinc ion is located at the point of convergence of two deep clefts in the surface of the subunit. One of the clefts has been identified as the coenzyme binding site.[59,70] The other cleft is believed to be the substrate binding site (since the X-ray structural studies show that competitive inhibitor phenanthroline occupies the cleft). The active zinc ion in the native enzyme is four-coordinate and exists in a warped tetrahedral ligand field geometry. Two of the four ligands are cysteinyl sulfhydryl residues (Cys 46 and Cys 174); the third is a histidyl residue (His 67). The fourth ligand appears to be a water molecule.[80] The water molecule bound to the zinc atom forms an internal hydrogen bond to Ser 48 which, in turn, is within hydrogen bonding distance of one of the nitrogen atoms of His 51 points towards the solution. This system of hydrogen bonds might provide a framework for proton release in the overall reaction which is induced by NAD^+ bonding.

In view of these structural results, the only plausible mechanism for alcohol oxidation is that of electrophilic catalysis mediated by the active center zinc atom. The structural determination strongly supports the mechanism proposed by Theorell and Chance, where the zinc-bound water molecule plays a crucial role. Binding of NAD^+ perturbs the pK_a value of this water molecule at lower pH with a concomitant proton release corresponding to the proton liberated in the overall reaction. Alcohol is then bound to zinc as the alcoholate ion, displacing the hydroxyl ion. The formation of alcoholate is mediated by the hydroxyl bound to zinc, that acts as a base and combines with the proton of the hydroxyl group of the alcohol to form a water molecule.[70]

Knowledge of the arrangement of the active center of plant ADH's is negligible and cannot be improved before sufficiently pure materials are available. Only indirect conclusions on the basis of similarities in the effect of inhibitors and substances forming complexes with the enzyme or the enzyme and coenzyme could be drawn concerning the arrangement of the active center of plant ADH's.

Pyrazole, pyridine and imidazole were bound to the LADH active center in an exactly determined way, related to the center structure. Pyridine was a competitive inhibitor with respect to NAD and ethanol with plant ADH's as well as with LADH, and the values of the inhibition constants were similar for plant ADH's and LADH (Table XI). The qualitative effect of pyrazole was similar for the plant and animal enzymes. However, its inhibition constant was about 100 times higher for plant ADH's than for LADH; in contrast to LADH, the effect on the rate of the enzymatic reaction catalyzed by plant ADH's was immediate. Imidazole exhibited CIS-type inhibition with respect to ethanol and NAD for both plant ADH's and LADH; hence it was either an inhibitor or an activator, depending on its concentration. Imidazole probably formed EOI and ERI complexes with plant as well as with animal ADH. Since the effect of the inhibitors tested depended on the structure of the active center, it can be assumed that there was a certain similarity between the arrangement of the catalytic center in plant ADH's and in liver enzyme;[38,46] as these substances form ERI ternary complexes, this finding also suggested a similarity in the active center structures. However, here, as well as with pyrazole, berberine and its derivatives and chloride ions, the sensitivity of plant ADH's was substantially lower. This finding could be related to the fact that K_m values for plant ADH's were substantially higher compared with those for LADH.

Intermediates of the carbohydrate metabolism exhibited very different behavior toward the plant and animal enzymes, as follows from Table X. Inhibition by lactate, acetate and succinate had no analog with LADH and, on the other hand, the inhibitive effect of malate and pyruvate was substantially smaller with plant ADH's than with LADH.

With LADH, it is assumed that bonding of the coenzyme to the enzyme protein takes place through the adenosine part of NAD, relatively far from the central zinc molecule, close to which the nicotinamide part was located. The coenzyme bond to the protein was hydrophobic. The competitive character of NAD inhibition by ATP, ADP and AMP suggested that the coenzyme was also bound to the protein through the adenosine part with plant ADH's. The hydrophobicity of the bond was then indicated by:

1. derivatives of the alkaloid berberine which exhibited an increasing inhibitive effect toward NAD when the hydrophobicity of the substituent in their side chains increased; a similar effect was observed for LADH (Table XIV);
2. aminonaphthalenesulfonic acid was bound to ADH and could be displaced from this bond by NAD or NADH.

In LADH, the substrate-enzyme bond was considered to be hydrophobic. Propanol and butanol were oxidized by liver ADH better than ethanol and the inhibition by fatty acids, competitive with respect to ethanol, increased with increasing chain length in the acid. With plant ADH's, butyric acid was a stronger inhibitor than propionic acid, and fatty acids also competed with ethanol for the bonding site.

CONCLUSIONS

In this work, the results obtained are given for ADH isolated from the germinating seeds of broad bean, lentil, pea, bean, soybean, maize, cucumbers, wheat, rice, rape and sunflower. The isolation procedure indicated that the protein parts of plants' ADH's were not identical. The enzymes were relatively labile and lost their activity during dialysis, preparative electrophoresis, incubation at $0°C$ for several days (where cysteine, mercaptoethanol and NAD exhibited a protective effect, while ethanol and methanol did not), heating above $60°C$ and in acidic media. The decrease in the activity followed first-order kinetics.

The molecular weights of ADH's from the plants studied were from 52,000-80,000. The enzymes were metalloproteins inhibited by chelating agents. All the ADH's contained sulfhydryl groups essential to the catalysis, although the quantitative effect of substances known as SH-poisons was not identical for all the ADH's. Alkylation of sulfhydryl groups by iodoacetate was blocked by some substances, which were bound close to the active site of plant ADH's. These were ATP with an inhibition constant of 1250 μM; ADP, 250 μM; AMP, 70 μM; NAD, 100 μM; phenanthroline, 8 μM; and chloride ions, 11 μM. All the ligands tested, except for phenanthroline, were competitive with respect to iodoacetate.

The specificity for alcohols and aldehydes was not clearly defined. In addition to ethanol, plant ADH's also oxidized analogs with longer chains, unsaturated faster than saturated. Diols, cyclic alcohols, aliphatic secondary alcohols and carbohydrate or terpenic alcohols were not oxidized. There were not only quantitative, but also qualitative differences among individual ADH's in the oxidation of alcohols, as follows from Table VI. The substrate specificity was also not clearly defined for aldehydes.

All ADH's oxidized alcohols most rapidly in a pH region of 8.2-9 and reduced aldehydes around pH 7. The Michaelis constant values were one order of magnitude lower for aldehydes than for alcohols. The rate of ethanol oxidation depended on the temperature, and the kinetic coefficients obeyed the Arrhenius equation; the Michaelis constants for ethanol and NAD increased with temperature.

Fatty acids and their thio derivatives were inhibitors competitive with respect to ethanol, with inhibition constants of about $10^{-2}M$. These substances were uncompetitive with respect to acetaldehyde, as they did not form the E-NADH-I complex, but the E-NAD-I complex with the product to acetaldehyde reduction. Two conclusions follow from the interaction of fatty acids with plant ADH. First, the binding side for ethanol probably has a rather strongly hydrophobic character. Second, the binding side for ethanol is not identical with the binding side for acetaldehyde and only on this assumption can the competition with ethanol and the uncompetitive behavior toward acetaldehyde be explained. The nonequivalence of the binding sites for ethanol and acetaldehyde can be accounted for by a conformational change of the protein backbone caused, *e.g.*, by some amino acid residue acting as an acid-base catalyst of the redox reaction catalyzed by ADH.[9,58]

Chloride ions behaved similarly with an inhibition constant of $10^{-1}M$. Hence, the inhibition of plant ADH by chloride anions is not only a quantitative expression of the affinity of anions for the site binding the pyrophosphate group of coenzyme but at the same time also a phenomenon which supports the hypothesis of nonequivalence of the binding sites for ethanol and acetaldehyde.

Amides (butyramide and acetamide), acetoxime and dimethylsulfoxide were inhibitors competitive with respect to acetaldehyde and noncompetitive with respect to ethanol, with inhibition constants of $10^{-2}M$ to $10^{-3}M$, as well as intermediates in the carbohydrate metabolism with inhibition constants of 10^{-1} to $10^{-2}M$.

ATP was a competitive inhibitor of coenzyme NAD and NADH, with $K_{0.5}=10^{-3}M$. The inhibition constants decreased with increasing pH. A similar inhibitive effect was exhibited by ADP and AMP. The alkaloid berberine and its derivatives and heterocycles, such as imidazole, 2-benzyl-4, 5-dihydroimidazole, pyrazole and pyridine were effective inhibitors of plant ADH's, with a K_i of 10^{-2}-$10^{-3}M$. Hydrophobic substances, *e.g.*, aminonaphthalenesulfonic acid or auramine, were bound to plant ADH's. Aminonaphthalenesulfonic acid was displaced from the bond with ADH by NAD, NADH, phenanthroline, pyrazole and mercuric ions.

As for the bonding of the coenzyme to the enzyme protein, it seems that the coenzyme is bonded through the ADP-ribosyl residue and also through the nicotinamide part, which probably interacts with the metal ion present in the enzyme molecule. This interaction might be of the ionic character, in view of the carboxyl group polarization. This hypothesis is supported by the finding that nicotinic acid, whose polarization is little pronounced, does not act as an inhibitor of plant ADH, even at a concentration of $0.5M$.[63]

The bonding site for the substrate is probably located close to the metallic component of ADH, as follows from the mixed inhibition by o-phenanthroline and nicotinamide with respect to ethanol.[61,62]

Analysis of the kinetics of the ethanol oxidation and the acetaldehyde reduction at various pH values indicated that the first reaction step was bonding of NAD or NADH to the enzyme and the last dissociation of NAD or NADH from the binary complex. The last step was rate-determining for the oxidation of ethanol and the reduction of acetaldehyde. The K_m values for all substrates and the dissociation constants for the E-NAD and E-NADH complexes were determined from the kinetic point of view.[68]

It can be concluded from the knowledge gained so far that ADH's isolated from the 11 kinds of germinating seeds were very similar, definitely more similar than to YADH or LADH. However, the ADH's will have a certain specificity in the individual plants, due to different behavior of the enzyme protein part, differences in the reaction rates, substrate specificity and sensitivity toward inhibitors, which suggest certain differences in the active centers of the ADH's studied.

REFERENCES

1. Leblová, S., D Ehlichová and J. Barthová. "Výskyt ethanolu a alkoholdehydrogenasy v klíčních rostlinách," Ústav vědeckotechnických informací, *Rostlinná výroba, Praha* 19(46):1209 (1973).
2, Leblová, S., J. Zima and E. Perglerová. "Conversion of Pyruvate under Natural and Artificial Anaerobiosis in Maize," *Aust. J. Plant Physiol.* 3(3):755 (1976).
3. Cossins, E. A., L. C. Kopala, B. Blawacky and A. M. Spronk. "Some Properties of Higher Plant Alcohol Dehydrogenase," *Phytochemistry* 7(7):1125 (1968).
4. Pattee, H. E. and H. E. Swaisgood. "Peanut Alcohol Dehydrogenase. I. Isolation and Purification," *J. Food Sci.* 33(3):250 (1968).
5. Davies, D. D., K. D. Patil, E. N. Ugochukwu and G. H. N. Towers. "Aliphatic Alcohol Dehydrogenase from Potato Tubers," *Phytochemistry* 12(3):523 (1973).
6. Davies, D. D., E. N. Ugochukwu, K. D. Patil and G. H. N. Towers. "Aromatic Alcohol Dehydrogenase," *Phytochemistry* 12(3):531 (1973).
7. Hatanaka, A. "Leaf Alcohol: NAD Oxidoreductase from Tea Seeds," *Bull. Inst. Chem. Res.*, Kyoto University 50(3): 135 (1972).
8. Hatanaka, A. and T. Harada. "Purification and Properties of Alcohol Dehydrogenase from Tea Seeds," *Agric. Biol. Chem.* 36(11):2033 (1972).
9. Lapka, R. "Pea Alcohol Dehydrogenase," Unpublished dissertation, Charles University, Prague, 1976.
10. Leblová, S. and P. Mančal. "Characterization of Plant Alcohol Dehydrogenase," *Physiol. Plant.* 34(3):246 (1975).

11. Theorell, H. and K. Tatemoto. "Thermal Stability of Horse Liver Alcohol Dehydrogenase and its Complexes," *Arch. Biochem. Biophys.* 143(2):354 (1971).
12. Wiseman, A. "Allosteric Binding of Coenzymes to Alcohol Dehydrogenases," *Biochem. J.* 124(5):78 (1971).
13. Wiseman, A. and N. J. Williams. "Thermal Inactivation of Alcohol Dehydrogenase in the Presence of NAD$^+$ or NADP$^+$," *Biochem. Biophys. Acta* 250(36):1 (1971).
14. Eriksson, C. F. "Alcohol: NAD Oxidoreductase from Peas (*Pisum sativum*)," *Acta Chem. Scand.* 21(1):304 (1967).
15. Eriksson, C. F. "Alcohol: NAD Oxidoreductase (E.C.1.1.1.1.) from Peas," *J. Food Sci.* 33(5):525 (1968).
16. Suzuki, Y. and K. Kyuwa. "Activation and Inactivation of Alcohol Dehydrogenase in Germinating Pea Cotyledons," *Physiol. Plant.* 27 (2):121 (1972).
17. Suzuki, Y. "Alcohol Dehydrogenase from Pea Seedlings," *Phytochemistry* 5(4):761 (1966).
18. Duffus, J. H. "Alcohol Dehydrogenase in Barley Embryo," *Phytochemistry* 7(7):1135 (1968).
19. Leblová, S. and J. Hlochová. "Rice and Pea Alcohol Dehydrogenase," *Coll. Czech. Chem. Commun.* 40(10):3220 (1975).
20. Leblová, S. and D. Ehlichová. "On the Alcohol Dehydrogenase of Zea Mays," *Phytochemistry* 11(4):1345 (1972).
21. Leblová, S., I. Zimáková, J. Barthová and D. Ehlichová. "On Plant Alcohol Dehydrogenase," *Biol. Plant.* 13(1):33 (1971).
22. Leblová, S., P. Mančal, D. Sofrová and J. Barthová. "Isolation of Alcohol Dehydrogenases from Germinating Seeds of Pea, Broad-bean, Lentil and Kidney-bean," *Biol. Plant.* 15(6):405 (1973).
23. Leblová, S. and M. Stiborová. "Investigation of the Properties of Alcohol Dehydrogenase from Bean, Rape, Wheat and Broad-bean," *Biol. Plant.* 17(4):268 (1975).
24. Leblová, S. "Isolation and Partial Characterisation of Alcohol Dehydrogenase from Broad-bean," *Aust. J. Plant. Physiol.* 1(2):579 (1974).
25. Leblová, S. and E. Perglerová. "Isolation and Characterization of Alcohol Dehydrogenase from Germinating Sunflower Seeds," *Coll. Czech. Chem. Commun.* 41(11):3482 (1976).
26. Leblová, S. and E. Perglerová. "Alcohol Dehydrogenase from Cucumber Seeds," *Biochem. Physiol. Pflanzen* 171(1):1 (1977).
27. Leblová, S. and E. Perglerová. "Alcohol Dehydrogenase from Soya-bean," *Phytochemistry* 15(5):813 (1976).
28. Leblová, S. and Mustapha El Ahmad. "Substrate Specificity of Cucumber and Broad-bean Alcohol Dehydrogenase," *Biol. Plant.* (in press.)
29. Stiborová, M. and S. Leblová. "Purification and Properties of Rape Alcohol Dehydrogenase," *Physiol. Plant.* 38(3):176 (1976).
30. Hedrick, J. L. and A. J. Smith. "Size and Charge Isomer Separation and Estimation of Molecular Weights of Proteins by Disc Gel Electrophoresis," *Arch. Biochem. Biophys.* 126(1):155 (1968).

31. Fisher, M. and D. Schwartz. "Dissociation and Reassociation of Maize Alcohol Dehydrogenase: Allelic Differences in Requirement for Zinc," *Molec. Gen. Genet.* 127(1):33 (1973).
32. Sedlak, J. T. and R. H. Lindsay. "Estimation of Total, Protein-bound and Nonprotein Sulfhydryl Groups in Tissue with Ellman's Reagent," *Anal. Biochem.* 25(1-3):192 (1968).
33. Goksöyr, J., E. Boeri and R. K. Bonnichsen. "The Variation on Alcohol Dehydrogenase During the Germination of Green Pea (*Pisum sativum*)," *Acta Chem. Scand.* 7(4):657 (1953).
34. Stafford, H. A. and B. Vennesland. "Alcohol Dehydrogenase of Wheat Germ," *Arch. Biochem. Biophys.* 44(2):404 (1953).
35. App, A. A. and A. N. Meiss. "Effect of Aeration on Rice Alcohol Dehydrogenase," *Arch. Biochem. Biophys.* 77(1):181 (1958).
36. Cossins, E. A. and E. R. Turner. "Losses of Alcohol Dehydrogenase Activity in Germinating Seeds," *Ann. Bot.* 26(104):591 (1962).
37. Lapka, R. and S. Leblová. "The Structure of Pea Alcohol Dehydrogenase Using Inactivation by Iodoacetate," *Physiol. Plant.* 39(1):86 (1977).
38. Yonetani, T. and H. Theorell. "Studies on Liver Alcohol Dehydrogenase Complexes. III. Multiple Inhibition Kinetics in the Presence of Two Competitive Inhibitors," *Arch. Biochem. Biophys.* 106(1):243 (1964).
39. McManmon, M. and R. M. M. Crawford. "A Metabolic Theory of Flooding Tolerances. The Significance of Enzyme Distribution and Behavior," *New Phytol.* 70(2):299 (1971).
40. Mazelis, M. and B. Vennesland. "Carbon Dioxide Fixation into Oxaloacetate in Higher Plants," *Plant. Physiol.* 32(6):591 (1957).
41. Streeter, J. G. and J. F. Thompson. "Anaerobic Accumulation of Gammabutyric Acid and Alanine in Radish Leaves," *Plant. Physiol.* 49(4):572 (1972).
42. Streeter, J. G. and J. F. Thompson. "*In Vivo* and *In Vitro* Studies on Gammaaminobutyric Acid Metabolism with the Radish Plant," *Plant. Physiol.* 49(4):579 (1972).
43. Arslanian, M., J. Pascoe and J. G. Reinhold. "Rat Liver Alcohol Dehydrogenase," *Biochem. J.* 125(4):1039 (1971).
44. Perlman, R. L. and J. Wolff. "Dimethyl Sulfoxide: An Inhibitor of Liver Alcohol Dehydrogenase," *Science* 160(3825):317 (1968).
45. Theorell, H., T. Yonetani and G. Sjöberg. "On the Effect of Some Heterocyclic Compounds on the Enzymic Activity of Liver Alcohol Dehydrogenase," *Acta Chem. Scand.* 23(1):255 (1969).
46. Reynolds, C. H., D. L. Morris and J. S. McKinley-McKee. "Complexes of Liver Alcohol Dehydrogenase. Further Studies on the Rate of Inactivation," *Eur. J. Biochem.* 14(1):14 (1970).
47. Theorell, H. and J. S. McKinley-McKee. "Liver Alcohol Dehydrogenase. II. Equilibrium Constant of Binary and Ternary Complexes of Enzyme, Coenzyme and Caprate, Isobutyramide and Imidazol," *Acta Chem. Scand.* 15(9):1811 (1961).
48. Theorell, H. and J. S. McKinley-McKee. "Liver Alcohol Dehydrogenase. III. Kinetics in the Presence of Caprate, Isobutyramide and Imidazol," *Acta Chem. Scand.* 15(9):1834 (1961).

49. Winer, A. D. and H. Theorell. "Dissociation Constants of Ternary Complexes of Fatty Acids and Fatty Acids Amides with Horse Liver Alcohol Dehydrogenase–Coenzyme Complexes," *Acta Chem. Scand.* 14(8):1729 (1960).
50. Yonetani, T. "Studies on Liver Alcohol Dehydrogenase Complexes. II. The Interaction of the Enzyme with *o*-Phenanthroline and Crystallization of Complexes of Phenanthroline-Enzyme, Enzyme-Adenosine Diphosphate Ribose, and Phenanthroline-Enzyme-Adenosine Diphosphate Ribose," *Biochem. Zeit.* 338:300 (1963).
51. Yonetani, T. "Studies on Liver Alcohol Dehydrogenase Complexes. I. The Coenzyme-Binding Sites and Effects of Adenosine Diphosphate Ribose and *o*-Phenanthroline," *Acta Chem. Scand.* 17:Suppl. 1,96 (1963).
52. Coleman, P. L. and H. Weiner. "Interaction of Chloride Ion with Horse Liver Alcohol Dehydrogenase-Reduced Nicotinamide Adenine Dinucleotide Complexes," *Biochemistry* 12(9):1702 (1973).
53. Theorell, H., A. P. Nygaard and R. Bonnichsen. "Studies on Liver Alcohol Dehydrogenase. III. The Influence of pH and Some Anions on the Reaction Velocity Constants," *Acta. Chem. Scand.* 9(7):1148 (1955).
54. Kovář, J. "Direct Fluorometric Determination of Liver Alcohol Dehydrogenase with Berberine," *Eur. J. Biochem.* 49(1):179 (1974).
55. Skurský, L., J. Kovář and J. Michalský. "Interaction of Liver Alcohol Dehydrogenase with Neuroleptics Chlorprothixene and Chlorpromazine," *FEBS Letters* 51(1):297 (1975).
56. Kovář, J. and L. Skurský. "Fluorescence Study of Liver Alcohol Dehydrogenase Complexes with Berberine and Other Ligands," *Eur. J. Biochem.* 40(1):233 (1973).
57. Pavelka, S. and J. Kovář. "Interaction of Liver Alcohol Dehydrogenase with Protoberberine Alkaloids," *Coll. Czech. Chem. Commun.* 40(3):753 (1975).
58. Leblová, S., R. Lapka and J. Kovář. "Binding of Various Ligands to Pea Alcohol Dehydrogenase," *Coll. Czech. Chem. Commun.* 42(3):1082 (1977).
59. Brändén, C. I., H. Eklund, B. Nordström, T. Boiwe, G. Süderlund, E. Zeppezauer, I. Ohlsson and Å. Åkeson. "Structure of Liver Alcohol Dehydrogenase at 2,9 Å Resolution (Crystallographic Structure Zinc Coenzyme Binding)," *Proc. Natl. Acad. Sci. U.S.A.* 70(8):2439 (1973).
60. Einarsson, R., H. Eklund, E. Zeppezauer, T. Boiwe and C. I. Brändén. "Binding of Salicylate in the Adenosine-binding Pocket of Dehydrogenases," *Eur. J. Biochem.* 49(1):41 (1974).
61. Stiborová, M. and S. Leblová. "Rape Alcohol Dehydrogenase. A Study of the Active Centre," *Biochem. Physiol. Pflanz.*,(in press.)
62. Stiborová, M. and S. Leblová. "A Study of the Bonding Site of Alcohol Dehydrogenase Isolated from Germinating Rape Seeds," *Coll. Czech. Chem. Commun.*
63. Stiborová, M. "A Study of the Functional Mechanism and the Structure of Plant Alcohol Dehydrogenases," PhD. Thesis, Charles University, Prague (1977).

64. Dalziel, K. "Initial Steady-State Velocities in the Evaluation of Enzyme-Coenzyme-Substrate Reaction Mechanism," *Acta Chem. Scand.* 11(10):1706 (1957).

65. Theorell, H. and B. Chance. "Studies on Liver Alcohol Dehydrogenase and Reduced Diphosphopyridine Nucleotide," *Acta Chem. Scand.* 5(11):1127 (1951).

66. Dalziel, K. and F. M. Dickinson. "The Kinetics and Mechanism of Liver Alcohol Dehydrogenase with Primary and Secondary Alcohols as Substrates," *Biochem. J.* 100(1):34 (1966).

67. Bäcklin, K. "The Equilibrium Constant of the System Ethanol, Aldehyde, DPN^+, $DPNH^+$ and H^+," *Acta Chem. Scand.* 12(6):1279 (1958).

68. Lapka, R. and S. Leblová. "Kinetics of Redox Reaction Catalyzed by Pea Alcohol Dehydrogenase," *Coll. Czech. Chem. Commun.* 42 (4):1262 (1977).

69. Bonnichsen, R. and J. Wassén. "Crystalline Alcohol Dehydrogenase from Horse Liver," *Arch. Biochem. Biophys.* 18(2):361 (1948).

70. Eklund, H., B. Nordström, E. Zeppezauer, G. Söderlund, I. Ohlsson, T. Boiwe and C. I. Brändén. "The Structure of Horse Liver Alcohol Dehydrogenase," *FEBS Letters* 44(2):200 (1974).

71. Iweibo, I. and H. Weiner. "Role of Zinc in Horse Liver Alcohol Dehydrogenase. Coenzyme and Substrate Binding," *Biochemistry* 11(6):1003 (1972).

72. Jörnvall, H., C. Woenckhaus, E. Schättle and R. Jack. "Modification of Alcohol Dehydrogenase with Two NAD Analogues Containing Reactive Substituents on the Functional Side of the Molecule," *FEBS Letters* 54(2):297 (1975).

73. Hayes, J. E. and S. F. Velick. "Yeast Alcohol Dehydrogenase: Molecular Weight, Coenzyme Binding and Reaction Equilibria," *J. Biol. Chem.* 207(1):225 (1954).

74. Jörnvall, H. "Horse Liver Alcohol Dehydrogenase. The Primary Structure of an N-terminal Part of the Protein Chain of the Ethanol-active Isoenzyme," *Eur. J. Biochem.* 14(3):521 (1970).

75. Jörnvall, H. "Horse Liver Alcohol Dehydrogenase. The Primary Structure of the Protein Chain of the Ethanol-active Isoenzyme," *Eur J. Biochem.* 16(1):25 (1970).

76. Pietruszko, R., K. Crawford and D. Lester. "Comparison of Substrate Specificity of Alcohol Dehydrogenases from Human Liver, Horse Liver and Yeast towards Saturated and 2-enoic Alcohols and Aldehydes," *Arch. Biochem. Biophys.* 159(1):50 (1973).

77. Winer, A. D. "A Note on the Substrate Specificity of Horse Liver Alcohol Dehydrogenase," *Acta Chem. Scand.* 12(8):1695 (1958).

78. Veillon, C. and A. J. Sytkowski. "The Intrinsic Zinc Atoms of Yeast Alcohol Dehydrogenase," *Biochem. Biophys. Res. Com.* 67(4):1494 (1975).

79. Sytkowski, A. J. and B. L. Vallee. "The Catalytic Metal Atoms of Cobalt Substituted Liver Alcohol Dehydrogenase," *Biochem. Biophys. Res. Com.* 67(4):1488 (1975).

80. Dunn, M. F., J. F. Biellman and G. Branland. "Roles of Zinc Ion and Reduced Coenzyme in Horse Liver Alcohol Dehydrogenase Catalysis. The Mechanism of Aldehyde Activation," *Biochemistry* 14 (14):3176 (1975).

81. Yonetani, T. and H. Theorell. "On the Ternary Complex of Liver Alcohol Dehydrogenase with Reduced Coenzyme and Isobutyramide. Effect of p-Chloromercuriphenyl Sulfonate and Stability of the Complex," *Arch. Biochem. Biophys.* 99(3):433 (1962).

82. Jörnvall, H., G. Woenckhaus and G. Johnscher. "Modification of Alcohol Dehydrogenase with a Reactive Coenzyme Analogue. Identification of Labelled Residues in the Horse Liver and Yeast Enzyme after Treatment with Nicotinamide 5-bromacetyl-4-methyl Imidazole Dinucleotide," *Eur. J. Biochem.* 53(1):71 (1975).

SECTION II

RECENT CONTRIBUTION ON ANAEROBIOSIS

15

LACTATE DEHYDROGENASE FROM GERMINATING PLANTS

Jana Barthová and Sylva Leblová

Department of Biochemistry
Charles University
Prague, Czechoslovakia

ENZYME ISOLATION

Purified enzymes were obtained from germinating plants of soybean (*Glycine soja* L.) and peas (*Pisum sativum* L.) during natural anaerobiosis. A procedure involving fractionation with ammonium sulfate, chromatography on a DEAE-cellulose column and preparative electrophoresis on a polyacrylamide gel for soybean dehydrogenase and gel filtration on a Sephadex G-150 column for the pea enzyme was employed.[1] Electrophoresis represents an especially effective part of the purification procedure—the specific activity of the preparation increased about 20 times (Table I). The purified enzyme exhibited an activity approximately 1200 times higher than the crude extract and was electrophoretically homogeneous.[2,3] So far we have not succeeded in finding suitable conditions for application of this method to lactate dehydrogenase from other plants; their enzymes were considerably or completely inactivated during this relatively long operation (approximately 8 hr).

Isolation of lactate dehydrogenase from plant material has so far been mentioned only by King,[4] who isolated the enzyme from soybean cotyledons and purified it by a factor of approximately 110 and by Davies and Davies,[5] who obtained a purification factor of 60 during the isolation of lactate dehydrogenase from potato tubers.

Table I. Isolation of Lactate Dehydrogenases from Soybean and Pea Seedlings

Fraction	Soybean LDH		Pea LDH	
	Specific Activity (μ/mg)	Degree of Purification	Specific Activity (μ/mg)	Degree of Purification
Crude Extract	2.3	1	1.8	1
Ammonium Sulphate Fraction	6.7	3	9.5	5
DEAE-Cellulose Chromatography	150.0	65	138.8	77
Preparative Electrophoresis	2643.0	1160	–	–
Sephadex G-150 Chromatography	–	–	337.2	187

PROPERTIES OF LACTATE DEHYDROGENASES FROM GERMINATING PLANTS

Structural Study of Lactate Dehydrogenases

It has long been known that animal lactate dehydrogenase has a molecular weight of about 140,000 and that its molecule consists of four subunits, which can be of two types. Consequently, this enzyme can occur in the form of five isoenzymes which have also been detected in animal tissues.[6]

To our knowledge, no study of plant lactate dehydrogenase has been undertaken from this point of view. Only the occurrence of two to three isoenzymes of lactate dehydrogenase in legume tissues,[7] five isoenzymes in *Vicia faba* roots[8] and two to three in potato tubers (*Solanum tuberosum*)[9,10] have been mentioned. The isoenzymes were only qualitatively detected in raw extracts, and no data are available concerning their structure or their physiological function in the plant organism.

By gel filtration on Sephadex G-150, we have determined approximate molecular weights for purified lactate dehydrogenases from soybean and pea seedlings: 140,300 for the soybean enzyme and 145,500 for the enzyme from peas. Using electrophoresis in a sodium dodecylsulphate medium we were further able to prove that even the plant enzyme was composed of subunits. Their molecular weights were also determined. According to our experiments, the soybean lactate dehydrogenase consisted of a single subunit type with a molecular weight of 36,000. In contrast to the papers cited, we only detected a single lactate dehydrogenase isoenzyme in the seedlings.

Structural–Kinetic Study of Lactate Dehydrogenase

It seems that most dehydrogenases are SH-enzymes. Sulfhydryl groups are usually essential for the enzyme activity, as they are a part of the active site. Moreover, they participate in the maintenance of the quaternary structure of the enzyme molecule.[6]

We were able to determine the content of sulfhydryl groups in the electrophoretically pure lactate dehydrogenase isolated from soybean seedlings; one molecule contained 24 of these groups. We further investigated the effect of some SH-poisons, namely, p-chloromercuribenzoate (PCMB), iodoacetamide (IAA) and N-ethylmaleimide (NEMI), on the activity of the two enzymes. While all three substances acted as strong inhibitors of pea lactate dehydrogenase, their effect on the soybean enzyme was relatively weaker. Both enzymes were inhibited noncompetitively by p-chloromercuribenzoate with an inhibition constant of the order of 1 mM for pyruvate as a substrate and 0.1 mM for NADH. N-ethylmaleimide and iodoacetamide inhibited both enzymes irreversibly, the inhibition being proportional to the inhibitor concentration and increasing with time. The pea lactate dehydrogenase was inhibited by 65% in the presence of 10 mM N-ethylmaleimide in 5 minutes and by 60% in the presence of iodoacetamide of the same concentration. On the other hand, the inhibition of soybean lactate dehydrogenase was only 15% under identical conditions, not exceeding 30% even after one hour.

We assume that the sulfhydryl group also plays an important role in the active site of the plant enzymes, in view of the strong inhibition of both enzymes by p-chloromercuribenzoate, but the function of the SH- groups in the molecular structure will be somewhat different for the pea lactate dehydrogenase than for soybean lactate dehydrogenase.

With pea lactate dehydrogenase, which was strongly inhibited by the substances mentioned above, the protective effect of enzyme preincubation with the coenzyme (NADH) or substrate (pyruvate) against inhibition by SH-poisons was tested. The results are summarized in Table II.

Enzyme preincubation with both NADH and pyruvate led to a decrease in inhibition by all the inhibitors tested. However, protection of lactate dehydrogenase by the coenzyme was effective against p-chloromercuribenzoate inhibition, only. On the basis of these results, it can be assumed that pea lactate dehydrogenase contains a sulfhydryl group in the active site and that this group, analogous to the animal enzyme,[6] takes part in the formation of the enzyme-NADH binary complex, which is the first step in the reaction catalyzed by this enzyme, as follows from the reaction mechanism of animal lactate dehydrogenase.

N-ethylmaleimide and iodoacetamide are inhibitors preferentially bound to the lactate dehydrogenase structural sulfhydryl groups, which are not essential for enzyme activity.[11] The substantially weaker protective effect

Table II. Influence of Preincubation of Lactate Dehydrogenase of Pea Seedlings with NADH (A) and Pyruvate (B) on the Inhibition of Enzyme by Agents Reacting with Sulfhydryl Groups

Inhibitor	Conc. (mM)	% Inhibition	
		No Preincubation	5-min Preincubation
A (NADH)			
PCMB	0.1	15.3	2.5
	1	65.0	12.5
NEMI	1	18.5	7.5
	10	44.2	27.5
IAA	1	20.3	12.5
	10	64.0	35.0
B (Pyruvate)			
PCMB	0.1	6.9	3.3
	1	46.3	35.5
NEMI	1	16.2	6.5
	10	65.6	38.7
IAA	1	13.8	9.7
	10	58.7	45.2

PCMB — p-chloromercuribenzoate.
NEMI — N-ethylmaleimide.
IAA — iodoacetamide.

of the coenzyme and substrate against inhibition by these substances probably occurs because a group in the active site is not attacked directly. Preincubation of enzyme, especially with NADH, led to the formation of the enzyme-NADH binary complex, whose protein molecule stearic structure was probably altered; in this complex the SH- groups were less accessible to alkylation. The results cited indirectly indicated that the kinetic of the reaction catalyzed by plant lactate dehydrogenase exhibited an ordered ternary complex mechanism.

Kinetic Character of Lactate Dehydrogenase

The K_m values were determined for the enzymes from soybean and pea seedlings for lactate, pyruvate, NAD and NADH. As can be seen from Table III, the values of constants differ very little for the two enzymes. The lactate dehydrogenases from both sources also catalyzed conversions of some other α-hydroxy- and α-ketoacids. The lactate dehydrogenase affinity toward these substrates was usually much lower than that toward the proper substrates (the K_m values were by at least one order of magnitude higher, compared with those for the proper substrates).

Table III. Values of Michaelis Constants (K_m) of Lactate Dehydrogenases of Soybean and Pea Seedlings for Different Substrates

Substrate	K_m (mM)	
	Soybean LDH	Pea LDH
Lactate	35	50
Pyruvate	0.63	0.33
NAD	3.1	1.0
NADH	0.014	0.028
Malate	120	660
Citrate	560	–
Glycolate	38	73
α-Ketoglutarate	8.7	3.4
Glyoxylate	–	16

It was interesting, however, that the affinity, especially that of the soybean lactate dehydrogenase, for glycolate was almost the same as for the lactate. The high activity of the glyoxylate cycle was typical for germinating seeds, especially fatty seeds. We assumed, in agreement with King,[4] that lactate dehydrogenase may participate in the fat metabolism in this phase of seed germination. Potato lactate dehydrogenase[5] also metabolized some of the abovementioned substrates; on the other hand, the enzyme isolated by Rothe[9] from potato tubers exhibited no activity toward glycolate, glyoxylate and β-hydroxybutyrate as substrates.

We found that the kinetic behavior of soybean lactate dehydrogenase was also a function of the enzyme concentration. The hyperbolic shape of the plot of reaction rate vs the substrate concentration changed to sigmoid when the enzyme concentration exceeded 25 μm/ml. The allosteric character of animal lactate dehydrogenase was observed several times.[12,13] Markert and Massaro[14] explain the enzyme allostery by its dissociation to the dimer at lower enzyme concentrations. An association-dissociation equilibrium between the dimer and tetramer may constitute a molecular basis for the allosteric behavior. On the other hand, Monod et al.[15] and Koshland et al.[16,17] assume that the allosteric behavior of the enzyme is caused by conformational changes in the subunit structure. Lactate dehydrogenase from soybean seedlings apparently exhibited a sigmoid saturation curve when its allosteric site was unoccupied, i.e., with a high enzyme concentration and a relative substrate deficiency. When the allosteric site was occupied, either by substrate or by other compounds, e.g., citrate cycle intermediates,[2] the reaction rate dependence on the substrate concentration changed to give a hyperbolic shape (Figures 1 and 2). Lactate dehydrogenase from the latex tree (Hevea braziliensis), studied by D'Anzac and Jacob,[18,19] and potato lactate dehydrogenase

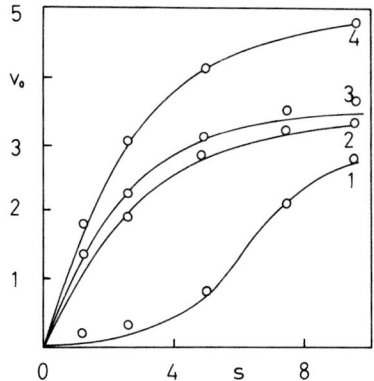

Figure 1. Dependence of the initial rate of reaction catalyzed by soybean lactate dehydrogenase (v_0) on the concentration of substrate (s), *i.e.,* pyruvate (M x 10^{-4}) - (1), and the same dependence in the presence of effectors in concentration 5 x $10^{-5} M$: malate (2), succinate or α-ketoglutarate (3), citrate (4).

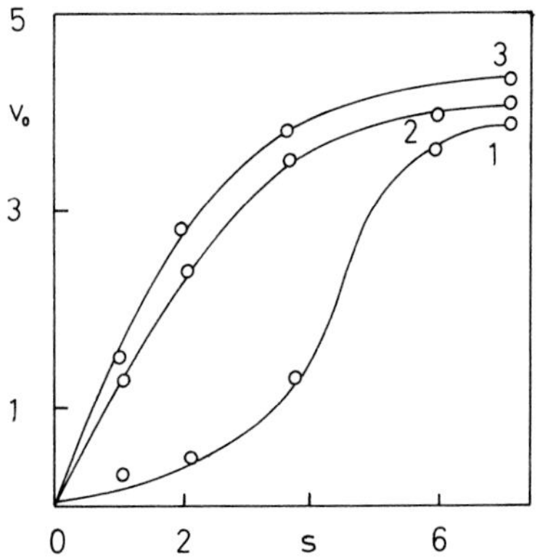

Figure 2. Dependence of the initial rate of reaction catalyzed by soybean lactate dehydrogenase (v_0) on the concentration of substrate (s), *i.e.,* NADH (M x 10^{-5}) - (1), and the same dependence in the presence of effectors in concentration 5 x 10^{-5} M: α-ketoglutarate or citrate (2), malate (3).

studied by Davies and Davies,[5] also exhibited allosteric character under certain conditions.

The pH optimum values for the two studied enzymes were different for the opposite directions of the reaction catalyzed by lactate dehydrogenase, similar to the animal enzyme (Figures 3 and 4). The maximum rate of pyruvate reduction was attained at pH 7.2 and 7.1 for soybean and pea lactate dehydrogenase, respectively; the maximum rate for the oxidation of lactate was attained at pH 9.0 and 9.1 for the former and the latter enzymes, respectively. The optimum pH values for the seedling lactate dehydrogenases were very close to the values found for the animal enzyme, but differed greatly from the data obtained for potato lactate dehydrogenase. Davies and Davies[5] found for the latter that the optimum for the reduction of ketoacids lay at pH 6.5, and for the oxidation of lactate, around pH 7.8. The different pH optima for the activity for the opposite reaction directions apparently play a significant role in the plant and animal organism during regulation of the pyruvate metabolism.

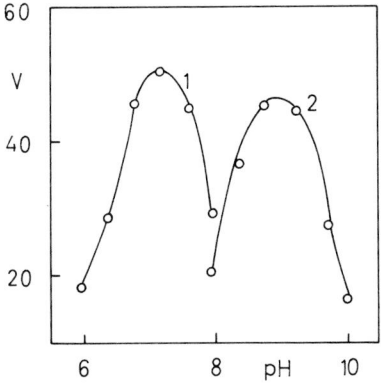

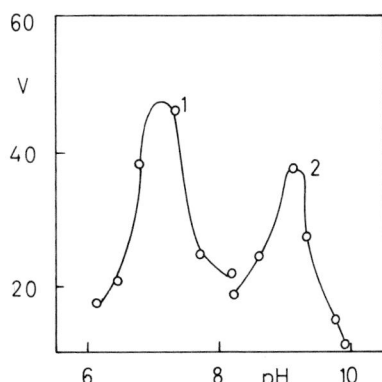

Figure 3. Dependence of the maximum reaction rate (V) for soybean lactate dehydrogenase on pH as regards reduction of pyruvate (1) and oxidation of lactate (2).

Figure 4. Dependence of the maximum reaction rate (V) for pea lactate dehydrogenase on pH as regards reduction of pyruvate (1) and oxidation of lactate (2).

LACTATE DEHYDROGENASE ACTIVITY REGULATION

We found that soybean lactate dehydrogenase was inhibited by some compounds analogous to the coenzyme. We studied the effect of adenine, adenosine, AMP, ADP and ATP. All these compounds inhibited the lactate dehydrogenase activity, the inhibition had a competitive characteristic with respect to NAD and NADH, and the inhibition constant values (K_i) decreased from adenine to ATP (Table IV). The phosphate or pyrophosphate group in position $5'$ apparently played a role in the inhibition mechanism. Consequently, it can be assumed that also the coenzyme binding to the protein was effected through this part of the molecule.

We have also found that inhibition of lactate dehydrogenase by the physiologically most important compound of this series, *i.e.*, ATP, was dependent upon the pH of the medium. The lower the pH, the stronger was the inhibition. On a pH change from 7.2 to 6.5, the inhibition constant for ATP decreased to almost one half (K_i was 10.5 μM at pH 7.2 and only 6.0 μM at pH 6.5). In this respect, the seedling lactate dehydrogenase behaved analogously to the potato enzyme[5] and the enzyme from the latex tree.[19] This property of lactate dehydrogenase was very important for living organisms. Lactate was produced in the cell by the activity of lactate dehydrogenase. Its accumulation led to a decrease in the cytoplasma pH, which would be dangerous for the

Table IV. Inhibition of Soybean Lactate Dehydrogenase by Compounds Analogous to the Coenzyme

Inhibitor	K_i (mM)	
	NADH as a Substrate	NAD as a Substrate
Adenine	0.75	3.4
Adenosine	0.26	2.3
AMP	0.96	4.4
ADP	0.0027	0.012
ATP	0.00056	0.0025

cell if it exceeded a certain limit. However, there are ways of dealing with this situation in the cell, which also involve the regulation properties of lactate dehydrogenase. Due to the decrease in the cytoplasma pH, the inhibition of lactate dehydrogenase by ATP increased and consequently, the lactate production was retarded.

Certain other intermediates in the carbohydrate metabolism also exerted an effect on the lactate dehydrogenase activity. The enzyme was inhibited, e.g., by fructose diphosphate. However, it followed from the inhibition constant (K_i = 64 mM) that this effect would not be as important in the lactate dehydrogenase activity regulation as the effect of ATP, even though the concentration of this compound in the organism is probably higher.

By the experiments presented in this review, we confirmed the important role of lactate dehydrogenase in the plant metabolism, above all during the change of the conditions from aerobic to anaerobic. In this situation, the enzyme functions in restoration of oxidized NAD required for the surviving of the plant. The enzyme undoubtedly takes part in removing products of anaerobic metabolism after the termination of anaerobiosis.

The possibility of regulation of the lactate dehydrogenase by the pH in cytoplasma and by some intermediates, mainly ATP, assures that the concentration of undesirable products of anaerobic metabolism will not exceed their toxic level in plant tissue.

In our experiments, we continue in the study of structural and kinetic properties of plant lactate dehydrogenase supposing that elucidation of these enzyme properties will contribute to the understanding of processes in the plant organism in which the lactate dehydrogenase takes part.

REFERENCES

1. Barthová, J., J. Borvák, and S. Leblová. "Isolation and Properties of Lactate Dehydrogenase from Germinating Pea Plants," *Phytochem.* 15, (1):75 (1976).
2. Barthová, J., P. Hrbas, and S. Leblová. "Isolation and Properties of Plant Lactate Dehydrogenase," *Colection Czech. Chem. Commun.* 38, (7):2174 (1973). (In Czechoslavakian.)
3. Barthová, J., P. Hrbas, and S. Leblová. "Some Structural and Kinetic Characteristics of Lactate Dehydrogenase from Soybean Seedlings (*Glycine max.* L.)" *Colection Czech. Chem. Commun.* 39, (11):3383 (1974). (In Czechoslovakian.)
4. King, J. "The Isolation, Properties, and Physiological Role of Lactic Dehydrogenase from Soybean Cotyledons," *Can. J. Bot.* 48, (3):533 (1970).
5. Davies, D. D., and S. Davies. "Purification and Properties of L/+/ −Lactate Dehydrogenase from Potato Tubers," *Biochem. J.* 129, (4): 831 (1972).
6. Everse, J., and N. O. Kaplan. "Lactate Dehydrogenase: Structure and Function," *Adv. Enzymol.* 37, (1):61 (1973).
7. Fottrell, P. F. "Dehydrogenase Isoenzymes from Legume Root Nodules," *Nature* 210, (5032):198 (1966).
8. Hadačová, V., and J. Švachulová. "Isoenzymes of Lactate Dehydrogenase in the Root Growth Zones of *Vicia faba* L.," *Biol. Plant, 14,* (2):170 (1972).
9. Rothe, G. M. "Catalytic Properties of Three Lactate Dehydrogenases from Potato Tubers (*Solanum tuberosum*)," *Arch. Biochem. Biophys.* 162, (1):17 (1974).
10. Brinkman, F. G., and L. J. van der Meer. "Dehydrogenases in the Potato Tubers (*Solanum tuberosum*). Identity, Coenzyme-Specificity and Isoenzyme Composition of Malic Enzyme, Malate Dehydrogenase and Lactate Dehydrogenase," *Z. Pflanzenphysiol.* 75, (4):322 (1975).
11. Holbrook, J. J., and R. A. Stinson. "Reactivity of the Essential Thiol Group of Lactate Dehydrogenase and Substrate Binding," *Biochem J.* 120, (2):289 (1970).
12. Hathaway, G., and R. S. Cridle. "Substrate-Dependent Association of Lactic Dehydrogenase Subunits to Active Tetramer," *Proc. Nat. Acad. Sci., U.S.* 56, (2):680 (1966).
13. Fritz, P. J. "Rabbit Muscle Lactate Dehydrogenase 5; A Regulatory Enzyme," *Science* 150, (3694):364 (1965).
14. Markert, C. L., and E. J. Massaro. "Lactate Dehydrogenase Isoenzymes: Dissociation and Denaturation by Dilution," *Science* 162, (3854):695 (1968).
15. Monod, J., J. Wyman, and J. P. Changeux. "On the Nature of Allosteric Transition: A Plausible Model," *J. Mol. Biol.* 12, (1):88 (1965).
16. Koshland, Jr., D. E., G. Némethy, and D. Filmer. "Comparison of Experimental Binding Data and Theoretical Models in Proteins Containing Subunits," *Biochemistry* 5, (1):365 (1966).

17. Koshland, Jr., D. E., and K. E. Neet. "The Catalytic and Regulatory Properties of Enzymes," *Ann. Rev. Biochem.* 37:359 (1968).
18. D'Anzac, J., and J. L. Jacob. "Inhibition par l'ATP de la malate-déshydrogénase, de l'alcool-déshydrogénase et de la lactate-déshydrogénase du latex d'Hevea braziliensis," *Bull. Soc. Chim. Biol.* 50, (1): 143 (1968).
19. Jacob, J. L., and J. D'Anzac. "La l'alanine, inhibiteur allosterique de la lactate déshydrogénase du latex d'Hevea braziliensis. Influence du pH, par MM," *C. R. Acad. Sci. Ser. D* 266:631 (1968).

16

REGULATIVE INTERACTION BETWEEN ANAEROBIC CATABOLISM AND NITROGEN ASSIMILATION AS RELATED TO OXYGEN DEFICIENCY IN MAIZE ROOTS

Johannes-Gunter Kohl, Jaroslava Baierova,
Gisela Radke and Konrad Ramshorn

Section Biology
Humboldt-University of Berlin
Berlin, German Democratic Republic

INTRODUCTION

With its root system, the higher plant is adapted to the soil conditions with regard to its oxygen demand, for the oxygen supply in the soil may be lower than that in the above-ground atmosphere. The share of oxygen in the soil air varies between about 10 and 20% by volume, depending on the structure and type of soil concerned. Oxygen uptake through the root is, however, controlled by the water content or by the water films around the root, respectively, through which the dissolved oxygen would diffuse to the organ. The immediate oxygen supply to the root will then be much lower (0.6% by volume in H_2O at 24°C), not to mention the fact that the oxygen supply from the above-ground atmosphere for replacement of the soil oxygen spent on respiration would be more or less retarded due to the structure of the soil. Further reduction of oxygen supply resulting from the large amount of water contained in the soil will lead to more or less distinct anaerobiosis bringing about a certain change of metabolism. The same will also occur in hydroponic culture of higher plants, if the usual nutrient solution is not aerated. Without aeration, only about 60% of the maximal soluble oxygen quantity will be present.[1]

Moreover, the root morphology gives rise to a different mode of dissimilation in the various parts of the root. Even under "aerobic" conditions, *i.e.,* in normal atmosphere, fermentation processes take place together with oxidative respiration in the meristem of the root tip. This may be ascribed to an insufficient oxygen level in that compact tissue and would then lead to increased anaerobic catabolism and to the accumulation of ethanol and lactate.[2-6]

On the level of metabolic regulation, however, this change of metabolism proceeds also through specific ADH and LDH induction.[3,6-9]

However, the ADH activity rises only in so-called flood-intolerant plants. In flood-tolerant plants, there will be no appreciable increase in the glycolytic rate and no or only slight ADH induction. This ecologically different behavior has been observed in a large number of plants.[6,11-13]

The presence or absence of nitrate, respectively, plays an important part in the regulation of the fermentation component. Arnon[14] referred to the fact that nitrate supply partly relieves the growth depression arising from anaerobiosis in the root zone. Willis and Yemm[15] found the RQ to decline in the presence of nitrate. The delay of the nitrate effect after application, as established in their experiments, may well be explained by the occurrence of a lag-phase on NR induction.[16] This indicates that nitrate reduction or nitrogen assimilation, respectively, would bring about a physiological compensation of the effects of oxygen deficiency. The regulative interactions between nitrogen metabolism and anaerobic dissimilation should therefore be the focus of increased research when studying the metabolic adaptation of the root to anaerobic conditions. This chapter concentrates on clarifying the regulation of the ADH activity and the ADH isoenzyme pattern (as indicators for satisfying the oxygen demand of the root) through the oxygen tension and the interactions with nitrate reduction and amino acid synthesis in maize. Moreover, for the purpose of clarifying the problem of the presence and causality of a cultivar-specific adaptability to oxygen deficiency in the root tissue of different maize cultivars, comparative investigations are presented regarding the dependence of respiratoric rate, ADH activity, seedling growth and mobilization of storage material on the oxygen tension.

INTERACTION OF OXYGEN AND NITRATE IN THE REGULATION OF ADH AND LDH ACTIVITIES IN MAIZE ROOTS

As related to the specific activities of some representative enzymes of the oxidative carbohydrate breakdown routes, the ADH activity is higher in the meristem of the seedling root than in the other parts of the root.

With increasing root length and decreasing root diameter, this gradient of the relative ADH activity within the root is offset with simultaneous activity decline in all parts (Figure 1). These different levels of ADH activity occurring at equal exogenous oxygen supply seem to be due to a

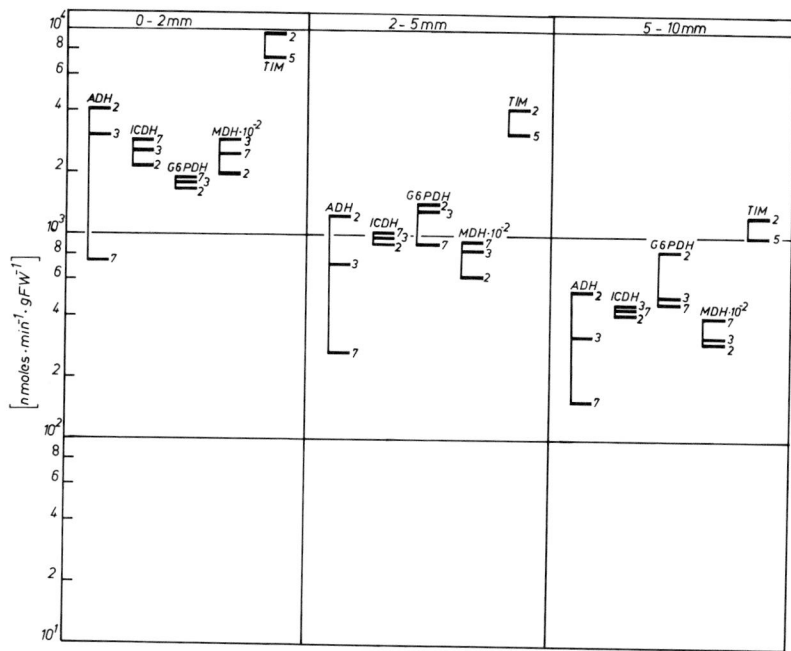

Figure 1. Enzyme activities of different zones of 2-, 3-, 5- and 7-cm roots of maize seedlings (*Zea mays* L.) (sand culture, DD, $\bar{x}$, n=5).

different "degree of aerobiosis" in the tissue. This degree of aerobiosis is essentially dependent upon oxygen consumption (respiration intensity) and on the rate of oxygen diffusion into the tissue. Internal diffusion of oxygen is essentially dependent upon certain morphological characteristics: root diameter, relative size of intercellular space, packing density of cells or degree of vacuolization. Similar relations are found also when comparing the enzyme patterns of roots of different plant species having highly different root diameters: In *Vicia faba* L., the ADH activity reaches about the same level as in *Zea mays* L., but ADH levels are much lower in the thinner roots of *Triticum aestivum* L. and in *Hordeum vulgare* L. (Figure 2).

In case of oxygen deficiency, with no aeration of nutrient solution, oxygen saturation in the nutrient solution is about 60% and the ADH

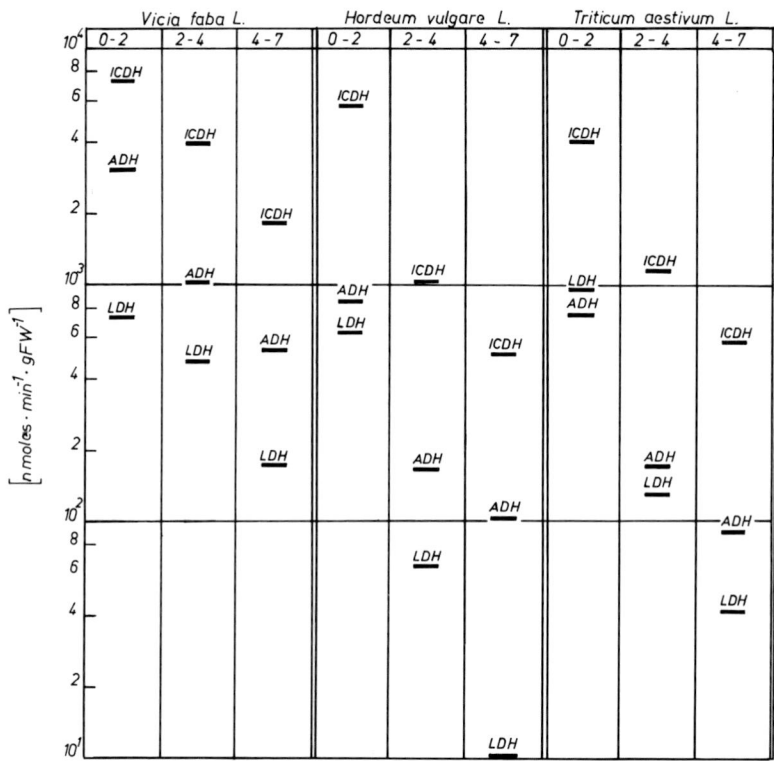

Figure 2. Enzyme activities of different sections of the root tips of 3-cm primary roots of different crop plant species (sand culture, DD, $\bar{x}$, n=6).

activity goes up considerably. The increase, however, is much less pronounced in the presence of nitrate than in a nitrate-free nutrient medium (Figure 3). In case of an adequate oxygen level (aerated nutrient solution), nitrate supply will lead to a considerable decline in ADH activity in the meristem only. This, too, indicates that under "aerobic" conditions morphologically induced oxygen deficiency existed in the meristem. The effect of this oxygen deficiency on the ADH activity may, however, be compensated by nitrate as well.

The LDH activity, too, goes up in case of oxygen deficiency, but to a lesser extent (Figure 3). Moreover, it is somewhat higher in the presence of nitrate. All other enzymes investigated (Figure 3) were relatively insensitive to an oxygen deficiency. As compared with results obtained on other objects,[13,17] the maize ME is not repressed under oxygen deficiency.

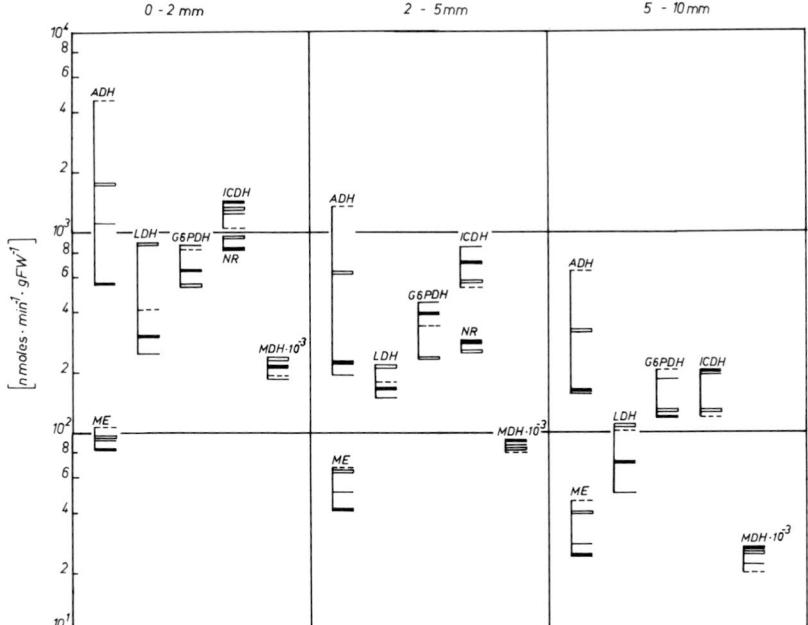

Figure 3. Enzyme activities of different zones of the primary root of 4-day-old maize seedlings at different degrees of aerobiosis in the nutrient solution and in dependence on nitrate supply (KNOP's nutrient solution without nitrogen or with 10.9 mM nitrate, 18 hr with or without aeration: ■■■ with aeration, with nitrate; ☐ without aeration, with nitrate; ——— with aeration, without nitrate; - - - -without aeration, without nitrate; $\bar{x}$, n=5).

Unlike the conditions found in facultatively anaerobic bacteria,[18] the NR activity in maize is not stimulated by oxygen deficiency (Figure 3). This is inconsistent with results obtained in experiments with whole roots of *Vicia faba* L.[19] and with several flood-tolerant plants.[6,20,21] On the one hand, this discrepancy may be due to the methical approach, as a number of sources of error could be traced in NR activity determination.[22] On the other hand, the high NR activity in meristems may lead to increased NR activity in the whole root in connection with changes in root morphology (higher degree of branching, formation of lateral root initials) without the activity being changed in the same tissues of treated and untreated plants.

It is shown that the NR activity is not influenced by oxygen deficiency in maize. Therefore, the degree of oxygen deficiency compensation by nitrate reduction is dependent upon the actual NR activity in the tissue, *i.e.,* the higher the NR level, the smaller the increase of ADH activity following oxygen deficiency. This becomes quite obvious from the different rates of increase of the ADH activity found in root tips of 4- and 10-day old maize plants, which differ significantly from each other as to their NR activity (Table I). The ADH activity is increased 2.8 fold in plants aged 4 days (NR activity: 43.2 nm min^{-1} gFW^{-1}) and about 8 fold in plants aged 10 days (NR activity: 20.7 nm min^{-1} gFW^{-1}). Such a high rate of increase is found in 4-day-old plants only in the absence of nitrate (Figure 3). In the root, the ability to compensate for oxygen deficiency is therefore essentially dependent upon the rate of nitrate assimilation.

Table I. Influence of Culture Aeration on ADH Activity
(nm min^{-1} gFW^{-1}) of Root Tip (1 cm) of 4- or 10-Day-Old Maize Plants
Under Different Concentrations of Nitrate (Knop-solution, LL, x̄, n=5)

Age (days)	Conc. (mM)	ADH Activity	
		With Aeration	Without Aeration
4	10.9	220	616
10	10.9	125	1063
10	1.0	183	1346
10	0.1	202	1473

REGULATION BY OXYGEN LEVEL OF THE ADH ISOENZYME PATTERN

Considering the difference in ADH activity between the various zones of the root tip of maize seedlings, it was tested whether these zones would have different ADH isoenzyme spectrums as well.[1,2,3] However, no qualitative differences were found between the isoenzyme patterns of the root sections tested. The pattern is identical with that of the root tip as a whole.

Different relations are found when comparing the three maize cultivars 'Hd 405', 'Sz 71' and 'Schindelmeiser' and in case of different rearing conditions (Figure 4). Cultivar-specific isoenzyme spectrums are found in 3-day-old seedlings reared in moist sand. 'Hd 405' shows both the largest number of isoenzyme bands and the highest ADH activity (Table II). This feature agrees well with the outstanding vigor of 'Hd 405,' which manifests itself, for example, in the rate of RNA synthesis in the root tip.[24] A certain relation between these parameters might exist inasmuch as more intensive root metabolism requires more ATP and induces more drastic oxygen deficiency. In all the cultivars tested, the

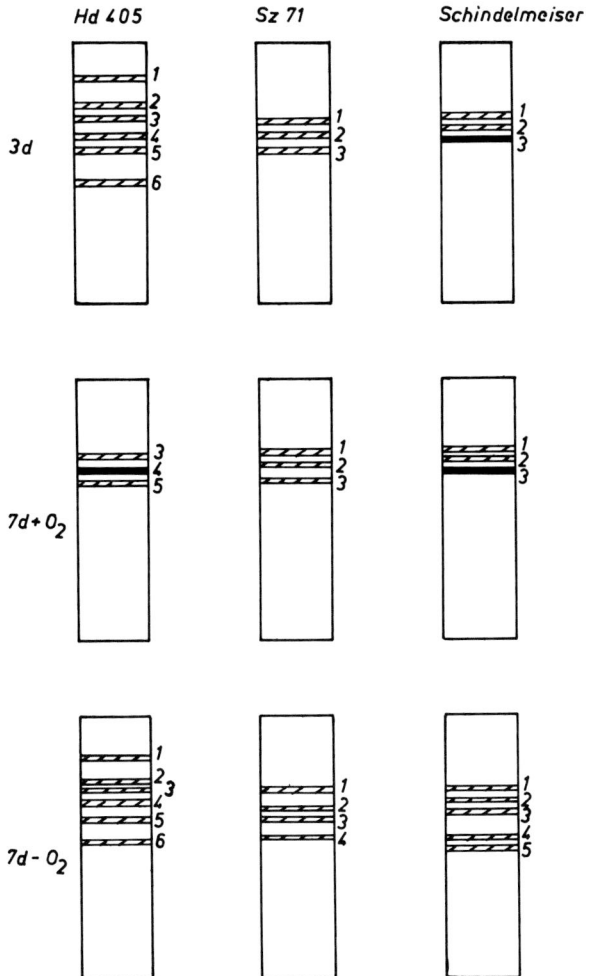

Figure 4. The ADH isoenzyme pattern (protein separation in 7.5% polyacrylamide gel) of the 1-cm root tip of three maize cultivars in dependence on plant age and on the aeration of the nutrient solution (3 days: sand culture; 7 days + O_2: aerated hydroponic culture according to KNOP; 7 days - O_2: hydroponic culture, no aeration during the last 48 hr) (n=6).

Table II. ADH Activity (a) (nm min^{-1} gFW^{-1}, $\bar{x}$) and Number of Bands in Isoenzyme Pattern (b) of Root Tips of Three Different Maize Cultivars (n=6)

| Cultivar | Plant Age | | | | | |
| | 3 Days Sand | | 7 Days Hydroponics With Aeration | | 7 Days Hydroponics the Last 48 hr Without Aeration | |
	a	b	a	b	a	b
Schindelmeiser	812	1-3	160	1-3	1710	1-3
Sz 71	1001	1-3	148	1-3	1450	1-4
Hd 405	1480	1-6	235	3-5	2150	1-6

increase in ADH activity found in 7-day-old seedlings after 48 hr of partial anaerobiosis is connected with the occurrence of new isoenzyme bands, but also in that case, there appear both varietal differences and a certain correlation between the number of bands and the ADH activity (Table II).

It may be easily concluded from these comparisons that the increase in ADH activity up to a certain degree of anaerobiosis would first be accomplished through the increase in the activity of the same isoenzymes (turnover regulation) and that the "induction" of the additional iso-enzymes would be triggered off only when exceeding that limit. This possibility of interpretation is also well supported by the equally missing effect of the nitrate level on the isoenzyme spectrum.

The experiments described here were performed only at 60% oxygen saturation of the nutrient solution. A further increase in ADH activity and a further rise in the number of bands obtained under full anaerobiosis is therefore not to be excluded.[25]

In addition, the important question as to whether the occurrence of the new ADH bands would be reversible was tested in the 'Schindelmeiser' and 'Hd 405' cultivars. In both, aeration of the nutrient solution follow-ing 48 hr of partial anaerobiosis caused the additional bands to disappear. This corroborates the idea that the oxygen concentration was the deci-sive factor influencing the variation of the ADH isoenzyme spectrum.

Inhibitor studies (cycloheximide, ug^{-1}ml^{-1}) have shown that both the increase in activity and the appearance of new bands (Table III) depend on protein synthesis.

The molecular weights of the individual ADH isoenzymes of the 'Schindelmeiser' cultivar were compared by means of electrophoresis according to Hedrick and Smith.[26] This comparison revealed that the ADH proteins newly formed during partial anaerobiosis had the same molecular weight as the proteins of the "constitutive" bands.

Table III. ADH Activity (a) (nm min^{-1} gFW^{-1}, $\bar{x}$) and Number of Bands in Isoenzyme Pattern (b) of Root Tips of Maize (cv. Schindelmeiser) Under Influence of Cycloheximide (1 μg/ml) (n=3)

With Aeration		7 Days the Last 48 hr Without Aeration + Cycloheximide		the Last 48 hr Without Aeration	
a	b	a	b	a	b
160	1-3	429	1-3	1710	1-5

Considering that the formation of new ADH bands may be inhibited by cycloheximide and that the information for ADH synthesis is coded by two genes,[27-29] several explanations may be derived as to the formation of the anaerobic pattern of the ADH bands, a *de novo* synthesis being the most probable.[1]

INFLUENCE OF OXYGEN DEFICIENCY ON THE TRANSLOCATION OF STORAGE MATERIAL AND GROWTH OF SEEDLINGS

The varietal specificity of the increase in ADH activity and of the variation of the isoenzyme patterns during anaerobiosis suggested varietal differences in the growth susceptibility to oxygen deficiency in the root zone. The extent to which the relations between the oxygen content of the surrounding medium and the enzyme activity and enzyme structure of the ADH become effective in the range of total metabolism may be recorded by the analysis of the storage material mobilization and of biomass growth. In general, in all three cultivars tested, partial anaerobiosis had an inhibiting effect on the rate of storage material decline in the endosperm as well as on the dry matter increase of root and shoot (Figure 5). Varietal differences were found with regard to the response to nonaeration of the nutrient solution. In 'Hd 405' the dry matter growth of the axis was inhibited by oxygen shortage; during the first 24 hr of partial anaerobiosis, the effect of the reduced oxygen supply of the root system on the dry matter growth was less pronounced in 'Hd 405' than in 'Sz 71' and 'Schindelmeiser' (Figure 5). The latter two cultivars were distinguished by their almost complete stagnation of dry matter growth during the first 24 hr of partial anaerobiosis. However, this stage was followed by a rapid increase in the rate of dry matter growth. Regardless of the cultivar-specific response to the changed conditions, in all three cultivars under investigation, the root dry matter growth was more affected by nonaeration of the nutrient solution than

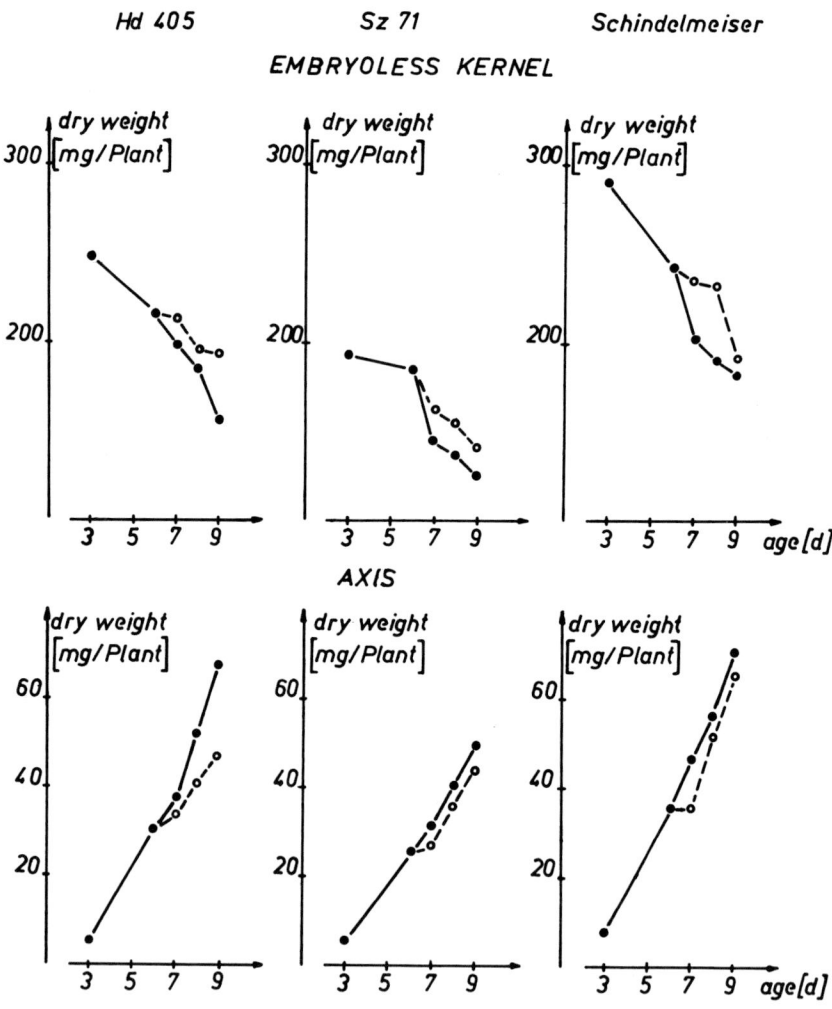

Figure 5. Changes in dry matter of axis and residual caryopsis in mg/plant in aerated ($\bullet$ O_2) and nonaerated ($\circ$ O_2) nutrient solution ($\bar{x}$, n=5).

the dry matter growth of the above-ground part of the plant (Table IV). By removing some of the plants from the nutrient solution, oxygen

Table IV. Changes in Oxygen Saturation (%) of Hydroponic Solution of Aerated Maize Cultures Under Nonaerated Conditions of 24, 48 and 72 hr, Respectively, with Plant Density Unchanged (a) and Decreasing Plant Density (b)

Duration of Culture Without Aeration (hr)	Relative Oxygen Saturation (%)	
	a	b
24	59.8	62.2
48	63.1	73.4
72	56.8	75.3

saturation may rise from 60-70% and thereby reach the threshold value where cultivars 'Sz 71' and 'Schindelmeiser' shift to aerobic growth velocity (Figure 6; Table V). However, it appears that cultivar 'Hd 405' may have a threshold value higher than 70% oxygen saturation because at this oxygen level, the growth velocity was much less than under aerated conditions.

Table V. Dry Weight of Root and Shoot and Amino Acid Content of Meristematic (M) and Differentiated Tissues (D) of Roots of 6-Day-Old Maize Plants, Cultivated Under Different Supply of Oxygen and Mineral Nitrogen (Figure 7)

	Culture Conditions				
	-N		$+ NO_3^-$		$+ NH_4^+$
	$+ O_2$	$-O_2$	$+O_2$	$-O_2$	$+O_2$
			mg of DW		
Shoot	54.1	35.0	75.6	42.8	55.7
Root	26.1	12.7	25.8	11.3	20.9
			Amino Acids nm/100 mg FW		
M	3139	2684	4631	3447	4586
D	1185	1434	813	3937	4001

INFLUENCE OF OXYGEN DEFICIENCY ON THE ABILITY OF THE RESPIRATORY APPARATUS

In connection with the investigation of ADH regulation, there arose the question whether the "potential respiration" of the root tip, as

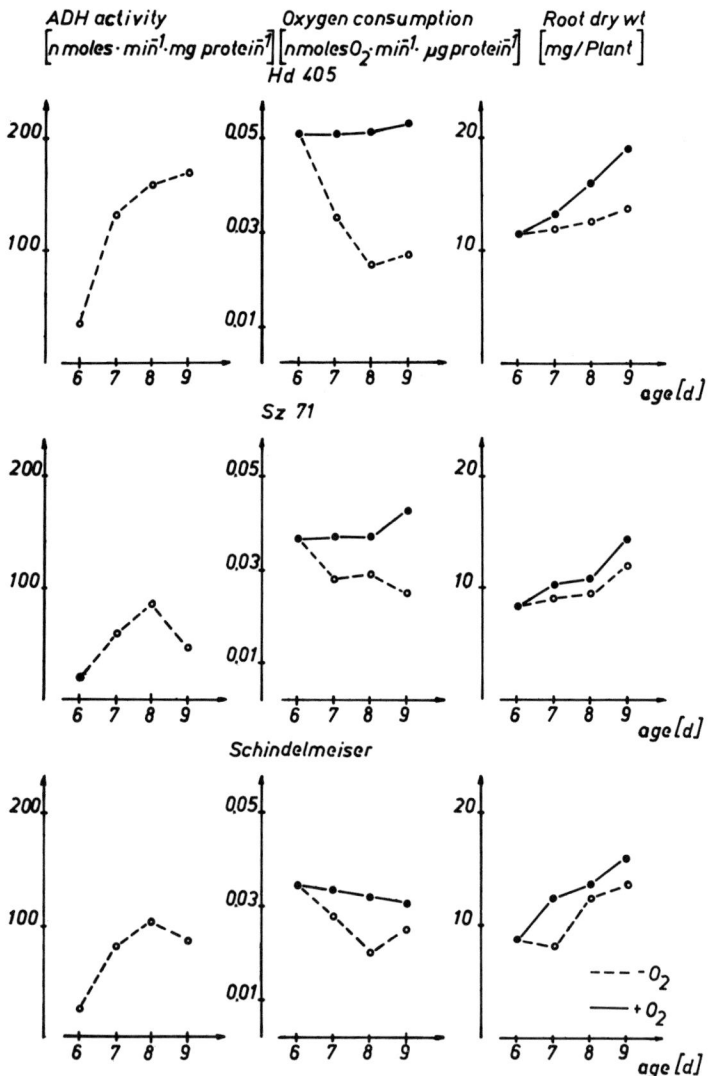

Figure 6. ADH activity (nm mg prot[-1] min[-1]), oxygen uptake by means of oxygen electrode after Clark (nm O_2 μg prot[-1] min[-1]) of the 1-cm root tip, and the change in dry matter of the main root (mg/plant) in seedlings of the three maize cultivars tested, reared under different conditions: aerated nutrient solution (————); nonaerated nutrient solution (- - - - -); ($\bar{x}$, n=5).

measured at oxygen saturation of the nutrient solution, would change during anaerobiosis or remain the same.[1] The results of these investigations showed that after 24 hr of partial anaerobiosis, oxygen consumption, as measured under atmospheric oxygen saturation of the nutrient solution, declined in all the cultivars tested. This amount of decline was, however, dependent upon the cultivar concerned (Figure 6) and amounted to 24.3% in 'Sz 71', 37.7% in 'Hd 405' and 17.4% in 'Schindelmeiser'. After 96 hr of partial anaerobiosis, the "potential respiration" decline further in 'Hd 405' and 'Schindelmeiser,' whereas no changes were observed in 'Sz 71.'

Figure 6 illustrates the effect of oxygen deficiency on ADH activity, potential respiration of the root tip, and dry matter growth of the main roots of the three maize cultivars tested.

Under anaerobic conditions, potential respiration declined as the ADH activity went up, *i.e.,* the respiration intensity of roots grown under oxygen deficiency was much lower than that of roots continuously grown in an aerated nutrient solution when measured under oxygen saturation. The responsiveness of potential respiration and of ADH activity to changed oxygen levels was most pronounced in 'Hd 405.' Also, the root dry matter increase of 'Sz 71' and 'Schindelmeiser' were less sensitive to oxygen shortage than 'Hd 405,' and respiration potential, after an initial decrease, increased with slightly raised oxygen levels. It is interesting that the rise in growth started before a rise in respiration intensity could be detected. A possible explanation of this lag may be that upon exposure to oxygen after its absence, an adaptive process in the mitochondria must take place before it becomes fully functional.[30] During the adapting stage, lack of energy supply from oxidative respiration seems to be compensated by fermentation. In the coleoptile of the rice plant—the latter being very well adapted to anaerobic conditions—oxidative respiration becomes fully effective within 30 minutes of the onset of aeration.[31] Thus, the regulatory mechanism in the rice coleoptile is different from that in the root. It therefore seems feasible that besides the metabolic regulation of oxidative respiration and fermentation, and the epigenetic regulation of ADH, still other regulatory mechanisms (*e.g.,* variations in the enzyme pool of the mitochondria through varying membrane bonding of the enzymes of the respiratory electron transport system etc.) were present, which cause metabolism to be dependent on the oxygen level.

INFLUENCE OF OXYGEN DEFICIENCY ON THE LEVEL AND PATTERN OF THE AMINO ACIDS IN DIFFERENT ROOT ZONES

Insufficient oxygen supply causes fundamental changes in the intermediary levels of glycolysis, TCC, etc. Particularly remarkable are the changes of the pools of the keto acids pyruvate and oxalacetate and of the reduction products ethanol, lactate and malate derived therefrom.[5,32-35] As the effect of oxygen deficiency was compensated by nitrate supply and as the levels of major amino acid precursors rose under anaerobiosis, it appeared to be particularly urgent under these conditions to investigate the change of the amino acid level in the roots. The amino acid contents were determined by a single-column technique using an amino acid analyzer.[36]

Reduced oxygen tension (nonaerated nutrient solution according to KNOP) has, in many respects, significant effects on the amount of free amino acids contained in the meristem and differentiated tissue of the roots of 6-day-old maize plants. A comparison of the roots from the nonaerated culture with the aerated one showed that the sum of free amino acid contents in the differentiated tissue increased 4.8 fold in case of nonaeration with nitrate (Table V). The total amount of amino acids in the meristem declined by 25.6%. The background of changes in total contents of amino acids were clarified by analyzing the absolute quantities as well as the percentages of the individual free amino acids (Figures 7 and 8).

The amino acid pattern in the meristem changed in relation to the quantitative and percentage decline of almost all amino acids except glutamate; the absolute quantity of glutamate remained almost unchanged as compared with the aerated culture, while its relative quantity increased as a result of that.

An absolute rise in the alanine content without aeration but with nitrate and an increase of the percentage from 20.8-44.6%, as compared with the aerated culture, is obvious. Under partial anaerobiosis, the amounts of the individual amino acids contained in the differentiated tissue showed a more or less pronounced increase, and the rise in alanine content to about 35 times the original value was particularly striking. This outstanding change was expressed also in the relative contents of the individual amino acids. Except for alanine, their shares in the total content tended to decline. Alanine increased strikingly from 8 to 59.1%. Among the major quantitative variations of the absolute amino acid contents that occurred from anaerobiosis, the decline in the asparagine and glutamine contents of the meristem was quite remarkable. On the other hand, these amides were found to increase in the differentiated tissue.

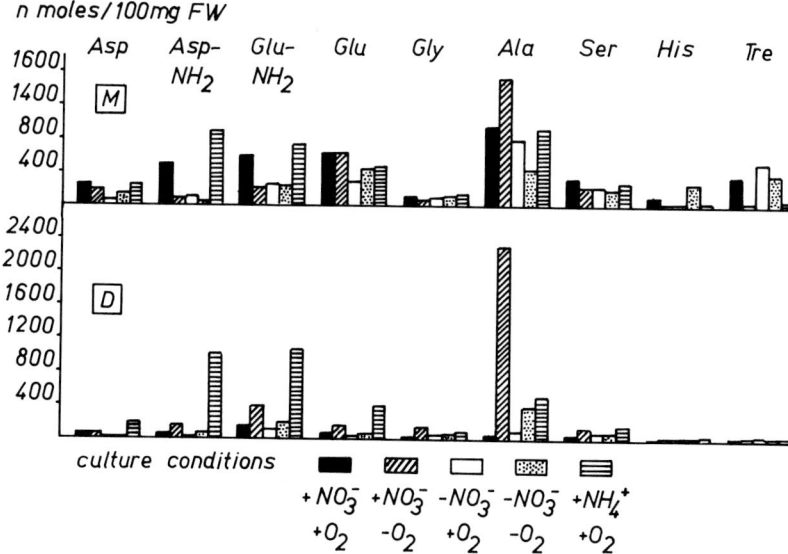

Figure 7. Amounts of free amino acids contained in meristem (M) and in differentiated tissue (D) of 6-day-old maize seedling roots as affected by cultural conditions (KNOP's nutrient solution without nitrogen, with 10.9 mM NO$_3^-$ or with 11 mM NH$_4^+$, each with (+ O$_2$) and without (-O$_2$) aeration, $\bar{x}$, n=5).

In the differentiated root section, the vascular tissues held a remarkable share of the total content, especially with regard to those amino acids that were translocated from the endosperm to the growing tip of the root. With a declining growth rate under partial anaerobiosis, an increase in the amide contents in the vascular tissues seemed to be the most probable reason for the rise in total amide content.

The large alanine quantities which accumulated under anaerobic conditions were included in the metabolism on subsequent aeration of the nutrient solution or transported off, respectively, for after 24-hr aeration the alanine content declined to 9.9%, while at the same time the amide content increased.[36]

No alanine accumulated in the absence of nitrate, *i.e.,* alanine accumulation was the result of the incorporation of the assimilated nitrate nitrogen into the alanine pool with simultaneous reduction of amide synthesis. This diversion becomes obvious if one considers at the same time the results of the test variants without nitrate supply or with ammonium supply, respectively (Table VI). When comparing the amino acid contents of plants reared on aerated nutrient solution, each without nitrogen and with NO$_3^-$ or NH$_4^+$, respectively, the high meristem glutamine contents,

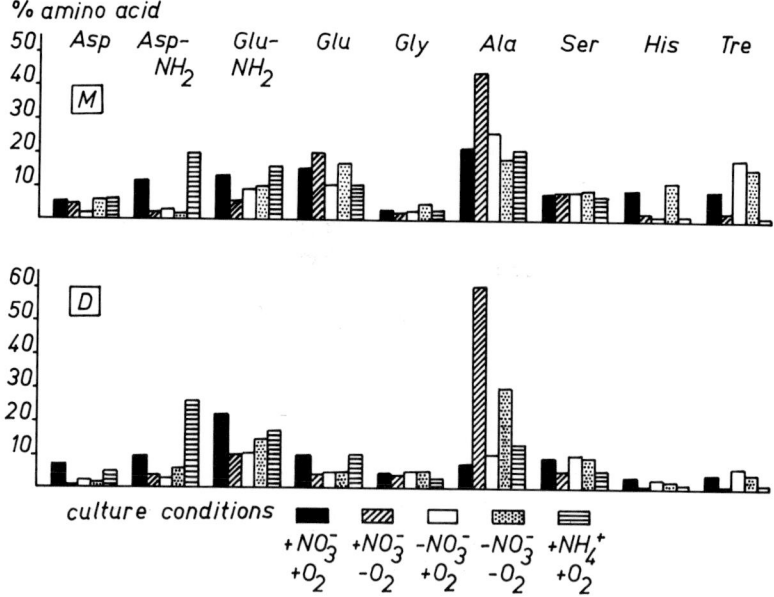

Figure 8. Percentages of the individual amino acids in total amino acid content. See Figure 7 for legend.

Table VI. Content of Amino Acids (nm/100 mg FW) in Meristematic (M) and Differentiated Tissues (D) of Roots of 7-Day-Old Maize Plants, Cultivated for the Last 24 hr Under Different Supply of Oxygen and Mineral Nitrogen (x, n=5)

Duration of Culture		Culture Conditions					
		$+ O_2$			$- O_2$		
0-6 days 6-7 days		$+ NO_3^-$	- N $+ NH_4^+$	- N	$+ NO_3^-$	- N $+ NH_4^+$	- N
Glu - NH_2	M	926	1116	180	505	450	211
	D	880	1593	170	524	290	309
Asp - NH_2	M	119	204	37	74	411	24
	D	78	294	22	54	160	77
Ala	M	510	205	558	952	985	406
	D	249	167	91	469	344	268
Total	M	3364	3167	3649	2972	3724	2023
Amino	D	1910	2857	1388	1629	1630	1441
Acids							

in the case of nitrate supply, indicate the reduction or assimilation of nitrate to be remarkably high in that tissue. In the case of ammonium supply, compared with the nitrate variant, the glutamine values in the meristem were not much higher, whereas the high glutamine contents that are found under these conditions, in the differentiated tissue, produced evidence for the almost equal participation of all root sections in the assimilation of NH_4^+.[37]

In case of oxygen deficiency, the amide levels were much lower in the presence of NO_3^- and NH_4^+, but nevertheless somewhat higher than in the control variant (without nitrogen). On the other hand, there appeared to be an extremely intensive alanine accumulation with both NO_3^- and NH_4^+ supply (Table VI). While in an aerated nutrient solution, glutamine would be the quantitatively predominant amino acid; under anaerobic conditions alanine was dominant even in the presence of NH_4^+. This was probably because under anaerobic conditions glutamine synthesis was reduced at low energy charge and alanine synthesis was increased by a higher pyruvate level.[5,32,38]

The increase in the amount of nitrogen contained in the alanine pool was, however, not equivalent to the decline of the nitrogen content in the amide pool; rather, it was lower. On the other hand, the alanine quantities contained in the differentiated tissue were many times higher. This may point to an intensive translocation to the shoot of the alanine developed under anaerobic conditions. However, the indications available are not yet sufficient for such interpretation. Experimental proof of this assumption would, therefore, be of very great importance as it would provide a practical biochemical mechanism of overcoming anaerobiosis. This mechanism corresponds to the classical fermentation processes. The "oxidative fermentation reaction" ($NADH_2$ reoxidation, e.g., due to the formation of ethanol) seems to correspond to the reductive amination of α-ketoglutarate, for in spite of intensive efforts no alanine dehydrogenase activity can be detected in maize roots.[39] In this process, glutamate amino nitrogen will be secondarily transmitted to the pyruvate via alanine aminotransferase, and alanine will develop as a fermentation product, and possibly be translocated to the shoot. Thus, instead of the toxic ethanol, the nontoxic alanine is produced, which is of great importance to the anabolic metabolism of the plant (Figure 9). Such alanine accumulation under anaerobiosis was also found in buckwheat seedlings[33] as well as in radish leaves.[40] Under the same conditions (100% nitrogen) in these objects, as in pea seedlings,[41] there is also some GABA accumulation. In maize roots, too, the GABA content increases during partial anaerobiosis in presence of nitrate,[36] but only in the presence of nitrate (Figure 7). The negligible amount of this amino acid in roots (below 5%), however, excludes significance with regard to a function as fermentation product.

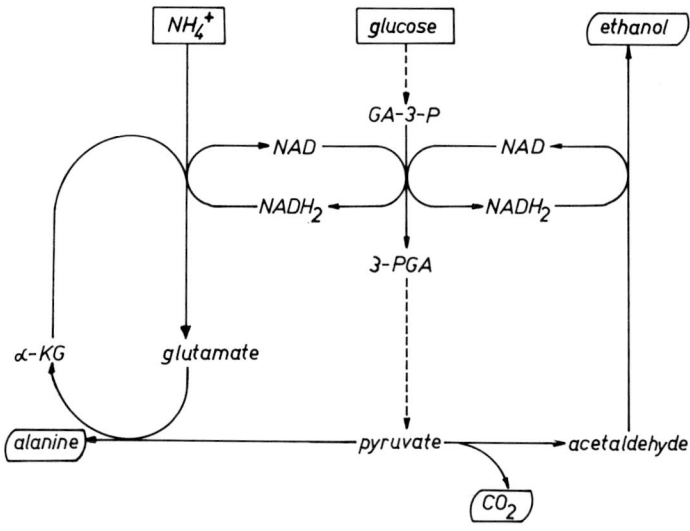

Figure 9. Schematic representation of routes of anaerobic metabolism in the maize root.

As on the change to anaerobic conditions, there is no increase in NR activity; root adaptability to anaerobic conditions is essentially dependent upon the actual NR activity. On the other hand, in case of NH_4^+ supply the adaptability is not limited by that factor. It is, therefore, not yet clear why, under anaerobiosis, alanine accumulation within 24 hr would not be higher in the presence of NH_4^+ than in the presence of NO_3^- (Table VI). Due to the recent findings regarding the alternative route of NH_4^+ assimilation (route of glutamate synthetase[42-44]), however, the regulating relations have become even more involved, for both NH_4^+ concentration and energy charge play a certain role in the system. In case of high NH_4^+ concentrations, glutamine synthesis is achieved first of all by glutamate dehydrogenase, while at low NH_4^+ concentrations, glutamate synthesis seems to be accomplished primarily by glutamine synthetase and glutamate synthetase (Figure 10), as the K_m-values for NH_4^+ are lower in glutamine synthetase than in glutamate dehydrogenase.[45-47]

However, under anaerobic conditions the route of glutamate synthetase is inhibited via the control of glutamine synthetase by the energy charge.[38] On the other hand, the NR will be inactivated at high NH_4^+ levels.[48-50] It cannot yet be reliably rated in what manner a high nitrogen assimilation rate will be realized within these limits at low NH_4^+ supply or with a nitrate supply (which lead also to low NH_4^+ levels within the cells). To

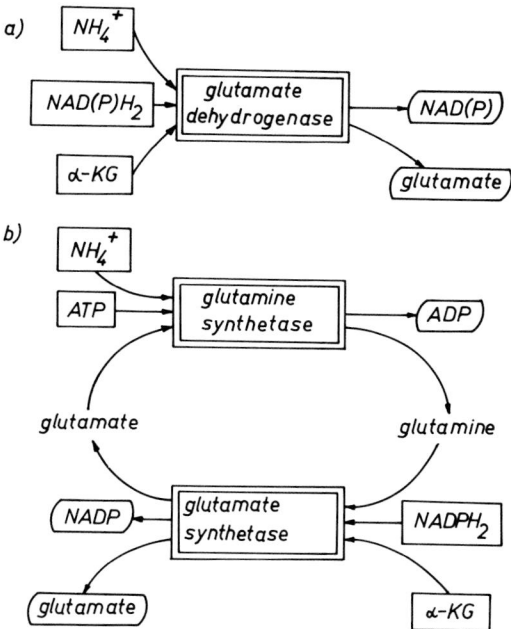

Figure 10. Schematic representation of the routes of ammonium assimilation and of glutamate synthesis, respectively.

clarify the question about which special biochemical features allow intensive nitrogen assimilation under anaerobic conditions and thus provide a possibility for compensating the effects of anaerobiosis, it will be necessary henceforth to systematically examine these regulating mechanisms.

CONCLUSIONS

The anaerobic metabolism of the maize root is characterized by an increased rate of ethanol formation which may be regulated on an epigenetic basis as well. Increased ADH activity is achieved, in part, through the development of further ADH-active proteins having about the same molecular weight as the "constitutive" ADH isoenzymes whose formation is bound to protein synthesis. The rate of ADH activity increase is the result of the complex action of endogenous and exogenous factors that are directly or indirectly related to the oxygen consumption in the tissue or to the oxygen level, respectively; for example, the morphological structure of the root, the growth intensity, and also the intensity of reductive syntheses (*e.g.*, of lipid synthesis[51]), of nitrate reduction, or of NH_4^+

assimilation. Moreover, the latter two processes should be considered as processes of anaerobic metabolism that are alternatives to ethanol fermentation. From the energetic point of view, alanine formation is as important to cell physiology as is ethanol formation, since by this pathway a limited ATP production is possible under anaerobic conditions. However, the nontoxic alanine is formed instead of ethanol, which is toxic in larger quantities. The good usability of this "fermentation product" in places of intensive protein synthesis (*e.g.,* in the aerobic horizons of the root or in the shoot) means at the same time that this anaerobic fermentation reaction represents an essentially more valuable mechanism of adaptation to anaerobic conditions. On the other hand, inclusion of ethanol in the metabolism does not seem to be certain in maize, as the ADH cannot be induced in the leaf.[10] Ethanol elimination via the transpiration flow[52,53] or via lenticels[54] has to be regarded as a less suitable method regarding wastage.

Other reduced intermediates that are formed under reoxidation of the pyridine nucleotides and that may be accumulated to a certain degree have to be considered less important from that point of view as well. Lactate formation seems to have quantitative limits,[5] while malate is formed avoiding pyruvate kinase via PEPCO and therefore does not lead to net ATP synthesis.[55] On the other hand, oxygen supply of the root via the shoot is an important adaptive mechanism within crop plants only in seedlings or in rice, respectively.[19,53,56-60]

Therefore, the formation of alanine seems to represent an effective and physiologically relevant mechanism of adaptation to anaerobic conditions. That the rise in ADH activity under oxygen deficiency is reduced in the presence of nitrate is strong evidence that this metabolic route would be predominant in a regulative way even over ethanol fermentation. The genetic control of both ADH and NR activity perhaps offers starting points for discovering by way of the genotypic patterns of the respective enzymes, those species or cultivars of crop plants that are particularly suitable with regard to their adaptability to short-term oxygen deficiency.

ABBREVIATIONS

ADH	alcohol: NAD oxidoreductase
G-6-PDH	D-glucose-6-phosphate: NADP oxidoreductase
GS	L-glutamate: ammonia ligase (ADP)
ICDH	L-isocitrate: NADP oxidoreductase (decarboxylating)
LDH	L-lactate: NAD oxireductase
MDH	L-malate: NAD oxireductase

ME	"malic enzyme"
	L-malate: NADP oxireductase (decarboxylating)
NR	$NADH_2$: nitrate oxireductase
TIM	D-glyceraldehyd-3-phosphate ketolisomerase
LL	continuous light
DD	continuous dark
TCA	tricarboxylic acid cycle

REFERENCES

1. Baierova, J. "Zum Einfluβ partieller Anaerobiose auf die Alkoholde-hydrogenase und andere Stoffwechselcharakteristike der Wurzelspitze verschiedener Maissorten unter dem Aspekt unterschiedlicher Toleranz gegenüber Sauerstoffmangel im Boden," Unpublished dissertation (A), Humboldt University, Berlin, 1975.
2. Ruhland, W. and K. Ramshorn. "Aerobe Gärung in aktiven pflanz-lichen Meristemen," *Planta* 28:471 (1938).
3. App, A. A. and A. Meiss. "Effect of Aeration on Rice Alcohol Dehydrogenase," *Arch. Biochem. Biphys.* 77:181 (1958).
4. Betz, A. "Die aerobe Gärung in aktiven Meristemen höherer Pflanzen," in *Handbuch der Pflanzenphysiologie,* W. Ruhland, Ed. Vol. XII, No. 2 (Berlin: Springer–Verlag, 1960), p. 88.
5. Kohl, J. G. and U. Matthaei. "Zur Rolle der Milchsäure im Stoffwechsel der Keimpflanzenwurzeln von Zea mays L.," *BPP* 162:119 (1971).
6. Chirkova, T. V. "Role of Anaerobic Respiration in Adaptation of Trees to Temporary Anaerobiosis," *Vestn. Leningrad. Uni.* 3:88 (1973).
7. Hageman, R. H. and D. Flesher. "The Effect of an Anaerobic Environment on the Activity of Alcohol Dehydrogenase and Other Enzymes of Corn Seedlings," *Arch. Biochem. Biophys.* 87:203 (1960).
8. Crawford, R. M. M. and M. McManmon. "Inductive Responses of Alcohol and Malic Dehydrogenases in Relation to Flooding Tolerance in Roots," *J. Exp. Bot.* 19:435 (1968).
9. Kolloffel, C. "Activity of Alcohol Dehydrogenase in the Cotyledons of Peas Germinated under Different Environmental Conditions," *Acta Bot. Neerl.* 17:70 (1968).
10. Kohl, J. G., G. Radke and J. Schade. "Untersuchungen zur Differenzierung des Enzymsystems der wachsenden Wurzelzelle. II. Interaktion von Sauerstoff und Nitrat bei der Induktion der ADH und LDH," *Biol. Zbl.* 92:151 (1973).
11. Crawford, R. M. M. and P. D. Tyler. "Organic Acid Metabolism in Relation to Flooding Tolerance in Roots," *J. Ecol.* 57:235 (1969).
12. Chirkova, T. V. and I. V. Chasova. "Respiratory Metabolism of *Glyceria Aquatica* Wahlb under Conditions of Full and Partial Flood-ing," *Vestn. Leningrad. Uni.* 21:96 (1973) (In Russian).

13. Chirkova, T. V., I. V. Chasova and T. P. Astafurova. "On Metabolic Regulation of Plant Adaptation to Temporal Anaerobiosis," *Fiziol. Rast.* 21:102 (1974) (In Russian).

14. Arnon, D. I. "Ammonium and Nitrate Nitrogen Nutrition of Barley at Different Seasons in Relation to Hydrogenion Concentration, Manganese, Copper and Oxygen Supply," *Soil Sci.* 44:91 (1937).

15. Willis, A. J. and E. W. Yemm. "The Respiration of Barley Plants. VIII. Nitrogen Assimilation and the Respiration of the Root System," *New Phytol.* 54:163 (1955).

16. Beevers, L. and R. H. Hageman. "Nitrate Reduction in Higher Plants," *Ann. Rev. Plant Physiol.* 20:495 (1969).

17. McManmon, M. and R. M. M. Crawford. "A Metabolic Theory of Flooding Tolerance: The Significance of Enzyme Distribution and Behaviour," *New Phytol.* 70:299 (1970).

18. Showe, M. K. and J. A. De Moss. "Localization and Regulation of Synthesis of Nitrate Reductase in *Escherichia Coli,*" *J. Bacteriol.* 95:1305 (1968).

19. Chirkova, T. V. and G. N. Benko. "Role of Leaves in Respiration and Uptake Activity of Roots under Different Conditions of Aeration," *Agrochem.* 6:74 (1973) (In Russian).

20. Garcia-Novo, F. and R. M. M. Crawford. "Soil Aeration, Nitrate Reduction and Flooding Tolerance in Higher Plants," *New Phytol.* 72:1031 (1973).

21. Lambers, H. "Respiration and NADH-Oxidation of the Roots of Flood-Tolerant and Flood-Intolerant Senecio Species as Affected by Anaerobiosis," *Physiol. Plant.* 37:117 (1976).

22. Kohl, J. G., E. Sommerfeld, G. Dudel and I. Irmler. "Der Einfluß der exogenen Nitratkonzentration auf die Nitratreduktaseaktivität in der Wurzel und im Blatt von *Zea mays.* L.," *BPP* 165:123 (1974).

23. Kohl, J. G., and J. Baierova. "Veränderungen des ADH-Isoenzymspektrums der Maiswurzel bei partieller Anaerobiose," *BPP* 164: 624 (1973).

24. Nicklisch, A. "Vergleichende Untersuchungen an Chromatinkraktionen aus Wurzelspitzensegmenten unterschiedlichen Differenzierungsgrades zweier Maishybridsorten unter besonderer Berucksichtigung der Histone," Unpublished dissertation (A), Humboldt University, Berlin, 1972.

25. Freeling, M. and D. Schwartz. "Simultaneous Induction by Anaerobiosis or 2,4-D of Multiple Enzymes Specified by Two Unlike Genes: Differential ADH 1-ADH 2 Expression in Maize," *Mol. Gen. Gen.* 127:215 (1973).

26. Hedrick, J. L. and A. J. Smith. "Size and Charge Isomerseparation and Molecular Weight of Protein by Acryl Amide Gel Electrophoresis," *Arch. Biochem. Biophys.* 126:155 (1968).

27. Schwartz, D. and T. Endo. "Alcohol Dehydrogenase Polymorphism in Maize - Simple and Compound Loci," *Genetics* 53:709 (1966).

28. Scandalios, J. G. "Genetic Control of Alcohol Dehydrogenase in Maize: Genetic Basis for Enzymes," *Science* 166:623 (1969).

29. Efron, Y. "Alcohol Dehydrogenase in Maize: Genetic Control of Enzyme Activity," *Science* 170:751 (1970).

30. Grineva, G. M. and L. A. Frolova. "Cytochrome Components of Respiration Chain of Corn Root Mitochondria under Conditions of Hypoxia," *Dokl. Acad. Nauk. USSR* 209:746 (1973) (In Russian).
31. Vartapetian, B. B. and A. I. Maslov. "Respiration of Coleoptile of Anaerobically Grown Rice Seedlings," *Fiziol. Rast.* 21:807 (1974) (In Russian).
32. Wager, H. G. "The Effect of Anaerobiosis on Acids of the Tricarboxylic Acid Cycle in Peas," *J. Exp. Bot.* 12:34 (1961).
33. Effer, W. R. and S. L. Ranson. "Respiratory Metabolism in Buckwheat Seedlings," *Plant Physiol.* 42:1042 (1967).
34. Effer, W. R. and S. L. Ranson. "Some Effects of Oxygen Concentration on Levels of Respiratory Intermediates in Buckwheat Seedlings," *Plant Physiol.* 42:1052 (1967).
35. Khan, M. and A. Aziz. "Changes in the Tricarboxylic Acid Cycle in Potatoes after Anaerobiosis," *J. Biol. Sci.* 26:1081 (1973).
36. Radke, G. "Untersuchungen zur Abhängigkeit der Aminosäure-Muster der Keimpflanzenwurzel von endogenen und exogenen Faktoren bei *Zea mays* L. und zum Aminosäure-Efflux aus dem Endosperm von einigen Winterweizengenotypen," Unpublished dissertation (A), Humboldt University, Berlin, 1975.
37. Kohl, J. G. "Die Regulation der Aktivitäten ausgewählter Schlusselenzyme des Intermediär- und N-Stoffwechsels in verschieden Organen junger Getreidepflanzen durch endogene und exogene Faktoren," Unpublished dissertation (B), Humboldt University, Berlin, 1976.
38. O'Neal, D. and K. W. Joy "Pea Leaf Glutamine Synthetase," *Plant Physiol.* 55:968 (1975).
39. Dudel, G. "Untersuchungen zur Regulation der Nitratreduktase und Glutamatdehydrogenase durch endogene und exogene Faktoren unter Berücksichtigung ertragsphysiologischer Aspekte," Unpublished dissertation (A), Humboldt University, Berlin, 1974.
40. Streeter, J. G. and J. F. Thompson. "Anaerobic Accumulation of γ-Aminobutyric Acid and Alanine in Radish Leaves (*Raphanus Sativus*)," *Plant Physiol.* 49:572 (1972).
41. Zemlianukhin, A. A., A. M. Makeev, B. F. Ivanov and A. N. Ershova. "Investigation of γ-Aminobutyric Acid Metabolism of Pea Seedlings under Conditions of Different Composition of Atmosphere," *Fiziol. Rast.* 21:1025 (1974) (In Russian).
42. Dougall, D. K. "Evidence for the Presence of Glutamate Synthetase in Extracts of Carrot Cell Cultures," *Biochem. Biophys. Res. Com.* 58:639 (1974).
43. Fowler, M. V., W. Jessup and G. S. Sarkissian. "Glutamate Synthetase Type Activity in Higher Plants," *FEBS Letters* 46:340 (1974).
44. Miflin, B. J. "The Location of Nitrite Reductase and Other Enzymes Related to Amino Acid Biosynthesis in the Plastides of Roots and Leaves," *Plant Physiol.* 54:550 (1974).
45. Pahlich, E. and K. W. Joy. "Glutamate Dehydrogenase from Pea Roots: Purification and Properties of the Enzymes," *Can. J. Biochem.* 49:127 (1971).

46. Lea, P. J. and D. A. Thurman. "Intracellular Location and Properties of Plant L-Glutamate Dehydrogenases," *J. Exp. Bot.* 440 (1972).

47. Lea, P. J. and B. J. Miflin. "Alternative route for Nitrogen Assimilation in Higher Plants," *Nature* 251:614 (1974).

48. Palacian, F., F. de la Rosa, F. Castillo and C. Gomez-Moreno. "Nitrate Reductase from Spinacea Oleracea: Reversible Inactivation by NAD(P)H and by Thiols," *Arch. Bioch. Biophy.* 161:441 (1974).

49. Castillo, F., F. de la Rosa and E. Palacian. "Nitrate Reductase from Spinacea Oleracea, Effects of Sulfhydryl-group Reagens on the Activities of the Complex and the Inactivation by NADH," *Biocehm. Biophys. Res. Com.* 64:546 (1975).

50. Wallace, W. "Effects of a Nitrate Reductase Inactivating Enzyme and NAD(P)H on the Nitrate Reductase from Higher Plants and Neurospora," *Biochem. Biophys. Acta* 377:239 (1975).

51. Buder, E., and K. Ramshorn. "Fettsäuresynthese im Wurzelmeristem von *Vicia faba* L. - eine mögliche Folge der Partiellen Gärung," *Biol. Zbl.* 88:357 (1969).

52. Fulton, J. M. and A. E. Erickson. "Relation between Soil Aeration and Ethyl Alcohol Accumulation in Xylem Exudates of Tomatoes," *Soil Sci. Soc. Am. Proc.* 28:610 (1964).

53. Grable, A. E. "Soil Aeration and Plant Growth," *Adv. Agron.* 18: 57 (1966).

54. Chirkova, T. V. and T. S. Gutman. "Physiological Role of Branch Lenticels of Willow and Poplar under Conditions of Root Anaerobiosis," *Fiziol. Rast.* 19:352 (1972) (In Russian).

55. Davies, D. D., K. H. Nascimento and K. D. Patil. "The Distribution and Properties of NADP Malic Enzyme in Flowering Plants," *Phytochem.* 13:2417 (1974).

56. Chirkova, T. V. "The Delivery of Additional Oxygen to Root System," *Selskochos. Biol.* 3:350 (1968) (In Russian).

57. Vartapetjan, B. B., I. N. Andreeva, N. G. Davtjan and I. P. Maslova. "Ultrastructure of Root Cells of *Cucurbita pepo* in Connection with Oxygen Transport," *Fiziol. Rast.* 15:19 (1968) (In Russian).

58. Vartapetjan, B. B., I. N. Andreeva and I. P. Maslova. "Ultrastructure of Root Mitochondria under Conditions of Anaerobiosis and Enhanced Temperature," *Fiziol. Rast.* 19:1105 (1972) (In Russian).

59. Greenwood, D. J. "Studies on the Distribution of Oxygen Around the Roots of Mustard Seedlings," *New Phytol.* 70:97 (1971).

60. Kordan, H. A. "Patterns of Shoot and Root growth in Rice Seedlings Germination under Water," *J. Appl. Ecol.* 11:685 (1974).

17

STRUCTURAL AND CYTOENZYMOLOGICAL ASPECTS OF THE MITOCHONDRIA IN EXCISED ROOTS OF OXYGEN-DEPRIVED *LYCOPERSICUM* CULTIVATED *IN VITRO*

Christiane Morisset

Universite Pierre et Marie Curie
Laboratoire de Biologie Vegetale VI
75230 Paris Cedex 05, France

INTRODUCTION

Root anoxia is a major problem, from both the agronomical and the ecological points of view. Consequently, research has been undertaken on the behavior of plants submitted to oxygen-impoverished surroundings. Diverse substances derived from the fermentations have been revealed: ethanol,[1] malate,[2] ethanol and/or lactate,[3] gamma-aminobutyrate[4] and ether products.[5] The behavior of the enzymatic systems linked to the fermentory metabolism have been investigated in natural,[6] or in induced anoxic conditions.[7-9] In some plants that are well adapted to prolonged flooding, new metabolic paths appear, different from those in plants that are less resistant to flooding, *i.e.*, products are formed that are less toxic and more easily metabolized when plants return to an aerated medium.[10,11]

The possibility of an oxygen transfer from the aerated aerial plant system to the oxygen-deprived root system has been considered by several researchers. The results depend on the techniques that have been used and upon the age and size of the plant. According to Greenwood[12] and Kordan,[13] oxygen transfer allows normal root growth whereas Healy and Armstrong[14] think that it limits growth because of a rather low extension of the root system. By contrast, the method of chemiluminescence, shows the absence of any oxygen transport from leaf to root.[15]

Other research has related the conditions of germination,[16,17] of organo-
genesis[18,19] and of energetic charges,[20,21] to the conditions of growth,
respiration and capacity of oxydoreduction of the electron transport chain
of the mitochondria[22,23] in hypoxia. Ferron et al.[24] point out the modification in the circulation of water under similar conditions.

It was proved in 1969 that mitochondria exist in yeast cells grown
under anaerobiosis. Their ultrastructure is altered (cristae are less numerous and less well developed than in aerobiosis), and it requires the
enrichment of the culture medium with unsaturated fatty acids and
ergosterol. Such mitochondria are devoid of the cytochromes of the
electron transfer chain.[25]

In higher plants, two orientations of research coexist with regard to
the ultrastructures. Anoxia is used as a means of inducing the phenomena
and observing for concentric layers of saccules of the endoplasmic reticulum[26-29] and cell autophagic processes connected with the formation of
phytolysosomes.[30,31] Other works are more concerned with the effects
of anoxia. Among different materials, the rice coleoptile is chosen for its
ability to grow in anaerobiosis, whereas neither leaves nor roots develop.
Ueda and Tsuji,[32] Vartapetian et al.[33,34] and Opik[35] showed that during
the first hours of germination, there are not ultrastructural differences
between the mitochondria of coleoptiles grown in a normal atmosphere
and those grown in an oxygen-deprived atmosphere. Growth of the coleoptile decreases but continues. After 6 days of anaerobiosis, mitochondria
vesiculate and their matrix becomes electron transparent. If the germinations are then placed in an aerated medium, the mitochondria of the
coleoptile become normal again, and roots and leaves emerge.

Contrary to what happens in anaerobic yeast cells, the cristae of rice
coleoptile develop normally, and there is no obligatory correlation between
cristae number and oxidative metabolism. Analyses show that these mitochondria possess all typical cytochromes of plants, but in lesser quantities
than in reference cultures;[36] hence, cytochrome activity is low.[35]

At present, research at the ultrastructural level in higher plants has
principally been carried out, either during short-term anoxia without considering the growth possibilities after returning to an aerated atmosphere,
or during longer oxygen deprivation, but with organs which grow and
continue cell division in anaerobiosis, although at a slower rate.

What happens to an isolated organ that survives a certain period of
anoxia, but stops growing? That is what I tried to determine, using excised tomato roots, cultivated in vitro as experimental material.

MATERIAL AND METHODS

Techniques of Culture and of Anoxia

Roots originating from the germination of tomato seedlings (variety Super Marmande produced by Vilmorin) were sterilized and put into culture vessels according to White[37] and in an optimum culture medium established by Street et al.[38] The media contained glycocolle, nicotinic acid, vitamin B_1 and B_6, 2% sucrose, and mineral salts (including EDTA chelated iron). Roots were subcultured every 7 days for several weeks before being used for experimentation. They were kept in the dark in a culture cupboard at 25°C.

Anoxia was achieved in 300-ml flasks with two roots in 80-ml medium (Figure 1). The bubbling of nitrogen (with less than 5 volumes O_2 per million) was maintained for at least 4 hours. The locking device with mercury allowed for the maintenance of a slight positive pressure of nitrogen in the flask, the atmosphere of which remained confined during the experimental period. To observe the consequences of returning to a normal atmosphere, roots were transferred to a new culture medium, in the presence of air.

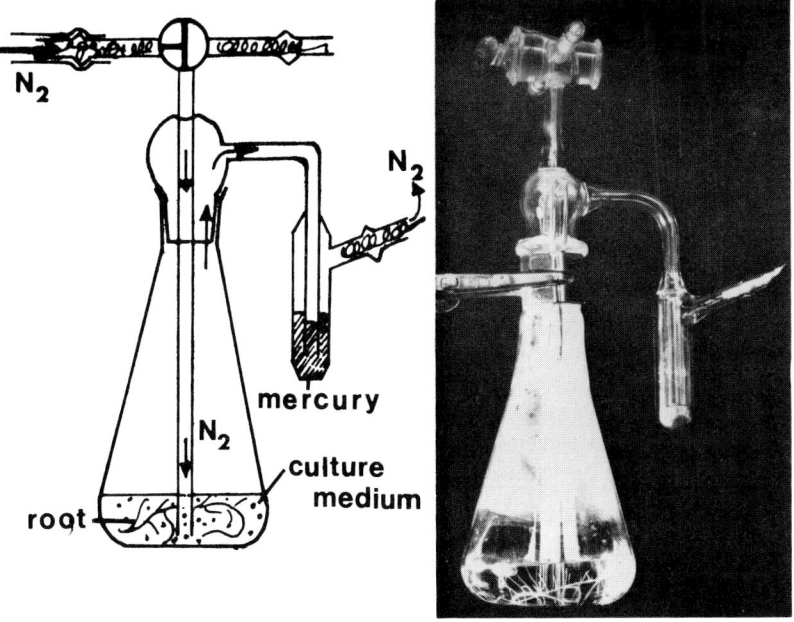

Figure 1. Experimental apparatus allowing for the culture of excised roots, in enclosed nitrogen atmosphere.

Electron Microscopy

Fixation, Embedding and Post-Staining

Fixing was carried out a 4°C. The cooled plant material was rapidly transferred into glutaraldehyde, 4 or 6% in cacodylate buffer, 0.1 M at pH 7.4 and chilled before excision. It was postfixed in 1% osmium tetroxide in veronal buffer, 0.1 M at pH 7.4.[39] After dehydration in an ethanol series they were transferred to propylene oxide and embedded in a mixture of araldite-epon resin.[40]

Sections were cut on a Porter Blum ultramicrotome and taken when interference colors were silver or pale yellow. They were post-stained in uranyl acetate and lead citrate, according to Reynolds.[41] Specimens were examined in a Hitachi HU 11E electron microscope, at 75 kV.

Cytochrome Oxidase Activity

Cytochrome oxidase activity was tested in mitochondria by the diaminobenzidine (DAB) method. The DAB, oxidized by the departure of hydrogen, polymerizes into large, water-insoluble, and osmiophilic molecules. During postfixation in the osmium tetroxide, they formed a complex opaque to electrons, called "osmium black."[42]

Samples were prefixed for 2 hr at 4°C with glutaraldehyde (4%) in the cacodylate buffer (pH 7.4). To avoid the diffusion of the enzymes, 10% sucrose was added to the cacodylate buffer which was used as a washing medium.

Prefixed samples were incubated in the dark for 1 hr at room temperature and for 1 hr at 37°C. The reaction mixture contained DAB as the hydrochloride (Sigma) 10 mg/ml; phosphate buffer, 0.05 M; 6% sucrose; hydrogen peroxide, 0.001%; after addition of the DAB, the final pH was 7.3. This medium was prepared immediately before use. For control, fixed and rinsed root segments were kept for 2 hr at room temperature, in a phosphate buffer containing 0.01 M potassium cyanide. Then they were incubated in the complete reaction mixture, containing 0.01 M KCN.

After incubation, samples were washed, postfixed with osmium tetroxide, dehydrated and embedded using the standard techniques. Thin sections were not post-stained, as areas of enzymatic activity were revealed by electron-opaque deposits.

Respiration Rate

The respiratory gas exchanges were measured by the Warburg technique.[43]

Measurements were made in sterile conditions. Each Warburg flask contained one root segment, 10-15 cm long, the fresh weight of which was

15-25 mg. This was steeped in 5 ml of the usual culture medium, to which was added either mannitol or 2-4 dinitrophenol, depending on the experiments. One flask containing 5 ml of culture medium, without any root segments, was kept as a proof of sterility. The unchanging gaseous volume showed that no bacterial pollution of the sweetened medium took place during the time of measurement. On the other hand, measures were often made 24 hr later to check root vitality.

The graphs represent the mean of two series of measures, realized each time with five reference roots and five experimental ones; the mean-gaps represented on the graphs show the homogeneity of response of the roots used for the measures. Sometimes root segments were fixed and embedded to control the state of the cells and especially of the mitochondria, in the electron microscope. For growing excised tomato roots, two types of segments are used: either a segment of the main root axis, 4-6 cm long, with its rootlets (Figure 2a) or the terminal part of the root (Figure 2b).

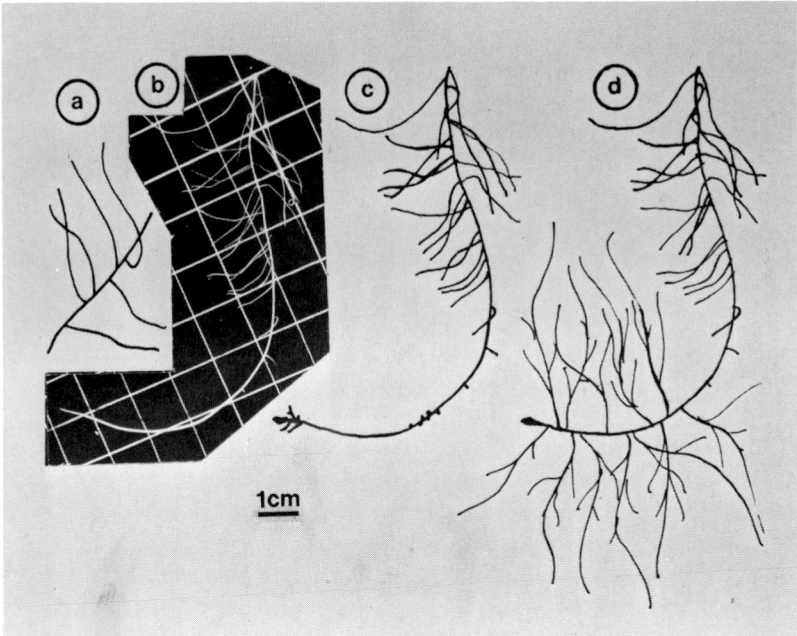

Figure 2. (a) diagram; (b) photograph of the segments used in the subculturing of excised roots, cultivated *in vitro*; (c) and (d) evolution of the explant "b." It has been submitted to 3-day-long anoxia and then 6 (c) and 14 days (d) aerated culture; x 0.75.

RESULTS

Growth after Anoxia

Both excised root segments lengthened 10-15 mm/day in air and branched at the same time in aerated media. When air was replaced by nitrogen, their growth stopped.

When the organs were placed in aerated conditions after 3 days of anaerobiosis, their behavior was quite different. The length of segment "a" did not increase. On the contrary, some parts of segment "b" grew again, but only after 4-6 days rest time. During that time the end meristem became more or less completely cankered.

Apical dominance, which occurred in aerated media, disappeared after three days of anoxia treatment. In reference roots, the second-order rootlets appeared only at a distance of several centimeters (4-6 cm) from the meristem; they bore third-order rootlets only when they were more than 10 cm distant from the meristem.

After the necrosis of the root apex, two types of growth appeared.

Numerous rootlets that were already being formed before experimentation, appeared simultaneously. They lengthened all in the same way. So, their way of growth was different from the usual one (length increasing with the distance from the main root meristem).

Rootlets appeared immediately above the dying meristem, in a place where they were normally never formed (Figure 2c).

After several days, all new rootlets had the same length (Figure 2d), all of them bore rootlets of second order, but the phenomenon of dominance was not expressed.

Rootlets, with or without rootlets of second order formed before experimentation, remained alive, with the exception of the main-end meristem. However, they did not continue to grow, and the cells did not function normally.

So it seems that only cells that divided to build up a new meristematic massif were able to maintain their potentialities intact, after a prolonged anaerobiosis. Only the cells of the new forming rootlet meristems and the cells of the pericyclic layer met these requirements.

On the contrary, the metabolism of functioning meristems, the cellular multiplication of which was related to the lengthening of already existing rootlets, was disturbed in such a way that they kept living up to 12 days, at least, after the period of anoxia, but did not proliferate.

Anaerobiosis and Cell Ultrastructure

Semblance of Plasmolysis

A noteworthy effect of anoxia was the important alteration of the osmotic pressure of cells, which was expressed through a loss of stiffness, obvious when touching the roots, and appeared in the microscope as a semblance of strong plasmolysis, in the meristematic cells as well as in the differentiated cells of all tissues.

The results of anoxia appeared precociously: 4 hr after the moment at which roots were put in an oxygen-deprived culture medium, in ultra-thin sections, the plasmalemma had a sinuous outline, and it separated from the pecto-cellulosic cell wall, mainly in the corners of the cells (Figure 3c). The separation between the plasmalemma and the cell wall increased with the duration of anaerobiosis. After 3 days of anoxia, the cytoplasm of the cells was considerably contracted (Figure 4b). The organelle sections were weakly contrasted and difficult to recognize at low magnification.

After returning the roots to a normal atmosphere, the semblance of plasmolysis disappeared gradually; after 1-hr aeration, the cytoplasm was

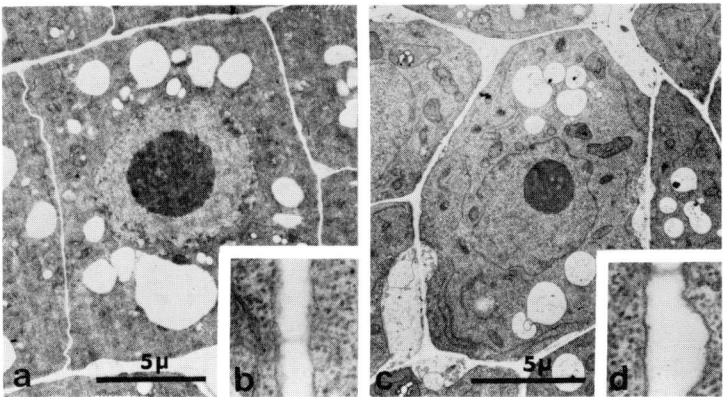

Figure 3. (a) reference meristematic cortical cells of the main end meristem; x 4,000; (b) the plasmalemma adhere tightly to the pecto-cellulosic cell wall; x 48,000; (c) resemblance to a slight plasmolysis after 4-hr anaerobiosis, in a cell of the same region; x 4,000; (d) sinuous plasmalemma more or less detached from the cell wall, after 4-hr anoxia; x 48,000.

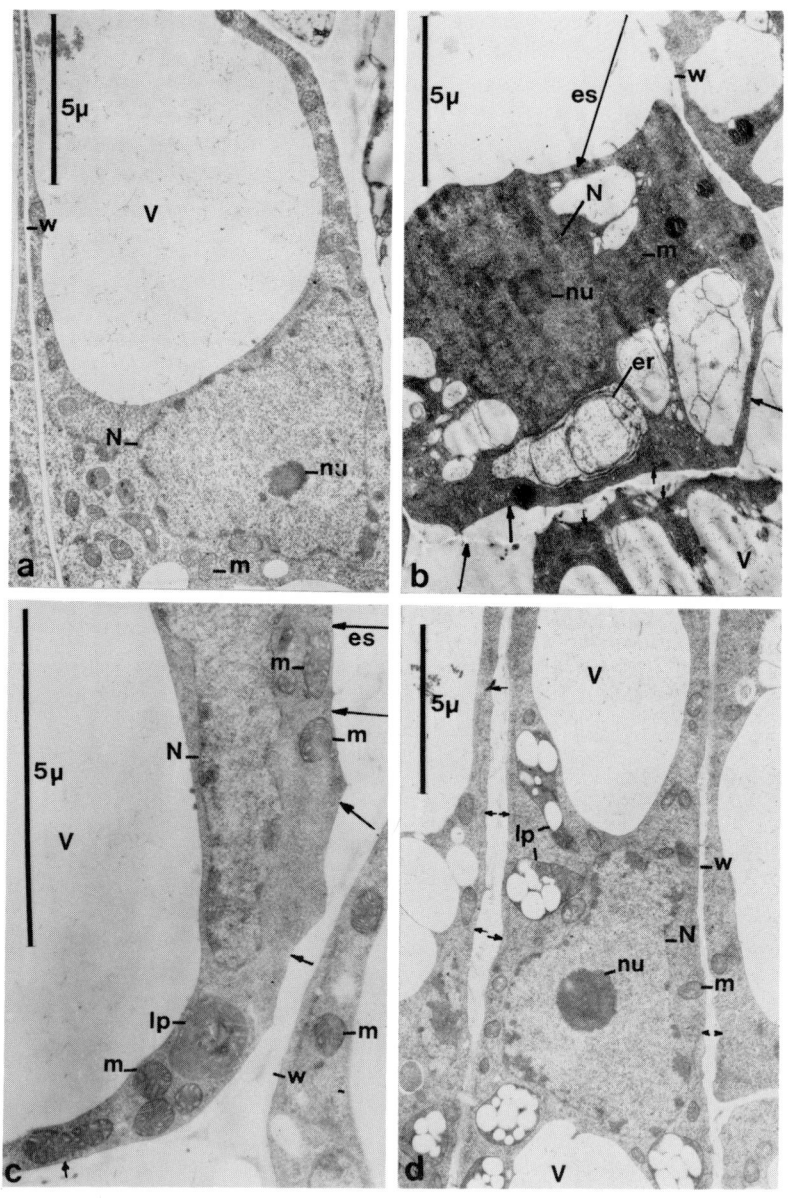

Figure 4. Sections of differentiating cortical cells: (a) reference root; x 6,000; (b) 3-day-long oxygen-deprived root, dense cytoplasm, separated from the cell wall (w) by wide, electron-transparent spaces (es). Organelles are difficult to distinguish; x 6,000; (c) 3-day-long oxygen-deprived root, aerated one-half hour, large shrinkage spaces subsist, mitochondria are still unchanged (m); x 12,000; (d) root oxygen deprived during 3 days and aerated during 15 hr, the cells again become similar to the reference, although the plasmic membrane is still separate from the cell wall in some places (arrows) x 6,000. N, nucleus; Nu, nucleolus; W, pecto-cellulosic cell wall; V, vacuole; m, mitochondria; er, endoplasmic reticulum; lp, leucoplasts. Arrows indicate the places where the plasmic membrane does not adhere to the cell wall.

still shrunken (Figure 4c), but the contrast of the organelles was similar to the reference ones. After 15-hr aeration, thin shrinkage spaces persisted between the cell wall and the plasmalemma, fringed with numerous poly-ribosomes (Figure 4d). The presence of fibrils in these spaces of clear vesicles or of crushed cytoplasmic vesicles, demonstrated that in the living cell, during anaerobiosis, the plasmalemma did indeed separate from the cell wall, the fixation process possibly increasing this phenomenon.

We use the term of "semblance of plasmolysis," because this aspect which was seen after fixation, can hardly be attributed to an altering of the osmolarity of the culture medium, which was voluminous (80 ml) with regard to the roots (2 times 10-15 mg). It was more likely due to changes of the cellular functioning, which could start an evolution of the osmolarities of the different inner compartments of the cells.

The Organelles

Under the influence of anaerobiosis, certain organelles showed more important ultrastructural modifications than others. These were best seen in differentiating cells in which the cytoplasm was still abundant. All photomicrographs came from cortical cells. The behavior of organelles was, apart from a few exceptions, similar in the different tissues and in the differentiated cells.

After 72 hr of anoxia, the ultrastructure of the nucleus did not seem changed, although during the shrinking of the cell, the nucleolus and the chromatin lost their contrast (Figure 4b).

Fibrillar structures aggregated in the cytoplasm (Figure 5, f) sometimes in the nucleoplasm. Leucoplasts were generally devoid of starch granules (Figure 5, lp). Dictyosomes did not appear altered. In some mitochondria, the cristae tended to form a parallel arrangement (Figure 5, m).[43],[44]

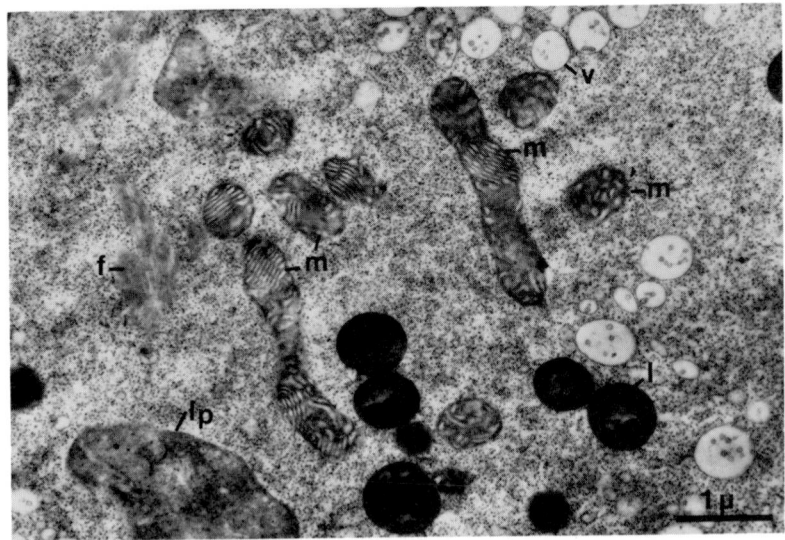

Figure 5. Differentiating cortical cell from root cultivated for 3 days in anaerobiosis: the mitochondria (m) possess either cristae disposed in parallels, or non-oriented and slightly dilated; leucoplasts are devoid of starch granules (lp); fibrillar structures (f) appear in the cytoplasm; the vesicles (v) are perhaps of phytolysosomal origin; 1, lipidic granule; x 15,000.

In differentiating cells, the osmium impregnation technique revealed the behavior of the endoplasmic reticulum; it presented three characteristics.[45] In reference cells, short saccules were disposed parallel to the cell walls. They were not longer seen in cells deprived of oxygen for 3 days. Long saccules formed regular parallel or concentric stacked layers (Figure 6, er), numerous impregnated vesicles, more or less anastomosed, existed in the cytoplasm all around the nucleus (Figure 6, vn).

After returning to a fresh culture medium and atmosphere, all these alterations disappeared gradually; starch accumulated again in the leucoplasts of all cells; the endoplasmic reticulum saccules were mostly found in the peripheric cytoplasm of the cells; the cristae of the mitochondria had no particular orientation. Cytologically, cells again became similar to reference cells, although their metabolism was deeply impaired, no longer allowing normal growth.

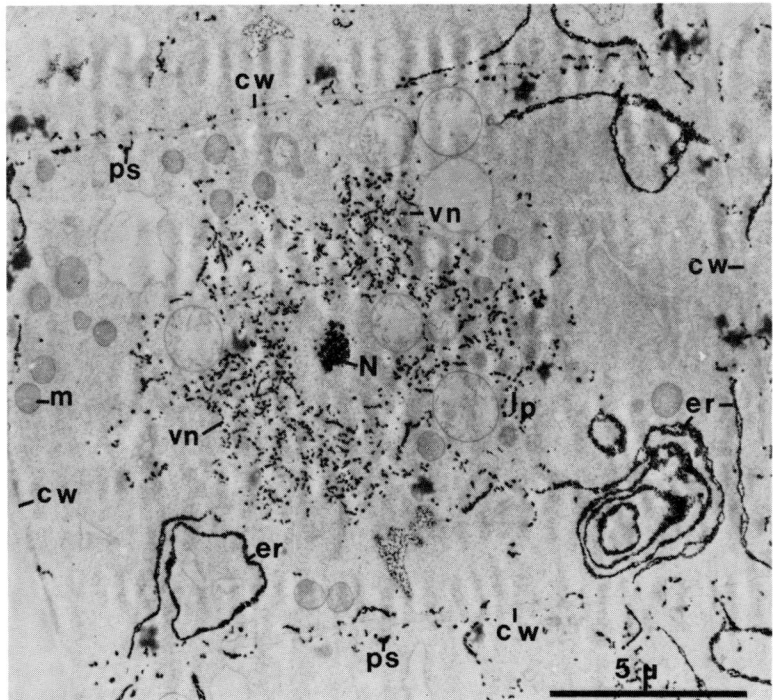

Figure 6. Osmium-impregnated structures after 2- day-long anoxia. cw, pecto-
cellulosic cell wall; N, tangential section of the nuclear envelope; m, mitochondria;
lp, leucoplast; er, endoplasmic reticulum, constituted of long concentric saccules;
vn, impregnated vesicles in the perinuclear cytoplasm (they do not appear in reference cells); ps, peripheric saccules, shorter and less numerous than in reference
cells; x 6,000.

Respiratory Intensity, Ultrastructure of Mitochondria and Cytochrome Oxidase Activity

Respiratory Rate

The values for oxygen uptake in reference roots and in oxygen-deprived
roots were compared during short-term anoxia (2 hr) and during longer
periods (1-6 days).

The oxygen uptake of roots that had been deprived of oxygen for 2 hr was
the same as that of reference roots. This value was considerably diminished in
roots after 3 days anaerobiosis (Figure 7). But if roots were left for 24 hr in
an aerated medium after 3 days anaerobiosis, before measures were made, the
oxygen uptake then reached 80% of the reference root (Figure 8).

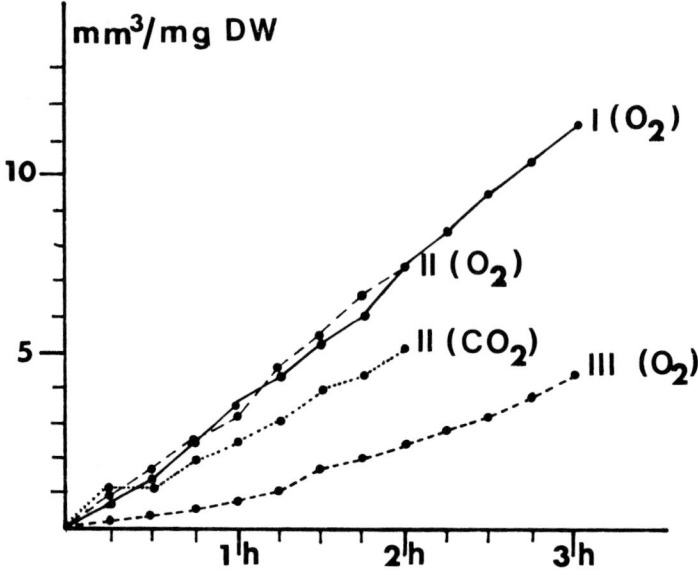

Figure 7. Graph representing the intake of oxygen and the output of carbon dioxide, expressed in mm^3 per mg of dry weight (mm^3/mg DW), in relation to the time (hr). I (O_2), reference graph, O_2 absorption. II (O_2), O_2 absorption after 2-hr anaerobiosis. II (CO_2), CO_2 output during 2-hr anaerobiosis. III (O_2), O_2 intake after 3 days anaerobiosis and 1-hr aeration.

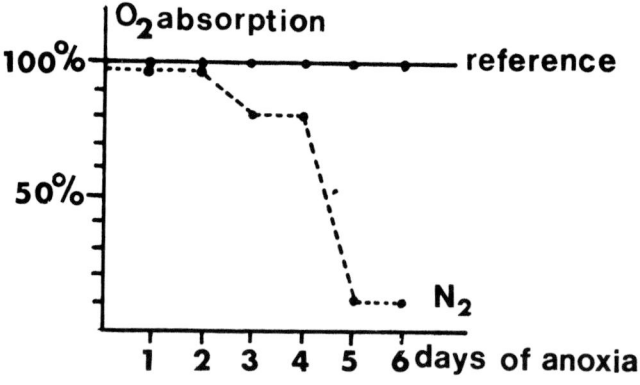

Figure 8. Respiratory intensity values of anoxied roots that were placed for 24 hr in an aerated medium, represented in % of the reference values, for increasing periods of anoxia.

The carbon dioxide output during 2-hr anaerobiosis showed the existence of a fermentory metabolism. The volumes of CO_2 sent out per mg of dry weight, per hour, were, respectively, 2.9 μl for reference roots and 2.5 μl for oxygen-deprived ones. The respiratory quotient was close to unity in reference roots. If it is considered that 6 CO_2 molecules are released per mole of glucose utilized and only 2 CO_2 during anaerobiosis, the ratio of CO_2 released, ought to be 2.9 μl for reference and 1.0 μl for anoxied roots. It seemed that the utilization of glucides was speeded up. The consequences of the Pasteur effect appeared also through the decrease of the amount of starch, measured by titration after hydrolysis.[46] The decrease of starch granules, as can be seen with the electron microscope, was variable from one tissue to another. The use of intracellular starch allowed for lesser energy consumption than would be necessary for an increased absorption of sucrose from the culture medium.

We also measured the oxygen uptake after increasing periods of anoxia, when roots had stayed 24 hr in an aerated medium, after the anoxia treatment. If anoxia lasted only 1-2 days, the tissues recovered normal respiratory exchanges. If oxygen deprivation was maintained during 3 or 4 days, roots again attained 80% of the normal respiratory rate. Observations with the microscope did not make it possible to determine if the decrease was due to either the death of 20% of the cells, or if all cells underwent a decrease of the respiration rate. For longer periods of anoxia the decrease was dramatic (Figure 8); this was stressed since the tomato plant is known to be "anoxia sensitive."[47] Nevertheless, even after 6 days anaerobiosis, some tissues were still alive. After the return to normal atmosphere, rootlets appeared in the subapical zone, near the main root meristem, after a period of latency.

To study the ultrastructural features, we chose the period of 3 days anaerobiosis, the effects of which were already perceptible, but not completely irreversible.

The Behavior of the Mitochondria

The behavior of mitochondria during anaerobiosis was noteworthy. They reacted in different ways, in the various regions of the root, depending on their ability to tolerate anoxia.

In the cells of reference roots (Figure 9), the development of the mitochondrial cristae was connected with cellular differentiation. Mitochondria had various traits, possibly depending on their physiological activity.

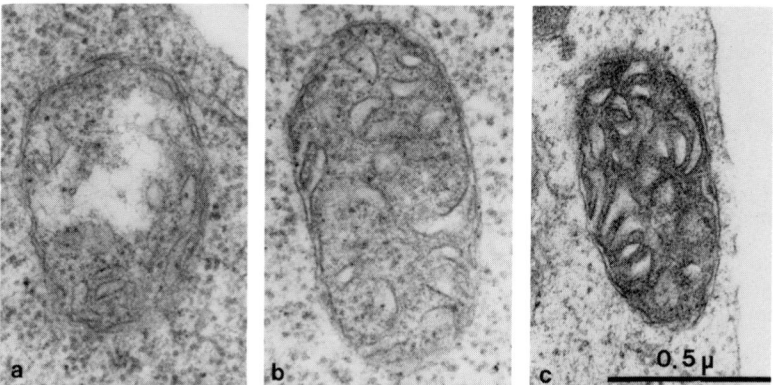

Figure 9. Mitochondria of reference roots, diversity of the ultrastructural (a) mitochondria of meristematic cell; (b) differentiating cell; (c) highly differentiated cortical cell of the main root. x 60,000.

Mitochondria during Anaerobiosis

After 3 days of oxygen deprivation, the mitochondria of the differentiating cortical cells and also those of more differentiated tissues of roots, exhibited two main aspects.

In some mitrochondria, the cristae built up one (Figure 10 a,c) or several (Figure 10 b,d) groups where they were disposed parallel one to another, giving a polarized aspect to the organelle. These regular stacks were made of 5-10 cristae and may or may not have been divided into compartments. The gaps between the cristae (about 75-175Å) were narrower than the inner space of the cristae (250-350Å). The intra-cristae volume seemed to be higher than the volume left for the matrix in this part of the mitochondria. In the region devoid of these groups, the electron-dense matrix contained only a few dilated cristae.

In the same cells, other mitochondria did not exhibit such notable alterations. Their periphery was sinuous and the observed deformations might have been due to the shrinkage of the cytoplasm (Figure 10 e, f). In the ultrathin sections of the roots we observed, the ratio of each type of mitochondria was variable. The frequency of mitochondria with parallel-disposed cristae was unequal from one cell to another, and also from one root to another. Generally, highly differentiated cells, with a thin parietal cytoplasm (0.5 μ broad) had mitochondria in which the parallel cristae were disposed all over the section. It can be said that, on the average, modified mitochondria were a little less common than mitochondria with nonpolarized cristae.

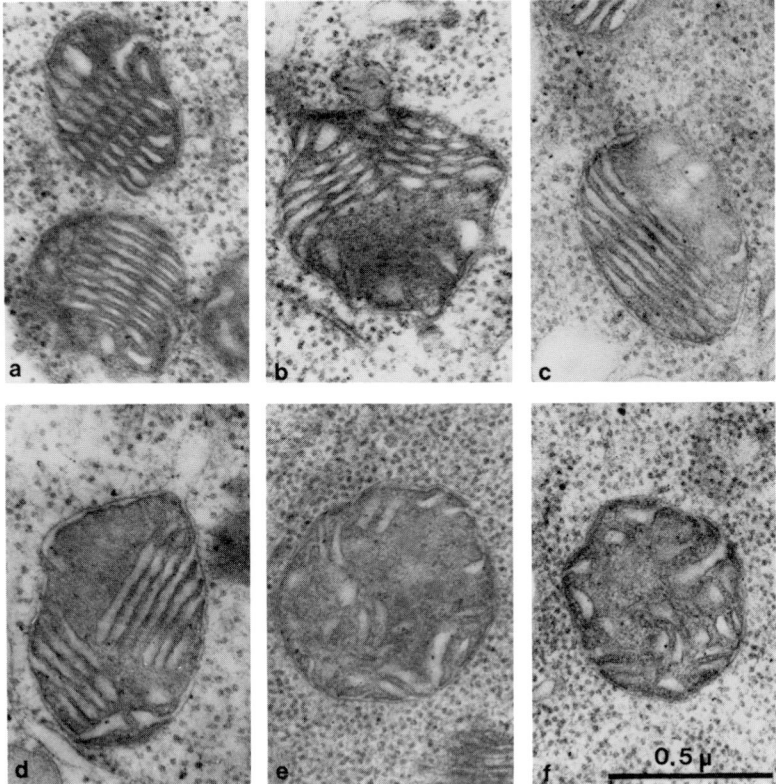

Figure 10. Mitochondria of 3-day-long anoxied roots, cortical cell. (a) (b) (c) and (d) cristae arranged in parallels between them, a few dilated cristae in the dense matrix; (e) and (f) slightly modified mitochondria, with irregular rounded shape. x 60,000.

On the contrary, in meristematic cells, where mitochondria possessed only a few short cristae, the response to anoxic conditions was more uniform:

1. In the shrunken cytoplasm of the meristematic cells of the main root meristem, all the mitochondria were considerably swollen (contrary to the leucoplasts, which were not swollen), and their matrix was electron translucent (Figure 11a).
2. The mitochondria of rootlet meristems were smaller in section; their matrix was dense and electron opaque, and their cristae were somewhat dilated and sometimes annular in shape (Figure 11b).

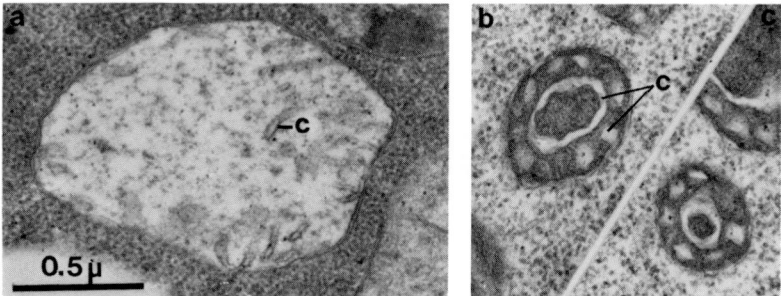

Figure 11. Three-day-long anoxied roots. (a) swollen mitochondria of the main end meristem, the cristae (c) dispersed in the electron transparent matrix; (b) and (c) mitochondria of newly formed rootlets. The cristae are occasionally annular in section (compare with Figure 9a). x 40,000.

Mitochondria during the Return to an Aerated Culture Medium

The mitochondria of the main root meristem were even more swollen than previously; some were contracted and did not recover a normal ultrastructure. The characteristic behavior of the mitochondria of this part of the root might have been connected with the fact that irreversible disturbances appeared in this meristem, destined to necrosis.

In roots after 1 hr of exposure to an aerated medium, the parallel oreintation of the cristae in the mitochondria persisted temporarily (Figure 12a). Both the inner and the outer membranes formed loops, which occasionally contained a part of the matrix (Figure 12 b, mw).

After maintaining the roots 15 hr in aerated conditions, the mitochondria regained their former ultrastructure; many were elongated and seemed to divide (Figure 12c).

Cytochrome Oxidase Activity Revealed by Reaction to DAB

The activity of cytochrome c[48] and cytochrome c oxidase[42] may be located by use of DAB, either with acid or neutral pH. Electron-dense deposits were located on the membranes of the cristae, and possibly on the inner membrane of the mitochondrial envelope. The "osmium black" was deposited on the side of the membrane, which was turned towards the inner space of the cristae. When reactions were very intensive, the whole inner cristae space was darkened with the osmium deposit. Treatment of the tissues with potassium cyanide, before and during incubation in the reaction mixture, completely inhibited reactions in the mitochondrial structures.

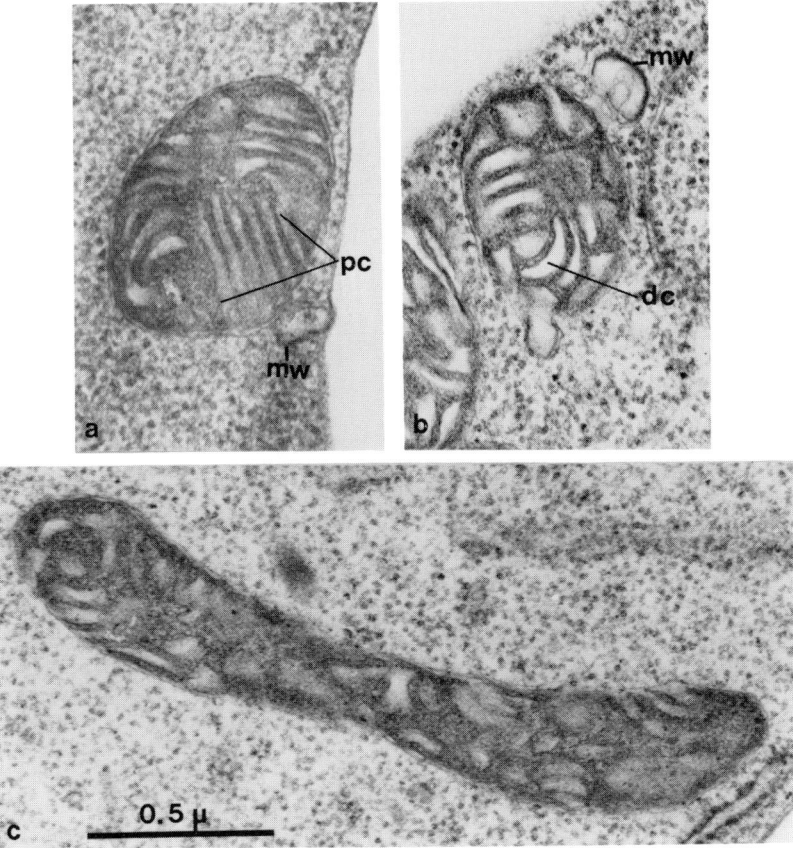

Figure 12. Mitochondria during the return of the roots to aerated conditions after 3 days anoxia: (a) 1/2-hr aeration; (b) 1-hr aeration; (c) 15-hr aeration. x 60,000.

To make comparisons, we always fixed and incubated reference samples and samples coming from anoxied roots at the same time. DAB enters slowly and in small quantities in the tissues; so we mainly observed the mitochondria, which presented the most intensive reactions in the cortical parenchyma cells.

In reference roots (Figure 13a), the positive reaction emphasized the outlines of relatively short cristae, the distribution of which was irregular. If potassium cyanide was dissolved in the incubation medium, the reaction was inhibited.

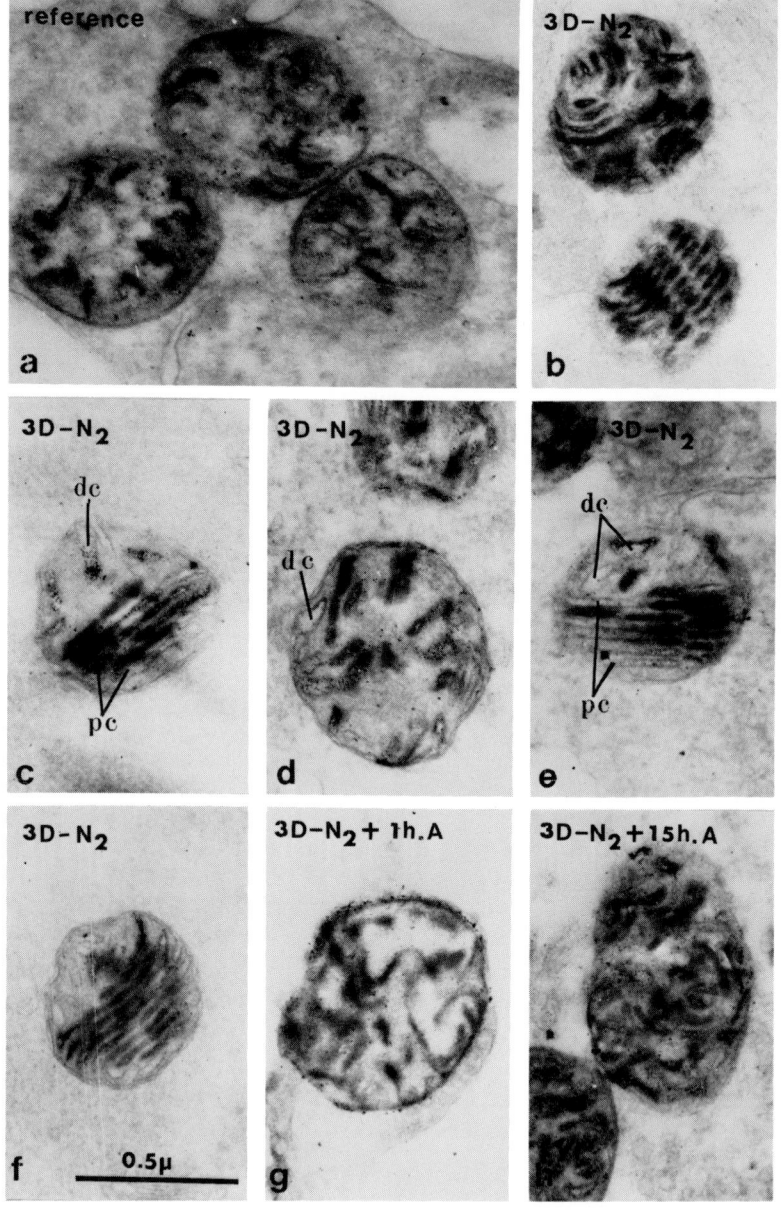

Figure 13. The cytochrome oxidasic activity revealed by the reaction to DAB, in differentiating cortical cells: (a) reference cell; (b) (c) (d) (e) and (f) mitochondria of 3 days anoxied roots (3D-N_2). The "osmium black" deposit can be seen in the inner space of compact parallel cristae (pc) and to a lesser degree in the few dilated cristae (cd) situated in the area where the matrix is the most abundant. Staining is less in nonpolarized cristae (d); (g) heavy DAB staining in cristae which are still partially oriented, when 1–hr aeration has followed 3 days anoxia (3D-N_2 + 1 hr A); (h) after 3 days anoxia and 15–hr aeration, the polarization disappears. The cristae remain heavily stained by the DAB (3D-N_2 + 15 hr A); x 60,000.

When roots had undergone 3 days of anaerobiosis, "osmium black" deposits could still be seen in the mitochondria. The reaction took place, no matter how the cristae were distributed; its intensity was comparable to that obtained in the reference roots. The absence of oxygen for 72 hr provoked the interruption of the functioning of the respiratory chain. In spite of this, the cytochemical reaction to DAB proved that oxidasic enzymes remained functional in the same areas as in normal conditions.

The cristae of polarized mitochondria appeared particularly long (Figure 13 b, c, e, f) and they remained connected to the inner mitochondrial membrane. Deposits were clearly more abundant in polarized cristae (Figure 13 c, e, pc) than in short, dilated cristae (Figure 13 c, d, e, dc). The product of the reaction formed a fine deposit, which lined almost completely the inside of the cristae space.

In the swollen mitochondria of the main meristem, the electron-transparent matrix enclosed a few short cristae with osmium deposits.

In mitochondria submitted to 3-day-long anoxia, and then 1–hr aeration (Figure 13 g), the reaction to the DAB was intensified; the deposit in the cristae was more dense than in reference roots. Some mitochondria still possessed parallel cristae. In other mitochondria, the "osmium black" filled the enlarged cristae. The abundance of the deposit sometimes induced artifacts when making the thin sections or during the observation; the cristae where the electron absorbing material accumulates were often torn away.

After a longer aeration of the anoxied roots (15 hr, Figure 13 h), in the specimens treated with the DAB, mitochondria were very similar to the reference ones, not only in the disposition of the cristae but also in intensity of the reaction.

The polarized structure involved a spatial rearrangement of the different components of the mitochondria. The connection between the membranes of the cristae and the matrix was modified. In reference cells, the cristae were surrounded by the matrix on all sides, except the zone of continuity with the inner mitochondrial membrane. In mitochondria altered by the anoxia treatment, cristae were stacked, and the contacts with the matrix seemed to be reduced. The spaces between the parallel-lying cristae were very narrow. Only the lateral cristae of the group maintained contact with the matrix, and on one side only.

It was possible that this new arrangement was connected with an inhibition of the enzymatic functioning. It could be a system of regulation, making it possible to preserve the integrity of the mitochondrial membranes, while the oxygen deprivation prevented the functioning of the respiratory chain. Nevertheless the process of penetration and oxidation continued to be accomplished when a substrate DAB was dissolved in the incubation medium.

Effects of Plasmolysis upon Respiratory Rate and Ultrastructures

Two criteria led us to experiment with the effects of plasmolysis. First, it appeared necessary to determine whether the structures of mitochondria during anoxia could be explained by a simple osmotic phenomenon. The rearrangement of the cristae then would be an artifact, determined by the fixation in a nonisotonic fixative solution. On the other hand, it seemed useful to determine if plasmolysis, by decreasing the respiratory intensity (as it does in potato slices),[49] could induce, in an indirect way, ultrastructural modifications.

Respiratory Intensity

Roots were placed in their usual culture medium, to which mannitol had been added, its concentrations being 0.47 M or 0.75 M. Roots remained 2 hr in such a medium before measurements were made.

When plasmolysis was weak (0.47 M), the average oxygen uptake was slightly depressed, compared to that of reference roots (Figure 14); however, average variations were slight and occasionally overlapped. The differences between the two curves was not significant. If plasmolysis was intensified (0.75 M), the average oxygen uptake by the root tissues was reduced significantly.

Ultrastructural Aspects

Excised roots were placed in hypertonic solutions, with spaced-out and increasing osmolarities. Plasmolysis was maintained for 24 hr. Such a long treatment, although shorter than the anoxia treatment, might induce structural modifications of the mitochondria, which could be compared to

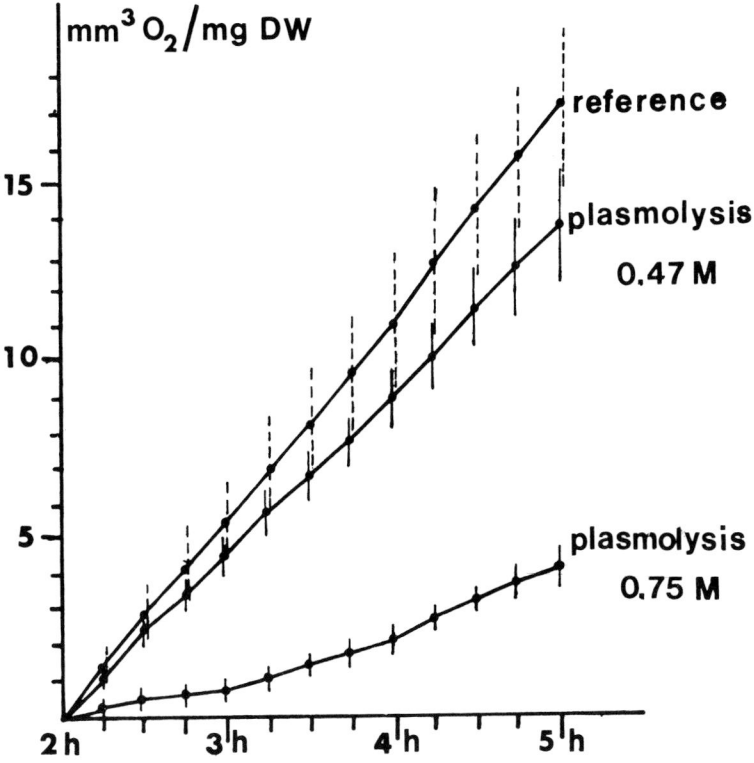

Figure 14. Respiratory rates (mm³ oxygen absorbed per mg of dry weight) of reference roots and roots plasmolyzed in culture medium to which mannitol had been added to a concentration of 0.47 M and 0.75 M (with indication of mean gaps). Measures carried out between 2 and 5 hr plasmolysis.

those produced by lack of oxygen. Fixation of the specimens was done with glutaraldehyde solutions to which mannitol had been added, in the same concentration as in the plasmolyzing culture medium.

When roots were immersed in a 470- (Figure 15 a) or 600-milliosmolar solution (Figure 15 b), the mitochondria had a dense matrix and long, sinuous, slightly dilated, electron-transparent cristae. A more intense plasmolysis (1000 milliosmolar) during a short time only (a half hour, Figure 15 f) did not modify the ultrastructure of the mitochondria in a significant way, although the cytoplasm was more contracted. They resemble those observed in plasmolyzed onion protoplasts[50] or cauliflower bud cells.[51]

If plasmolysis had been severe during a longer period, 1000 milliosmols during 24 hr (Figure 15 c), the mitochondria had diversified structure.

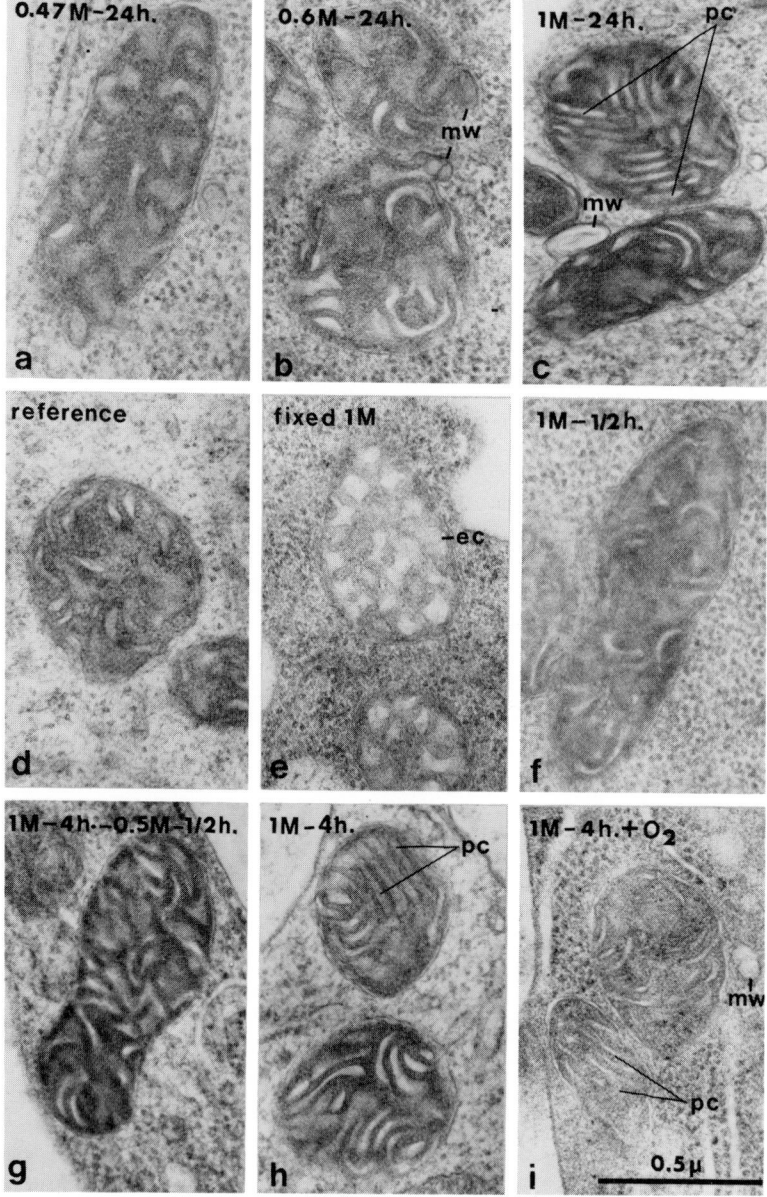

Figure 15. *In situ* mitochondria of differentiating cortical cells, in varying conditions of plasmolysis. (a) (b) and (c) plasmolysis of increasing intensity, 0.47 *M*, 0.60 *M*. 1 *M* mannitol added to the culture medium during 24 hr. The cristae are at first slightly dilated (a), then the mitochondria contract, membranous whorls (mw) are formed, cristae appear longer (b). At 1 *M* concentration, long cristae are arranged in parallel groups (c); (d) reference mitochondria; (e) the utilization of a plasmolyzing fixative (1 mole mannitol in glutaraldehyde) provokes an intense dilation of the short and irregularly disposed cristae; (f) and (g) both during a strong plasmolysis (1 *M*) of short length (1/2 hr and during its regression (4 hr in a molar mannitol solution and 1/2 hr in a 0.5 *M* solution), the inner mitochondrial membranes do not appear clearly. In the first case (f) there has been no time for the orientation of the cristae to become established. In the second case (g), it is **disappearing**; (h) and (i) the effect of oxygen bubbling during 4 hr (i) in a plasmolyzing medium (1 *M*). Cristae are thin, the parallel arrangement is hardly visible. x 60,000.

Some were only slightly modified—their cristae appeared more clearly than in reference specimens. This phenomenon was attributed to the reduction of the volume of each mitochondria, which makes the matrix more dense and the sinuous cristae more distinct. In other mitochondria, the whole section was covered with numerous, stacked cristae (Figure 15 c, h). Their thickness was very regular, as well as their intercristae distance, which measured about 170 Å.

A strong plasmolysis, during a long enough period, was necessary to induce a dramatic reorganization of the mitochondrial ultrastructure; moreover, it affected only a part of the mitochondrial population.

Such modifications were reversible. We fixed samples of roots which had stayed in a molar mannitol solution for 4 hr and then 1 hr in a 500 milliosmolar solution. Under such conditions, cytoplasmic vesicles were crushed between the plasmic membrane and the pecto-cellulosic cell wall. These vesicles might have been torn away from the cytoplasm during the period of strong plasmolysis and had been crushed when plasmolysis diminished and the plasmic membrane drew nearer to the cell wall. The cristae of the mitochondria appeared long and rather thin (Figure 15 g) and were rarely arranged parallel to one another. So the polarization of the cristae seemed to depend upon the intensity and duration of the plasmolysis. But does plasmolysis interfere with the ultrastructures directly, or indirectly by means of the observed decrease in the respiratory intensity?

Reference roots were fixed by means of buffered glutaraldehyde solution to which mannitol had been added up to 1 *M* concentration. This experiment was made to determine if in oxygen-deprived roots, the observed ultrastructural modifications were due only to the hypertonicity of the fixatives. The cytoplasm of the cells was indeed locally separated from the cell wall and nucleus and nucleolus were poorly contrasted. But these endoplasmic reticulum remained formed of short isolation vacuoles and did not build the long stacks of vacuoles usually seen in anaerobic cells. The mitochondria possess short porific dilated cristae (Figure 15 e).

Modifications existed, provoked by the osmotic phenomena during fixation, but to a limited extent. In particular, the reorganization of the mitochondrial cristae, was indeed due to oxygen deficiency during a prolonged anaerobiosis, and not to a simple plasmolytic effect during fixation.

In roots submitted to a prolonged and intense plasmolysis (1 M, 24 hr; Figure 15 h), which results also in a decrease in the oxygen uptake (Figure 14), mitochondria resembled strongly the mitochondria of oxygen-deprived roots. If the oxygen concentration can be artificially increased in the culture medium where roots are steeped, it might be possible that transfer of the oxygen molecules through the cytomembranes (plasmic membrane, mitochondrial membranes) is improved. If this were so, it would become possible to distinguish the effects of plasmolysis and those of oxygen deprivation.

When oxygen was bubbled through the plasmolyzing culture media for 4 hr (1 M, mannitol), the cellular ultrastructures were obviously modified. The saccules of the endoplasmic reticulum extended to the whole cytoplasm; the nucleus was not highly contrasted, nor were the organelles, which were discerned with great difficulty. The mitochondria (Figure 15 i) differed considerably from those observed in roots coming from a plasmolyzing medium without oxygen bubbling (compare Figures 15 h and 15 i). The cristae were very thin and faintly visible, although long. The matrix was seen mainly in the peripheric region of the organelles, widely separating the cristae. In the central part of the mitochondria, an electron transparent zone existed with anastomosed filaments, which probably contained the DNA. It was noteworthy that the oxygen bubbling applied to nonplasmolyzed roots did not provoke any of the modifications described previously.

The observed ultrastructures showed that the fine structure of the mitochondria was modified when the oxygen uptake was increased in plasmolyzed cells. The piling up of the mitochondrial cristae thus seems to be due to the direct or indirect effects of a local oxygen deprivation connected with a strong plasmolysis. The disposition appears to decrease the contact areas between the cristae and the matrix. This variation in the connection between matrix and cristae may be related to the oxygen deficit.

Effects of an Uncoupling Agent, 2-4 DNP, upon the Respiratory Rate and upon the Ultrastructures

According to the ultrastructural observations of anoxied or plasmolyzed roots, it seems that the modification of the mitochondrial structures are caused by a more or less complete absence of oxygen. This can be a direct or indirect effect, such as the stopping of electron transfer, which suppresses phosphorylation. In the presence of oxygen, we tested an uncoupling agent to try to determine the process that starts the reversible morphological alterations.

A 2-4 dinitrophenol solution was added to the culture medium, and the final concentration varied between 1 and 100 μM. Roots remained in this medium for 5 hr. The dissolving of the 2-4 DNP in the culture medium was slow, and the pH was slightly modified: it increased from 4.8 to 5.2, but it remains within the usual limits of the pH variation during one week of culture (it increased from 4.8 to about 5.6).

Respiratory Intensity

In a concentration equal to 10 μM of 2-4 DNP, the oxygen uptake increased noticeably in comparison to the reference values (Figure 16). However, a concentration of 50 μM caused the uptake to decrease significantly, as it was a toxic solution. When the 2-4 DNP concentration was 100 μM, all cells were killed. These results obtained with excised tomato roots are in accordance with the previous experiments with Tobacco callus[52] and yeast[53] respiration response to 2-4 DNP.

Ultrastructural Observations

The 2-4 DNP solutions, used at the concentration of 5 and 10 μM, induced modifications similar to those produced by anoxia: cells appeared plasmolyzed, fibrillar structures appeared in the cytoplasm, some mitochondria had irregularly dilated cristae (Figure 17 a), and other mitochondria had cristae arranged in compact and polarized groups (Figure 17 b, d, f, g). Nevertheless, this aspect was less frequent than in anoxied roots. The mitochondria of all meristems became electron transparent, but not in a synchronous way (Figure 17 c); they were not swollen during anaerobiosis, contrary to the mitochondria of the main root meristem. The poisoning due to the 2-4 DNP was revealed through the frequent presence of disturbed mitochondrial membranes in the cells.

The concentration of 50 μM was toxic. Cells were probably not killed, since they still took up oxygen, but their physiological condition was unsound and they were disorganized during fixation. Only a few cells kept an

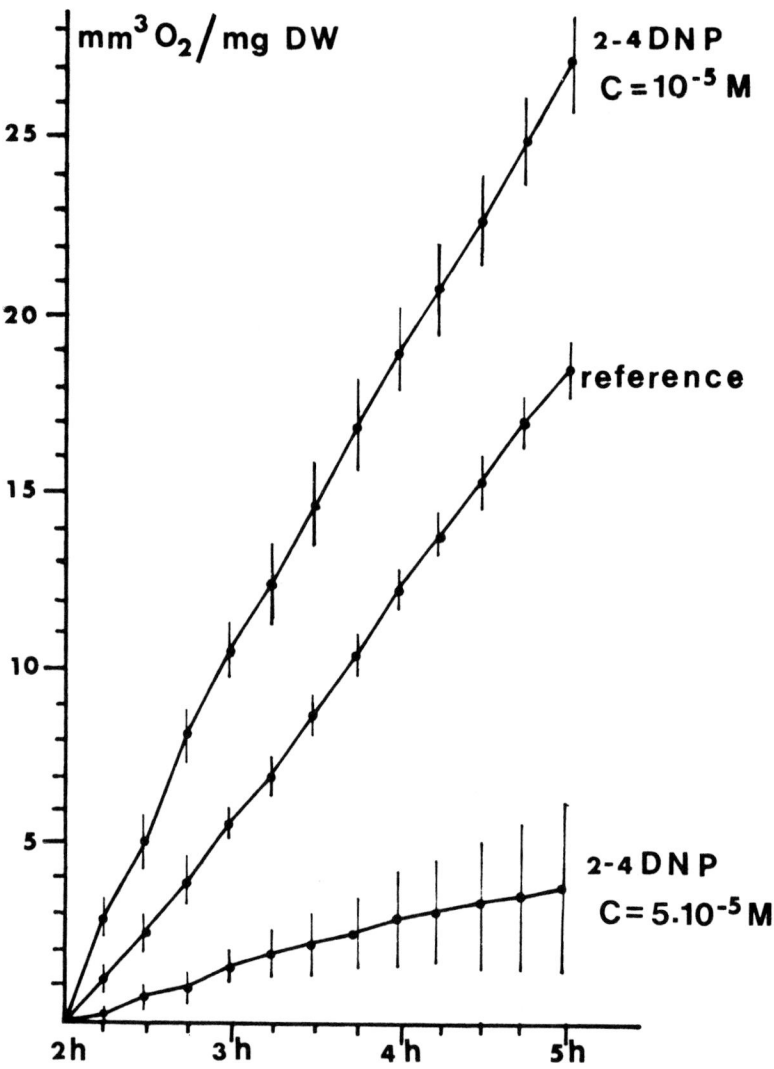

Figure 16. Respiratory rate (mm^3 oxygen absorbed per mg of dry weight) of roots steeped for 2-5 hr in 2-4 DNP solutions, at concentrations of 10 μM and 50 μM in the culture medium.

intact plasmic membrane, though their cytoplasm appeared extremely condensed.

When the uncoupling agent was utilized at a concentration that provoked an increase in the oxygen absorption, a part of the mitochondrial population showed structural modifications similar to those induced by anoxia. In both experiments, it seemed to be the stopping of the energetic coupling, rather than the oxygen deprivation, which induced the modifications of the mitochondria. In this hypothesis, the lack of oxygen would act indirectly by impeding the energetic coupling, because of interference with the electron transfer.

DISCUSSION

1. Through the use of culture techniques, it was possible to determine precisely the effects of oxygen deficiency upon the root, an organ which is normally underground, and subjected to wider perturbations in gas circulation than above-ground organs. Moreover, the utilization of excised roots eliminated the correlations between organs, and made it possible to constitute root clones and to experiment upon them in germ-free conditions.

2. The tissues of the root behaved quite differently during anaerobiosis. Although it was difficult to draw conclusions concerning the behavior of tomato roots in general, a certain analogy with the known activity of adventitious roots[54] could not be excluded. When differentiation was sufficiently advanced for different tissues to be distinguished, a short time before the formation of the quiescent center cells lost their ability to divide after prolonged oxygen deprivation.

Most of the cells remained alive in the rootlets, as well as in the older tissues of the main root. This was verified by the accumulation of starch granules in the leucoplasts upon aeration.

The initiating rootlet meristems and the pericycle cells of the subapical zone at the main end meristem were most resistant to anoxia, *i.e.*, they were still able to divide.

3. The aspects of cells, which appeared plasmolyzed, after anoxia or after a treatment by 2-4 DNP, received considerable attention. It was likely that the living cells were indeed plasmolyzed during anoxia (roots having lost their stiffness, cytoplasmic vesicles being crushed between the plasmalemma and the cell wall, in roots placed back into aerated medium). The fact that anoxia damage caused a contraction of the cytoplasm, showed that the osmotic pressure of the cells was already modified before fixation. This change was attributed to variations in the permeability of the plasmic membrane, and also

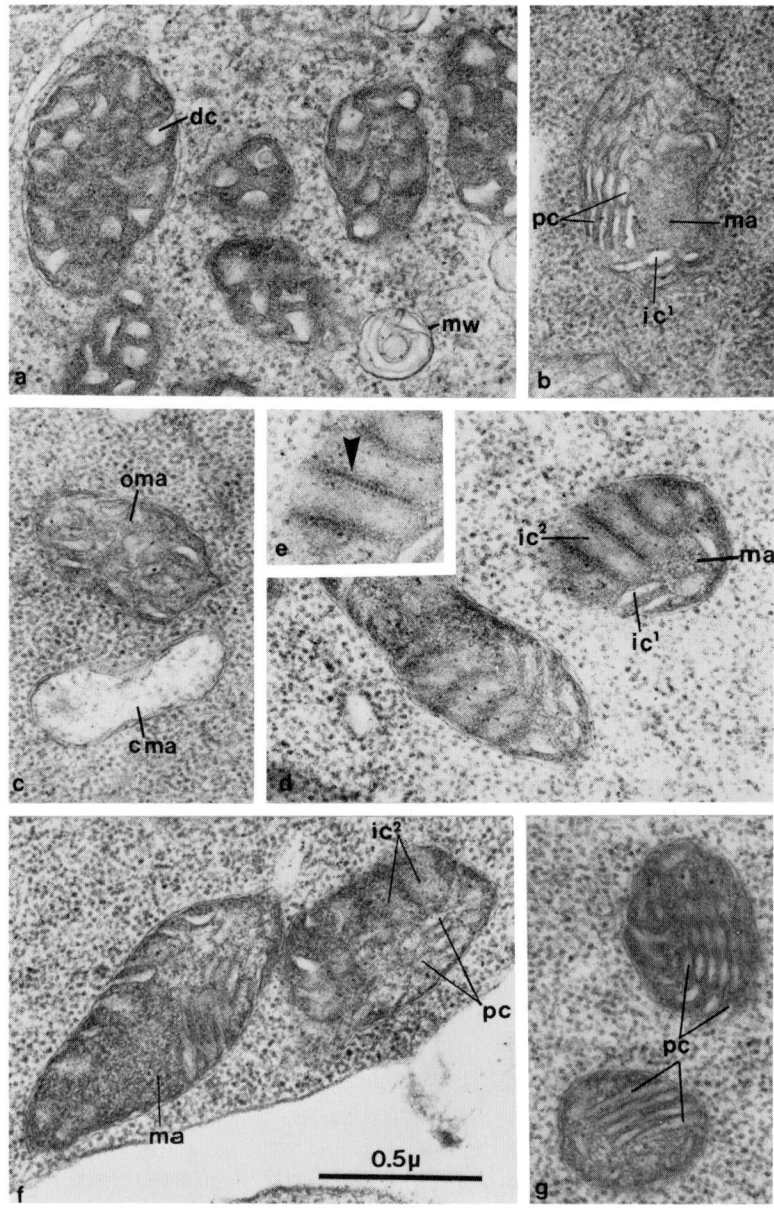

Figure 17. The ultrastructure of mitochondria treated with the 2-4 DNP in a concentration which increases the respiratory rate (10 M). In the subterminal area of the root certain mitochondria (a) have cristae more dilated (dc) than those of reference roots. Others (b) (d) (f) and (g) present cristae that are regularly disposed (pc); the matrix forms zones devoid of cristae (ma). The inner space of the cristae appears clear if it is cut across (ic^1); (b, d) slightly opaque if the cut is tangential (ic^2); (d, f) to the broad flattened cristae. The periodic structure may be formed in the narrow space between two consecutive cristae [arrows, (d) and inset x 100,000]. In the meristematic cells some mitochondria have a clear content [cma (c)] and a few cristae, while others do not present such a drastic change [oma (c)] x 60,000.

to the cytomembranes surrounding the different organelles, in experimental anoxic conditions, which caused a decrease of the phosphorylating process.

On the other hand, the modifications of the organelles could not be explained by an altering during the fixation process. In reference roots fixed with a plasmolyzing fixative, the ultrastructural modifications of endoplasmic reticulum and of mitochondria never occurred.

The osmotic pressure of the vacuole is necessary for the entry of water in plant cells, but its transfer to the central cylinder cells against the natural gradient of turgor pressure, regains an energy consumption on the part of the cells. So the active maintenance of the turgescence of cells can be disturbed if the quantity of ATP decreases. The concentration of ions and small molecules in the vacuole might be diminished, by simple passive diffusion towards the external medium, if the selective permeability is altered through energetic shortage. This phenomenon was certainly one of the reasons why cells looked plasmolyzed after a treatment by anoxia or 2-4 DNP.

Moreover, the molecular composition of the membranes may vary under the influence of both experimental treatments, *i.e.*, oxygen deprivation and uncoupling agents. Hiatt and Lowe[55] attributed the loss of organic acids, amino acids and of ions to an impairment of the cellular membranes in barley roots deprived of oxygen or treated with 2-4 DNP or NaCl. MacLaughlin[56] suggested that the 2-4 DNP, adsorbed by the membranes (formed from a neutral lipid), determined the electronegative potential and thus had a direct effect upon their permeability.

The quantitative analyses of the lipid and phospholipid composition showed that they differed appreciably from the check sample. Results differed, according to the experimental material. In yeast,[57,58] the amount of ergosterol decreased, and the oleic fraction increased at the expense of the unsaturated fatty acids. In rice coleoptile, on the contrary,[59] phospholipids were more abundant, saturated residues (palmityl) decreased, in correlation with an increase of the unsaturated residues. Since the results were contradictory in rice and yeast, it could be interesting to determine the lipid composition of our own material (excised tomato roots) which, contrary to yeast and rice, did not manifest any growth during anoxia. However, in both cases, quantitative analyses of materials revealed that anoxia altered the chemical components of the membranes and, consequently, may have modified their selective permeability.

These phenomena were complicated by the different reactions of the plasmic membrane and the organelles. Thus, in the shrunken cytoplasm of the cells of the main root meristem during anoxia, the mitochondria were notably swollen (leucoplasts were not). On the contrary, in rootlet meristems and in more differentiated cells, mitochondria were slightly contracted.

Our experimentation was conducted under confined conditions. Therefore the observed phenomena could have been the consequence of diverse factors. In the absence of oxygen, CO_2 may accumulate, and diverse toxic substances such as ethanol[60] may appear in the culture medium and in the nitrogen atmosphere of the incubation flask. If we assumed that the CO_2 output was the same during 72 hr of anaerobiosis, the maximal CO_2 concentration reached in the 300-ml flask was 0.7%. (We have not yet determined the toxicity at this concentration.) It is commonly recognized that roots are very sensitive to ethanol. However, 1% ethanol in the culture medium during 24 hr did not provoke any ultrastructural modifications in excised tomato roots. If other toxic substances were produced by the fermentory metabolism in the oxygen-deprived tissues, they could be diffused into the liquid culture medium and the surrounding atmosphere. Both were voluminous (80 ml and 220 ml) with regard to the roots (two roots of about 10-15 mg fresh weight each).

4. *Structural reactions of the mitochondria.* To distinguish the main methods of classifying anaerobiosis we use herein the terminology of Vartapetian.[34] When plants were entirely grown in anerobic conditions, it was called primary anoxia (yeast cells, rice coleoptile). If plants were first developed in a normal atmosphere and then submitted to oxygen deprivation, it was called secondary anoxia (rice or tomato roots,for example).

Our results were apparently contrary to those of Coulomb and Coulomb.[61] The mitochondria of gourd meristems, deprived of oxygen for 2 hr, were considerably enlarged. When the germinations were returned to aerated conditions, the mitochondria became a condensed structure; their matrix was dense and the cristae were dilated. In tomato roots, we observed the swelling of the mitochondria of the main root meristem, but not their restoring. On the contrary, the mitochondria of more differentiated cells did not swell during anoxia; their matrix remained opaque to electrons. These contradictory results can possibly be explained through differences in the experimental modalities, applied to two different species (gourd, tomato). Anoxia lasted only 2 hr. The germinations were steeped in oil, or were in nitrogen-enriched conditions; the excised tomato roots were cultured in a liquid medium saturated with nitrogen for 72 hr. The age of the organs was not the same: 1-week germination of the case of gourds, 8-10 weeks subculturing, at least, for the tomato roots.

We hypothesized that the mitochondria of gourd roots, submitted to a short-term anoxia, would not undergo such an important and irreversible alteration as those of the main meristem of a tomato root system, after 72-hr oxygen deprivation. After a short aeration, they were still able to recover an aspect, which was similar to the one we observed in more differentiated cells after 72-hr anoxia and 1-hr aeration, *i.e.*, they had typical enlargement of the cristae and the opacity of the matrix.

Mitochondria from rice coleoptile cells, germinated in anaerobiosis responded considerably differently from those of anoxied tomato roots. They developed more or less normally at first, and then vesiculated more and more with the lengthening of anaerobiosis.[32,33,35] But it did not seem entirely justifiable to compare the alterations provoked by a primary (rice) or a secondary (tomato) anoxia.

Rice roots, contrary to the coleoptile, only develop in normally aerated conditions, so they can only be submitted to a secondary anoxia in conditions similar (although roots were not isolated from the other organs) to those we applied to tomato roots. Resistance of rice roots to anoxia was low; after 24-hr anoxia, the cellular structures were already disturbed.[34,62] On the contrary, the grouping of the cristae in coleoptile mitochondria, when germinates were submitted to a prolonged secondary anoxia, were comparable to our own results, obtained with tomato roots. Results of our experiments with excised roots cultivated *in vitro* proved that it was oxygen deprivation that determined the process of spatial redistribution of the cristae.

A comparable reorganization of mitochondrial cristae was observed in Pteridophtes[63,64] in certain cells at a particular period of their evolution. In the oosphere stage at the end of its maturation, cristae became disposed in parallels, were somewhat enlarged, and the matrix was more opaque. In all surrounding cells, mitochondria remained unchanged. Such response was attributed to a possible oxygen deficiency in the cells.[64] Under these conditions it seems difficult to imagine how oxygen transfer could be impeded. We suggest, at this particular stage of the egg cell maturation, that the respiratory intensity was in fact modified, but as a consequence of the reorganization of the mitochondrial cristae.

5. *Oxidase activity.*   Quantitative analysis and cytochemical reactions
have demonstrated that in rice coleoptile, contrary to the results obtained
with yeast, the enzymes of the oxidative phosphorylating chain develop
in anaerobiosis, although they were less abundant than in reference measurements.[36,65] The oxidative polymerization of the DAB, due to the
activity of the cytochrome oxidase in the cristae of rice coleoptile cells
entirely gown in anaerobiosis, was much slower;[66] these observations were
hence in accordance with the previous measurements.

On the contrary, in our materials submitted to 72-hr anoxia, the reaction with the DAB seemed to be more intensive than in reference cells.
The osmium deposits made the orientated cristae opaque, whereas it was
less abundant in the dilated cristae of the same mitochondria, or in the
less-modified mitochondria. It looks as if the tight disposition of the
parallel cristae in some way protects the activity of the enzymes of the
respiratory chain. In our experiment, it was impossible to analyze the
cytochromes quantitatively because of the small amount of fresh weight of
the root material used (10-15 mg from each root).

When the period of anoxia had been followed by 1 hr of aeration, the
oxygen uptake was reduced. Nevertheless, the reaction with the DAB was
very strong in the parallel-orientated cristae, which were less tightly disposed, as well as in the enlarged and nonorientated cristae. After a longer
period of aeration (15 hr), the ultrastructure and intensity of the cytochemical reaction became similar to that observed in reference cells.

In 1958, Simon[67] attributed the increased activity of the cytochromes,
when that of the succino-oxidasic system decreased, to the action of digi-
tonine and other detergent upon the membrane structures. In the same
way Nir *et al.*[68] observed in water-deprived roots, an increase of "osmium
black" deposit. They suggested that the modification of the spatial distribution of the enzymatic complexes was due to the loss of water, therefore
resulting in a greater reactivity.

The demonstration with the DAB of the cytochrome oxidasic activity
was not quantitative, but it still showed that if a turnover of the cytochromes occurs, 3 days anaerobiosis did not lead to their disappearance.
The reasons for the high reactivity to the DAB during anoxia, and during
a short period after the return to aerated culture conditions, must be due
to either a modification of the molecular structure of the membranes,
allowing for a better diffusion of the substrate, or to a partial release of
the cytochromes out of the enzymatic complex where they are normally
found.

6. *Ultrastructure and energy state.* Is it possible, as some researchers
have tried to do in other animal or plant cells,[69-72] to relate the ultrastructure of the mitochondria of isolated tomato roots, to their energetic
state?

The respiratory states of isolated mitochondria have been defined by Chance and Williams.[73] State III is the state of oxidative phosphorylation, when the respiratory rate is fast, the levels of O_2, ADP, Pi and substrate are high, the rate-limiting component being the respiratory chain itself. State IV corresponds to the exhaustion of ADP, with a coincident decrease in respiration, the levels of O_2 and substrate remaining high; the rate-limiting component is the phosphate acceptor. In state V, on the contrary, ADP, Pi and substrate are at high level, but the respiratory rate is slow or nonexistent as oxygen is rare or absent (anoxia).

The ultrastructural configurational changes in mitochondrial membranes from animal tissues have been attributed to energy-linked reactions associated with oxidative phosphorylation, not only in isolated mitochondria[69,74] but also in situ.[70] The orthodox configuration which approaches the structure of the mitochondria in situ, is associated with the respiratory state IV, and the condensed configuration is associated with metabolic state III. The ultrastructural changes of the mitochondria would seem to reflect configurational changes in the proteins of the oxysomes, and so appear to be the real energy transformer making the respiratory coupling possible.

Another interpretation is proposed, derived from experiments carried out on mitochondria isolated from animal[71,75] or plant[72,76] tissues. The active and energy-dependent penetration of ions into the matrix causes an osmotic swelling of the mitochondria. So the different configurational states could "represent metabolically triggered osmotic adjustments rather than mechano-coupling events associated with oxidative phosphorylation."[72]

In our own experiments, the mitochondria in situ, of roots cultivated in vitro, were oxygen deprived. But as the ADP and substrate levels have not yet been determined, we cannot affirm that they actually were in respiratory state V. Some mitochondria presented an "orthodox configuration," close to that observed in reference cells. Others reacted differently and had a quite original ultrastructure. Cristae grouped in several areas in which they were parallel to one another, while other areas contained only the matrix. This represented a spatial segregation of the two components of the mitochondria: membranous system on one hand, and matrix on the other hand.

A similar structure took place under the influence of an uncoupling agent (2-4 DNP), at a concentration that increased the oxygen uptake. Some mitochondria were hardly modified, and their ultrastructure remained "orthodox." This observation was similar to that of Hackenbrock et al.[70] In the presence of oxygen, the electron transfer goes on, the respiratory rate increased. Although all required conditions seem assembled for the appearance of a "condensed configuration," this was not realized in the absence of the energy coupling.

Other mitochondria possessed thin cristae arranged in parallels, with little space left for the matrix. The matrix space was very narrow and filled with an osmiophilic substance with a periodic structure (arrow, Figure 17 e). This aspect had also been found in the mitochondria of anoxied roots. This structured system lies in the matrix space, between two closely juxtaposed cristae. Thus, it was quite different from that observed by Green et al.[77] in the inner space of the cristae and in the space between the inner and the outer mitochondrial membrane.

In situ mitochondria of excised tomato roots, treated by anoxia or with an uncoupling agent, showed the same ultrastructural modifications. So it seemed that the observed reaction was more specifically due to the absence of phosphorylation than to the oxygen shortage.

Plasmolysis led to the same alterations. It was not inconsistent with the previous conclusion, where a 750 or 1000 milliosmol plasmolysis provoked temporarily the appearance of a "condensed configuration" (Figure 12 b). The inner cristae space was enlarged and the matrix was electron opaque. This was realized during a period when oxygen uptake was still depressed compared to the reference values, but evidently higher than during anaerobiosis. ADP or AMP may have been abundant as well as the reduced substrates. In Ascite tumor cells, Hackenbrock et al.[70] reported that the orthodox configuration of the mitochondria changed to a condensed configuration when the ADP level was increased *in situ*. In the same way, in excised tomato roots cultivated *in vitro*, we might observe *in situ* the transformation of the "orthodox configuration," and of the specific ultrastructure of mitochondria in the absence of phosphorylation, into the "condensed configuration" characteristic of metabolic state III.

Thus, it seemed possible to correlate the different typical ultrastructures to energetic events. But we must not forget that important osmotic phenomena accompanied the ultrastructural modifications. So the osmotically triggered events, more difficult to determine *in vivo*, were always added to the experimental conditions. We showed that important osmotic alterations appeared during anoxia, different from one cell compartment to another (the cytoplasm shrunk, the mitochondria were swollen in the main root meristem cells, for example). After a period of anaerobiosis, when roots were replaced in an aerated medium, cells gradually regained their normal osmotic pressure, the cristae of the mitochondria broadened out, and then this aspect disappeared when cells became turgescent again. In our experimentation, the structural modifications seemed to be highly determined by the osmotic readjustments. Thus, they were linked to the transport of ions which is dependent on the energetic phenomena.

7. *Heterogeneity of the mitochondrial population.* All the mitochondria of the same cell did not react in a uniform way. Some mitochondria were slightly modified (as a consequence of the variations of the osmotic pressure, as it seems); others exhibited a highly polarized structure, in the differentiating cells of roots deprived of oxygen for 3 days.

Kerpel-Fronius and Hajos[78] studied the accumulation of Sr^{2+} in mouse liver mitochondria *in vivo* as well as *in vitro*. It was energy-dependent and the energy was provided either by the hydrolysis of added ATP or by the oxidative phosphorylation. All mitochondria did not possess Sr^{2+} precipitate; "this points to possible differences in the substrate utilization of mitochondria located in the same cells."[78] On the other hand, in an isolated mitochondria pellet submitted to a determined treatment, not all, but usually only a fraction of mitochondria behave in a uniform way. Thus, the mitochondrial population seems to be heterogeneous for several functions. An explanation for these phenomena could be that in the same cell, mitochondria either are in different functional states or undergo an evolution and are at various stages in the course of evolution.

CONCLUSIONS

Our research related to an anoxia-sensitive plant at a particular period. After long anaerobiosis, the ultrastructural alterations were still reversible. The experimental results with excised tomato roots, cultivated *in vitro*, point out that:

1. *Meristems behaved differently* according to whether they belonged to well developed rootlets or to the group of meristematic cells of initiating rootlet meristems. After 3 days of anaerobiosis, meristems that had already been functioning for a certain time, were no longer able to regain their property of cell division. Only the meristems in the process of being formed were able to proliferate again, as were the pericyclic cells of the subapical zone of the main root meristem, which cells can and do differentiate and initiate new rootlets.

2. *A modification of the cell permeability took place,* expressed by a resemblance to plasmolysis, when roots were deprived of oxygen, or submitted to the action of an uncoupling agent (2-4 DNP). In both experiments, the variations of the osmotic pressure of the cell were related to the decreased production of ATP, which started a reduction of the active transport of ions and molecules on one hand, and altered the molecular structure of the membranes, which impaired their properties of selective permeability.

3. *Mitochondria underwent specific structural and functional modifications.* Similar alterations were triggered by other factors: uncoupling and plasmolysis. The characteristic reaction of the mitochondria revealed either reversible alterations or a true adaptation to a new regime, during which some mechanism—electron transfer, uncoupling and ATP synthesis —were more or less completely inhibited. If the different results were

compared, it seemed that the stopping of ATP synthesis was more
a determining factor in the reorganization of the mitochondrial cristae
than the oxygen deficiency.

Because of the regularity and accuracy of the reaction, we hypothesized
that the phenomenon corresponds to an inhibition of the respiratory function of the organelle. Inhibition mechanisms exist at the molecular level,
and it may be possible that they exist at the ultrastructural level also.
The assembling in a tight group of the cristae, which bear the electron
transfer chain, and the enzymes of the oxidative phosphorylation, leads
to their isolation, at least partly, from the matrix, which normally has
wide surfaces of contact with them.

It is possible to compare the grouping and the polarization of the cristae
in the mitochondria to the behavior of the endoplasmic reticulum, which
forms dense groups of long saccules in the cytoplasm of oxygen-deprived
cells.[26,28,45] The proposed interpretation is supported by the fact that
the observed phenomenon was reversible. After the return to a normal
atmosphere, the distribution of the cristae became less regular and separated from one another. It would seem that the arrival of oxygen provoked
the spreading out of the cristae in the mitochondria, making maximum
contact areas possible again between the external side of the cristae and
the matrix, where most of the enzymes and the metabolites of the Krebs
cycle are located.

REFERENCES

1. James, W. O. *Plant Respiration* (London: Oxford University Press, 1953).
2. Mazelis, M., and B. Vennesland. "Carbon Dioxide Fixation into Ox-aloacetate in Higher Plants," *Plant Physiol.* 32:591-600 (1957).
3. Schneider, A. "Milchsäure in höheren Pflanzen," in *Handbuch der Pflanzenphysiologie*, Vol. 12 (Berlin: Springer-Verlag, 1960), pp. 1009-1022.
4. Streeter, J. G., and J. F. Thomson. "Anaerobic Accumulation of Aminobutyric Acid and Alanine in Radish Leaves (*Raphanus sativus* L.)," *Plant Physiol.* 49:572-584 (1972).
5. Turner, J. S. "Fermentation in Higher Plants; Its Relation to Respi-ration; the Pasteur Effect," in *Handbuch der Pflanzenphysiologie*, Vol. 12 (Berlin: Springer-Verlag, 1960), pp. 42-87.
6. Leblova, S. "Pyruvate Metabolism in Germinating Plants during Natural Anaerobiosis, Characterization of Lactate and Alcohol De-hydrogenase," in *Abstracts of the Twelfth International Botanical Congress* (Leningrad: Nauka, 1975).
7. Chirkova, T. V., E. L. Sokolovskaya and I. V. Khazova. "Activity and Isoenzyme Composition of Root Peroxydase as a Function of Temporary Anaerobiosis," *Fiziol. Rast.* 20(6):1236-1241 (1973). (In Russian).

75. Stoner, C. D., and H. D. Sirak. "Osmotically-Induced Alteration in Volume and Ultrastructure of Mitochondria Isolated from Rat Liver and Bovine Heart," *J. Cell Biol.* 43:521-538 (1969).
76. Lee, D. C., and R. H. Wilson. "Swelling in Bean Shoot Mitochondria Induced by a Series of Potassium Salts of Organic Anions," *Physiol. Plant* 27:195- 201 (1972).
77. Green, E., E. F. Korman, G. Vanderkooi and T. Wakabayashi. "Structure and Function of the Mitochondrial System," in *Autonomy and Biogenesis of Mitochondria and Chloroplasts,* N. K. Boardman, A. W. Linnan and R. M. Smillie, Eds. (Amsterdam: North-Holland Publishing Company, 1970), pp. 1-17.
78. Kerpel-Fronius, S., and F. Hajos. "Electron Microscopic Demonstration of Energy Production and Coupled Respiration of *In Situ* Mitochondria," *J. Histochem. Cytochem.* 81:740-745 (1970).

18

LIPIDS IN RICE SEEDLINGS GROWN UNDER
ANAEROBIC AND AEROBIC CONDITIONS

B. B. Vartapetian

Timitiazev Institute of Plant Physiology
Academy of Sciences
Moscow 127106 USSR

R. Bazier and C. Costes

Institut National Agronomique
Thiverval - Grignon, France

INTRODUCTION

During recent years, three teams of investigators have demonstrated that mitochondria are formed and preserved without notable signs of destruction in rice (*Oryza sativa* L.) coleoptiles grown under strictly anaerobic conditions.[1-3] Moreover, anaerobically cultivated rice seedlings when transfered into an ordinary atmosphere, can absorb oxygen from the environment,[3-5] thus escaping the stage of oxygen adaptation.[6]

Mitochondria isolated from anaerobically grown coleoptiles contain cytochromes a, b and c[3,7,8] and are able to transport electrons of the Krebs cycle substrates to molecular oxygen, hence releasing the energy accumulated in ATP.[6]

Rice seedlings can easily tolerate long-term secondary anoxia, as compared to other higher plants, *i.e.,* anoxia induced not from the very beginning of seed germination but after a preliminary cultivation of the seedlings in the presence of molecular oxygen. In this case their mitochondria exhibit a very dense and packed membranous system[9] similar to the mitochondria in animal cells performing very active work (cells of flight muscles in insects, myocardium of mammals). Finally, the rice coleoptile cells grown under anaerobic conditions contain not only mitochondria but all subcellular organelles, which are typical for the plants grown under aerobic conditions.

The existence of such well developed organelle membranes without oxygen led us to study the effect of anoxia on total lipids of such anaerobic coleoptiles compared to aerobic rice coleoptiles.

MATERIALS AND METHODS

Rice coleoptiles (Arpa-Shali variety) were grown in darkness for 5 days at 27°C, in distilled water bubbled with nitrogen (anaerobic conditions) or with air (aerobic conditions). In the anaerobic treatment, water was previously flushed for 1 hr with a flow of gaseous nitrogen (O_2 content did not exceed 30 vpm). Sterilization of seeds, when used, was performed by dipping them in mercuric chloride solution (1% W/v). In both aerated and anaerobic treatments, coleoptiles were white and about 4-5 cm long.

Coleoptiles were cut off, freeze-dried, ground and extracted with a mixture of chloroform-methanol, 1/1, v/v to test for total lipids. Lipid analysis was performed as described by Costes et al.[10]: thin-layer chromatography of the extract on Silicagel H with $CHCl_3$ - CH_3OH - H_2O, 65/25/4, v/v for polar lipids and petroleum ether-ethyl ether-acetic acid, 80/20/1, v/v for neutral lipids, and identification of each class of lipids through specific reactions. Fatty acids were separated by gas-liquid chromatography after scraping of silicagel and *trans*-esterification by the mixture CH_3OH - C_6H_6 - H_2SO_4, 20/10/1, v/v for 2 hr at 77°C.

RESULTS

The results arising from four series of repetitive experiments are given in Figures 1-4. Individual differences between experiments were mainly due to differences between samples as variations were taken into account in averaging values.

1. On the dry matter weight basis, the total lipid content increased under anaerobic development of coleoptiles (Table I): the increase in anaerobiosis was almost twice that in aerobiosis, with or without sterilization. Therefore, the lipid increase was due to coleoptile production and not to contaminant bacteria.

ACYL RESIDUES DISTRIBUTION IN TOTAL LIPIDS

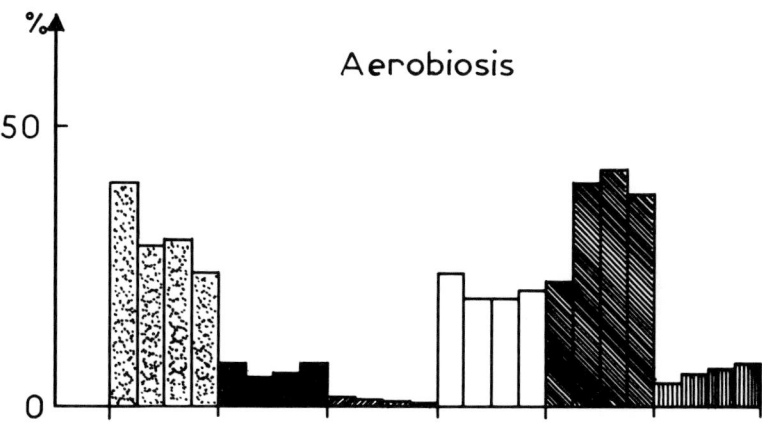

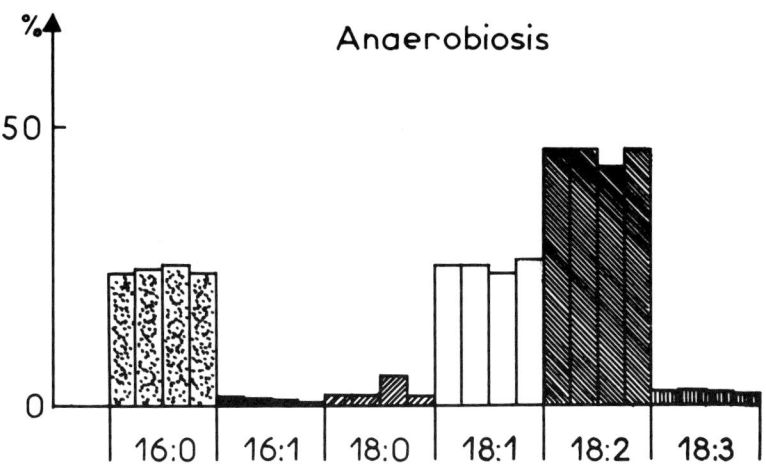

Figure 1. Acyl residue distribution in total lipids. Each bar represents the result of one experiment in anaerobiosis coupled with one experiment in aerobiosis.

ACYL RESIDUES DISTRIBUTION IN PHOSPHATIDYLCHOLINE

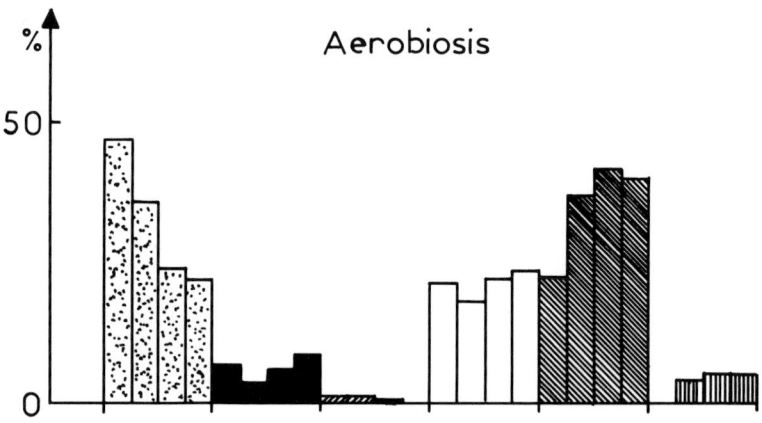

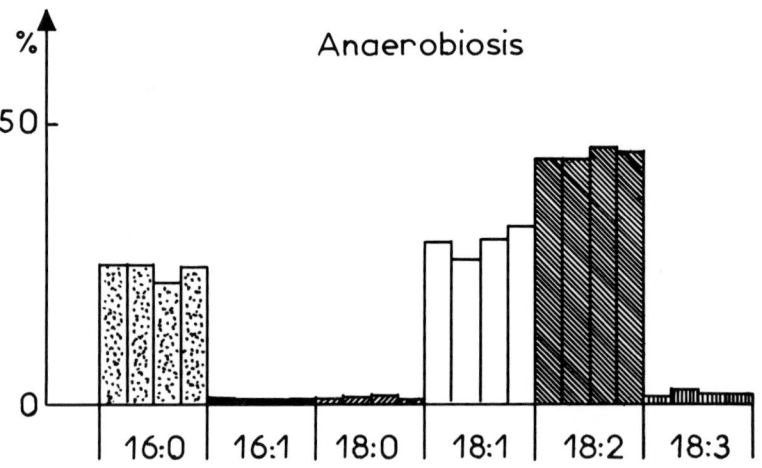

Figure 2. Acyl residue distribution in phosphatidylcholine. (Expression of results as in Figure 1.)

ACYL RESIDUES DISTRIBUTION IN PHOSPHATIDYLETHANOLAMINE

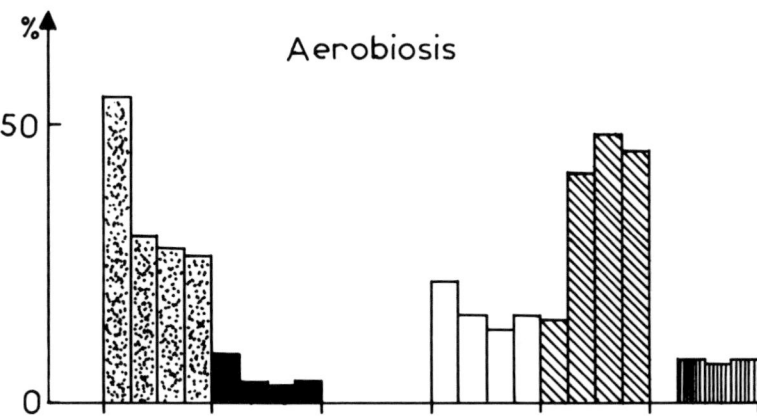

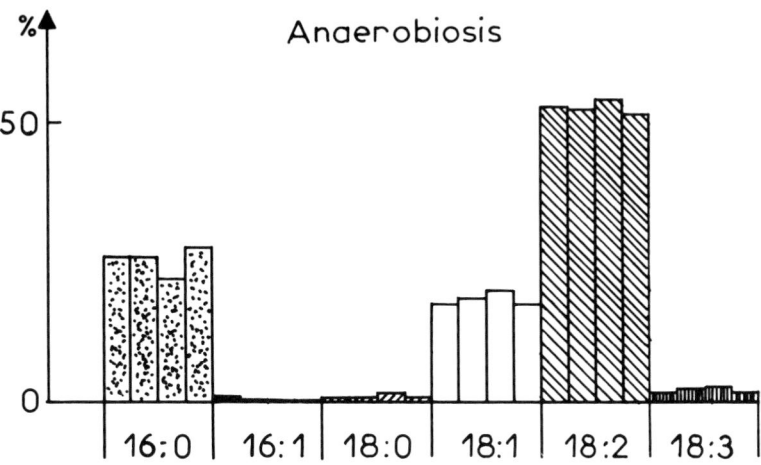

Figure 3. Acyl residue distribution in phosphatidylethanolamine. (Expression of results as in Figure 1.)

ACYL RESIDUES DISTRIBUTION IN NEUTRAL LIPIDS

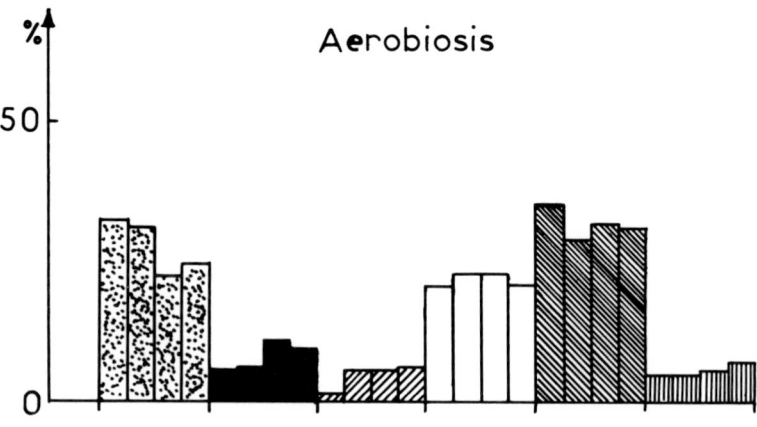

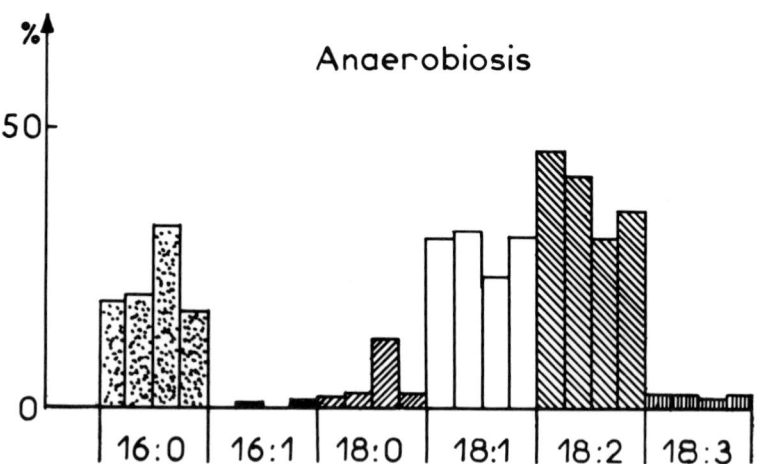

Figure 4. Acyl residues distribution in neutral lipids. (Expression of results as in Figure 1.)

Table I. Total Fatty Acids in Lipids
(μmol g^{-1} of dry weight)

Sterilization	Anaerobiosis (An)[a]	Aerobiosis (Ae)[a]
-	143.6	94.8
-	168.0	
+	111.8	71.0
+	140.5	73.7
+	119.0	77.0
mean value	136.6	79.1
	100%	58%

[a]An/Ae = 1.7.

2. In both treatments, the same lipids appeared: phosphatidylcholine (PC), phosphatidylethanolamine (PE), neutral lipids containing mainly triglycerides. In anoxia, the contents of PC, PE, neutral lipids and fatty acids increased strongly on the dry weight matter basis (Table II): the main variation occurred in neutral lipids and fatty acids which exhibited a three-fold increase.

Table II. Fatty Acids in Each Lipid
(mean values, μmol g^{-1} of dry weight)

	Anaerobiosis (An)	Aerobiosis (Ae)	$\dfrac{An}{Ae}$
Phosphatidylcholine	39.4	21.6	1.8
Phosphatidylethanolamine	30.5	15.5	2.0
Neutral lipids and free fatty acids	40.1	13.4	3.0

3. On total lipids, the acyl residues distribution varied significantly with anaerobic treatment. In rice coleoptiles grown under anaerobiosis, the ratios of palmityl ($C_{16:0}$), of palmitoleyl ($C_{16:1}$) and linolenoyl ($C_{18:3}$) residues decreased, with a significant increase in oleyl ($C_{18:1}$) and a marked increase in linoleyl ($C_{18:2}$) residues (Figure 1, Table III).

4. The same variations occurred in each class of lipids. In phosphatidylcholine, $C_{16:0}$ residue ratio decreased in anaerobiosis; a slight increase of $C_{18:1}$ residues ratio was observed (21%-27%) but the strongest variation

Table III. Acyl Residues Distribution in Total Lipids

		$C_{16:0}$ (%)	$C_{18:1}$ (%)	$C_{18:2}$ (%)
Seeds		17.8	40.9	38.3
Coleoptiles	Aerobiosis	29.1	20.6	35.9
	Anaerobiosis	23.6	25.1	45.1

occurred in $C_{18:2}$ residues ratio, which changed from 35% with molecular oxygen to 45% under anoxia (Figure 2). Phosphatidylethanolamine exhibited parallel changes: a decrease of $C_{16:0}$ residues and a higher ratio of $C_{18:2}$ residue in anaerobiosis (52%) than in aerobiosis (38%), without significant variation on $C_{18:1}$ residues (Figure 3).

In neutral lipids, the same variations were observed as in phosphatidylcholine (Figure 4). In all cases, $C_{16:1}$ and $C_{18:3}$ residue ratios were lower under anaerobiosis.

DISCUSSION AND CONCLUSION

These results can be summed up by three points. The growth of rice coleoptiles without molecular oxygen, induced:

1. an accumulation of lipids;
2. a decrease in ratios of all C_{16} residues and of the most unsaturated C_{18} residue; and
3. an increase in the distribution of $C_{18:1}$ and $C_{18:2}$.

The high content of lipids obtained under anaerobiosis may be explained by an increase in the biosynthesis rate or by an exportation of lipids from the seeds during germination. In fact, in anaerobiosis, the total amount of lipids in coleoptiles was 10 times lower (0.6% of the lipids of seeds) than in aerobic coleoptiles (6.8% of seed lipids–Table IV). So a small transfer of lipids from seeds to coleoptiles may occur only under anaerobiosis. But if one considers the fatty acid distribution, a substantial difference appeared between lipids of seeds and total lipids of anaerobic coleoptiles, mainly for $C_{18:1}$ and $C_{18:2}$ residues ratios (Table III). So lipids were not transfered as a whole, but the translocation of a small part of fatty acids from embryos to coleoptiles cannot be excluded, indicating that lipids are renewed.

Table IV. Total Fatty Acids In:

	μmol	%
Rice Seeds (100 ml)	7234	100
Coleoptiles in Anaerobiosis (356 mg dry weight)	48	0.6
Coleoptiles in Aerobiosis (5185 mg dry weight)	491	6.8

The biosynthesis of fatty acids during the development of coleoptiles cannot be excluded either. In this case, the unexpected fact is the hypothetical biosynthesis of unsaturated fatty acids as $C_{18:1}$ and $C_{18:2}$ under anaerobiosis. One may assume that a small biosynthetic flow of such acids is possible even with the very small content of molecular oxygen which may remain in the anaerobiosis experiments. In fact, two more probable explanations must be considered.

First, without molecular oxygen, biosynthesis of some new fatty acids would be depressed; only C_{16} and $C_{18:3}$ ratios are affected, for these fatty acids are known to have parallel biosynthetic pathways, different from those of $C_{18:1}$ and $C_{18:2}$ fatty acids.[11-13]

Secondly, elongation from C_{16} to C_{18} would still be possible without molecular oxygen. In this hypothesis, it is easy to understand the anaerobic formation of monounsaturated acid $C_{18:1}$, as it was demonstrated in bacteria.[14,15] But it is difficult to explain the desaturation giving $C_{18:2}$ acid, which needs molecular oxygen: the formation of long-chain unsaturated fatty acids occurs by aerobic desaturation, which is catalyzed by enzymes of oxygenase type.[14,15] This mechanism seems to be the rule in Eukaryotes: nevertheless, $C_{18:2}$ appears in membrane cells of anaerobically grown yeast.[17]

So the problem remains open. Even if unsaturated fatty acid biosynthesis is impossible in such anaerobic conditions, at least total biosynthesis and turnover of lipids is possible in this material without molecular oxygen. Anaerobically grown rice coleoptiles could be a good system for studying the biosynthetic pathways of unsaturated fatty acids in lipids of higher plants.

REFERENCES

1. Vartapetian, B. B., I. N. Andreeva and I. P. Maslova. "Ultrastructures Cellulaires des Coléoptiles de Riz en Conditions Anaérobies et Aérobies," *Dokl. Akad. Nauk SSSR* 196(5):1231-1233 (1971).

2. Ueda, K., and H. Tsuji. "Ultrastructural Changes of Organelles in Coleoptile Cells During Anaerobic Germination of Rice Seeds," *Protoplasma* 73:203 (1971).

3. Öpik, H. "Effect of Anaerobiosis on Respiratory Rate, Cytochrome Oxidase Activity and Mitochondrial Structures in Coleoptiles of Rice (*Oryza Sativa* L.)," *J. Cell Sciences* 12:725 (1973).

4. Tsuji, H. "Respiratory Activity in Rice Seedlings Germinated Under Strictly Anaerobic Conditions," *Bot. Mag. Tokyo* 85:207 (1972).

5. Vartapetian, B. B., I. P. Maslova and I. N. Andreeva. "Mitochondria of Coleoptiles of Rice (*Oryza sativa*) Grown Under Anaerobic Conditions," *Fiziol. Rast.* 19(1):106-112 (1972).

6. Vartapetian, B. B., A. I. Maslov and I. N. Andreeva. "Cytochromes and Respiratory Activity of Mitochondria in Anaerobically Grown Rice Coleoptiles," *Plant Sci. Letters* 4:1-8 (1975).

7. Vartapetian, B. B., A. I. Maslov, I. N. Andreeva and G. I. Kozlova. "Cytochromes of Mitochondria of Rice Coleoptiles Grown Under Conditions of Strict Anoxia," *Fiziol. Rast.* 20(6):1279-1284 (1973).

8. Tsuji, H., T. Katoh and K. Ueda. "Growth and Metabolism in Plants Under Anaerobic Conditions," in *Abstracts of the Twelfth International Botanical Congress* (Leningrad: Nauka, 1975), p. 373.

9. Vartapetian, B. B., I. N. Andreeva and A. L. Kursanov. "Appearance of Unusual Mitochondria in Rice Coleoptiles at Conditions of Secondary Anoxia," *Nature,* 248:259 (1974).

10. Costes, C., R. Bazier and D. Lechevallier. "Rôle Structural des Lipides dans les Membranes des Chloroplastes de Blé," *Physiol. Veg.* 10(2):291-317 (1972).

11. Stumpf, P. K. Personal communication (1974).

12. Jaworski, J. G., E. E. Goldschmidt and P. K. Stumpf. "Fat Metabolism in Higher Plants. Properties of the Palmityl Acyl Carrier Protein: Stearyl Acyl Carrier Protein Elongation System in Maturing Sofflower Seed Extracts," *Arch. Biochem. Biophys.* 163:769-776 (1974).

13. Packter, N. M., and P. K. Stumpf. "Fat Metabolism in Higher Plants. The Effect of Cerulenin on the Synthesis of Medium and Long-Chain Acids in Leaf Tissue," *Arch. Biochem. Biophys.* 167:655-667 (1975).

14. Bloch, K., P. Baronowsky, H. Goldfine, W. J. Lennarz, R. Light, A. T. Norris and G. Scheuerbrandt. "Biosynthesis and Metabolism of Unsaturated Fatty Acids," *Federation Proc.* 20:921 (1961).

15. Fulco, A. J., R. Levy and K. Bloch. "The Biosynthesis of Δ^9 and Δ^5 Monounsaturated Fatty Acids by Bacteria," *J. Biol. Chem.,* 239: 998 (1964).

16. Talamo, B., N. Chang and K. Bloch. "Desaturation of Oleyl Phospholipid to Linoleyl Phospholipid in *Torulopsis utilis*," *J. Biol. Chem.* 248(8):2 38-2742 (1973).

17. Jollow, D., G. M. Kellerma and A. W. Linnane. "The Biogenesis of Mitochondria. III—The Lipid Composition of Aerobically and Anaerobically Grown *Saccharomyces cerevisiae* as Related to the Membrane Systems of the Cells," *J. Cell Biol.* 37:221 (1968).

INDEX